Short-Term Bioassays in the Analysis of Complex Environmental Mixtures III

ENVIRONMENTAL SCIENCE RESEARCH

Recent Volumes in this Series

A Continuation Order Plan is available for this series. A continuation order will bring
delivery of each new volume immediately upon publication. Volumes are billed only upon
actual shipment. For further information please contact the publisher.

Short-Term Bioassays in the Analysis of Complex Environmental Mixtures III

Edited by

MICHAEL D. WATERS
SHAHBEG S. SANDHU
JOELLEN LEWTAS
LARRY CLAXTON
NEIL CHERNOFF

and

STEPHEN NESNOW

U.S. Environmental Protection Agency
Research Triangle Park, North Carolina

PLENUM PRESS • NEW YORK AND LONDON

Library of Congress Cataloging in Publication Data

Main entry under title:

Short-term bioassays in the analysis of complex environmental mixtures III.

(Environmental science research; v. 27)
Includes bibliographical references and index.
1. Pollution—Toxicology—Congresses. 2. Toxicity testing—Congresses. 3.
Biological assay—Congresses. 4. Environmental chemistry—Congresses. I.
Waters, Michael D. II. Series.

RA566.S473 1983	615.9'07	82-22323

ISBN-13: 978-1-4613-3613-6   e-ISBN-13: 978-1-4613-3611-2
DOI: 10.1007/978-1-4613-3611-2

This report has been reviewed by the Health Effects Research Laboratory, U.S.
Environmental Protection Agency, and approved for publication. Approval does
not signify that the contents necessarily reflect the views and policies of the U.S.
Environmental Protection Agency, nor does mention of trade names or commercial
products constitute endorsement or recommendation for use.

INTRODUCTION

In the four years since the 1978 Symposium on the Application of Short-Term Bioassays in the Fractionation and Analysis of Complex Environmental Mixtures the use of short-term bioassays to evaluate potential health hazards of complex environmental mixtures has substantially increased.

Increased research activity has been particularly noticeable in mobile source emissions, where initial observations on the mutagenic activity of diesel particulate extracts reported at the 1978 symposium stimulated the development of major research programs in government and industry. In the absence of appropriate reference materials, the U.S. Environmental Protection Agency initiated comparative genotoxicity studies to determine the relative mutagenic and carcinogenic activity and, ultimately, the potential human health risk due to exposure to various complex emission products. Among the materials investigated were those of known health risk, such as coke oven and roofing tar emissions and cigarette smoke condensates, and those of unknown hazard, such as exhaust from diesel-and gasoline-powered vehicles.

Studies on diesel emission products proved useful in short-term bioassay development, as the diesel exhaust extracts were genetically active with low cellular toxicity and could be obtained in relatively large quantities. Availability of such samples aided chemical characterization, and it was eventually determined that the nitro-polynuclear aromatic hydrocarbons were among the mutagenic components of diesel exhaust particulate.

The productivity of airborne emissions research had its counterparts in the other environmental media. The success of these efforts attested to the general utility of the combined approach, which couples short-term bioassays with the state-of-the-art techniques of chemical analysis. Perhaps more importantly, the baseline data developed in comparative evaluation of numerous complex mixture samples have provided, in short order, valuable information for use in estimating potential human health hazards.

While the methodologies described above are new, and in many cases unconventional, they hold substantial promise. Careful performance of the biological and chemical procedures and prudent interpretation of the results have yielded significant public health and environmental benefits. The prospects for the future are for greater use of short-term bioassay data, especially to estimate human health hazards.

The present volume continues to describe state-of-the-art techniques to collect and prepare environmental samples for bioassay and to develop short-term in vitro and in vivo bioassays for mutagenicity, cytotoxicity, carcinogenicity, and teratogenicity. Current efforts to integrate biological and chemical data to assess human health hazards are described in the final section. The ultimate questions of human health risk from exposure to complex environmental mixtures remain largely unanswered, but important new direction has been given to the field and much of the technology described in this volume will be used to answer these questions.

Michael D. Waters, Ph.D.

ACKNOWLEDGMENTS

Many fine papers were presented at this symposium, which, because of space limitations, could not be included in the proceedings volume. We would like to express our sincere appreciation to the following people for their excellent contributions to the sympoisum: M. Maskarinec, F.W. Larimer, C.W. Francis, T. Atherholt, G. McGarrity, D. Schuetzle, T.D. Traylor, T.J. Hughes, T.J. Wolff, A.R. Kolber, C.M. Sparacino, E.D. Pellizzari, A.L. Brooks, P.O. Zamora, R.F. Henderson, F.F. Hahn, W.Z. Whong, J. Stewart, T.M. Ong, D. Spangenberg, S. Scher, C.Y. Ma, C. Ho, T.K. Rao, J.L. Epler, E.L. Loechler, J. King, A.R. Jones, D.M. DeMarini, A.W. Hsie, A. Austin, L. King, K. Loud, S. Tejada, R. Young, C. Hutchinson, D.R. Jagannath, D.J. Brusick, S.H. Toney, L.D. Claxton, A.G. Braun, B.B. Nickinson, P.B. Horowicz, J.A. Bantle, B.H. Erickson, J.C. Chuang, B.A. Paterson, A.F. McFee, M.N. Sherrill, E.H. Jackim, G.G. Pesch, A.R. Malcolm, H.H. Chen, T.S. Banerjee, R.H. Allen, B.L. Berquist, R. Simmons, C.S. Woo, A.F. Hanham, B.P. Dunn, H.F. Stich, R.H. Stevens, D.A. Cole, H.F. Cheng, D.D. Gay, J. Santodonato, M. Neal, P. Durkin, F. Stoss, L.J. Schiff, S.F. Elliot, S.J. Moore, M.S. Urcan, J.A. Graham, and S.G. Banerjee.

We wish to thank Olga Wierbicki and Leslie Silkworth of Northrop Services, Inc. for coordinating the symposium and also, with Barbara Elkins, Linda Cooper, and Michele Mann, for editing the proceedings and preparing the final manuscript.

CONTENTS

---

*Invited paper
+Poster paper

SESSION 1

COLLECTION AND PREPARATION OF ENVIRONMENTAL SAMPLES FOR BIOASSAYS

SOME ASPECTS OF MUTAGENICITY TESTING OF THE PARTICULATE PHASE AND THE GAS PHASE OF DILUTED AND UNDILUTED AUTOMOBILE EXHAUST

Ulf Rannug,[1] Annica Sundvall,[1] Roger Westerholm,[2] Tomas Alsberg,[2] and Ulf Stenberg[2]

[1]Department of Toxicology Genetics, Wallenberg Laboratory University of Stockholm, and [2]Department of Analytical Chemistry, Arrhenius Laboratory, University of Stockholm S-106 91 Stockholm, Sweden

INTRODUCTION

Mutagenicity testing of complex chemical mixtures always involves difficulties not encountered when testing individual chemicals. This is especially evident when testing automobile exhausts for mutagenicity in vitro, since many factors may affect the results. Some of these factors will be discussed.

The emission from an internal combustion engine is distributed between gas phase and particles. This distribution is a function of temperature as well as amount (surface) of particles. This is particularly significant in the comparison of liquefied petroleum gas (LPG) and diesel vehicles because of the difference in particle emission.

It is necessary to concentrate the samples prior to biological tests in vitro and chemical analysis, and this may alter the composition. Furthermore, automobile exhaust contains oxides of nitrogen (NO, NO_2, NO_3) which, together with water, produce acidic extracts.

All these factors may change the sample substantially, thereby obstructing the evaluation. These "artifacts" have recently become significant when sensitive in vitro tests are used. In this paper, the risk of nitration of polycyclics, which cause potent mutagens, is discussed.

MATERIALS AND METHODS

Vehicles and Driving Cycles

The results discussed in this paper are based on experiments carried out with light-duty vehicles. Different car makes and year models as well as different types of fuel-engine combinations were included, although the major part of the work was performed with cars fueled with gasoline or alcohol/gasoline fuels. The vehicles were put at the disposal of the Air Pollution Research Laboratory, Motor Vehicle Section at Studsvik, by Volvo Car Corp., Saab-Scania AB, and Swedish Motor Fuel Technology Co. All the chassis dynamometer tests were carried out by the Motor Vehicle Section at Studsvik. The vehicles were driven according to the U.S. driving cycle, average speed 31.7 km/h (FTP 1972), using cold (20 to 30°C) starts, if nothing else is stated. For comparison, test series with cold starts at a lower (about 0°C) ambient temperature were also carried out. Another driving cycle, the ECE-driving cycle adopted by the ECE (United Nations Economic Commission for Europe), was also used for comparison. With the ECE driving cycle, the maximum and average speed is 50 km/h and 18.7 km/h, respectively. The vehicles were conditioned by road driving at ~ 90 km/h for 45 min before each test series. No conditioning was performed between tests (three repeated cycles).

Fuels

Commercially available leaded gasoline and diesel fuel were used. Other fuels were also tested, such as M 15 (a methanol/gasoline fuel with 15% methanol) and LPG.

Sampling Methods

Exhaust samples were taken both directly at the end of the tail pipe and after dilution in a dilution tunnel. Proportional sampling was used with undiluted exhausts, i.e., the amount of air passing through the carburetor is used as reference and as a measure of the amount of emitted exhaust. The dilution tunnel was constructed at the Motor Vehicle Section at Studsvik according to the specifications in the Federal Register (1980). The approximate diluting ratio was 1:10 for the gasoline vehicles and 1:7 for the diesel vehicles. Each run was performed three times. Two different tunnels were used, one for the diesel vehicles and another for the other vehicles.

Particulate phase. The particulate matter was trapped on glass-fiber filters (Gelman A/E). These were washed in ethanol and heated for 1 h at 120°C before being used. After sampling, the filters were either extracted within two days or stored at -20°C and extracted within three weeks. Each filter was extracted with 250 ml acetone in a Soxhlet apparatus for 16 h. Siphoning time was 15 min. The volume was then reduced at 37°C under a stream of nitrogen and adjusted to correspond to a given volume of raw exhaust, normally 1 ml of extract was equivalent to either 50 1 of exhaust (diesel) or 100 or 150 1 of exhaust (gasoline). When chemical analysis (see below) was carried out, parallel filters from the same run were extracted and cleaned up according to the procedure given elsewhere (Stenberg et al., 1981a).

Gas phase. Parallel to the particulate sampling, gas phase trapping was also carried out in some experiments. The cryo-gradient technique and procedure used in these experiments have been described in detail by Stenberg et al. (1981b). Briefly, the following equipment was used. The sampling probe connected to the dilution tunnel permits particulate sampling on three identical filters. Normally, however, a simple sampling probe with one filter holder was used. For the gas phase enrichment, three consecutive condensers were used: one ice/water condenser, one CO_2/ethanol condenser, and one liquid nitrogen condenser (Figure 1). The cooling area of the condensers is ~ 0.5 m^2. The condensers are connected with a pump system including flow elements, temperature/pressure sensors, valves and a dry-gas meter. The pump unit can either be used as a proportional sampler, using a reference signal from the carburetor, or as a constant flow sampler compensating for the pressure drop. Maximum sampling velocity is 100 1 gas/min. The extracts from the nitrogen condenser contained no PAH or PAH derivatives. These extracts also showed higher toxic effects than the extracts from the other two condensers. Therefore, only the extracts from the ice/water and CO_2/ethanol condenser, respectively, will be discussed in this context.

After each run, the condensers were extracted with acetone in a special Soxhlet apparatus (Stenberg et al., 1981b). The volume of the extracts was reduced and adjusted as described for the particulate extract. In some experiments, the condensers were extracted with dichloromethane (DCM), together with a buffer. In these experiments, DCM was evaporated to dryness under nitrogen at reduced pressure and the samples were redissolved in acetone.

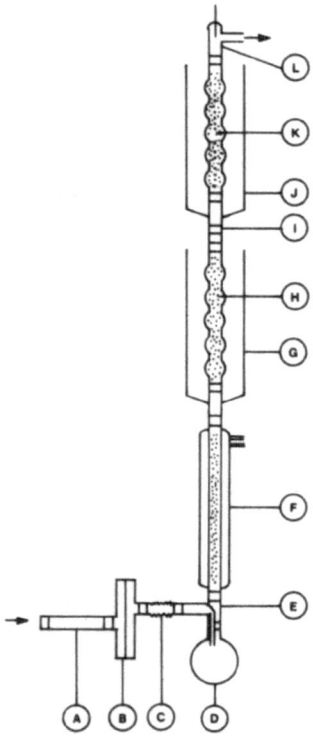

Figure 1. Sampling equipment:
  A. Flexible metal tubing, 250 mm.
  B. Filter holder, ⌀ 127 mm, stainless steel.
  C. Teflon bellows.
  D. Round-bottom flask, 1000 ml.
  E. Normal glass joint.
  F. Ice/water condenser.
  G. Cryo vessel, dry ice/ethanol.
  H. Condenser cartridge.
  I. Ball glass joint.
  J. Cryo vessel, liquid nitrogen.
  K. Condenser cartridge.
  L. Teflon/glass joint, temperature sensor.

Test Procedure

The <u>Salmonella</u> <u>typhimurium</u> strains TA98 and TA100 were kindly provided by Dr. B.N. Ames, University of California, Berkeley, CA. The nitroreductase-deficient strains TA98NR and TA100NR were kindly provided by Dr. H.S. Rosenkranz, Case Western Reserve University, Cleveland, OH. Basically, the plate incorporation assay according to Ames et al. (1975) was followed, with minor modifications. Thus, histidine (0.1 μmol/plate) and biotin (0.1 μmol/plate) were added to the minimal medium instead of to the soft agar. Series for determining the toxic effects were also included. In these series, the overnight culture was diluted 2×10^6 times and the soft agar mixture was poured onto nutrient agar plates (antibiotic medium 3, Difco) instead of minimal agar plates. Liver preparations (S9 fractions) from Aroclor-pretreated Sprague-Dawley male rats were used as metabolizing systems.

The extracts were tested at three different concentrations normally corresponding to 2.5, 5.0, and 7.5 l of exhaust in the case of gasoline, or 1.25, 2.5, and 5.0 l in the case of diesel. All plates were incubated at 37°C for 48 h before scoring. The results were analyzed by regression analysis from the linear portion of the curves and the results shown in tables and figures (i.e., numbers of revertants per liter exhaust or per kilometer were calculated from the slope of the regression line.

RESULTS

Particulate Phase

Diluted and undiluted exhaust were sampled simultaneously and the results from two mutagenicity experiments are shown in Table 1. With the particulate extracts, no major difference in the mutagenic effects was seen between diluted and undiluted exhaust gases, although the diluted samples showed lower effects in most cases. However, the relative distribution of PAH between particulate phase and gas phase was different in the diluted compared to the undiluted samples. Diluting the exhaust increases the amount of PAH found on the particles (Table 2). For 3- and 4-ring PAH, the adsorption onto particles is enhanced by dilution, but a quantitative collection by filtering only is not sufficient. Since we noted toxic effects of the gas phase extracts from undiluted exhausts, we omitted the use of this method for the subsequent experiments.

To test the influence of temperature on the mutagenicity of the extracts from the particulate phase, cold starts at ~ 0°C were compared with cold starts at 23°C. The results from different gasoline and diesel cars are shown in Figure 2. In most cases, a signficantly higher mutagenicity was noted at 0°C. The increase was seen with both tester strains and varied between 20 and 130%.

Table 1. The Mutagenic Effect of Particulate Extracts on
          Salmonella typhimurium TA98 and TA100 With and
          Without a Metabolizing System (S9)[a]

| | | No. of Revertants/1 Exhaust | | | |
| | | Strain TA98 | | Strain TA100 | |
Series	Samples	-S9	+S9	-S9	+S9
1	Undiluted	6.0	16.1	5.0	25.6
1	Diluted	4.9	11.8	1.7	42.9
2	Undiluted	26.0	52.0	10.2	69.9
2	Diluted	3.0	30.6	1.2	52.8

[a]Samples from diluted and undiluted exhaust were collected
simultaneously.

Table 2. Particle-Associated PAH from Diluted and Undiluted
          Gasoline Exhaust Sampled Simultaneously[a]

| Type of PAH | Molecular Weight | Amount of PAH ($\mu g/m^3$ exhaust) | |
		Undiluted	Diluted
Phenanthrene	178	0.6	6.6
Pyrene	202	3.9	64.6
Benzo(a)anthracene	228	2.9	7.9
Benzo(a)pyrene and 6-H-benzo(cd)pyrene-6-one	252,254	4.3	5.0
Benzo(ghi)perylene	276	12.2	10.6
Coronene	300	10.8	8.2

[a]FTP-72 driving cycle with cold, 23°C, starts.

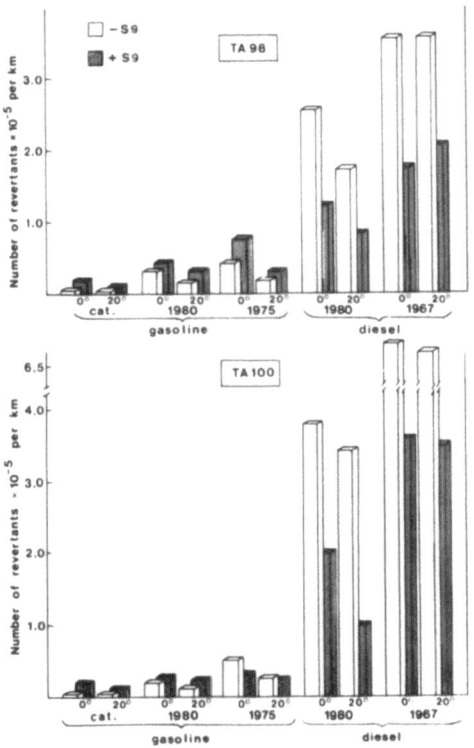

Figure 2. A comparison of the mutagenic effects of particulate
extracts from cold starts at 0 and 23°C, respectively,
from gasoline and diesel exhaust on Salmonella
typhimurium TA98 and TA100 in the presence and absence
of a metabolizing system (S9).

The 1967 diesel car, however, did not show this temperature
dependence in these experiments.

    The influence of the driving cycle on the mutagenicity of the
particulate extract was also investigated. A comparison of the
mutagenic effect of particulate extracts from gasoline- and
diesel-fueled light-duty vehicles driven according to the FTP-72
and the ECE driving cycle, respectively, is shown in Figure 3. In
all cases except the 1975 gasoline car, cold starts at about 0°C
were used. For the 1975 car, cold starts at about 23°C were
employed. The two diesel vehicles both gave lower mutagenic
effects with the ECE driving cycle. The most significant
reduction was seen in the absence of S9, e.g., a reduction from

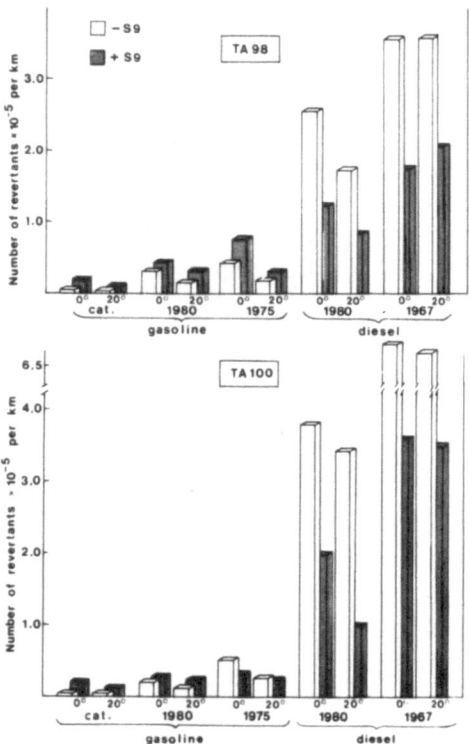

Figure 3.   A comparison of the mutagenic effects of particulate
            extracts from the emission of light-duty vehicles
            driven according to driving cycle FTP-72 (U) and
            ECE (E) on Salmonella typhimurium TA98 and TA100.   The
            mutagenicity was tested both in the presence and in the
            absence of a metabolizing system (S9).   All starts were
            at about 0°C except for the 1975 gasoline car where
            starts at about 23°C were used.

250,000 to 130,000 and from 380,000 to 230,000 rev/km on strain
TA98 and TA100, respectively, for the diesel 1980 model.   The
catalyst-equipped car also showed lower mutagenic effects with the
ECE driving cycle, i.e., no significant mutagenic effects were
seen with this driving cycle.   With the other gasoline cars, no
consistent difference was noted between the two driving cycles if
both tester strains are considered.   With the 1975 model gasoline
car (starting temperature 23°C instead of 0°C), a higher mutagenic
effect was seen with the ECE driving cycle in the presence of S9.

Although the number of samples is relatively small, these results, together with results presented elsewhere (Egebäck et al., 1982; Rannug, 1982), make a rough classifiction of the mutagenic effects possible. Thus, based upon the mutagenicity of particulate extracts of the exhaust emission from light-duty vehicles with different fuel-engine combinations, three main groups can be seen. Light-duty diesel vehicles constitute the high mutagenicity group, giving between 50 and 250 rev/l exhaust (Figure 4) or 100,000 and 700,000 rev/km (Figures 2 and 3). The medium mutagenicity group, giving between 20,000 and 100,000 rev/km (Figures 2 and 3), consists of different types of gasoline fuels, i.e., leaded and lead-free gasoline, and alcohol/gasoline fuels (Figures 2, 3, and 4). Two other fuels, methanol (Egebäck et al., 1982; Rannug, 1982) and LPG (Figure 4), constitute the low mutagenicity group. In addition, when gasoline-fueled cars are equipped with a three-way catalyst in combination with a closed loop (i.e., oxygen sensor), the mutagenicity of the particulate extracts is reduced to the low mutagenic effect significant for the third group.

Gas Phase

So far, only the mutagenicity of the particulate phase has been discussed. Experiments have also been carried out with the gas phase from different types of light-duty vehicles. Also in these cases the tests have been carried out with acetone extracts. Some experiments have also been done with DCM extracts. As mentioned previously, extracts from automobile exhausts are acidic, partly due to formation of nitric/nitrous acid during combustion. Since there is a possible risk for nitration during the condensation process, we conducted a series of parallel sampling experiments using different buffers (pH 3, 7, or 11) in the flask for collecting the condensed water (Figure 1). In these experiments we used DCM and thus obtained a two-phase extraction system. The DCM phase was tested for mutagenicity. Only minor differences in mutagenicity were noted between the extractions at different pH. The chemical analysis, however, showed that DCM did not extract the organics quantitatively from the condenser (Stenberg et al., 1981b). Consequently, acetone was chosen as solvent.

In Figure 4, the mutagenicity of extracts from the H_2O condenser and the CO_2 condenser can be compared with the corresponding particulate extract from light-duty vehicles fueled with four different fuels. The extracts from both the condensers gave a significant mutagenic effect on strain TA100 with the following fuels: diesel, gasoline, and leaded M 15. The same

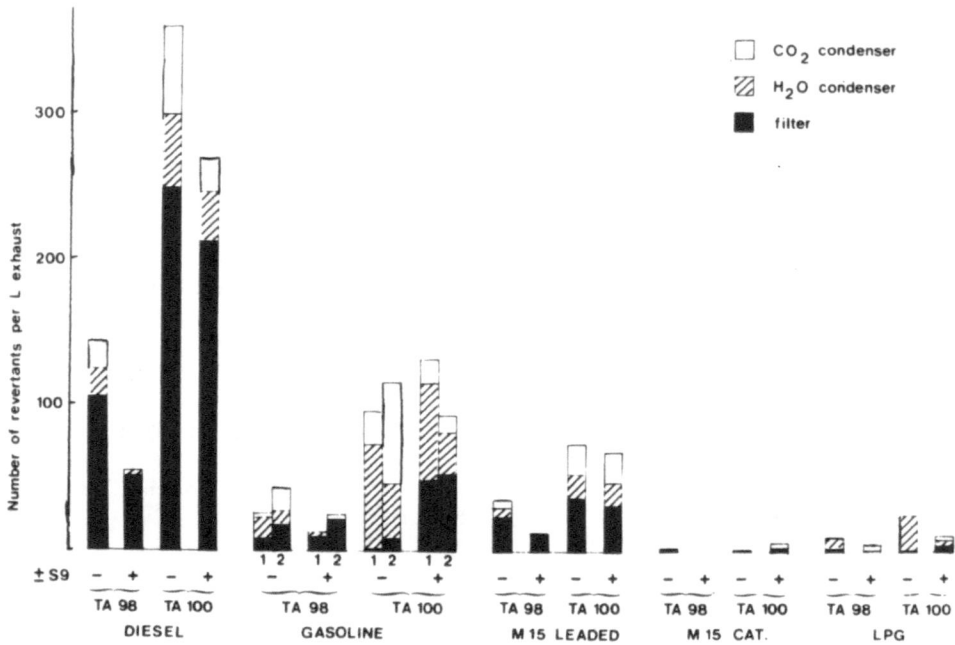

Figure 4. The mutagenic effects (revertants per liter exhaust) on
Salmonella typhimurium TA98 and TA100 of the
particulate phase (acetone extracts) and the gas phase
(acetone extracts from H_2O and CO_2 condensers,
respectively) of exhaust emission from diesel-,
gasoline-, M 15-, and propane-fueled cars. FTP-72
driving cycle (cold start at about 0°C). One of the
M 15-fueled cars was equipped with a three-way catalyst
with closed loop (cat). All samples were tested with
and without the addition of a metabolizing system (S9).

extracts gave a lower mutagenic effect on strain TA98. For both
strains a lower mutagenic effect was noted for the extracts in the
presence of S9 than in absence of S9. With the M 15- (lead-free)
fueled car, equipped with a three-way catalyst and closed loop, no
significant ($p > 0.05$) mutagenic effect was found in either of the
two condensers. With the propane-fueled car (LPG, Figure 4),
mutagenic effects were seen with extracts from either the H_2O
condenser or the CO_2 condenser. Again, the mutagenic effect was
higher in the absence of S9. This significant ($p < 0.001$) direct
mutagenic effect was found only with extracts from the H_2O
condenser. Strain TA100 showed the highest effect, 25 rev/l.

DISCUSSION

The mutagenic effect of automobile exhaust in the Ames Salmonella assay is dependent on both sampling technique and test procedure. For a comparison of the mutagenicity of particulate and gas-phase extracts for different fuel-engine combinations it is, therefore, important to discuss some of these factors.

The results discussed in this context are derived from a rather small number of vehicles driven according to a specific driving cycle (FTP-72). Changing the driving cycle changes the mutagenic effect. In our case, the ECE driving cycle decreased the difference in mutagenicity between the high-effect group, i.e., diesel exhaust, and the medium-effect group (gasoline and alcohol/gasoline) in strain TA98 (Figure 3). It is still possible, however, to distinguish between the two groups.

Other factors such as cold start temperature change the mutagenicity of all partiuclate extracts similarly, i.e., lower temperatures give rise to increased mutagenicity (Figure 2).

Another question of importance is to what extent the sampling procedure influences the composition of the samples and thereby the mutagenicity. In this context the role of nitration of PAH has been discussed, and one crucial question is whether nitroarenes are the significant contributors to the mutagenicity of all types of exhaust samples. In the case of diesel, there are both chemical and biological data that indicate that approximately 20% of the mutagenicity of particulate extracts can be attributed to 1-nitropyrene (Gibson et al., 1981; Pederson and Siak, 1981a,b; Salmeen et al., 1981). There are also indications that dinitropyrenes are present, one of which, 1-8-dinitropyrene, is a very potent mutagen in bacteria (Löfroth et al., 1980; Rosenkranz et al., 1980; Mermelstein et al., 1981).

In the case of particulate extracts from gasoline exhaust, much less information is available. There are, however, some indications that 1-nitropyrene is present in these exhaust samples, for instance, a reduction in mutagenicity with the nitroreductase-deficient tester strain TA98NR (data not shown).

When it comes to the gas phase, no nitroarenes have been identified, although PAH that can be nitrated and thus give rise to potent mutagens are present in the extracts from the two condensers from gasoline exhaust (Stenberg et al., 1981a).

Preliminary experiments with extracts from the H_2O condenser from gasoline exhaust showed no direct effect on strain TA98 cr

TA100 of the fraction which should contain nitro-PAH if present (data not shown). No experiments have yet been carried out with the nitroreductase-deficient strains.

The experiment with DCM extractions at different pH (3, 7, and 11, respectively) showed no major differences in mutagenicity of the extracts from the water condenser. A difference would be expected if nitro derivatives are the major mutagenic components. On the other hand, the results indicated that DCM may not be the best solvent in this case and it is possible that minor differences could not be detected under these circumstances. Further experiments with addition of NO_2 to the exhaust, prior to filtering, showed a dose-dependent increase in the mutagenicity of particulate extracts of gasoline exhaust, while no such increase was noted with the extracts from the H_2O condenser. Our mutagenicity data on the gas phase indicate that nitro-PAH might not contribute significantly to the mutagenic effect of gasoline exhaust. Nevertheless, in the event nitro-PAH are present in samples from the particulate phase or the gas phase, it is essential to determine the role of the sampling technique for the nitration reactions. It should be mentioned, however, that experiments by others, using XAD-2 to trap the gas phase, have shown no mutagenicity above background levels in the Ames Salmonella test (Stump et al., 1981). Further studies are in progress to clarify these differences.

REFERENCES

Ames, B.N., J. McCann, and E. Yamasaki. 1975. Methods for detecting carcinogens and mutagens with the Salmonella/mammalian-microsome mutagenicity test. Mutation Res. 31:347-364.

Egebäck, K.E., ed. (in press). Chemical and biological characterization of vehicle exhaust emissions when using different fuel/engine combinations. Report to the Swedish Government Committee on Automotive Air Pollution. The National Swedish Environment Protection Board, Motor Vehicle Section.

Gibson, T.L., A.I. Ricci, and R.L. Williams. 1981. Measurement of polynuclear aromatic hydrocarbons, their derivatives, and their reactivity in diesel automobile exhaust. In: Chemical Analysis and Biological Fate: Polynuclear Aromatic Hydrocarbons. M. Cooke and A.J. Dennis, eds. Battelle: Columbus. pp. 707-717.

Löfroth, G., E. Hefner, I. Alfheim, and M. Møller. 1980.
    Mutagenic activity in photocopies. Science 209:1037-1039.

Mermelstein, R., D.K. Kiriazides, M. Butler, E.C. McCoy, and
    H.S. Rosenkranz. 1981. The extraordinary mutagenicity of
    nitropyrenes in bacteria. Mutation Res. 89:187-196.

Pederson, T.C. and J.-S. Siak. 1981a. The role of nitroaromatic
    compounds in the direct-acting mutagenicity of diesel
    particle extracts. J. Appl. Toxicol. 1:54-60.

Pederson, T.C. and J.-S. Siak. 1981b. Dinitropyrenes: their
    probable presence in diesel particle extracts and consequent
    effect on mutagenic activations by NADPH-dependent S9
    enzymes. Presented at the U.S. Environmental Protection
    Agency Diesel Emissions Symposium, Raleigh, NC.

Rannug, U. 1982. Data from short-term tests on motor vehicle
    exhausts. Presented at the Karolinska Institute Symposium on
    Biological Tests in the Evaluation of Mutagenicity and
    Carcinogenicity of Air Pollutants, Stockholm, Sweden.

Rosenkranz, H.S., E.C. McCoy, D.R. Sander, M. Butler,
    D.K. Kiriazides, and R. Mermelstein. 1980. Nitropyrenes:
    isolation, identification and reduction of mutagenic
    impurities in carbon black and toners. Science
    209:1039-1043.

Salmeen, I., A.M. Durisin, T.J. Prater, T. Riley, and
    D. Schuetzle. 1981. Contribution of 1-nitropyrene to direct
    acting Ames assay mutagenicities of diesel particulate
    extracts. Presented at the U.S. Environmental Protection
    Agency Diesel Emissions Symposium, Raleigh, NC.

Stenberg, U., R. Westerholm, T. Alsberg, U. Rannug, and
    A. Sundvall. 1981a. Enrichment of PAH and PAH derivatives
    from automobile exhausts, by means of a cryo-gradient
    sampling system. Presented at the Sixth International
    Symposium on Polynuclear Aromatic Hydrocarbons, Columbus, OH.

Stenberg, U., T. Alsberg, and R. Westerholm. 1981b. The
    applicability of a cryo-gradient technique for the enrichment
    of PAH from automobile exhaust, demonstration of methodology
    and evaluation experiments. Presented at the Karolinska
    Institute Symposium on Biological Tests in the Evaluation of
    Mutagenicity and Carcinogenicity of Air Pollutants,
    Stockholm, Sweden.

Stump, F., R. Bradow, W. Ray, D. Dropkin, R. Zweidinger,
    J. Sigsky, and R. Snow. 1981. Trapping gaseous hydrocarbons
    for mutagenic testing. Presented at the U.S. Environmental
    Protection Agency Diesel Emissions Symposium, Raleigh, NC.

SOURCE ASSESSMENT SAMPLING SYSTEM (SASS)

VERSUS DILUTION TUNNEL SAMPLING

Raymond G. Merrill, Jr.,[1] Joellen Lewtas,[2] Robert E. Hall[1]

[1]Industrial Environmental Research Laboratory, [2]Health Effects Research Laboratory, U.S. Environmental Protection Agency, Research Triangle Park, NC 27711

INTRODUCTION

Two sampling methods have received increased attention recently because of their ability to collect particulate and organic emissions from combustion sources (Huisingh et al., 1978; Lewtas, in press). One method of sampling, represented by the Source Assessment Sampling System (SASS), is designed to collect and size-classify particulate and to collect nonparticulate organic and inorganic materials at source conditions. The second method, represented by various types of dilution tunnel, is designed to collect total particulate and particulate-bound organic material at conditions which approximate ambient environment.

Both methods of sampling were evaluated by the U.S. Environmental Protection Agency (EPA) at its combustion research facility in Research Triangle Park, NC, using a typical residential oil furnace. The data acquired in these tests demonstrate the strengths and weaknesses of each method and the possible applications of the two methods to stationary source sampling.

17

DESCRIPTION OF THE SAMPLING SYSTEMS

SASS

The SASS was designed to collect sufficient quantities of
stationary source particulate and gaseous emissions to allow
particulate size characterization and inorganic, organic, and
bioassay analyses (Blake, 1978; Dorsey et al., 1978).

The SASS train consists of a stainless steel probe that
connects to three cyclones and a filter in an oven module, a gas
treatment section, and an impinger series (Figure 1). Size
fractionation is accomplished in the cyclone portion of the SASS
train, which incorporates the three cyclones in series to provide
large collection capacities for particulate matter nominally
size-classified into three ranges: > 10 μm, 3 to 10 μm, and
1 to 3 μm. By means of a standard 142-mm or 230-mm filter, a
fourth cut, < 1 μm, is also obtained. The gas treatment system
follows the oven unit and is composed of four primary components:
the gas cooler, the sorbent trap, the aqueous condensate
collector, and a temperature controller. Volatile organic
material is collected in a cartridge or "trap" containing XAD-2
sorbent, a micro-reticular resin with the capability of absorbing
a broad range of organic species. Volatile inorganic elements are
collected in a series of impingers that follow the condenser and
sorbent system. The last impinger in the series contains silica
gel for moisture removal. Trapping of some inorganic species also
may occur in the sorbent module.

Particulate material sampled by the SASS train in the furnace
tests is collected at 205°C and gas-phase organic material is
cooled and collected at 20°C in the gas conditioning module. To
obtain the 30 dscm SASS sample under conditions typical of home
furnace use, the furnace is operated for 25 cycles of
approximately 10 min on followed by 20 min off. The SASS train is
started 20 s before burner start-up and left on for 40 s following
burner shutdown. SASS sampling flow rate is held at
0.11 dscm/min.

Dilution Tunnel

Dilution sampling systems are designed to collect cooled
particulate and particulate-bound flue gas materials at
approximately ambient environment conditions. Dilution of flue
gas is accomplished with filtered ambient air which cools the
source sample to approximately 32°C and allows condensable
material to adsorb on the particulate material. The particulate

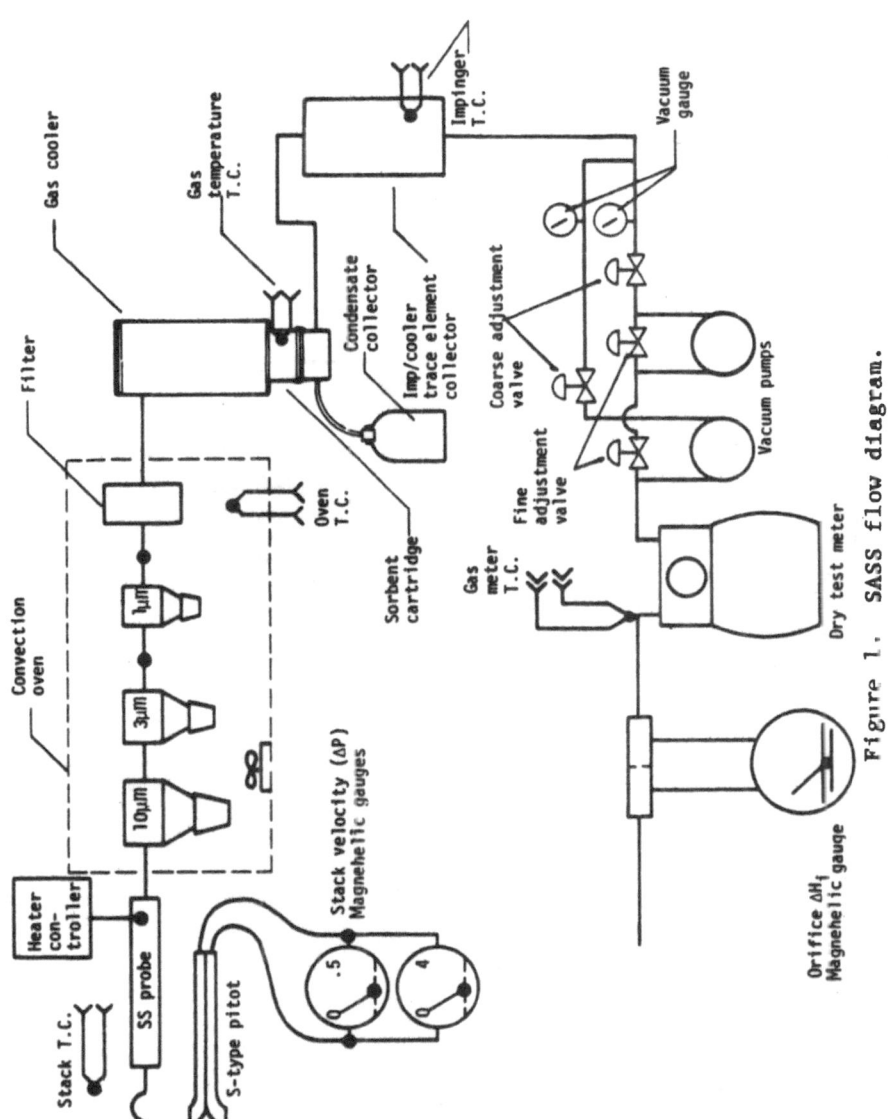

Figure 1. SASS flow diagram.

material is collected on an inert (Teflon) coated fiberglass
filter. No particulate size fractionation is accomplished in the
primary collection apparatus.

For the oil heater tests, a dilution tunnel consisting of a
12.7-m-long steel tube 0.45 m in diameter was used as shown in
Figure 2. The source stream and filtered dilution air are
admitted in one end of the tunnel and drawn through the tunnel to
the filter module. Dilution air enters through a laminar flow

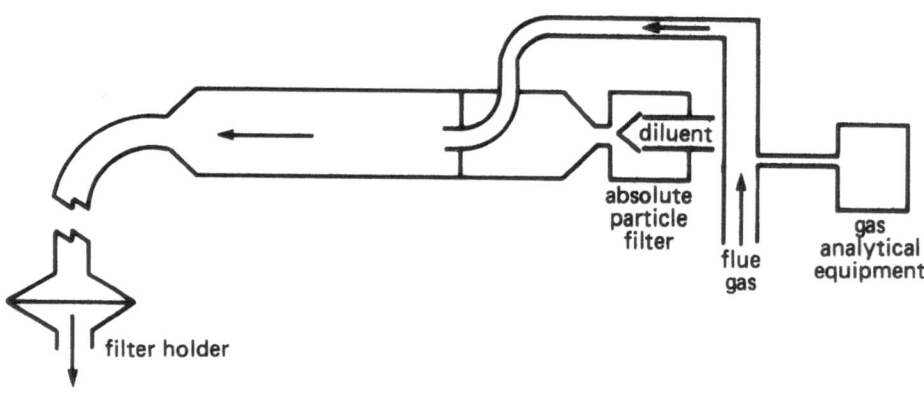

Figure 2.   Dilution tunnel system for collection of particulate.

element where the flow rate is measured. Cooled source sample and
dilution air are drawn through a 50- x 50-cm Teflon-coated filter
and pumped to an exhaust stack through another laminar flow
element where the final flow rate of source plus dilution air is
measured.

The dilution tunnel is operated on the same cycle as the SASS
train (i.e., 10 min on, 20 min off, starting 20 s before start-up,
and sampling 40 s after burner shutdown).

DESCRIPTION OF THE DOMESTIC FURNACE TEST

   The tests were conducted using a conventional (10-year-old)
commercial Williamson home furnace with a non-flame retention
burner fired with No. 2 fuel oil. Figure 3 is a schematic of the
residential warm air furnace. Source sampling and monitoring was
accomplished before and after the furnace barometric damper. Only
the results of sampling after the damper are presented here since
the source and dilution data are directly comparable for this
portion of the source stream.

   Flue gas analyzers and a SASS probe were inserted into the
stack above the atmospheric damper. All material in the flue gas
stream which passes these sampling probes was directed through the
dilution tunnel and filter. All samples were taken in duplicate,
and the results reported are the average of the two sampling runs.

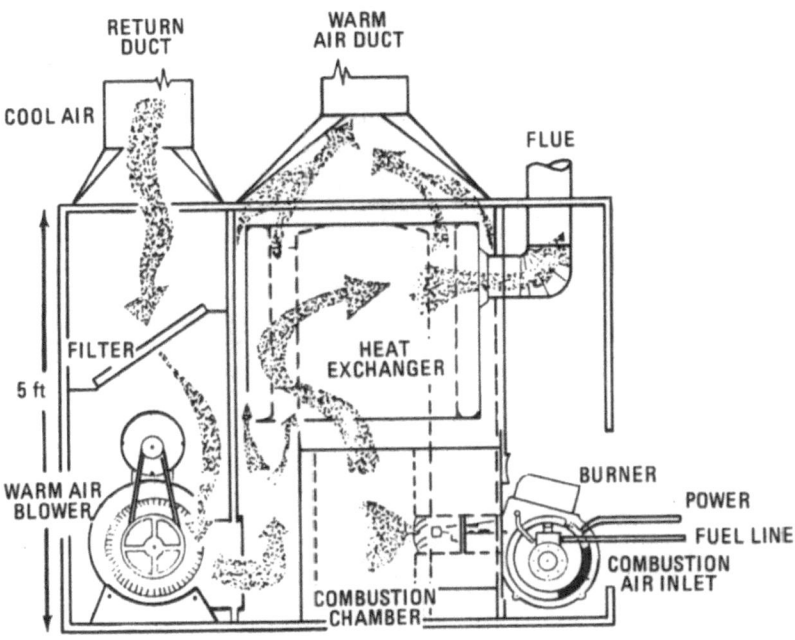

Figure 3. Residential warm air furnace.

SAMPLE ANALYSIS

Bulk Solids

Bulk solids from the SASS train were recovered from the probe, cyclones, and fiberglass filter. Solvent from probe and cyclone particulate washes were allowed to evaporate, and the residual was weighed. Filters were weighed after reaching constant weight in a desiccator.

Bulk solids on the dilution tunnel filters were determined by weighing after the filters had reached constant weight in a desiccator. No attempt was made to recover particles from the tunnel itself.

Extractable Organic Materials

In addition to total particulate mass, the total extractable organic material was recovered from both sampling systems. Limited bioassay and chemical analyses were performed on the organic extracts.

Organic samples from the SASS train were prepared by Soxhlet extraction with dichloromethane as described in the IERL-RTP Level 1 Procedures Manual (Lentzen et al., 1978). Since the amount of organic extract from the SASS particulate was extremely small, it was combined with the extract of the XAD-2 module.

Dilution tunnel filters (50 x 50 cm) were cut into 6 x 13 cm sections and extracted with dichloromethane in a Soxhlet extractor. These samples were filtered through a medium porosity fritted glass funnel to remove fine particles.

Total semiquantitative organic analysis of the samples was performed using selected Level 1 procedures shown in Figure 4. Samples were concentrated using Kuderna-Danish apparatus. Total chromatographic organic (TCO) analysis for components boiling between 100 and 300°C was performed by gas chromatography. Gravimetric analysis for components boiling above 300°C was performed by taking an aliquot of each sample to dryness in a preweighed pan and weighing the residue.

Ames reverse mutation assays were performed with and without S9 microsomal activation on organic extract samples. Only results from TA98 are reported. Results are reported as the best-fit computer-modeled slope of the experimentally observed response per microgram of organic material added to the plates.

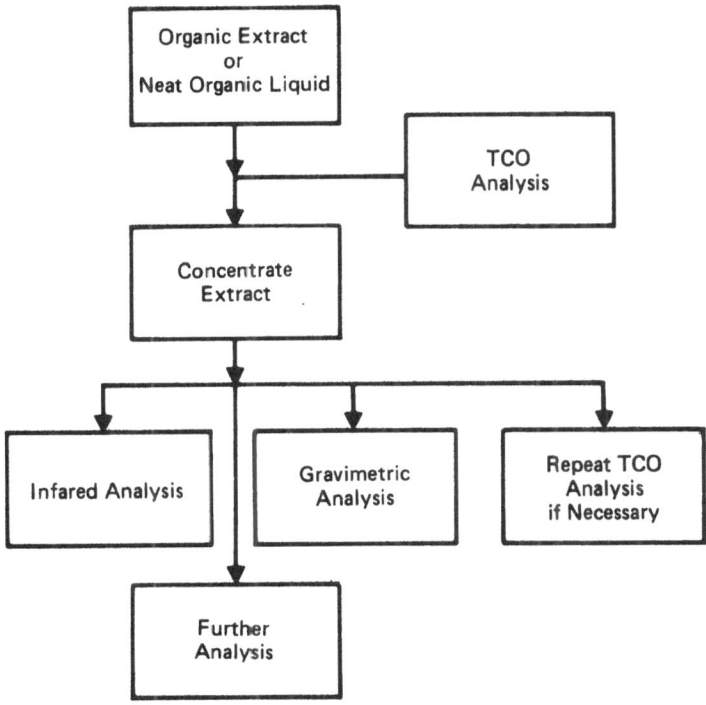

Figure 4.   Total extractable organic analysis of SASS and dilution
            tunnel samples.

RESULTS

    The first indication of a difference in the sampling
characteristics of these two systems appears in the mass of
particulate caught. For the residential home heater studies, the
dilution tunnel retained 1.5 times the weight of material per
cubic meter collected by the SASS (Table 1). The extra mass of
the particles collected in the dilution tunnel is due to the
presence of condensed organic material on the filtered particulate
present because the tunnel operates at ambient temperature.
Organic material in the SASS train remains in the vapor phase as

Table 1.   Total Particulate Mass -- SASS vs. Dilution Tunnel

	SASS[a]	Dilution Tunnel[b]
Condition 1, mg/dscm[c]	1.62	2.41
Condition 2, mg/dscm[d]	2.65	3.25

[a]Collected at 205°C.
[b]Collected at approximately 32°C.
[c]Equivalent to Bacharach Smokespot 1.
[d]Equivalent to Bacharach Smokespot 3.

it passes the 200°C oven cyclones and filter and condenses or
adsorbs on the XAD-2 after cooling to 20°C in the gas cooler.

A better example of the differences in the sampling system
can be seen by comparing total organic mass collection (Table 2).
The SASS method collected 17 to 20 times more total organic mass
than the dilution tunnel. SASS samples showed greater
concentration of organic material in the flue gas due to the
ability of the SASS train to collect and retain organic material
of lower boiling point (100 to 300°C) than the dilution tunnel.

Table 2.   Total Organic Mass Catch -- SASS vs. Dilution Tunnel

| | SASS | Dilution Tunnel |
	Filter and XAD-2	Filter
Condition 1, mg/dscm[a]	6.9	0.36
Condition 2, mg/dscm[b]	4.2	0.25

[a]Equivalent to Bacharach Smokespot 1.
[b]Equivalent to Bacharach Smokespot 3.

The dilution sampling method did not collect any of the 100 to 300°C organic material in these tests. In addition, the chemical analysis revealed that only 20% of the organic material boiling above 300°C was retained on the particulate in the dilution tunnel.

Most of the organic material collected by SASS and not by the dilution tunnel was chemically characterized as unburned fuel. The Ames reverse mutation assay results indicate that the organic material collected by both systems which boils above 300°C has similar biological activity (Table 3). The total SASS sample, including all organic material boiling above 100°C, is difficult to assay because it is toxic to bacteria; however, these samples seem to contain materials of higher biological activity than the dilution tunnel (Dorsey et al., 1978).

Table 3. Mutagenicity[a] of Organics

	SASS (Grav)		Dilution	Filter
	-S9	+S9	-S9	+S9
Condition 1[b]	2.96	2.51	2.15	5.11
Condition 2[c]	1.32	2.58	1.97	5.47

[a]Revertants per µg of organics in strain TA98.
[b]Equivalent to Bacharach Smokespot 1.
[c]Equivalent to Bacharach Smokespot 3.

CONCLUSIONS

The results from these tests confirm the design constraints of each of the sampling systems. The dilution tunnel collects particulate and material bound to particulate at temperatures near ambient. Samples taken this way retain less of the total organic emission by percent; but, because of the dilution tunnels' high sampling rate (4.3 dscm/min), the total material collected provides enough sample for chemical and bioassay analyses.

The SASS method collects the entire particulate and
condensable organic mass in separate parts of the sampling
apparatus. The SASS method is much more efficient than the
dilution tunnel at collecting organic material boiling between
100 and 300°C. The SASS method is much better than the dilution
method in accounting for the total condensable emissions from a
source, since the dilution tunnel collects only 5% of the organic
mass collected by the SASS.

REFERENCES

Blake, D. 1978. Source Assessment Sampling System: Design and
    Development. EPA-600/7-78-018. U.S. Environmental
    Protection Agency: Research Triangle Park, NC.

Dorsey, J.A., L.D. Johnson, and R.G. Merrill. 1978. A Phased
    Approach for Characterization of Multimedia Discharges from
    Processes. ACS Symposium Series No. 94. American Chemical
    Society: Washington, DC. pp. 29-48.

Huisingh, J.L., R. Bradow, R. Jungers, L. Claxton, R. Zweidinger,
    S. Tejada, J. Bumgarner, F. Duffield, M. Waters, U.F. Simmon,
    C. Hare, C. Rodriguez, and L. Snow. 1978. Application of
    bioassay to the characterization of diesel particle
    emissions. In: Application of Short-Term Bioassays in the
    Fractionation and Analysis of Complex Environmental Mixtures.
    M.D. Waters, S. Nesnow, J.L. Huisingh, S.S. Sandhu, and
    L. Claxton, eds. Plenum Press: New York. pp. 381-418.

Lentzen, D.E., D.E. Wagoner, E.D. Estes, and W.F. Gutknecht.
    1978. IERL-RTP Procedures Manual: Level 1 Environmental
    Assessment, 2nd edition. EPA-600/7-78-201.
    U.S. Environmental Protection Agency, Research Triangle Park,
    NC.

Lewtas, J. (in press). Comparison of the mutagenic and
    potentially carcinogenic activity of particle bound organics
    from wood stoves, residential oil furnaces, and other
    combustion sources. In: Proceedings of the International
    Conference on Residential Solid Fuels, Portland, OR, 1981.

PREPARATION OF HAZARDOUS AND COMPLEX SAMPLES FOR ECOLOGICAL

TESTING

Kenneth M. Duke and David J. Bean

Bioenvironmental Sciences Section, Biological Sciences
Department, Battelle Columbus Laboratories, 505 King Avenue
Columbus, OH 43201

INTRODUCTION

Releases of complex mixtures into the environment can result
in significant adverse ecological effects. These mixtures are
often gaseous, liquid, or solid wastes from energy and industrial
processes such as coal combustion and the manufacture of textiles.
They may be the result of a single chemical or combustion process
or a mixture of two or more waste streams generated by a facility.
Some effluents may have undergone various levels of waste
treatment before release. Of concern are those which are released
into the environment in sufficient quantities to have a potential
for adverse ecological effects.

The evaluation of the potential for a complex environmental
mixture to cause ecological effects is inherent in most
environmental legislation regulating the release of wastes into
the environment, e.g., the Clean Air and Clear Water Acts and the
Resource Conservation and Recovery Act (RCRA). These laws cite
the need for protection of fisheries, recreational resources, and
natural ecosystems from degradation or damage. While these laws
are enforced primarily through standards or discharge limits
governing the quantity of chemicals released, these limits are
usually based on toxicity data derived from health and ecological
effects bioassays. An excellent example of the research
techniques suitable for obtaining the necessary toxicity data is

the phased approach to environmental assessment developed by the
Industrial Environmental Research Laboratory of the
U.S. Environmental Protection Agency, Research Triangle Park, NC
(IERL-RTP) (Duke et al. 1977; Brusick et al., 1980). While the
data obtained by the phased approach are used primarily to define
control technology needs, they also are used in establishing
emission standards. There are three phases of data collection and
interpretation in this approach. The first and best developed
phase (at present) is Level 1 which is characterized by the use of
short-term bioassays. Of interest to this paper are the
ecological effects tests (Table 1). Other approaches to obtaining
ecological effects data, such as those used for the Toxic
Substances Control Act (TSCA), the Organization of Economic
Cooperation and Development (OECD), RCRA, and the Federal
Insecticide, Fungicide, and Rodenticide Act (FIFRA) are similar to
Level 1. Important to all approaches is proper preparation and
application of sample to each short-term ecological bioassay. The
complexity of many environmental mixtures often makes such
preparation and application difficult.

The objective of this paper is to present sample preparation
and application procedures and techniques recommended for EPA's
Level 1 ecological tests. Some problems encountered in
implementing the procedures and their solutions will be discussed.

Table 1.   EPA Level 1 Approach to Ecological Effects Testing of
           Complex Mixtures

Test	Sample Type	Sample Quantity Needed
Aquatic[a]		
Algae	Liquids, solids[b]	5 to 10 1
Invertebrate	Liquids, solids[b]	4 to 12 1
Fish	Liquids, solids[b]	75 to 150 1
Terrestrial		
Root elongation	Liquids, solids	5 1
Plant stress ethylene	Gases	1365 1
Insect toxicity	Liquids, solids	20 ml, 0.5 g

[a]Both freshwater and marine species may be tested.
[b]Solids must be leached to be compatible with the test.

SAMPLE PREPARATION AND APPLICATION

Sample preparation is required when the state of the raw
sample is incompatible with the ecological test. For example,
solid samples are incompatible with aquatic tests. Rarely are
solid wastes dumped directly into aqueous habitats. Yet through
the natural leaching process solid wastes have the potential to go
into solution, enter surface waters, and interact with aquatic
organisms. Therefore, to assess the effects of solid waste on
aquatic species, solid samples must be leached and the leachate
properly applied to the short-term Level 1 aquatic tests. Sample
preparation techniques are also useful when it is desirable to
test separately the various fractions (e.g., organic and
inorganic) of a complex mixture. In all cases, care must be taken
during the preparation not to introduce unnecessary contaminants
and to maintain sample integrity as much as possible. Similarly,
the application of the sample to the test must minimize
unnecessary alterations and contamination. Sample preparation and
application procedures can best be organized by the state of the
raw sample: gaseous, liquid, and solids/slurries. The Level 1
ecological test matrix for these types of samples is given in
Table 1.

Gaseous Samples

Gaseous samples generally require little or no preparation
after they have been collected. However, considerable
"processing" of a gaseous waste sample may be done during its
collection. IERL-RTP has developed a Source Assessment Sampling
System (SASS) which fractionates the gas stream into its organic,
inorganic, and particulate (solid) components during the
collection process. Since the volume of gas collected by the SASS
is limited, ecological effects testing is not usually recommended
for such samples. Because of the relatively large amount of
sample required for the plant stress ethylene test (the only
required Level 1 ecological test for gaseous samples), a grab
sample with a simple filter to remove coarse particulates, if
abundant, and a cooler in the sample line to reduce the gas
temperature to that which can be safely held by the sample
containers are sufficient. This sample is then applied directly
to the test without further preparation. Application is
accomplished by evacuating the plant exposure chamber until its
pressure is negative relative to that of the sample container,
eliminating the need to pump the gas. Inert materials such as
high density linear polyethylene and polypropylene are used to
connect the sample container to the exposure chamber.

Liquid Samples

Liquid samples are those containing less than 5% solids
(Brusick et al., 1980). They may be either aqueous or nonaqueous
(greater than 0.2% organics) (Brusick et al., 1980). Aqueous
samples are generally collected by simple grab techniques and
applied directly without further preparation to the appropriate
ecological test. Adjustment of the pH of liquid samples is also
of convern in the short-term biological testing of complex
mixtures. Some raw, untreated samples exceed the pH tolerance
range of the test organism. These will be toxic at least in the
higher test concentrations due to pH effects, masking the effects
of other chemical constituents. While the pH can be altered by
simple dilution, this may result on concentrations of the other
constituents below the potentially toxic effect levels. Sample
preparation procedures which incorporate pH adjustment have been
developed (U.S. EPA, 1978; Bause and McGregor, 1980; and Battelle
Columbus Laboratories, 1981). These procedures call for the
addition of reagent-grade acid or base to adjust the pH of a
sample to within the tolerance limits of the test organism
(Table 2). Acetic and hydrochloric acids have been proposed for
reduction in pH while sodium hydroxide has been used to increase
pH. Strong acids and bases offer the advantage of adjusting the
pH with the smallest quantity of salt formation, reducing the
potential for reaction between the sample constituents and the
added acid or base. Nevertheless, the sample will have been
altered both in pH and chemical composition as a result of the pH
adjustment process.

Application procedures involve serial dilutions with water of
appropriate quality followed by direct addition to the required
aquatic and terrestrial tests. In cases where foliar application
of liquids is desirable for the plant stress ethylene test, a
small atomizer permits direct application of the mixture to the
leaf surface. The quantity of sample needed if all marine,
freshwater, and terrestrial tests are implemented exceeds 150 l.
When the sample quantity is limited, the number of tests done may
be reduced. Alternatively, the number of serial dilutions for
each test can be reduced. The latter method may reduce the
reliability of the resulting effects calculations but it does
permit the collection of data on all the species deemed important
for assessing ecological effects.

Nonaqueous samples may be applied directly to the ecological
tests without preparation. They may also be extracted using an
XAD-2 resin column and both the aqueous and organic fractions
tested. Direct application is appropriate for the aqueous
fraction. Solvent exchange is necessary for the organic fraction

Table 2.   pH Tolerance Range for Ecological Test Organisms

Test Organism	pH Range
Selenastrum capricornutum	6 to 10[a]
Daphnia pulex	6 to 10[b]
Daphnia magna	6 to 10[c]
Pimephales promelas	4.1 to 9[d]

[a]Miller et al., 1978.
[b]David and Uzburn, 1969. Reproduction may be impaired at the
extremes of this range.
[c]Derived from experimental data from Battelle Columbus
Laboratories.
[d]Lower limits from Mount, 1973; reproduction is impaired at lower
limit. Upper limit derived from experimental data, Battelle
Columbus Laboratories.

since it is normally extracted in methylene chloride, a
biologically incompatible solvent. Dimethyl sulfoxide (DMSO) is a
suitable solvent for the insect and root elongation tests.
Acetone may be used in all the aquatic tests except for the
freshwater invertebrate, Daphnia, where it is too toxic (LC_{50} of
10 mg/l) (Dowden and Bennett, 1965). DMSO is not toxic to Daphnia
(Battelle Columbus Laboratories, 1981) and can be used in that as
well as the other aquatic tests. In all cases a solvent control
should be conducted confirming the compatibility of the solvent
with the test organism.

Solid/Slurry Samples

These samples offer the greatest challenge in terms of proper
preparation prior to testing. A variety of techniques may be
necessary to make these samples test-compatible, ranging from
grinding and sizing to leachate preparation designed to permit the
use of aquatic tests on solid samples.

Solid samples may be applied directly to terrestrial tests if
particle size is 5 μm or less in diameter. Larger particles must
be ground and sized to this dimension. A manual or automated

mortar and pestle is used for grinding. If the solid waste has
the potential of entering the aquatic environment, then the sample
must be leached and tested using marine and freshwater bioassays.
Numerous laboratory procedures for leaching are available
including two ASTM aqueous methods, the extraction procedure (EP)
used for RCRA (U.S. EPA, 1978), and a simple 30-min aqueous shake
procedure. The ASTM-B procedure and the EP technique involve pH
adjustment. An analytical chemistry study of four leachate
procedures was conducted by GCA/Technology Division (Bause and
McGregor, 1980) for IERL-RTP. The ASTM-A, ASTM-B, EP, and
carbonic acid extraction procedures were compared. The ASTM-A
technique was the most precise based on relative standard
deviations for several extracted ions. The two ASTM procedures
produced higher concentrations of ions although their efficiency
per gram of solid leached was lower than the other techniques
(Bause and McGregor, 1980). It is the ASTM-A procedure, with some
modification, which is used to prepare the leachates for Level 1
ecological tests (Table 3).

     The ASTM procedure was designed for small volumes (2.8 l or
less). The ecological tests may require as much as 165 l of
leachate. To meet this requirement, specialized equipment was
developed. It was deemed most effective to prepare the leachate
in a single large-volume container rather than numerous (up to 40)
4-l containers. Convenient volumes depending on the number of
aquatic tests to be performed were 2.8 l, 15 l, 90 l, and 165 l.
Solids are added to distilled water at a ratio of 1:4 by weight.
Solids are not dried before use to prevent the loss of volatile
constituents. Agitation of the sample is by rolling, with the
container in the horizontal position which exposes more particle
surface to the water. For the smaller volumes, 2.8 l and 15 l, a
bottle roller is used. The larger containers (90 l and 165 l)
require a heavy-duty drum roller modified to handle the stresses
incurred during the continuous 48-h agitation of a 200-kg
container (Figure 1). Glass containers with Teflon lid inserts
are used for the two smaller volumes. The larger volumes use high
density linear polyethylene drum liners. During agitation these
liners are placed in a reinforced steel 30- or 55-gallon drum to
prevent them from flexing and splitting (see Figure 1). In all
containers the air in the head space is replaced by high purity
nitrogen gas to prevent oxidation.

     At the end of the 48-h agitation period, the bulk of the
sample is decanted and filtered. Many samples may be filtered
through a large-diameter 0.45-μm membrane filter immediately.
Some require prefiltration in addition to decanting to remove
larger suspended solids and prevent the final filter from
clogging. Pressurized filtration using high purity nitrogen at

Table 3. EPA Level 1 Modifications to the ASTM-A Leaching
          Procedure for Use in the Preparation of Large-Volume
          Leachates for Ecological Tests

ASTM-A Procedure[a]	EPA Level 1 Modifications
Grind if necessary to pass 9.5-mm sieve.	Same.
Dry for 18 h at 104°C.	Do not dry.
Weigh 350- or 700-g portion for testing.	Weigh 700-g, 3.75-kg, 22.5-kg, or 41.2-kg portion for testing.
Add 1.4 1 or 2.8 1 distilled water.	Add 2.8, 15, 90, or 165 1 distilled water.
Agitate continuously for 48 h.	Same.
Prefilter to separate bulk of aqueous phase from solids.	Same.
Final filter through 0.45-μm membrane filter.	Same.

[a]"Proposed Method for Leaching Waste Materials," ASTM, 1916 Race
Street, Philadelphia, PA 19103.

pressures up to 80 psi is used in place of vacuum filtration to
speed the filtration rate. Some soil samples with high clay
content have proven essentially unfilterable in large volumes.
Particle size in such samples is in the submicron and greater
range. In one situtlon, rapid clogging of the 0.45-μm filter,
reducing its flow to 20 ml/h or less, could not be prevented by
prefiltration. In such cases it may be necessary to use only
prefiltration via centrifugation and dispense with the final
0.45-μm filter altogether. A large-volume centrifuge (4 to 8 1)
is needed to perform this operation efficiently.

The adjustment of pH may be desirable for leachates whose pH
is outside the normal tolerance range of the test organisms (see
Table 2) to prevent pH from masking the effects of the other
constituents of the leachate (see Liquid Samples). Acetic acid

Figure 1.   Apparatus for preparing large-volume leachates
             including modified drum roller and steel drum with
             114-1 polyethylene liner.

has been used for this purpose in the EP and ASTM-B procedures.
Problems encountered include interference with both chemical
analyses (Bause and McGregor, 1980) and the bioassay results
(Epler et al., 1980) caused by the large quantity of acetate
present (as much as 400 ml of acetic acid may be added to 1600 ml
of leachate in the EP procedure). In the Daphnia test, results
may be confounded by the fact that the acetate can be used as a
food source for bacteria which in turn are eaten by the Daphnia,
enhancing their survival. A comparative study between the
modified ASTM-A procedure (without pH adjustment) and the EP
procedure using two fluidized-bed combustion (FBC) solid waste

streams revealed some large differences in the Daphnia and algal test results between the two techniques (Table 4) (Eischen and Duke, 1980b). No pattern in the differences was discernible and the lack of detailed chemical analyses on the two leachates prevented definitive conclusions as to the cause. The raw data for Daphnia did reveal a marked increase in survival when the pH of the test solution fell within its tolerance range. Another problem concerning the use of a weak acid such as acetic acid was its inability to significantly alter the pH in highly buffered leachates. The pH of one FBC leachate was 12.2 before the addition of the acetic acid and 12.1 after (Eischen and Duke, 1980a,b). The net result was to add significant amounts of acetate to the leachate without altering the pH. Strong acids (HCl) and bases (NaOH) have been used to adjust the pH of complex mixtures successfully (Battelle Columbus Laboratories, 1981). Here considerably smaller quantities of acid are needed in most cases and the resulting ions (Cl^-, Na^+) are biologically compatible.

The leachate is added directly to the aquatic tests. It may also be applied in the plant tests to simulate root uptake of the leached material. No special techniques for sample application are necessary. In the marine tests, the test solution (leachate

Table 4. Comparison of Toxicity of Leachates of Two Fluidized-Bed Combustion Samples Prepared Using the Modified ASTM-A and EP Techniques

Sample/Test	Leachate Preparation Method	
	ASTM-A (%)	EP (%)
Regenerator cyclone		
Daphnia LC_{50}	15	11
Algal EC_{50}	4.6	0.55
Second cyclone		
Daphnia LC_{50}	5.5	67.4
Algal EC_{50}	6.3	0.35

plus dilution water) may require addition of marine salts (Rila mixture) to meet salinity requirements.

Slurries, containing more than 5% solids, may be added directly to the required ecological tests. Fractionation into aqueous and solid phases and testing both phases may facilitate testing and provide data on the toxicity of each phase. If significant organics are present, fractionation using an XAD-2 resin may be done. The resin column must be extracted and the resultant solution solvent exchanged into a biologically compatible solvent such as DMSO or acetone. Application may be made directly to the ecological tests.

SUMMARY

The preparation and application of complex environmental samples to ecological tests often requires specialized techniques. Most gaseous samples are "prepared" during their collection (particulate removal, cooling) and applied directly to the plant stress ethylene test. Many liquid samples may also be applied directly without preparation. Samples with significant organics may be extracted and the extract solvent exchanged and tested. Adjustments of pH using a strong acid or base may be required by samples with pH outside the tolerance range of the test organism. This permits the evaluation of the toxicity of the other chemical constituents of the sample whose effect might otherwise be masked by the pH effect. Special techniques for sample application, such as foliar application, are occasionally required for liquids. Solid and slurry samples can require more diverse preparation techniques including grinding and sizing, leaching and filtration, pH adjustment (of leachates), and extraction of organics and solvent exchange. In all the preparation and application procedures, care must be taken to make the sample compatible with the ecological test while altering it as little as possible.

REFERENCES

Battelle Columbus Laboratories. 1981. Ecological effects testing of Georgetown University fluidized bed combustion emissions. Battelle Columbus Laboratories: Columbus, OH. 47 pp.

Bause, D.E. and K.T. McGregor. 1980. Comparison of four leachate-generation procedures for solid waste characterization in environmental assessment programs. EPA-600/7-80-118. U.S. Environmental Protection Agency, Industrial Environmental Research Laboratory: Research Triangle Park, NC. 107 pp.

Brusick, D.J., R.R. Young, C. Hutchinson, A.G. Dilkas, and T.A. Gezo. 1980. IERL-RTP Procedures Manual: Level 1 Environmental Assessment Biological Tests. Litton Bionetics, Inc.: Kensington, MD. 178 pp.

Davis, P. and G.W. Ozburn. 1969. The pH tolerance of Daphnia pulex (Leydig, Emend., Richard). Can. J. Zool. 47:1173-1175.

Dowden, B.F. and H.J. Bennett. 1965. Toxicity of selected chemicals to certain animals. J. Water Pollut. Control Fed. 37(9):1308-1316.

Duke, K.M., M.E. Davis, and A.J. Dennis. 1977. IERL-RTP Procedures Manual: Level 1 Environmental Assessment Biological Tests for Pilot Studies. EPA-600/7-77-043. U.S. Environmental Protection Agency: Washington, DC. 117 pp.

Eischen, M.A. and K.M. Duke. 1980a. Health and ecological effects testing of Exxon miniplant fluidized bed combustion emissions. Battelle Columbus Laboratories: Columbus, OH. 102 pp.

Eischen, M.A. and K.M. Duke. 1980b. Comparison of two leachate preparation methods using selected Level 1 bioassays. Battelle Columbus Laboratories: Columbus, OH. 30 pp.

Epler, J.L., F.W. Larimer, T.K. Rao, E.M. Burnett, W.H. Griest, M.R. Guerin, M.P. Maskarinec, D.A. Brown, N.T. Edwards, C.W. Gehrs, R.E. Milleman, B.R. Parkhurst, B.M. Ross-Todd, D.S. Shriner, and H.W. Wilson, Jr. 1980. Toxicity of leachates. EPA-600/2-80-057. U.S. Environmental Protection Agency, Municipal Environmental Research Laboratory: Cincinnati, OH. 142 pp.

Miller, W.E., J.C. Green, and T. Shiroyama. 1978. The Selenastrum capricornutum printz algal assay bottle test: Experimental design, application, and data interpretation protocol. EPA-600/9-78-018. U.S. Environmental Protection Agency, Environmental Research Laboratory: Corvallis, OR. 132 pp.

Mount, D.I. 1973. Chronic effect of low pH on fathead minnow
    survival, growth and reproduction. Water Res. 7:987-993.

U.S. Environmental Protection Agency. 1978. Part 250--Hazardous
    waste guidelines and regulations. Fed. Regist.
    43(243):58954-59028.

SAMPLE COLLECTION AND PREPARATION METHODS AFFECTING MUTAGENICITY

AND CYTOTOXICITY OF COAL FLY ASH

Judy L. Mumford and Joellen Lewtas

Genetic Toxicology Division, Health Effects Research
Laboratory, U.S. Environmental Protection Agency, Research
Triangle Park, North Carolina 27711

INTRODUCTION

Reports by several investigators describing the biological
activity of coal fly ash have presented a variety of results which
in some cases (Fisher et al., 1979; Clark and Hobbs, 1980;
Kubitschek et al., 1980; Mumford and Lewtas, in press) are
conflicting. The biological activity of coal fly ash may differ
because of one or or more of the following factors: (1) the
samples studied were from different sources; (2) the samples were
prepared for bioassay differently; (3) the sampling method
differed, and, therefore, collected samples were different in
chemical or physical properties which affect the biological
activity. Several variables involved in coal fly ash
studies -- source, sample collection and preparation methods,
bioassay method -- are undoubtedly responsible for the diversity
of biological effects observed. The objectives of this study were
to examine the sample preparation and collection factors which may
affect the observed biological activity caused by coal fly ash and
to evaluate the mutagenicity and cytotoxicity of fluidized-bed
combustion (FBC) fly ash from experimental and commercial units.
The bioassays used in this study were the Ames Salmonella plate
incorporation test for mutagenicity and the rabbit alveolar
macrophage (RAM) system for cytotoxicity.

Worldwide use of coal as an energy source is predicted to increase substantially in the next two decades. The burning of coal by conventional combustion (CC) methods presents a number of environmental problems. Moreover, the addition of new control methods, such as flue gas scrubbing, may not always be economical in conventional plants. Therefore, it has been of great interest to develop alternative coal combustion technologies that are environmentally acceptable and economically feasible.

Of the new coal technologies under development, FBC is closest to large-scale commercialization. A limited number of commercial FBC units are now available. A commercial unit currently in operation at Georgetown University, Washington, DC, was the source of several of the emissions reported here. FBC systems for the production of steam or electricity have several advantages over CC systems:

(1) High heat transfer coefficients and volumetric heat release rates permit a reduced boiler size, which, in turn, lowers capital cost.

(2) Limestone is used as the bed sorbent material to remove SO_2, thus reducing SO_2 emissions and permitting use of high sulfur coal.

(3) The lower FBC temperature range (800 to 900°C, vs. the CC range of 1400 to 1600°C) can decrease NO_x emissions (Fennelly et al., 1977).

The low FBC combustion temperature, however, may decrease the combustion efficiency and increase the emission of mutagenic organics. Kubitschek et al. (1980) and Clark and Hobbs (1980) have reported mutagenic effects of coal fly ash from experimental FBC units. It has been speculated that the lower FBC combustion temperatures increase the emission of polycyclic organic matter (Kubitschek and Williams, 1980).

SAMPLE PREPARATION STUDIES

Two fly ash samples -- one from an FBC source and a reference sample from a CC source -- were used to study sample preparation methods for the mutagenicity bioassay. The CC fly ash sample was collected from a conventional coal-fired power plant burning Alabama Eastern bituminous coal with 1 to 2% sulfur and 11% ash. The combustion temperature was about 1400°C. The sample was collected downstream of an electrostatic precipitator (ESP) by the fabric filter of a high-volume particulate sampler. Sample

collection temperature was ~ 132°C. The fly ash particle size
was, in general, < 3 μm, with a geometric mean diameter of
1.05 μm. The FBC fly ash sample was collected from a pressurized
FBC miniplant (combustor diameter was 33 cm) at Exxon burning
Pennsylvania Eastern bituminous coal with 2% sulfur and 8% ash.
The combustor temperature was about 870°C. The sample was
collected at ~ 165°C by a fabric filter after three stages of
conventional cyclones. A detailed description of the Exxon FBC
miniplant and the sample collection has been reported (Kindya
et al., 1980). The particle size was < 3 μm with a geometric mean
diameter of 0.54 μm.

Mutagenicity Studies

    Two sample preparation methods were examined for Salmonella
mutagenicity assay: direct testing of particles and testing of
the extractable organics. The Ames Salmonella typhimurium plate
incorporation method (Ames et al., 1975) with the following
modification was employed. The standard minimal histidine was
added to the plating media instead of the soft agar overlay. The
plates were incubated 72 h (Claxton et al., 1979). For metabolic
activation a 9000-g supernatant (S9) of a liver homogenate was
prepared using Arochlor 1254-induced male rats (Charles River
CD 1; Ames et al., 1975). Negative and positive control tests
were conducted with each experiment. A sample was considered
positive if it showed a maximum response two times greater than
the spontaneous rate and if it gave a positive, linear, dose
response. The linear portion of the dose-response curve was used
to calculate a linear regression line. The slope of the linear
regression was reported as revertants per weight unit of samples
tested. Figure 1 presents a comparison of the mutagenicity of FBC
and CC fly ash samples, based on experiments in which the
particles are placed directly in agar overlay. The FBC fly ash
particles without metabolic activation gave a dose response for
mutagenic activity, whereas CC fly ash showed negative response
with and without activation. With activation, FBC fly ash also
showed a negative response for mutagenicity. The direct testing
of fly ash particles indicated the bioavailability of mutagens and
required much smaller sample size. This method, however, was less
sensitive than solvent extraction prior to bioassay.

    For solvent extraction studies, sonication and Soxhlet
extraction methods using a series of the organic solvents with
different polarity were examined to determine the most efficient
methods in removing the mutagens from the FBC and CC fly ash
samples. The four solvents examined were cyclohexane,
dichloromethane, acetone, and methanol, with polarity indices of

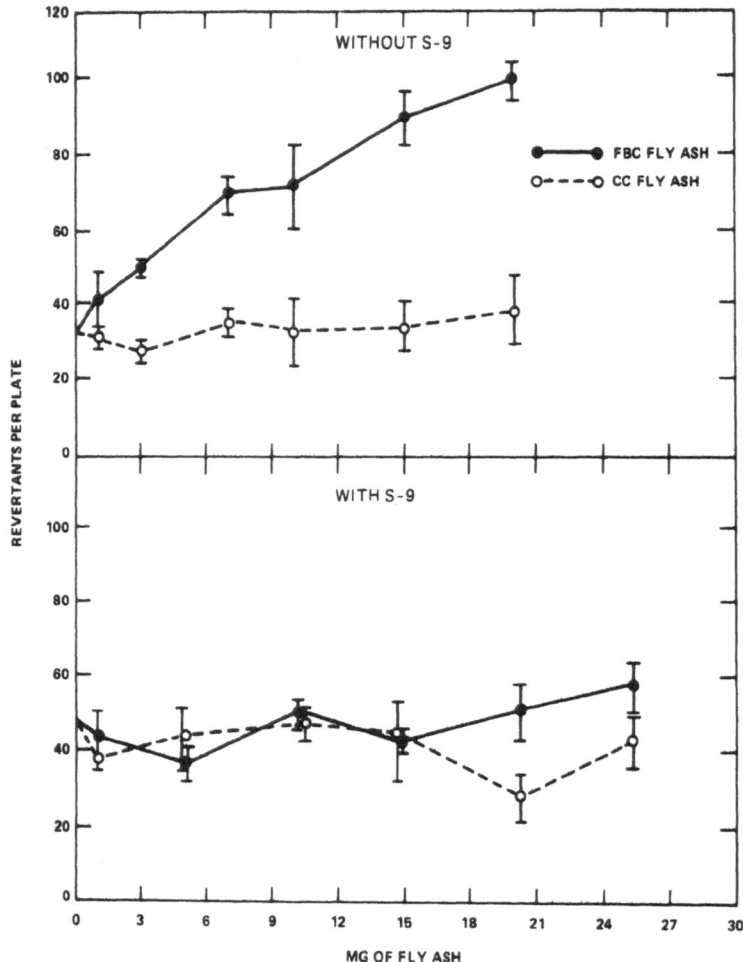

Figure 1.   Mutagenicity of fly ash particles in Salmonella
typhimurium TA98. Data presented are means ± standard
deviation of three replicates.

0.0, 3.4, 5.4, and 6.6, respectively. The detailed extraction
methods were reported previously (Mumford and Lewtas, in press).
The results of the solvent extraction studies for percent of
extractable mass are shown in Figure 2. These data show that
Soxhlet extraction and sonication yielded comparable extractable
mass for each solvent. Also shown is that the more polar solvents
acetone and methanol extracted more mass; extraction with the less
polar solvents cyclohexane and dichloromethane yielded lesser

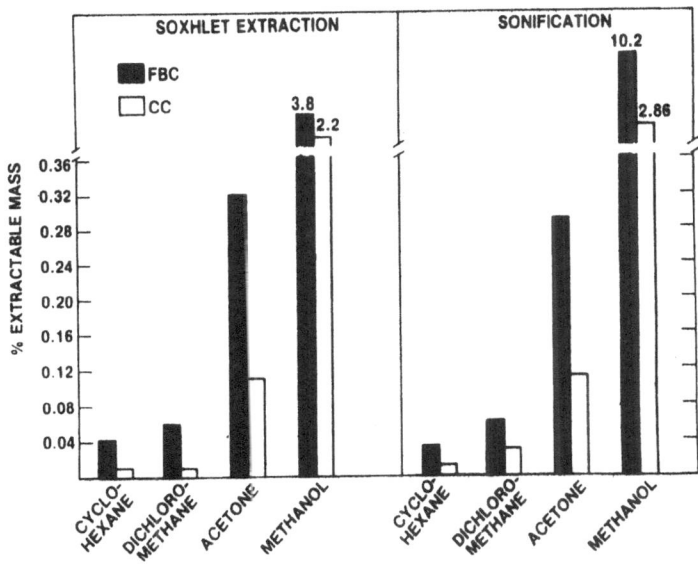

Figure 2.   Percent of extractable mass.

amounts of extractable mass. The FBC fly ash sample consistently contained more organic solvent extractable mass than CC fly ash, even though the FBC fly ash collection temperature was 33°C higher than the CC fly ash collection temperature.

The mutagenicity of the organic extracts of fly ash, determined with Salmonella typhimurium TA98 without S9 activation, is shown in Figure 3. Dichloromethane extraction resulted in the highest mutagenic activity per unit of extractable mass, followed by acetone and cyclohexane. The methanol extractable material produced a negative response in mutagenicity tests. These data show that this FBC fly ash sample has higher mutagenic activity than the CC fly ash sample for all the solvents which removed mutagens.

In Table 1, mutagenicity of the fly ash sample is expressed as revertants per microgram of extractable mass and as revertants per milligram of equivalent fly ash. Since acetone extraction yielded a relatively high extractable mass and moderate mutagenicity, this extraction gave the highest number of revertants per milligram of equivalent fly ash. These data show that addition of S9 generally decreased the mutagenic activity. This indicated the presence of direct-acting mutagens in both fly ash samples.

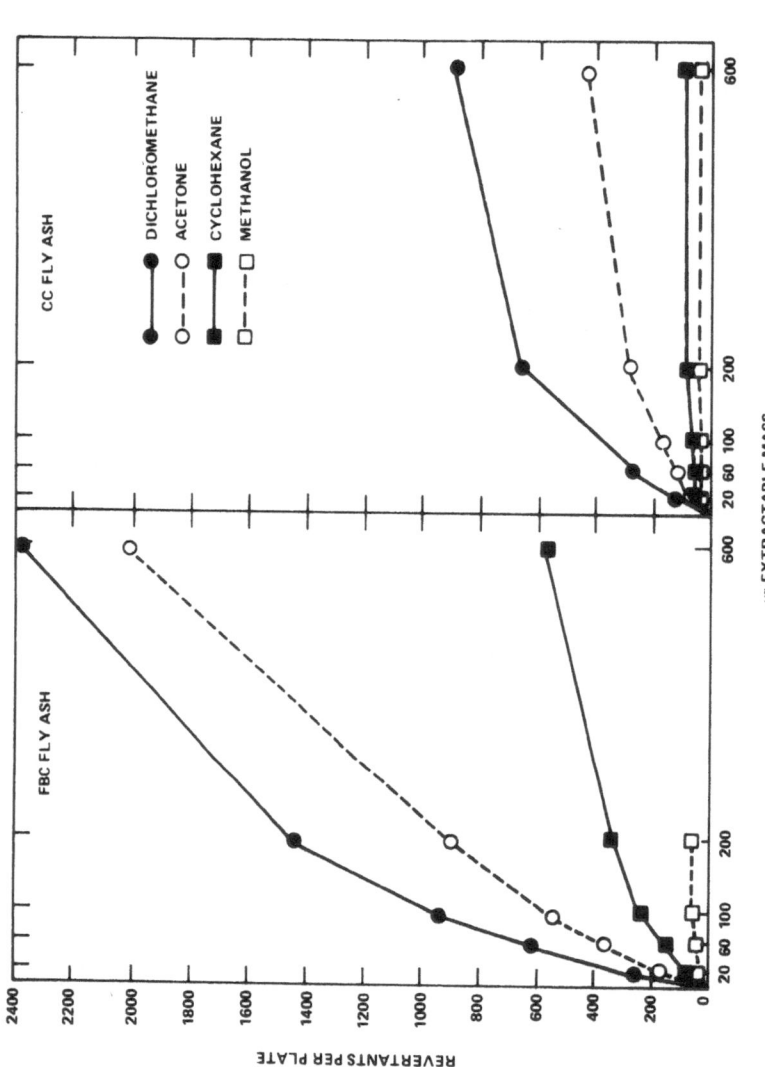

Figure 3. Mutagenicity of organic extracts of coal fly ash in Salmonella typhimurium TA98 without metabolic activation. Each point is an average of two replicates.

Table 1. Mutagenicity of Fly Ash in the <u>Salmonella typhimurium</u> Plate Incorporation Assay Using TA98

Sample	Extraction Solvent	% Extractable Mass	Revertants per microgram extractable mass[a]		Revertants per milligram equivalent fly ash[b]	
			With Activation	Without Activation	With Activation	Without Activation
FBC fly ash	Dichloromethane	0.06	5.21	9.43	3.13	5.66
	Acetone	0.30	1.33	4.45	3.99	13.35
	Methanol	7.0	neg	neg	neg	neg
	Cyclohexane	0.04	0.87	1.71	0.35	0.68
	None	--	--	--	neg	3.32[c]
CC fly ash	Dichloromethane	0.02	--	3.07	--	0.61
	Acetone	0.11	0.43	1.34	0.47	1.47
	Methanol	2.55	neg	neg	neg	neg
	Cyclohexane	0.01	--	0.10	--	0.01
	None	--	--	--	neg	neg

[a]Revertants per microgram of organics obtained by computing the slope of the linear regression in the linear portion of the curve.

[b]Revertants per milligram equivalent calculated by multiplying revertants per microgram of organics times the number of microgram organics per milligram of fly ash.

[c]Revertants per milligram of fly ash obtained by computing the slope of the linear regression from the testing of particles directly.

Since methanol extraction resulted in high yield of mass but showed no mutagenic activity, elemental and anion analyses were conducted to determine if methanol removed primarily inorganic matter. Indeed, a considerable amount of inorganics, especially the sulfates, was extracted by the methanol, as shown in Table 2. The methanol extract also showed a lower percentage of carbon in comparison to the acetone extract, which would indicate low organic content. The low organic content and the high inorganic matter of the methanol extract may account for the negative mutagenic response.

Cytotoxicity Studies

The alveolar phagocytic macrophages have been generally considered essential for defending the host against inhaled particles and other toxicants. The rabbit alveolar macrophage (RAM) system was used to evaluate the cytotoxicity of fly ash particles in this study. The technique for the assay has been reported previously (Mumford and Lewtas, in press). The RAM cells were exposed to FBC or CC fly ash particles for 20 h at 37°C in vitro. At the end of the incubation period, the cytotoxicity indices of viability, viability index, adenosine triphosphate, and protein synthesis were evaluated. The results are shown in Figure 4. These data show that FBC and CC fly ash samples both exerted dose-response toxic effects on the RAM cells. The FBC fly ash sample appears to be more cytotoxic than CC fly ash sample.

Table 2. Elemental and Anion Analysis

Sample	Extractable Mass (%)	C	H	F^-	Cl^-	$SO_4^=$
		(%)		($\mu g/g$)		
FBC fly ash						
Acetone extract	0.30	26.3	5.3	1.9	485.0	61.7
Methanol extract	7.0	5.1	3.7	178.3	672.7	3110.4
CC fly ash						
Acetone extract	0.11	38.3	5.9	0.8	22.6	12.7
Methanol extract	2.55	3.2	4.4	120.5	38.9	5777.7

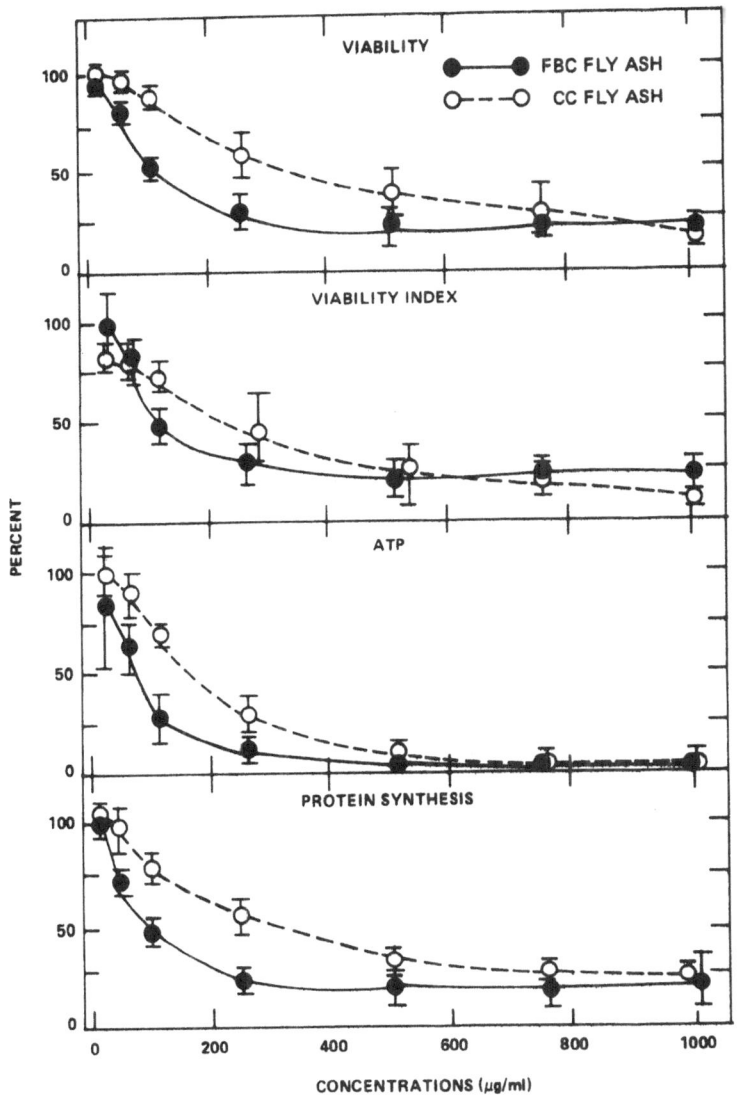

Figure 4. Toxic effects of fly ash on RAM cells after a 20-h
exposure _in vitro_. Data presented are means ± standard
deviation of three experiments (nine replicates).
Viability is calculated and expressed in percent viable
cells. Viability index was obtained by multiplying
cell viability by total cell numbers as a fraction of
the control. Adenosine triphosphate, ATP (fg/ml), and
protein synthesis (disintegrations per minute) are
expressed as percent of control.

SAMPLE COLLECTION STUDIES

The three devices most commonly used to remove particles from gas effluents are the cyclone, the electrostatic precipitator (ESP), and the baghouse filter. These collectors are based on different principles for removing particles from the air. Cyclones collect particles by centrifugal separation. Baghouse filters collect particles by forcing the gas effluent through the filters and trapping the solids. An ESP collects particles when gas effluent containing the particles passes through the device's electric discharge area, where ionization of the gas occurs. The ions produced will collide with the suspended particles, causing an electric charge within the particles. These charged particles drift toward an electrode with the opposite charge and are deposited on the plates. These different collection or sampling methods, therefore, result in the selective collection of particles with particular properties, which may produce different biological activity in bioassays.

At the Exxon FBC miniplant, the effluent stream passes through three stages of high efficiency, conventional cyclones in the plant and then through the ESP or baghouse filter outside the plant. Four FBC fly ash samples collected by the two cyclones (secondary and tertiary cyclones), the fabric filter, and the ESP at various temperatures were examined and compared for particle size, toxicity, and mutagenicity. The collection temperatures were 830°C for the secondary cyclone, 650°C for the tertiary cyclone, and 170°C for the filter and for the ESP. The median volume diameters of the four fly ash samples were 4.6, 3.0, 1.7, and 1.5 µm for the secondary cyclone, tertiary cyclone, ESP, and filter samples, respectively. Figures 5 and 6 present the results of cytotoxicity studies with the RAM assay and mutagenicity studies using the Salmonella assay, with and without activation. These data show that the two cyclone samples, collected at higher temperatures, contained larger particles which were very weakly cytotoxic or mutagenic. The filter and ESP samples, collected downstream of the cyclones at lower temperatures, contained finer particles that showed higher cytotoxic and mutagenic activity in the test systems. The ESP sample was the most mutagenic.

Biological activity vs. sample collection temperature was plotted (Figure 7), and, at lower collection temperatures, greater cytotoxic effects of the fly ash samples were observed. The increased cytotoxicity may be due to condensation of organics or volatile trace elements on the fly ash surface. When the biological activity vs. the particle size of the fly ash samples (Figure 8) was plotted, cytotoxicity correlated better with the particle size. In this case, the cytotoxicity of the fly ash

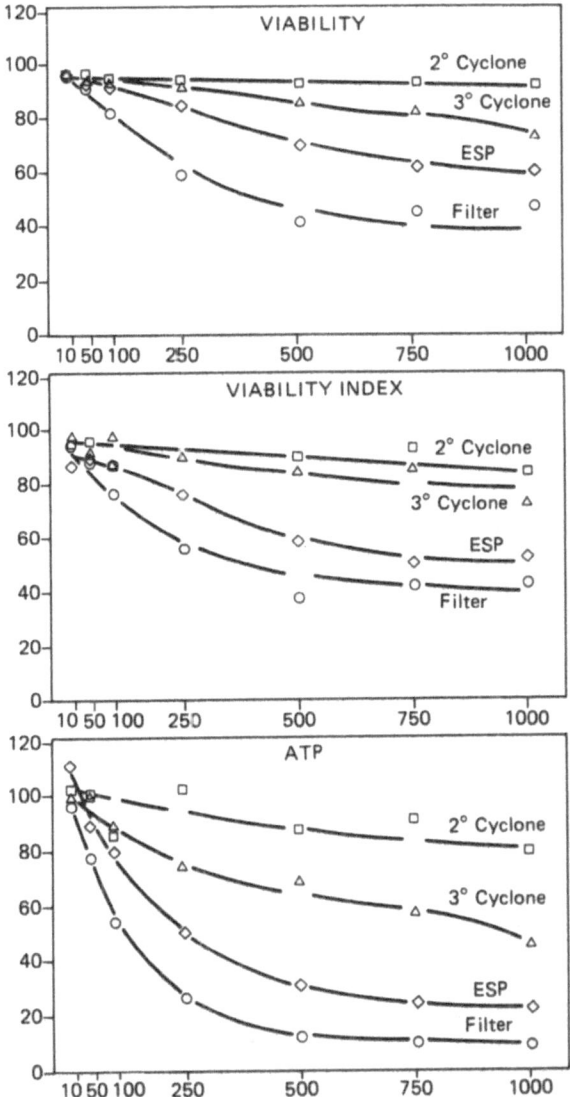

Figure 5. Toxic effects of fly ash on RAM cells after a 20-h
exposure in vitro. Data presented are means of six
replicates. Viability is calculated and expressed as
percent viable cells. Viability index was obtained by
multiplying cell viability by total cell numbers as a
fraction of the control. Adenosine triphosphate, ATP
(fg/ml), is expressed as percent control.

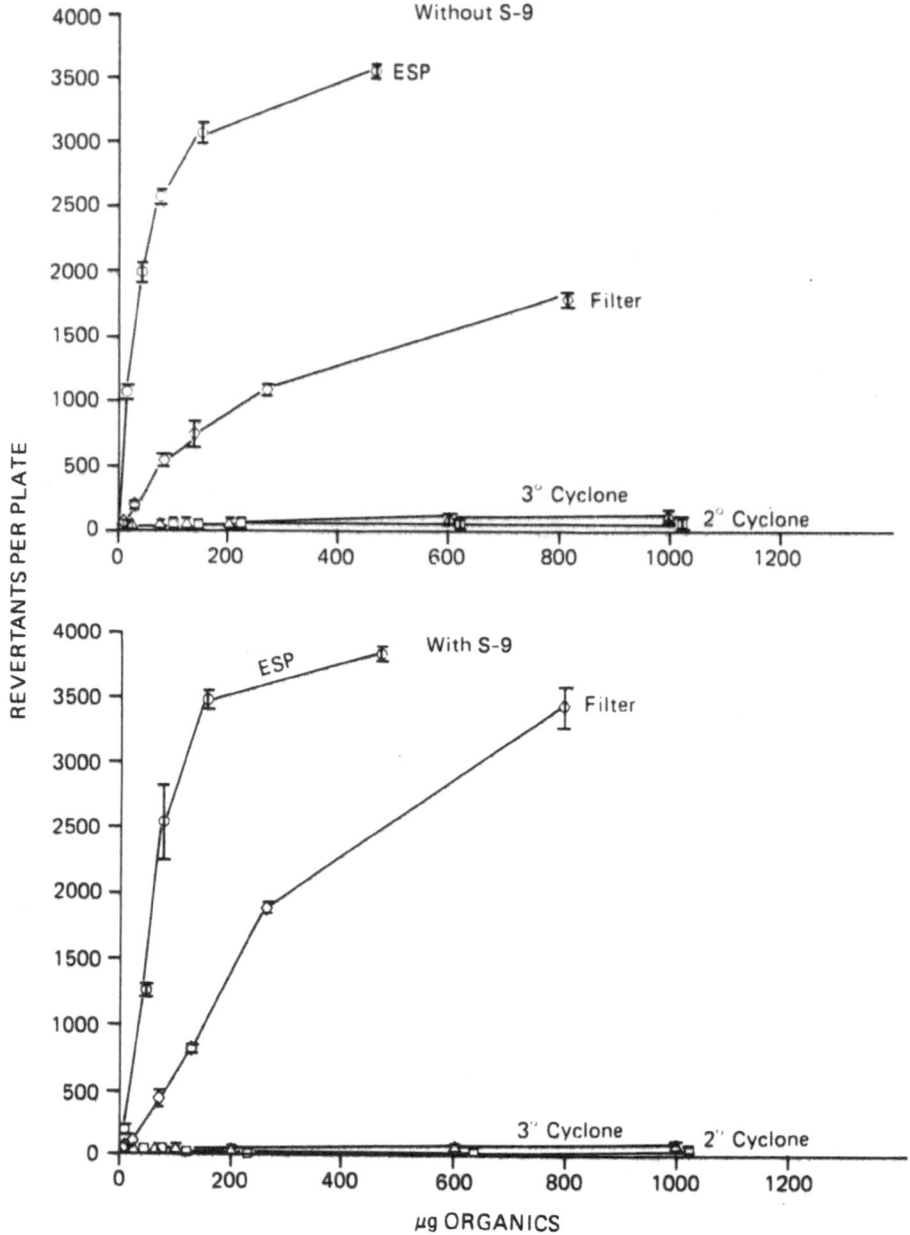

Figure 6. Mutagenicity of dichloromethane extracts of fly ash in
Salmonella typhimurium TA98. Data presented are means
± standard deviation of three replicates.

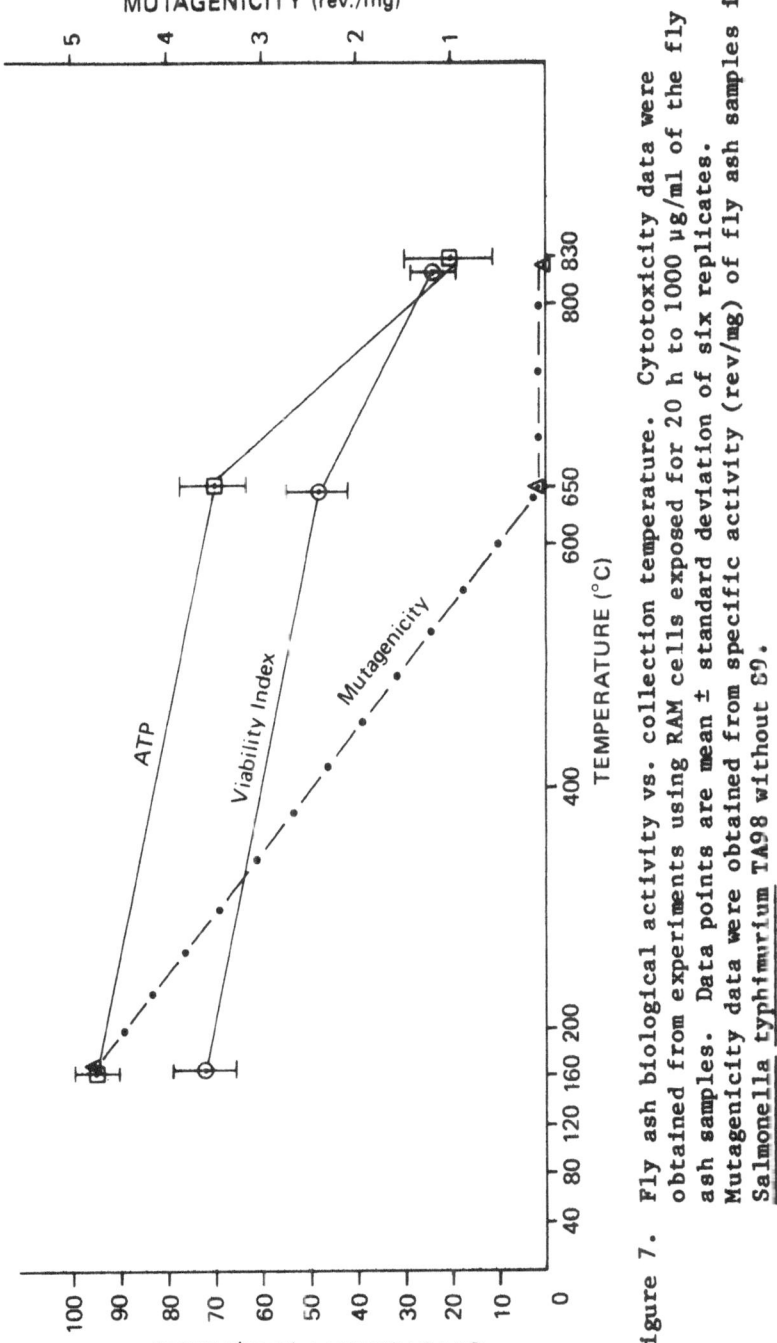

Figure 7. Fly ash biological activity vs. collection temperature. Cytotoxicity data were obtained from experiments using RAM cells exposed for 20 h to 1000 µg/ml of the fly ash samples. Data points are mean ± standard deviation of six replicates. Mutagenicity data were obtained from specific activity (rev/mg) of fly ash samples in Salmonella typhimurium TA98 without S9.

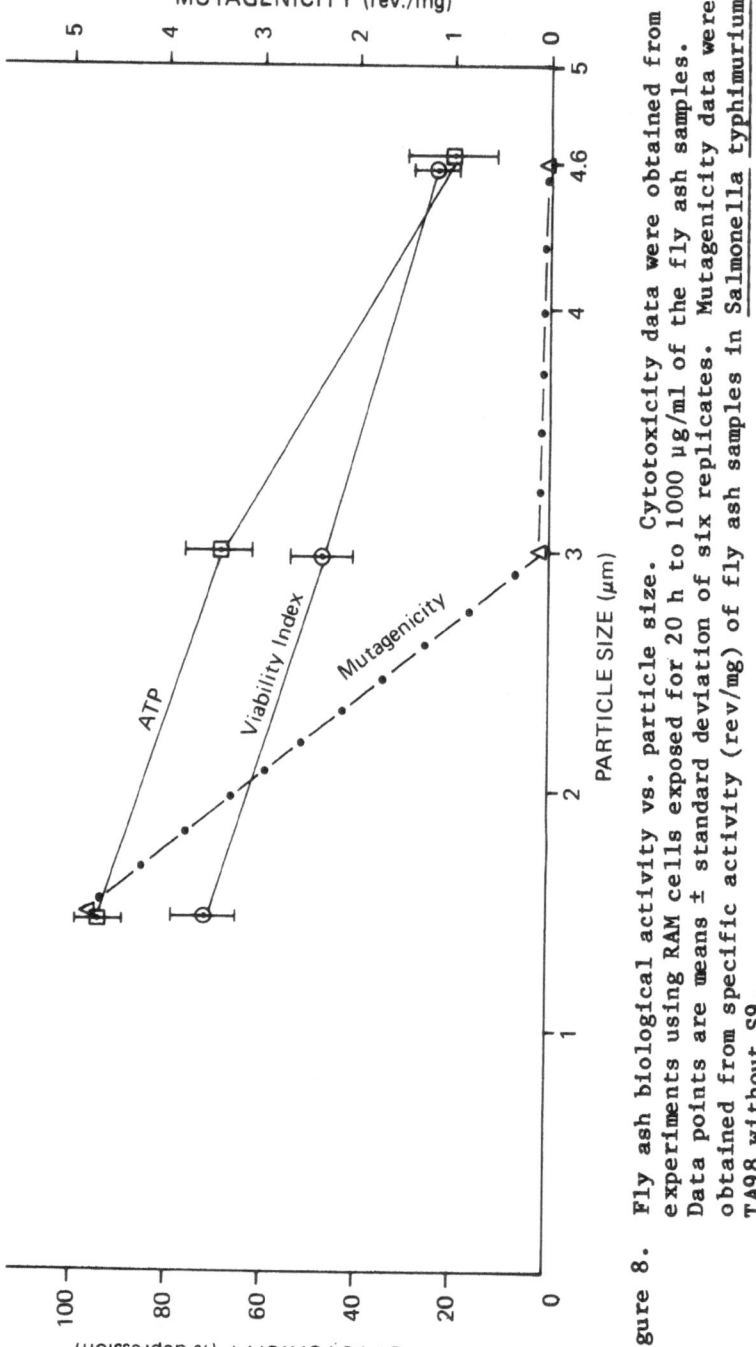

Figure 8. Fly ash biological activity vs. particle size. Cytotoxicity data were obtained from experiments using RAM cells exposed for 20 h to 1000 μg/ml of the fly ash samples. Data points are means ± standard deviation of six replicates. Mutagenicity data were obtained from specific activity (rev/mg) of fly ash samples in Salmonella typhimurium TA98 without S9.

appeared to be more size-dependent than temperature-dependent. The smaller particles have a larger specific area and can adsorb more potentially toxic trace elements and organics on their surface.

Mutagenicity of the samples, however, is more collection-temperature-dependent: The collection temperature must be lowered to a certain point before mutagenic effects are seen. When samples were collected in the high temperature region, mutagenicity did not increase with decreasing particle size. Thermal studies conducted by Fisher et al. (1979) showed that when coal fly ash was heated to 350°C or higher, mutagenicity was completely lost. Thus, the collection temperature, which can affect the degree of adsorption and the kind of organics deposited onto fly ash surface, appears to play an important role in determining the mutagenicity of the fly ash collected. When the sample is collected at a low temperature, in a range where adsorption of organics occurs, the particle size or specific area of particles will be important in determining the mutagenicity of fly ash.

For investigating the biological effects of coal fly ash, therefore, it is important to collect the samples at cooler temperatures after air dilution, if possible, or to evaluate the vapor phase of the gas effluent downsteam of the particulate collection device when particles are collected at a high temperature. In the Georgetown FBC unit, we examined fly ash samples collected at a fairly high temperature (163°C) and an XAD extract of gas effluent.

The Georgetown University atmospheric FBC boiler is one of the largest commercial applications of fluidized bed coal combustion in this country. Currently, it heats and cools 204,000 m^2 of space in the university. The boiler produces 45,360 kg of steam per hour. Five tons of Eastern bituminous coal containing 3.3% sulfur are burned each hour at ~ 840°C. Limestone is used for the bed material. The primary particulate control devices are mutlicyclones and a baghouse (Young et al., 1981).

Fly ash samples from the Georgetown FBC unit examined were baghouse particulate samples, collected at 163°C, and the vapor phase of the gas effluent, collected by the XAD-2 trap of a Source Assessment Sampling System (SASS) train at 23°C, downstream of the baghouse (Young et al., 1981). The results of the Ames Salmonella mutagenicity assay of these samples are shown in Figure 9. The data show that organic extracts of both gas and particulate effluents are mutagenic in the Salmonella typhimurium assay in strains TA98, TA1538, and TA100, without metabolic activation.

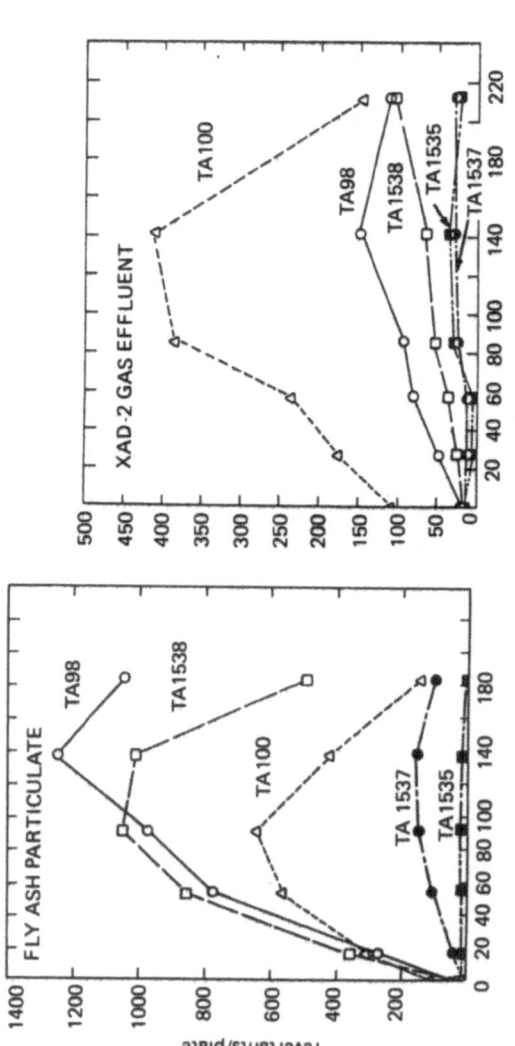

Figure 9. Mutagenicity of the dichloromethane extracts from the gas and particulate effluents of Georgetown FBC in Salmonella typhimurium without S9. Data presented are means ± standard deviation of three replicates.

The organic extract from fly ash was much more mutagenic than the organics obtained from the gas effluent. The mutagenicity of the fly ash extract was 9.26 revertants and 10.0 revertants per microgram of organics in TA98 and TA100, respectively, without activation. The mutagenicity of the extract of the XAD-2 gas effluent sample was 0.96 revertants and 3.92 revertants per microgram of organics in TA98 and TA100, respectively, without activation. The fly ash extract showed comparable high mutagenicity in TA98, TA1538, and TA100, and showed negative response in TA1535. The extract of XAD-2 gas effluent showed much higher mutagenic activity in TA100 than in TA98 or TA1538. Some mutagenic activity was observed in TA1535 for gas effluent. These results suggest that the gas and fly ash effluents may contain different mutagens. The higher activity of the gas effluent in TA100 suggests that it contains both base-pair substitution and frameshift mutagens, whereas the particulate effluent activity in the various tester strains is more consistent with the presence of primarily frameshift mutagens. Addition of S9 for metabolic activation to the fly ash organics generally decreased the mutagenic activity (from 9.26 revertants to 4.65 revertants per microgram of organics in TA98). This indicates that the mutagens in fly ash were direct-acting.

Polycyclic organic matter is known to be emitted from coal combustion. This study showed that the mutagens in coal fly ash are primarily direct-acting and frameshift. Some nitroaromatic compounds have been found to be potent, direct-acting, and frameshift mutagens in the Ames Salmonella assay (Rosenkranz et al., 1980) and are found in combustion products, for instance, diesel emissions (Schuetzle et al., 1981). One nitroaromatic compound, 1-nitropyrene, was detected in the coal combustion products. Table 3 shows the nitropyrene concentration of the particulate and gas phase effluents from the coal combustion sources that was determined by a reverse-phase HPLC fluorescence method (Tejada et al., 1982).

CONCLUSIONS

FBC and CC fly ash particles were found to be cytotoxic in vitro in mammalian lung cells. The organic extracts of FBC and CC fly ash samples contained direct-acting and frameshift mutagens. The direct-acting, frameshift mutagen 1-nitropyrene was detected in the coal combustion products. The fly ash particle size, collection temperature, and sample preparation method were shown to significantly affect the bioassay results. The fly ash particle size, which is affected by the collection site and device, is important in determining the toxicity of fly ash

Table 3.   1-Nitropyrene in Coal Combustion Products

Product	1-Nitropyrene Concentration	
	(ng/g fly ash)	(µg/g organics)
Fly Ash		
FBC--Exxon Miniplant[a]	130	161.1
FBC--Georgetown[b]	0.09	1.7
CC--Power Plant[a]	0.11	0.56
Gas Effluent		
FBC--Georgetown (XAD Extract)		8.0

[a]Respirable particles (< 3 µm).
[b]Larg particles (< 400 µm).

samples collected. The mutagenicity of fly ash can be affected by collection temperature. This finding is further substantiated by the presence of mutagenic activity in the XAD extract of the gas effluent following the collection of the particles at a high temperature. Careful consideration of the above factors is required before any meaningful comparison can be made of the biological data presented by the various investigators.

Fluidized-bed combustion of coal technology is new and still developing. Any improvement in technology such as combustion efficiency can change the biological effects of the fly ash. As the technology develops, fluidized-bed combustion may provide one of the best alternative coal technologies. These studies suggest that monitoring the mutagenicity and cytotoxicity of the emissions during the developmental stage of this technology may help to direct improvements so that the mutagenic and cytotoxic components of the emissions can be minimized.

REFERENCES

Ames, B.W., J.M. McCann, and E. Yamasaki. 1975. Methods for
    detecting carcinogens and mutagens with the Salmonella/
    mammalian microsome mutagenicity test. Mutation Res.
    31:347-364.

Clark, R.C., and C.H. Hobbs. 1980. Mutagenicity of effluents
    from an experimental fluidized bed coal combustor. Environ.
    Mutagen. 2:101-109.

Claxton, L.D., J. Huisingh, and M. Waters. 1979. The testing of
    environmental samples for mutagenicity and carcinogenicity
    using microbial assay systems. In: Symposium Proceedings:
    Textile Industry Technology, December 1978, Williamsburg, VA.
    F.A. Ayers, compiler. EPA-600/2-79-104. U.S. Environmental
    Protection Agency: Research Triangle Park, NC. pp. 231-233.

Fennelly, P.F., H. Klemm, and R.R. Hall. 1977. Coal burns
    cleaner in a fluid bed. Environ. Sci. Technol. 11:244-248.

Fisher, G.L., C.E. Chrisp, and O.G. Raabe. 1979. Physical
    factors affecting the mutagenicity of fly ash from a
    coal-fired power plant. Science 204:879-881.

Kindya, R.J., R.R. Hall, G. Hunt, W. Piispanen, and P. Fennelly.
    1980. Environmental Assessment: Source Test and Evaluation
    Report -- Exxon Miniplant Pressurized Fluidized-Bed Combustor
    with Sorbent Regeneration. GCA Corporation: Bedford, MA.
    pp. 8-25.

Kubitschek, H.E., and D.M. Williams. 1980. Mutagenicity of fly
    ash from a fluidized-bed combustor during start-up and steady
    operating conditions. Mutation Res. 77:287-291.

Kubitschek, H.E., D.M. Williams, and F.R. Kirchner. 1980.
    Biological monitoring of fluidized-bed combustion operations
    I. Increased mutagenicity during periods of incomplete
    combustion. In: Short-Term Bioassays in the Analysis of
    Complex Environmental Mixtures II. M.D. Waters, S.S. Sandhu,
    J. L. Huisingh, L. Claxton, and S. Nesnow, eds. Plenum
    Press: New York. pp. 411-420.

Mumford, J.L., and J. Lewtas. In press. Mutagenicity and
    cytotoxicity of coal fly ash from fluidized-bed and
    conventional combustion. J. Toxicol. Environ. Health.

Rosenkranz, H.S., E.C. McCoy, D.R. Sanders, M. Butler,
    D.K. Kiriazides, R. Mermelstein. 1980. Nitropyrenes:
    isolation, identification, and reduction of mutagenic
    impurities in carbon black and toners. Science
    209:1039-1043.

Schuetzle, D., F.S.C. Lee, and J.T. Prater. 1981. The
    identification of polynuclear aromatic hydrocarbon (PAH)
    derivatives in mutagenic fractions of diesel particulate
    extract. Int. J. Environ. Anal. Chem. 9:93-144.

Tejada, S.B., R.B. Zweidinger, and J.E. Sigsby. 1982. Analysis
    of nitroaromatics in diesel and gasoline car emissions.
    Paper presented at Society of Automotive Engineers Passenger
    Car Meeting, June 7-10, 1982. Troy, MI. Paper #820775.

Young, C.W., P.H. Anderson, R.J. Kindya, R.R. Hall, J.M. Robinson,
    and P.F. Fennelly. 1981. Environmental Assessment: Source
    Test and Evaluation Report -- Georgetown University
    Fluidized-Bed Boiler. GCA-TR-81-76-G. GCA/Technology
    Division, GCA Corporation: Bedford, MA.

THE USE OF SHORT-TERM BIOASSAYS TO MONITOR THE ENVIRONMENTAL IMPACT OF LAND TREATMENT OF HAZARDOUS WASTES

Kirby C. Donnelly,[1] Kirk W. Brown,[1] and Barry R. Scott[2]

[1]Soil and Crop Sciences Department, Texas A & M University College Station, TX 77843, and [2]Phoenix, Inc., P.O. Box 774 Smithville, TX 78957

INTRODUCTION

The feasibility of using a battery of short-term bioassays for the detection of mutagens and potential carcinogens released into the environment as a result of land treatment of hazardous waste is being investigated. Land treatment is the incorporation of a waste into the upper layer of soil resulting in the degradation or attenuation of hazardous waste constituents. EPA (1980) regulations state that a waste cannot be applied to the soil unless the waste is rendered less hazardous or nonhazardous by chemical or biological reactions in the soil. Biological analysis provides the only means of demonstrating a reduction in hazardous characteristics since chemical analysis fails to account for the interactions of the components of a complex mixture, the production of mutagenic metabolites via degradative pathways, and the chemical combination of nontoxic precursors to form mutagenic compounds.

The objectives of this research are to characterize the genotoxic constituents of three hazardous wastes, to monitor waste degradation in soil, and to determine the environmental fate of mutagenic waste constituents following land application. To meet these objectives and to develop a set of test protocols that can be used to monitor environmental contamination, it is necessary to account for the various environmental events that preclude the expression of genotoxic effects in the population. The collection of samples used for environmental monitoring should be conducted in such a manner as to ensure adequate information to define the background levels of mutagenic activity in the environment prior to waste application, the amount of mutagenic activity added to the soil with waste application, and the fate of soil-applied waste constituents following waste application. In addition, sample preparation and handling must be performed so that the mutagenic activity of a sample is neither increased nor decreased appreciably prior to biological analysis.

To develop the data base necessary to demonstrate the utility of biological analysis for regulatory and management activities, we have collected samples from several sources. Initially, three hazardous wastes were collected and characterized in the bioassays. Degradation has been monitored in undisturbed soil profiles to which two of these wastes were applied, and the movement of mutagenic waste constituents has been monitored in leachate and runoff water from waste-amended soil. While preliminary results indicate that short-term bioassays can be applied to environmental monitoring, previous research has demonstrated their utility for the detection of mutagens and potential carcinogens in samples of air (Hughes et al., 1980; Tokiwa et al., 1980; Wang et al., 1980; Talcott and Wei, 1977), water (Kool et al., 1981; Coleman et al., 1980; Cheh et al., 1980; Tabor and Loper, 1980; Rappaport et al., 1979), and complex mixtures (Rao et al., 1981; Wilson et al., 1980; Nestmann et al., 1980).

MATERIALS AND METHODS

Biological Analysis

The ability of samples to induce genetic damage was measured in three microbial systems (Table 1). The bioassay using an eukaryotic organism, Aspergillus nidulans (a fungus) can be used to detect point mutations and small deletions induced in a haploid genome, or can be used as a diploid organism to detect chromosome aberrations, mitotic recombination, gene mutation, nondisjunction, recombinogenic events, recessive lethals, and spindle poisons. These systems are capable of detecting changes in the genetic entity that are of relevance to the human species and are sensitive to compounds not detected in the Salmonella assay (Lilly, 1956; Scott et al., 1978; Scott et al., 1982). In the first phase of this study, the Aspergillus bioassay will be used to assess the mutagenic potential of the acid, base, and neutral fractions of hazardous industrial wastes by evaluating the

Table 1. Biological Systems Used to Detect Genotoxic Compounds in Hazardous Waste

Organism	Genotoxic Event Detected	Advantages	Disadvantages
Prokaryotes			
Bacillus subtilis[a]	Increased lethal damage in DNA repair-deficient strains	Sensitive to bacteriocides, inorganics; can evaluate several DNA repair mechanisms; large data base	Insensitive to volatiles; difficult to quantify
Salmonella typhimurium[b]	Point mutation	Well-validated; well-defined end point	Insensitive to volatiles, toxic mutagens, certain chlorinated hydrocarbons
Eukaryotes			
Aspergillus nidulans[c]	Point mutation; chromosome damage	Detects range of genetic damage including terratogens; good correlation in compounds tested; chromosome organization similar to humans	Limited data base; insufficient number of trained personnel

[a]Felkner et al., 1979; Kada et al., 1974; Tanooka et al., 1978.
[b]Ames et al., 1975; Skopek et al., 1978; Haroun and Ames, 1981.
[c]Scott et al., 1982; Scott et al., 1978; Bignami et al., 1981.

induction of mutations at the methionine-suppressor loci. Conidia
from four to five single colonies of the methG1 biA1 (requiring
methionine and biotin) Glasgow strain of Aspergillus nidulans
grown for 5 to 6 days on a complete medium at 37°C were used for
each experiment. Samples were tested at a minimum of three dose
levels and four exposure times. The procedures used were the same
as Scott et al. (1978). Mutant colonies were assayed by spreading
exposed cells on a methionine-free medium. Mutant colonies were
scored after incubation for 5 days at 37°C. Colonies were divided
by colony morphology into three classes, A, B, and C, and the
total number of mutant colonies. Each of these three classes is
believed to involve two genes. The morphology of Class A colonies
appears green; Class B, brown; and Class C, green with a white
hyaline edge. The frequency of mutations induced by a sample was
determined by subtracting from the total mutation frequency in
Classes A, B, C, and the total, the frequency of spontaneous
mutations that occurred in Classes A, B, C, or the total. A
sample was considered mutagenic if there was a positive slope on
the mutation-induction curve, or the induced-mutation frequency
for at least two exposure times was more than twice the
spontaneous mutation frequency. Positive controls included
8-methoxypsoralen (Sigma, St. Louis, MO) plus near UV light,
without activation, and benzo(a)pyrene (Aldrich, Milwaukee, WI),
with metabolic activation.

A microbial DNA-repair assay was used to measure the capacity
of a sample to produce increased lethal damage in DNA
repair-deficient strains. Six strains of B. subtilis deficient in
different recombination (Rec⁻) and/or excision (Exc⁻) repair were
used to test for lethal DNA damage. These included the
Rec⁻ strains rec A8, rec B2, rec E4, mc-1; Exc⁻ strain hcr-9; and
Rec⁻/Exc⁻ fh 2006-7. All of these strains are isogenic with
B. subtilis strain 168 which has all repair intact. These strains
were kindly supplied by Dr. I.C. Felkner of Clements Associates,
Washington, DC. Cultures were grown overnight in brain-heart
infusion broth (Difco, Detroit, MI) incubated at 37°C. Each
strain was streaked radially on a nutrient agar plate to a
centrally placed sensitivity disc containing 100 µl of the test
chemical. After incubation at 37°C for 18 h, the distance of
growth inhibition from the disc was measured in millimeters (Kada
et al., 1974). A response was considered positive if the distance
of growth inhibition was more than 2.5 mm greater in one of the
repair-deficient strains than in the repair-proficient strain 168.
Mitomycin C (Sigma), methylmethane sulphonate (Aldrich), and
sensitivity to ultraviolet light were used as appropriate positive
controls. Quadruplicate plates were run at each dose level for
all samples.

Fractional survival (N/N_o) was determined for those strains showing the greatest sensitivity (inhibition) to the test chemical. Brain-heart infusion broth was inoculated with the appropriate strain, incubated at 37°C for approximately 16 h until an OD_{540} of 0.3 was reached (approximately 1.2×10^9 cells/ml). The cells were then diluted with brain-heart infusion media to an OD_{540} of 1.16 (approximately 1.2×10^8 cells/ml), serially diluted with Spizizen's Minimal Salts from 10^{-1} through 10^{-6}, and plated onto nutrient agar (Felkner et al., 1979).

The Salmonella/microsome assay as described by Ames et al. (1975) was used to measure the ability of a sample to revert strains of bacteria to histidine prototrophy. The subfractions of three wastes were tested with strain TA98, a frameshift mutant, and TA100, a base-pair mutant, which were kindly supplied by Dr. B.N. Ames of University of California at Berkeley, Berkeley, CA. Cultures were grown overnight in Oxoid No. 2 (KC Biological, Lenexa, KS) at 37°C. Samples were tested in the standard plate incorporation assay at a minimum of four dose levels with and without enzyme activation (0.3 ml rat liver/1.0 ml S9 mixture; 0.5 ml S9 mixture/plate). Aroclor 1254-induced S9 rat liver was obtained from Litton Bionetics (Kensington, MD).

Soil

Two soils were selected to represent a range of soil textures. These soils were a Norwood sandy clay (Typic Udifluvent) and a Bastrop clay (Udic Paleustalf). The characteristics of these soils are given in Table 2. Soils were analyzed for mutagenic activity using the Salmonella/microsome assay only.

Waste

Ten wastes were initially selected for the project (Table 3). Criteria for the selection of the wastes were the wastes' amenability to land treatment and large volume of production. The acute toxic effects of each waste were evaluated in three bioassays.

Fractional survival was determined for the crude extract of all ten wastes using at least one strain in each of the microbial bioassays. Cultures of the appropriate strain were grown overnight to a cell density of approximately 1×10^9 cell/ml and serially diluted from 10^{-1} through 10^{-6}. To 2.0 ml of top agar, 0.1 ml of the microbial culture and 0.1 ml of the crude extract of

Table 2. Characteristics of the Unamended Soils Used in the Greenhouse and Lysimeter Study

| Soil Series | % | | | | pH | % Moisture | | | Mutagenic Activity (rev/mg) | |
	Sand	Salt	Clay	Organic Matter		WP[a]	FC[b]	SAT[c]	Organic Extract of Soil	Soil
Norwood	48.2	15.2	36.6	1.4	7.69	12	18	33	47	0.028
Bastrop	60.3	10.0	29.7	1.0	6.86	6	22	25	365	0.292

[a]Wilting point.
[b]Field capacity.
[c]Saturated.

Table 3. Characteristics of the Hazardous Waste

Waste	EPA No.	Physical State	% Oil & Grease	Suspected Genotoxic Constituents
Dissolved air flotation float (DAF)	K048	Sludge	5	Polynuclear aromatic hydrocarbons
Slop-oil emulsion solids (SOE)	K049	Slurry	89	Polynuclear aromatic hydrocarbons
Combo sludge (API separator and wastewater treatment)	K051	Sludge	41	Polynuclear aromatic hydrocarbons
Storm-water runoff silt (SWRI)	--	Sludge	21	Polynuclear aromatic hydrocarbons
Liquid stream from acetonitrile purification column (ACN)	K013	Liquid	1.8	Acetamide
Methyl ethyl ketone (MEK)	--	Slurry	97	Methyl cyanide
Biosolids-phenol	K022	Liquid	0.2	Phenols
Biosolids-ag. chem.	--	Liquid	0.2	Biocides
Wood-preserving bottom sediments (Penta-S)	K001	Sludge	27	Pentachlorophenol dioxin, polynuclear aromatic hydrocarbons
Petrochemical (Petro)	K051	Sludge	62	Polynuclear aromatic hydrocarbons, heterocyclic nitrogen compounds

each waste were added, mixed on a vortex mixer, and plated on a
complete medium. Cells were exposed to a minimum of four dose
levels of the crude extract. The plates were incubated for 24 h
at 37°C and fractional survival (N/N_o) was determined by comparing
cell counts on exposed plates to the cell count on plates exposed
to the solvent DMSO without the waste extract. Three of the
10 wastes were then selected for characterization of the chronic
toxic effects and for use in the greenhouse and lysimeter study.

Dichloromethane was selected from a group of agents to
extract the organic fractions of the wastes and the soil, since it
consistently provided the greatest extraction efficiency for the
wastes used (McGill and Rowell, 1980). Six volumes of
dichloromethane were added to the waste or waste soil mixture and
mixed in a Waring blender for 30 s. This extraction was repeated
twice or until the extracting solvent remained colorless. Solvent
extractions were then combined and taken to dryness on a
Brinkman-Bucci rotary evaporator. The residue from this
extraction was partitioned into acid, base, and neutral fractions
following the scheme outlined in Figure 1. The neutral fraction
of each waste was further separated into four subfractions using
sequential solvent extraction on a silica gel column. This
extraction approximately separated the neutral fraction into
saturated, aromatic, and condensed-ring fractions according to the
procedures of Warner et al. (1976).

Water

Leachate and runoff samples were collected from large,
undisturbed field lysimeters containing two soils. Soil
characteristics are given in Table 2. Each waste was incorporated
into the top 15 cm of the two lysimeters of each soil. A second
and third waste application was made to one lysimeter of each
soil, and one lysimeter to which no waste was applied served as a
control. Five monoliths of each soil were encased in the
150-cm deep lysimeters and brought to a central location. A
detailed description for the installation of lysimeters is given
by Brown et al. (1974). The leachate was collected from porous
ceramic suction cups (Coors Type No. 7001-P-6-C), which were
installed in the soil at the bottom of the profile. The suction
cups were connected by nylon tubing to a 20-1 glass bottle which
was connected to a vacuum manifold. Leachate samples were
collected once a week unless the rate of water flow required a
more frequent sampling schedule.

Runoff samples were collected by means of a peristaltic pump.
Glass beakers were placed in a small hole in the corner of each

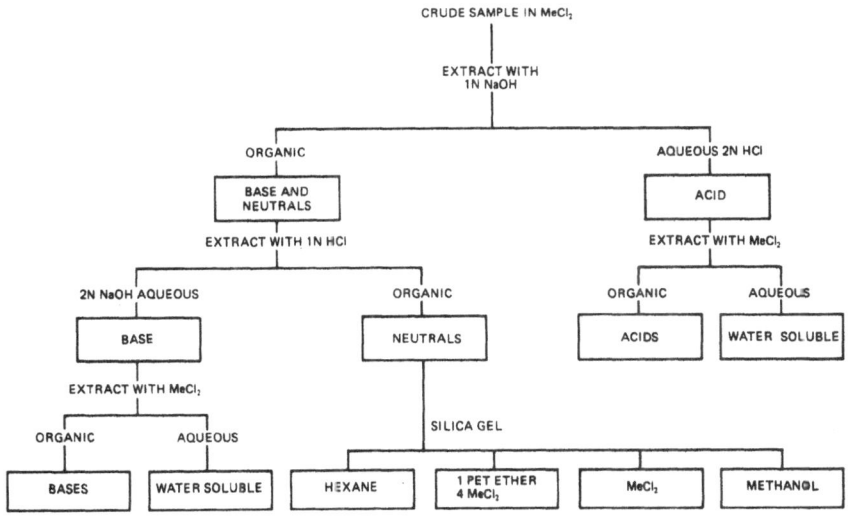

Figure 1. Fractionation scheme for waste-amended soils.

lysimeter; the soil surface in the lysimeter was sloped so that runoff water could collect in the beaker. A rainfall of greater than 1.27 cm activated a switch which supplied power to the pumps. The collected runoff was pumped through nylon tubing to a 20-1 glass bottle where the water was stored in an air-conditioned shed until collected. Water samples were collected in amber glass bottles and stored at 4°C until processed.

Two methods were employed for the extraction of water-bound organic compounds. Initially, a 20-cm^3 bed was made of nonpolar XAD-2 resin (Applied Science Laboratory, State College, PA) following the methods of Hooper et al. (1978). Later in the study, it was decided, due to the complex nature of the samples being analyzed, to supplement the XAD-2 with a moderately polar resin. Thus, samples that were extracted in the latter portion of the study were passed through a mixed bed of 4.0 g of XAD-2 and 6.3 g XAD-7 or approximately 20 cm^3 of each resin, as suggested by Rappaport et al. (1979). The resins were washed prior to use by swirling and decanting three times with ten volumes each of acetone, methanol, and distilled water. Washed resins were stored at 4°C prior to use. Glass econo-columns (Bio-Rad, Richmond, CA) 1.5 x 50 cm^2 were packed with 20 cm^3 of XAD-2 resin followed by 20 cm^3 of XAD-7 resin. A glass-wool plug was placed above the resin in order to trap soil particles. The columns were flushed with 1,200 ml of distilled water before loading the water sample. Leachate or runoff water was placed in a reservoir and allowed to pass through the column by gravity flow at about 50 ml/min. After loading the water sample, dry nitrogen was introduced into the column to remove the residual aqueous phase, and the column washed with 120 ml of distilled water to remove residual histidine. The adsorbed organic compounds were then eluted with 160 ml of acetone. The acetone extract was filtered through about 30 g of anhydrous Na_2SO_4 and Whatman No. 42 filter paper into a flat-bottom flask. Extracts were reduced to less than 10 ml on a Brinkman-Bucci rotary evaporator. The sample was then added to a large screw-capped glass culture tube filled one-third with sodium sulfate to remove residual water, and washed with acetone. The sample was then pipetted off into a smaller screw-capped glass culture tube and dried under a stream of nitrogen. Dimethyl sulfoxide (Sigma) was added to the dried extract at a rate of 0.5 ml DMSO/1 of unconcentrated water, and the resultant solution was passed through a 0.2-μm average pore diameter Teflon filter (Millipore-Fluoroporem, Bedford, MA). Samples were stored at 4°C prior to use.

RESULTS AND DISCUSSION

Waste Characterization

The results from preliminary testing and the acute toxicity study were inconclusive (data not shown). None of the crude waste extracts elicited a clear positive response in the spot tests using the Salmonella assay or Bacillus DNA-repair assay. In most cases, the acute toxicity of the crude waste extract was greatest in the haploid bacteria and least in the diploid Aspergillus. None of the wastes examined produced a definite increase in toxicity with increasing dose in the eukaryotic diploid Aspergillus. For the Salmonella and Bacillus systems, at least one dose level was determined for each waste at which less than 10% of the exposed prokaryotic cells survived. This dose will be used as the maximum dose level for all subsequent testing with the waste, as well as with the soil or water extract.

Three hazardous wastes have been characterized in one or more of the bioassays. The wood-preserving bottom sediment (Penta-S) has been examined in the Aspergillus, Bacillus, and Salmonella systems, and the acetonitrile (ACN) and petrochemical (Petro) wastes have been examined in the Salmonella system only. A summary of the results obtained from testing subfractions of each waste in S. typhimurium strain TA98 with metabolic activation is presented in Table 4. Both direct- and indirect-acting mutagens were detected in the acetonitrile waste, although the greatest amount of mutagenic activity was detected with enzyme activation. For the acetonitrile waste, the basic fraction induced the greatest amount of mutagenic activity; for the petrochemical and wood-preserving wastes, mutagenic activity was greatest in the acid fraction.

Table 4. Distribution of Mutagenic Activity in Fractions of
          Several Hazardous Wastes[a]

Waste	Mutagenic Activity (rev/mg)		
	Acid	Base	Neutral
Acetonitrile (ACN)	42	78	49
Wood-preserving (Penta-S)	500	370	280
Petrochemical (Petro)	498	278	246

[a]All assays performed with crude liver S9 from rats induced with Aroclor 1254.

The acid fraction of the wood-preserving waste also produced increased lethal damage in a recombinant repair-deficient strain of B. subtilis (Figure 2). The base and neutral fractions of the wood-preserving waste did not produce increased lethal damage in the repair-deficient strains. In A. nidulans without enzyme activation, the base and neutral fractions produced a positive response. The neutral fraction acted as might be expected for a pure compound or a group of compounds with equivalent kinetics, yielding straight-line mutation induction curves for the total, A, and B colonies (Figure 3); the base fraction acted more like a complex mixture, with different kinetics of mutation induction, yielding a straight-line for type C colonies only (Figure 4).

Soil

To evaluate the reduction of hazardous characteristics following land application of waste, it is necessary to define the level of mutagenic activity in the waste, the soil, and the

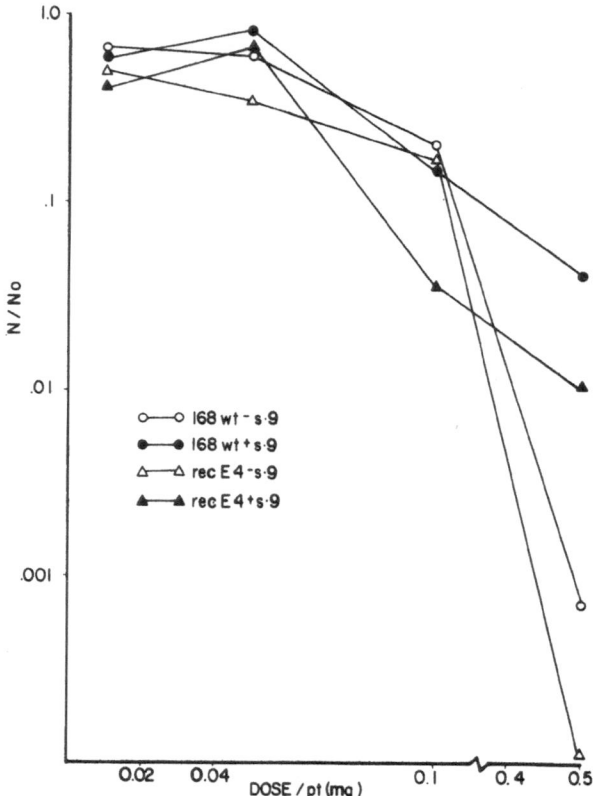

Figure 2.   Fractional survival of B. subtilis exposed to acid
            fraction of wood-preserving waste.

soil/waste mixture prior to degradation. The results obtained
from the analysis of the wood-preserving waste subfractions
immediately after the waste and soil were mixed are shown in
Figures 5, 6, and 7. These results demonstrate the ability to
recover approximately equal amounts of mutagenic activities from
the acid and base fractions of the waste and waste/soil mixture,
although greater amounts of mutagenic activity were recovered from
the neutral fraction of the waste/soil mixture than from the waste
or soil alone (Figure 7).

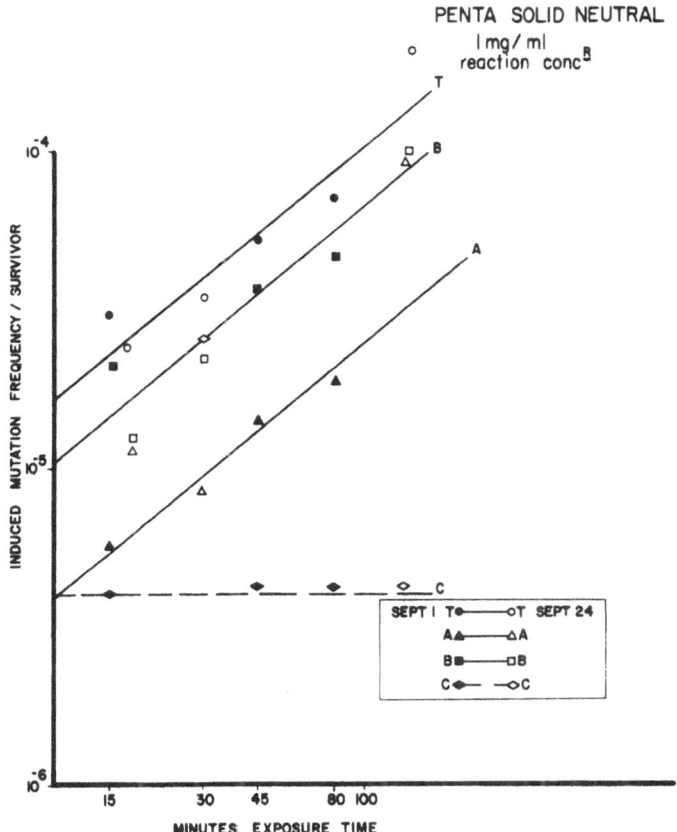

Figure 3. Mutation frequency induced by the neutral fraction of the wood-preserving waste without metabolic activation in _methG1_ _biA1_ strain of _A._ _nidulans_.

Water

Leachate and runoff samples were collected from waste-amended and unamended lysimeters in order to determine the potential for migration of mutagenic compounds from a hazardous waste land treatment facility. The analysis of leachate water indicated that only low levels of mutagenic activity could be detected in a few of the samples collected (Donnelly and Brown, 1981). While substantially greater amounts of mutagenic activity were detected

PENTA SOLID BASE

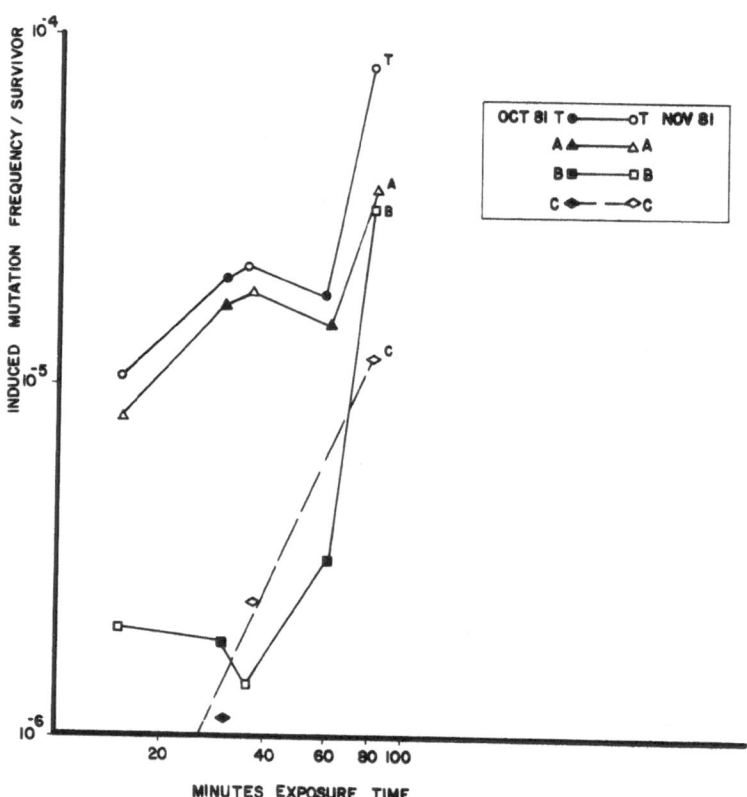

Figure 4. The mutation frequency induced by the basic fraction of
the wood-preserving waste without metabolic activation
in <u>methG1</u> <u>biA1</u> strain of <u>A</u>. nidulans.

in runoff water from waste-amended soil (Figure 8), the activity
returned to background levels within 5 months after the second
waste application. None of the leachate or runoff water collected
from unamended lysimeters induced a positive response in the
<u>Salmonella</u> assay.

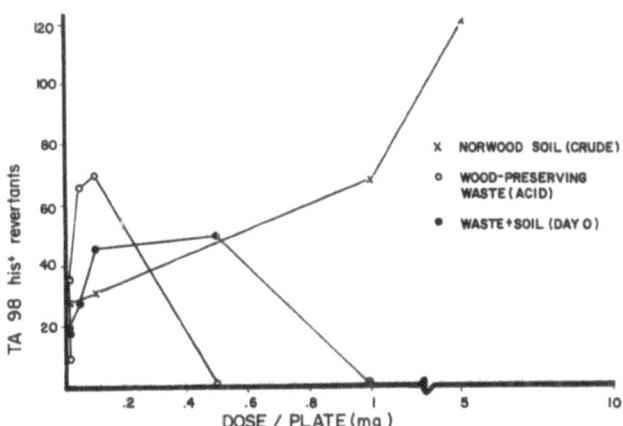

Figure 5. Initial effect of soil application on the mutagenic
activity of acid fraction of wood-preserving waste as
measured with S. typhimurium strain TA98 with metabolic
activation.

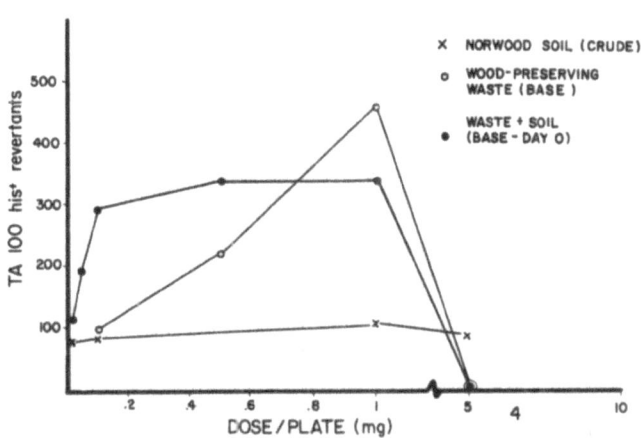

Figure 6. Initial effect of soil application on the mutagenic
activity of basic fraction of wood-preserving waste as
measured with S. typhimurium strain TA100 with
metabolic activation.

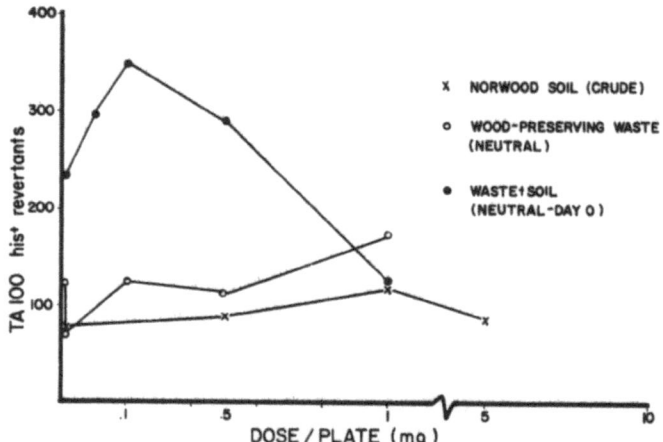

Figure 7.   Initial effect of soil application on the mutagenic
           activity of neutral fraction of wood-preserving waste
           as measured with S. typhimurium strain TA100 with
           metabolic activation.

CONCLUSIONS

     These preliminary studies have been conducted in order to
develop a series of biological test systems that can be used both
to evaluate the effect of land application on the hazardous
characteristics of a waste, and as a monitoring tool for a land
treatment facility. Although the initial results indicate that
biological analysis can be used to detect genotoxic compounds in a
hazardous waste, they have also demonstrated some of the
limitations of this combined testing protocol.

     The most important limitations of using a combined testing
protocol can be divided into three categories. First, a battery
of bioassays is needed so that a range of genetic damage and a
variety of genotoxic compounds can be detected. In addition,
adequate background samples must be collected in order to
accurately define the level of mutagenic activity present in soil,
water, and plant samples prior to waste application. Finally,
sample collection, preparation, and analysis must be conducted in
such a manner as to prevent the loss or generation of mutagenic
activity. The inability to detect the true mutagenic potential of

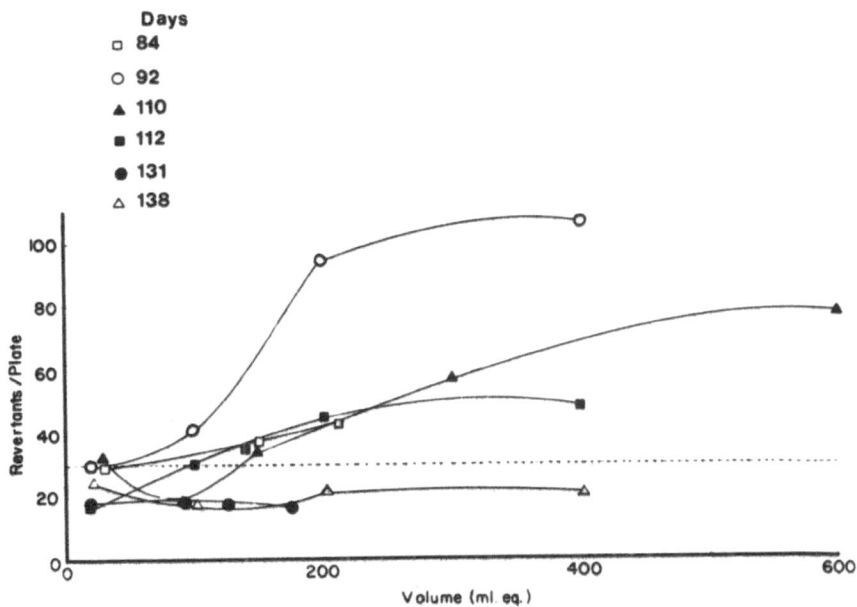

Figure 8.   Mutagenicity as measured with S. typhimurium strain
            TA98 without enzyme activation of runoff water
            collected various days after the second waste
            application from petrochemical waste-amended soil.

a sample can result from improper sample handling or processing
(samples may degrade due to time, temperature, or light), the
selection of a solvent that fails to completely dissolve a sample,
or the inability of the enzyme induction system to activate
proximate mutagens.

     Although these short-term tests can provide an accurate and
inexpensive means of assaying the hazardous nature of a complex
mixture, more research is needed to better define the limitations
of the various biological systems and to improve the chemical
analysis of the mutagenic subfractions. If these forms of
analysis can be refined and improved, then this type of combined
test protocol should provide a rapid and accurate risk assessment
for complex mixtures in the environment.

REFERENCES

Ames, B.N., J. McCann, and E. Yamasaki. 1975. Methods for
    detecting carcinogens and mutagens with the Salmonella/
    mammalian microsome mutagenicity test. Mutation Res.
    31:347-364.

Bignami, M., G. Conti, R. Crebelli, and A. Carere. 1981. Growth-
    mediated metabolic activation of promutagens in Aspergillus
    nidulans. Mutation Res. 68:31-40.

Brown, K.W., C.J. Gerard, B.W. Hipp, and J.T. Ritchie. 1974.
    A procedure for placing large undisturbed monoliths in
    lysimeters. Soil Sci. Soc. of Amer. 38:981-983.

Cheh, A.M., J. Skochdopole, P. Koski, and L. Cole. 1980.
    Nonvolatile mutagens in drinking water: production by
    chlorination and destruction by sulfite. Science 207:90-92.

Coleman, W.E., R.G. Melton, F.C. Kopfler, K.A. Barone,
    T.A. Aurand, and M.G. Jellison. 1980. Identification of
    organic compounds in a mutagenic extract of a surface
    drinking water by a computerized gas chromatography/mass
    spectrometry system (GC/MS/COM). Environ. Sci. Technol.
    14:576-588.

Donnelly, K.C., and K.W. Brown. 1981. The development of
    laboratory and field studies to determine the fate of
    mutagenic compounds from land applied hazardous waste.
    Presented at the U.S. Environmental Protection Agency Seventh
    Annual Research Symposium on Land Disposal: Hazardous Waste.
    EPA-600/9-81-002b. pp. 224-239.

Felkner, I.C., K.M. Hoffman, and B.C. Wells. 1979. DNA-damaging
    and mutagenic effects of 1,2-dimethylhydrazine on Bacillus
    subtilis repair-deficient mutants. Mutation Res. 68:31-40.

Haroun, L., and B.N. Ames. 1981. The Salmonella mutagenicity
    test: an overview. In: Short-Term Tests for Chemical
    Carcinogens. H.F. Stich and R.H.C. San, eds. Springer-
    Verlag: New York. pp. 108-119.

Hooper, K., C. Gold, and B.N. Ames. 1978. Development of methods
    for mutagenicity testing of wastewater and drinking water
    samples. Report to Water Resources Control Board, State of
    California, Sacramento, CA.

Hughes, T.J., E. Pellizzari, L. Little, C. Sparacino, and
    A. Kolber. 1980. Ambient air pollutants: collection,
    chemical characterization and mutagenicity testing. Mutation
    Res. 79:51–83.

Kada, T., M. Morija, and Y. Shirasu. 1974. Screening of
    pesticides for DNA interactions by REC-assay and mutagenic
    testing and frameshift mutagens detected. Mutation Res.
    26:243.

Kool, H.J., C.F. Van Kreijl, H.J. Van Kranen, and E. De Greef.
    1981. Toxicity assessment of organic compounds in drinking
    water in the Netherlands. Sci. Total Environ. 18:135–153.

Lilly, L.J. 1965. An investigation on the suitability of
    suppressors of methAl for the study of induced and
    spontaneous mutations. Mutation Res. 2:192–195.

McGill, W.B., and M.J. Rowell. 1980. Determination of oil
    content of oil contaminated soil. Sci. Total Environ.
    14:245–253.

Nestmann, E.R., E.G.H. Lee, T.I. Matula, G.R. Douglas, and
    J.C. Mueller. 1980. Mutagenicity of constituents identified
    in pulp and paper mill effluents using the Salmonella/
    mammalian-microsome assay. Mutation Res. 79:203–212.

Rao, T.K., B.E. Allen, D.W. Ramey, J.L. Epler, I.B. Rubin,
    M.R. Guerin, and B.R. Clark. 1981. Analytical and
    biological analyses of test materials from the synthetic fuel
    technologies, III: use of Sephadex LH-20 gel chromatography
    technique for the bioassay of crude synthetic fuels.
    Mutation Res. 85:29–39.

Rappaport, S.M., M.G. Richard, M.C. Hollstein, and R.E. Talcott.
    1979. Mutagenic activity in organic wastewater concentrates.
    Environ. Sci. Technol. 13:957–961.

Scott, B.R., E. Kafer, G.L. Dorn, and R. Staff. (in press).
    Aspergillus nidulans: systems and results of test for
    induction of mutation and mitotic segregation. Mutation Res.

Scott, B.R., A.H. Sparrow, S.S. Lamm, and L. Schairer. 1978.
    Plant metabolic activation of EDB to a mutagen of greater
    potency. Mutation Res. 49:203–212.

Skopek, T.R., J.L. Liber, J.J. Krowleski, and W.G. Thilly. 1978.
    Quantitative forward mutation assay in Salmonella typhimurium
    using 8-azaguanine resistance as a genetic marker. Proc.
    Natl. Acad. Sci. USA 75:410.

Tabor, M.W., and J.C. Loper. 1980. Separation of mutagens from
    drinking water using coupled bioassay/analytical
    fractionation. Int. J. Environ. Anal. Chem. 8:197-215.

Talcott, R., and E. Wei. 1977. Airborne mutagens bioassayed in
    Salmonella typhimurium. J. Natl. Cancer Inst. 58:449-451.

Tanooka, H., N. Munakata, and S. Kitahira. 1978. Mutation
    induction with UV- and x-radiation in spores and vegetative
    cells of Bacillus subtilis. Mutation Res. 49:179-186.

Tokiwa, H., S. Kitamori, K. Takahashi, and Y. Ohnishi. 1980.
    Mutagenic and chemical assay of extracts of airborne
    particulates. Mutation Res. 77:99-108.

U.S. Environmental Protection Agency. 1980. Hazardous waste
    management system: standards for owners and operators of
    hazardous waste treatment, storage, and disposal facilities.
    Fed. Reg. 45(98):33154-33258.

Wang, C.Y., M.S. Lee, C.M. King, and P.O. Warner. 1980. Evidence
    for nitroaromatics as direct-acting mutagens of airborne
    particulates. Chemosphere 9:83-87.

Warner, J.S. 1976. Determination of aliphatic and aromatic
    hydrocarbons in marine organisms. Anal. Chem. 48:578.

Wilson, B.W., R. Pelroy, and J.T. Cresto. 1980. Identification
    of primary aromatic amines in mutagenically active
    subfractions from coal liquefaction materials. Mutation Res.
    79:193-202.

MUTAGENICITY TESTING OF COMPLEX MIXTURES DERIVED FROM HUMAN BODY FLUIDS

Resha M. Putzrath[1] and Eric Eisenstadt[2]

[1]RMP Consulting, 3225 Grace Street, N.W., Washington, DC 20007 and [2]Laboratory of Toxicology, Department of Physiology and Department of Microbiology, Harvard University School of Public Health, Boston, MA 02115

Although short-term bioassays have usually been developed by testing individual chemicals, samples which need to be evaluated for potential human health hazard are often complex mixtures. These mixtures may contain many toxic substances, each present in small quantities. To detect such toxins in bioassays, sample preparation will involve concentration of the compounds of interest and removal of material which will interfere with the test system. If the toxicity is due to many compounds, isolation and concentration of the total biological activity will be useful either as a preliminary step for chemical analysis or as a method for characterizing some of the toxic and chemical properties of the mixture. Mutagenicity tests are useful for monitoring genotoxic activity during such isolation procedures, since these assays are relatively quick and simple. Moreover, a large number of carcinogens are mutagenic (McCann et al., 1975; Rinkus and Legator, 1979); therefore, isolating mutagenic activity may develop procedures which will be useful in isolating carcinogens. Using simple chemical separation techniques and mutagenicity testing, we have highly concentrated the mutagens in smokers' urine and have elucidated some of their chemical characteristics.

We elected to study smokers' urine for several reasons. First, smokers' urine had already been shown to be mutagenic in the Salmonella assay system by Yamasaki and Ames (1977). Therefore, an exposed population that was positive and a control

population that was negative were available for developing
preparative and analytical procedures for assaying urine samples.
Second, urine is one of the more easily sampled body fluids.
Third, characterization of the mutagens in urine due to smoking
might result in criteria that would allow these mutagens to be
distinguished from those due to other exposures. Finally, smokers
have been observed to have an elevated incidence of bladder cancer
(USDHEW, 1979). Although the correlation of mutagenic urine with
bladder cancer is not known, our studies might help to evaluate
the contribution of mutagens in urine to the risk of developing
bladder cancer.

Yamasaki and Ames (1977) showed that XAD-2 resin could be
used to concentrate the mutagenic activity in smokers' urine and
to separate the mutagens from histidine present in the urine. Our
protocol for characterizing the mutagens in smokers' urine
employed standard chemical separation procedures subsequent to
XAD-2 extraction (Figure 1). Details of this protocol have been
published (Putzrath et al., 1981).

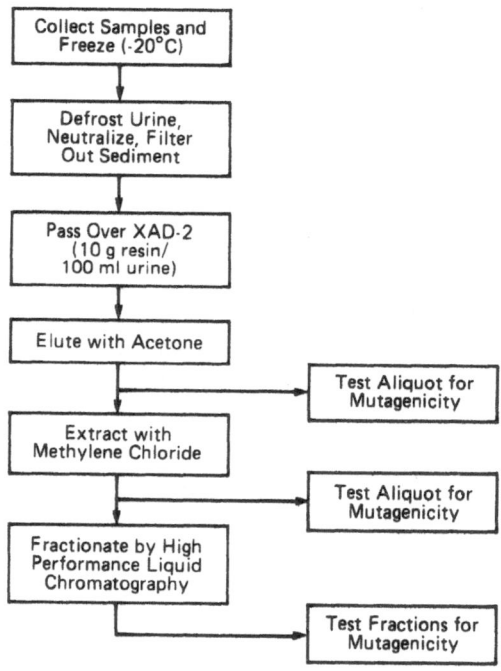

Figure 1.   Preparation of urine extracts for mutagenicity testing.

Since XAD-2 resin is widely used to prepare fluids for mutagenicity testing, it is worth mentioning some of the limitations of this procedure. The resin selectively adsorbs nonpolar molecules from aqueous solutions. Since human tissue or body fluid samples may contain conjugated mutagens which are highly polar, these mutagens may be systematically underestimated or lost from XAD-2 extracts. Even if conjugated compounds are extracted by XAD-2 (Bradlow, 1968), their mutagenicity could go undetected without either chemical or enzymatic hydrolysis. Our study examined the mutagens extracted by XAD-2 from untreated smokers' urine. We recognize, therefore, that we may be extracting a subset of the mutagenic activity in smokers' urine. Furthermore, some of the material from urine which was retained by the resin is not released by standard organic eluting solvents. A colored material remained on the column and could not be removed except by harsh cleaning treatments which would not be appropriate for sample preparation. Despite these limitations, XAD-2 remains a useful tool for rapidly concentrating nonpolar mutagens from aqueous solutions.

The acetone eluate from an XAD-2 column loaded with smokers' urine gives a positive dose-response curve with several of the Ames tester strains. Strain TA1538 was most frequently used in our studies to monitor mutagenic activity. We found that application of a small number of additional separation techniques subsequent to XAD-2 extraction can further purify and characterize the mutagens in smokers' urine. By resuspending dried XAD-2 eluate in water and extracting this solution with methylene chloride, the mutagens in smokers' urine could be isolated from nonmutagenic material. We recovered approximately 95% of the mutagenic activity in the methylene chloride fraction, while 95% of the dry weight remained in the aqueous phase. This purification procedure has an additional benefit since nonmutagenic materials that are toxic to the Salmonella system appear to be removed. Dolara et al. (1981) have recently observed a similar reduction in toxicity of urine extracts if the XAD-2 column is eluted with methylene chloride instead of acetone.

The mutagens purified by methylene chloride extraction of an acetone eluate from XAD-2 were sufficiently concentrated to be analyzed by reverse-phase high performance liquid chromatography (HPLC). Fractions from HPLC were assayed for mutagenic activity (Figure 2). The elution gradient was 35% to 100% methanol:water, changing at a rate such that 100% methanol was reached in fraction number 21. As could be expected from the preparative procedure, the mutagens were relatively nonpolar. A large number of the fractions were mutagenic, suggesting a complex mixture of mutagens

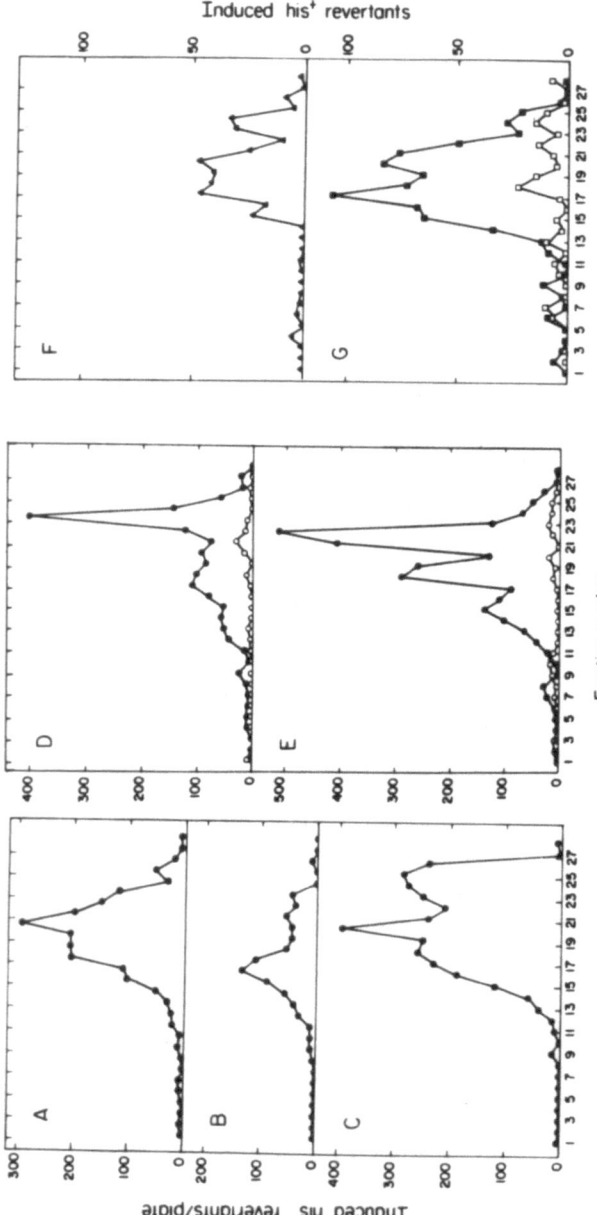

Figure 2.   HPLC fractionation of smokers' and nonsmokers' urine. Methylene chloride extracts of
XAD-2 concentrates were dried, resuspended in methanol, and fractionated by reverse-
phase HPLC using a Whatman Partisil Magnum-9 column and a Perkin-Elmer Series 2
programmed for an elution gradient from 35 to 100% methanol:water changing at 3%/min
with a flow rate of 5 ml/min. Fractions, collected at 1-min intervals, were tested
for mutagenicity. Volumes from each fraction, equivalent to 250 ml of fractionated
urine, were tested. B: a 5-l preparation; all other preparations are 1 to 2 l.
A-E: urine of a smoker (●) and pooled urines from nonsmokers (o) from Boston, MA.
F: pooled urines from smokers from Akron, OH (△); G: urine from a smoker (■) and
nonsmoking spouse (□) from the Republic of San Marino.   (From Putzrath et al., 1981,
©Elsevier/North Holland Biomedical Press, reprinted by permission.)

reminiscent of the multiple mutagens observed with cigarette smoke condensates (Kier et al., 1974).

   If this broad distribution of mutagenic activity is characteristic of smokers' urine extracts, HPLC analysis might be useful in distinguishing smokers' urine from other mutagenic urine. Although a general pattern of activity in the fractions emerged, we also observed variation between samples. Samples collected on different days from one smoker showed as much variation as samples collected from many different smokers. The differences observed in urine from one smoker from Boston, MA (Figures 2A through 2E) appeared to be no greater than differences between his urine and urine pooled from smokers from Akron, OH (Figure 2F) or urine from a smoker from the Republic of San Marino (Figure 2G). However, when a single sample of urine concentrate was analyzed twice by HPLC, the mutagenic activity of the two elutions was nearly identical. Therefore, the HPLC fractionation procedure is not responsible for the variation observed.

   To further characterize the multiple mutagens in the methylene chloride extract, extractions were performed under acidic, basic, and neutral conditions (Figure 3). Treatment with acid or base reduced the total mutagenic activity of the sample by 30 to 60%. Most of the remaining mutagenic activity was recovered in the aqueous phase of the acidic extraction and in the methylene chloride phase of both the basic and neutral extractions. These extracts were also analyzed by HPLC. The distribution of mutagenic activity in these HPLC fractions was similar, suggesting that much of the activity in methylene chloride extracts is attributable to basic organic material. Most of the mutagens in cigarette smoke condensates also appear to be basic (Kier et al., 1974). Although basic extraction removes additional nonmutagenic material (as demonstrated by the disappearance of major peaks in the absorbance trace), the loss of mutagenic activity made acidic or basic extraction inappropriate for standard preparation of smokers' urine. However, this simple procedure provided a useful characterization of the mutagenic activity in smokers' urine. Such procedures might also be useful in the preparation of other urine samples.

   Another method of characterizing mutagenic extracts is to compare their biological activities toward different bacterial strains and with different metabolic activation systems. Dolara et al. (1981) showed that, while strain TA100 is less efficient than TA1538 for detecting mutagens in smokers' urine, TA100 is more efficient in detecting occupational exposure to some agents. Using TA1538, urine from chemical workers who smoked had the same activity as urine from controls who smoked. Using TA100, urine

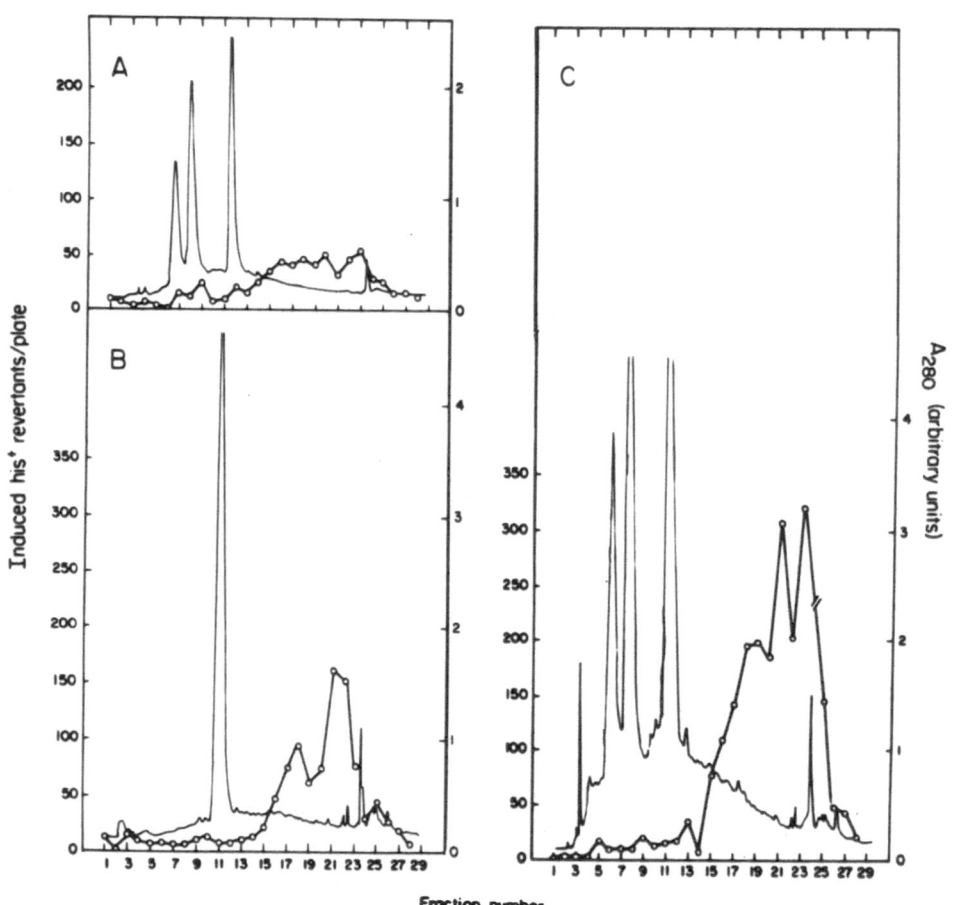

Figure 3.   HPLC fractionation after acidic or basic treatment. A
            methylene chloride extract of XAD-2 concentrate
            equivalent to 2.2 1 of smokers' urine was divided into
            thirds and partitioned between methylene chloride and
            either 1 N HCl or 1 N NaOH or distilled water. Those
            phases which contained mutagenic activity were
            chromatographed by HPLC as described in Figure 2.
            Mutagenicity (o) and A280 (solid line) of the fractions
            are shown. A: aqueous phase of acidic extraction;
            B: methylene chloride phase of basic extraction;
            C: same phase analyzed in Figure 2. (From Putzrath et
            al., 1981, ©Elsevier/North Holland Biomedical Press,
            reprinted by permission.)

from chemical workers who smoked had more activity than urine from controls who smoked. Furthermore, by varying the metabolic activation system, we found that the mutagenicity of smokers' urine extracts using Aroclor or 3-methylcholanthrene S9 was similar, while the activity with phenobarbital S9 was significantly lower (Table 1). Similarly, in each HPLC fraction assayed, more revertants were observed with Aroclor or 3-methylcholanthrene S9 than with phenobarbital S9.

Table 1.   Ability of Different Rat Liver S9 Preparations to Activate Urine Concentrates to Mutagens for Salmonella typhimurium[a]

		Induced histidine revertants per plate			
Strain	Experiment Number	Control (No S9)	Aroclor	3-Methyl- cholanthrene	Phenobarbital
TA1538	1	0	311	254	38
	2	2	432	439	102
TA98	1	0	428	427	48
	2	16	445	536	134
TA1535	2	0	11	5	8
TA100	2	0	154	203	64
TA1537	1	0	33	26	4
	2	1	27	48	11

[a]From Putzrath et al., 1981, ©Elsevier/North Holland Biomedical Press, reprinted by permission.

We were able to further concentrate the mutagens in smokers' urine by one additional procedure. Although the methylene chloride extract is quite soluble in methanol, we found that resuspending the extract in small volumes of methanol caused formation of a precipitate. About 50% of the dry weight of the extract could be precipitated, while all of the mutagenic activity appeared to remain soluble. Most of the mutagenic activity from

4 l of urine, therefore, could be concentrated into 200 μl of methanol. Preliminary characterization by HPLC of the methanol concentrate showed that the mutagenic activity eluted in a pattern like the methylene chloride extract, but that the large absorbance peaks which were nonmutagenic were decreased in size (Putzrath and Eisenstadt, unpublished results).

Our experiments have focused on the biological activity of a complex mixture and its components rather than its chemical composition. A few, comparatively simple concentration and fractionation procedures were sufficient to reveal information about the chemical and biological properties of mutagens in smokers' urine. Moreover, the mutagenic activity was isolated into relatively few fractions, thus limiting the number of samples that needed to be tested. This approach may be appropriate in the testing of other complex mixtures, for example, monitoring human body fluids for occupational exposure, evaluating water contamination, or assessing genotoxic hazards from uncontrolled waste sites.

ACKNOWLEDGMENTS

We would like to thank Jessica Brewster and Tom Burke for their important contributions to preliminary phases of this investigation and Donna Langley and Laura Jenkins for expert technical assistance.

This research was supported in part by grants from the National Institute of Occupational Safety and Health (1R01-OH-00856-01), the National Institutes of Health (1R01-ES-02021-02) and the Rita Allen Foundation. R.M.P. was supported by an Interdisciplinary Programs in Health fellowship funded by the Andrew W. Mellon Foundation.

The contributions of John Hermann are greatly appreciated.

REFERENCES

Bradlow, H.R. 1968. Extraction of steroid conjugates with a neutral resin. Steroids 11:265-272.

Dolara, P., S. Mazzoli, D. Rosi, E. Buiatti, S. Baccetti, A. Turchi, and V. Vannuci. 1981. Exposure to carcinogenic chemicals and smoking increases urinary excretion of mutagens in humans. J. Toxicol. Environ. Health 8:95-103.

Kier, L.D., E. Yamasaki, and B.N. Ames. 1974. Detection of
    mutagenic activity in cigarette smoke condensates. Proc.
    Natl. Acad. Sci. USA 71:4159-4163.

McCann, J., E. Choi, E. Yamasaki, and B.N. Ames. 1975. Detection
    of carcinogens as mutagens in the Salmonella/microsome test:
    Assay of 300 chemicals. Proc. Natl. Acad. Sci. USA
    72:5135-5139.

Putzrath, R.M., D. Langley, and E. Eisenstadt. 1981. Analysis of
    mutagenic activity in cigarette smokers' urine by high
    performance liquid chromatography. Mutation Res. 85:97-108.

USDHEW. 1979. Smoking and health, a report of the Surgeon
    General. DHEW 79-50066. U.S. Department of Health,
    Education, and Welfare: Washington, DC.

Yamasaki, E., and B.N. Ames. 1977. Concentration of mutagens
    from urine by absorption with the nonpolar resin XAD-2:
    Cigarette smokers have mutagenic urine. Proc. Natl. Acad.
    Sci. USA 74:3555-3559.

EVALUATION BY THE TRADESCANTIA-MICRONUCLEUS TEST OF THE

MUTAGENICITY OF INTERNAL COMBUSTION ENGINE EXHAUST FUMES FROM

DIESEL AND DIESEL-SOYBEAN OIL MIXED FUELS

T.H. Ma,[1] W.R. Lower,[2] F.D. Harris,[3] J. Poku,[2]
V.A. Anderson,[1] M.M. Harris,[1] and J.L. Bare[1]

[1]Institute for Environmental Management and Department of
Biological Sciences, Western Illinois University, Macomb, IL
and [2]Environmental Trace Substances Research Center and
[3]Department of Agricultural Engineering, the University of
Missouri, Columbia, MO

INTRODUCTION

A comparative study of the fuel efficiency of diesel and
diesel/soybean oil mixture for the stationary diesel engine has
been undertaken in the Department of Agricultural Engineering of
the University of Missouri. The present study was designed to
evaluate the mutagenicity of these internal combustion engine
emissions using the Tradescantia-Micronucleus (Trad-MCN) test with
Tradescantia clone #4430. A large number of control data obtained
from previous studies were analyzed to examine the underlying
distribution of the occurrence of micronuclei (MCN) in control
populations and to explore some methods of statistical analyses
appropriate to the data. Experiments were conducted by exposing
the early prophase I meiotic pollen mother cells of Tradescantia
plant cuttings to the diluted exhaust fumes and observing the
frequencies of MCN in the early tetrads of the same meiotic cycle.

The physiological, carcinogenic and mutagenic effects of
internal combustion engine exhausts have been subjects of
considerable investigation (Campbell et al., 1979; Chaudhart and
Dutta, 1979; Guerrero et al., 1979; Huisingh et al., 1979;
Misioroski et al., 1979; Orthoefer, 1979; Pereira et al., 1979a
and 1979b; Schuler and Niemeier, 1979) in recent years. The
present study with the Trad-MCN test is an extension of other
investigations and examines the mutagenic effects of primarily the

gaseous phase of the combustion products of diesel and diesel plus soybean oil mixed fuel.

Tradescantia inflorescences have been used for mutagenesis assays with both the MCN system (Ma et al., 1978; Ma, 1979; Ma et al., 1981) and the stamen hair system (Underbrink et al., 1973; Sparrow et al., 1974; Sparrow and Schairer, 1976; Schairer et al., 1978; Lower et al., 1978). Dose-response curves were established with Trad-MCN tests on 1,2-dibromoethane (Ma et al., 1978) and X rays (Ma et al., 1980). The versatility of the Trad-MCN test system is further demonstrated in this study.

MATERIALS AND METHODS

Tradescantia clone #4430 was used in the present study except those extended control data obtained from previous experiments using clone #03. Tradescantia plants were propagated in the greenhouse of the Environmental Trace Substance Research Center, University of Missouri, during December and January, with extended photoperiod using artificial lights. The detailed procedure of the Trad-MCN bioassay is described in earlier publications (Ma et al., 1978; Ma, 1979). The engine, air compressor, dilution chamber, and exposure chamber are illustrated in Figure 1.

Statistics and Experimental Design

Three major statistical analyses were carried out: an analysis for distribution of control data, a linear regression analysis, and the Jonckheere test.

Two separate groups of control data, one of 27 observations and one of 155 observations, were tested for normal distribution. The first group was composed of the control data, with Tradescantia clone #4430 reported in this study. The larger group was of control data previously generated from other experiments with clone #03. The analysis of both groups was included to establish the characteristics of the distribution of the frequency of MCN in control populations of Tradescantia plants.

Linear regression analysis was performed on the frequency of MCN and log frequency of MCN. The null hypothesis that the samples are from identical populations was tested against the alternative one-sided hypothesis that the value in the population tends to increase with length of time of exposure.

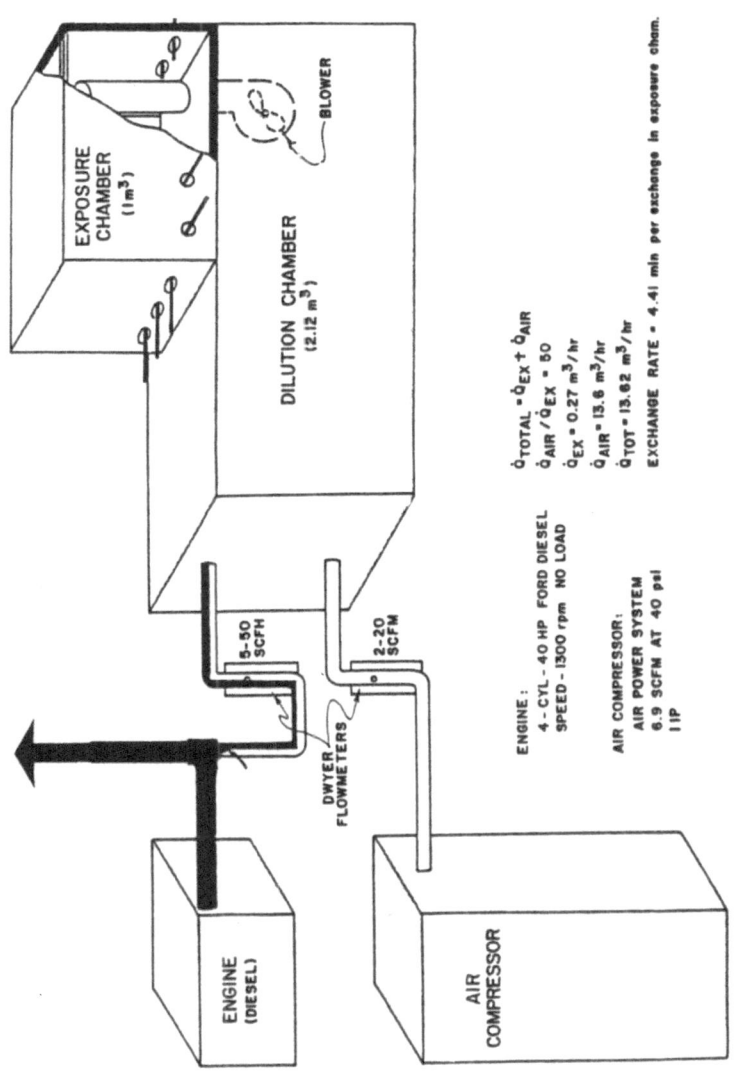

Figure 1. Production and measurement of engine exhaust.

The experimental design consisted of the following:

Engine: One 4-cycle, 4-stroke, 40-h.p. compression-
                     ignition engine, under no load, at 1300 rpm.

Fuels: No. 2 diesel fuel-soybean oil ratios 1:0; 3:1;
                     1:1.

Dilutions: 1:50 (exhaust:air); 1:80 (exhaust:air).

Exposures: 0, 20, 40, 60, 80, 100 min.

Controls: Laboratory control (used 0 exposure control);
                     Garage control (in garage room housing the
                     engine, air pump, dilution chamber, and exposure
                     chamber).

Number of
  Observations: 4 to 10, where one observation is the number of
                     MCN in approximately 300 tetrads from a single
                     bud; 4 to 10 buds analyzed per treatment.

RESULTS AND DISCUSSION

        The untransformed control data showed significant skewness
but no kurtosis in the samples of strain 4430 and both significant
skewness and kurtosis in strain 03. The log transformation
corrected for skewness in strain 4430 and somewhat reduced
skewness and kurtosis in strain 03. The skewness was
overcorrected in strain 03, going from significant positive
skewness to less significant negative skewness. It is apparent
that the two groups of data have different distributions. It is
unlikely that the differences were due to sample size, although it
has been observed that the distribution of the number of
Salmonella ceases to be Poisson at counts over 120. This suggests
the use of nonparametric statistics, in situations where there is
difficulty with the assumptions of distribution. The Wilcoxon,
Kurskal-Walter, and Jonckheere tests, as well as others, can serve
this purpose. The function of sample size, control data of
frequency of micronuclei in different strains of Tradescantia, and
the distribution of data of elevated frequencies of MCN of
different strains of Tradescantia are being investigated.

Strain 4430	(n = 27)		skewness (s)	kurtosis (k)
	MCN		s = 1.335	k = 1.056
			p(s) < 0.01	p(k) > 0.20
	Log MCN		s = -0.537	k = 0.290
			p(s) > 0.20	p(k) > 0.20
Strain 03	(n = 152)			
	MCN		s = 0.951	k = 1.615
			p(s) < 0.001	p(k) < 0.001
	Log MCN		s = -0.556	k = 1.016
			p(s) < 0.01	p(k) < 0.02

The transformation is of less importance in simple
description of data when the mean and an estimation of error,
i.e., S.E. or S.D. of regression, are employed. The log
transformation is of most value when statistical inferences are to
be made between treatment and controls between treatments.

In the first series of experiments on the exhaust fumes of
diesel/soybean oil (1:1) mixed fuel at the exhaust/air dilution
ratio of 1:50, linear response was indicated in Experiment No. 1
from 0 to 80 min of exposure. The MCN frequency and log MCN
frequency are shown in Table 1. The linear regression
dose-response curve is shown in Figure 2. The slopes of
regression and correlation coefficients are significantly
different from zero. The increase in frequency of MCN with length
of exposure is further indicated by the significant value from the
Jonckheere test. In Experiment No. 2 of this first series
(Table 1), linear response by the Jonckheere and log MCN was
indicated from 0 to 40 min of exposure, followed by evidence of
toxicity at 60 and 80 min of exposure and a sharp increase in MCN
frequency at 100 min. The untransformed data of Experiment No. 2
do not indicate that the slope or regression significantly departs
from zero.

In the second series of experiments, the dilution ratio of
exhaust/air was changed at 1:80 because of high toxicity at 1:50
dilution. The diesel/soybean oil mixture ratios were 1:0 and 3:1.
The slopes of regression for MCN frequency and log MCN frequency
were linear from 0 to 40 min at 1:0 ratio for diesel/soybean oil,
0 to 60 min at 3:1 ratio for diesel/soybean oil, and 0 to 80 min
at 1:1 ratio for diesel/soybean oil (Table 2). The Jonckheere
analyses of the data are in agreement with the regression analyses
of all three experiments. A comparison of the results of three
experiments of the second series indicates that increasingly
longer exposures are required to reach the toxicity level, as the
concentration of diesel in the mixed fuels is decreased. Further

Table 1. Percent of Micronuclei of Tradescantia Clone #4430 Exposed to Exhaust from an Engine Burning Diesel Fuel-Soybean Oil at 1:1 Ratio. Dilution of Exhaust-Air 1:50

Experiment Groups	Exposure (min)	Experiment #1			Experiment #2		
		MCN/100 Tetrads	S.E.	n	MCN/100 Tetrads	S.E.	n
Laboratory control	0	3.35[a]	0.69	7	5.30[a]	1.86	6
Garage control	0	4.12	0.69	8	3.56	0.75	5
Treated 1	20	5.70[a]	0.41	8	8.61[a]	2.04	7
Treated 2	40	7.99[a]	0.47	8	9.05[a]	1.26	6
Treated 3	60	10.59[a]	2.24	7	8.60	2.20	6
Treated 4	80	11.03[a]	2.55	10	8.02	1.72	7
Treated 5	100				18.23	5.41	6

MCN Frequency:

Experiment #1: $n = 40$
$b = 0.099$; $p(b) < 0.001$[b]
$r = 0.525$; $p(r) < 0.01$[c]

Experiment #2: $n = 19$
$b = 0.108$; $p(b) > 0.05$
$r = 0.383$; $p(r) > 0.05$

Log MCN Frequency:

Experiment #1:
$b = 0.00519$; $p(b) < 0.001$
$r = 0.579$; $p(r) < 0.01$

Experiment #2:
$b = 0.01113$; $p(b) < 0.05$
$r = 0.460$; $p(r) < 0.05$

Jonckheere Test ($p(z)$):

Experiment #1: < 0.001

Experiment #2: < 0.05

[a] MCN frequencies used in regression analysis and Jonckheere test.
[b] b = Slope.
[c] r = Correlation of regression.

Table 2. Frequencies of Micronuclei Induced by Exhaust Fumes from an Engine Burning Mixtures of Diesel/Soybean Oil at Ratios of 1:0, 3:1, and 1:1, and Exhaust/Air Dilution Ratio of 1:80.

Experimental Groups	Exposure (min)	Experiment #1 (1:0 D/S)			Experiment #2 (3:1 D/S)			Experiment #3 (1:1 D/S)		
		MCN/100 Tetrads	S.E.	n	MCN/100 Tetrads	S.E.	n	MCN/100 Tetrads	S.E.	n
Laboratory control	0	5.71[a]	0.73	7	5.71[a]	0.73	7	3.21[a]	0.93	5
Garage control	0	6.97	1.00	5	5.66	0.51	5	4.59	0.84	4
Treated 1	20	10.40[a]	2.48	5	7.09[a]	1.25	5	8.29[a]	1.12	4
Treated 2	40	13.50[a]	1.41	5	8.78[a]	2.32	5	9.84[a]	2.57	4
Treated 3	60	6.94	1.23	5	10.55[a]	0.91	5	12.37[a]	4.78	5
Treated 4	80	6.46	1.57	5	4.16	0.93	5	12.85[a]	4.67	5
Treated 5	100	2.71	2.71	5	6.63	0.47	5	10.11	1.25	5
MCN Frequency:		n = 17 b = 0.195; p(b) 0.01[b] r = 0.699; p(r) 0.01[c]			n = 22 b = 0.080; p(b) 0.01 r = 0.545; p(r) 0.01			n = 22 b = 0.117; p(b) 0.005 r = 0.488; p(r) 0.05		
Log MCN Frequency:		b = 0.0098; p(b) 0.001 r = 0.692; p(r) 0.01			b = 0.00449; p(b) 0.01 r = 0.482; p(r) 0.05			b = 0.0070; p(b) 0.005 r = 0.542; p(r) 0.05		
Jonckheere Test (p(z)):		0.005			0.05			0.01		

[a]Exposures used for regression analysis and Jonckheere Test.
[b]b = Slope.
[c]r = Correlation of regression.

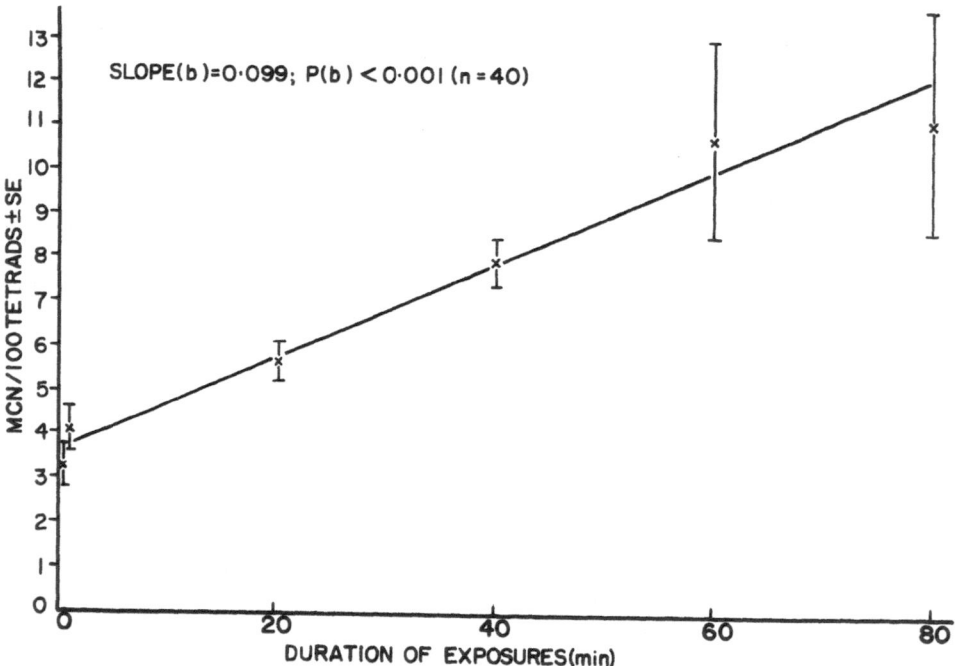

Figure 2.   Linear regression dose-response
            curve of diesel exhaust fumes.

experimentation is needed to determine whether this trend is true in reference to mutagenicity.

Results of the present study show agreement with that obtained from the Trad-MCN test on exhaust fumes generated by the standard Nissan (1970 Datsun) diesel automobile under a simulated city-driving cycle (Ma et al., 1981) using Tradescantia clone #03. The effective concentration of diesel automobile exhaust fumes was at 1:23- and 1:45-fold exhaust/air dilutions, while the present study was in the 1:50 and 1:80 dilution range. This difference could be due to a higher sensitivity of Tradescantia clone #4430 than that of Tradescantia clone #03.

ACKNOWLEDGMENTS

The authors wish to express their deepest appreciation to Mr. Clifford Mongler, Engineering Research Technician of the Department of Agricultural Engineering, University of Missouri, for his technical assistance in operating and maintaining the diesel engines to provide the exhaust fumes.

Research was partially supported by the U.S. Environmental Agency research grant #R80-7497-01.

REFERENCES

Campbell, K.I., E.L. George, and I.S. Washington, Jr. 1979. Enhanced susceptibility to infection in mice after exposure to diluted exhaust from light-duty diesel engines. Presented at the International Symposium on Health Effects of Diesel Engine Emissions, Poster Session, Cincinnati, OH.

Chaudhart, A., and S. Dutta. 1979. Effect of exposure to diesel exhaust on pulmonary prostaglandin dehydrogenase (PGDH) activity. Presented at the International Symposium on Health Effects of Diesel Engine Emissions, Poster Session, Cincinnati, OH.

Guerrero, R.R., D.E. Rounds, and J. Orthoefer. 1979. Sister chromatid exchange analysis of Syrian hamster lung cells treated in vivo with diesel exhaust particulates. Presented at the International Symposium on Health Effects of Diesel Engine Emissions, Cincinnati, OH.

Huisingh, J., S. Nesnow, R. Bradow, and M. Waters. 1979.
    Application of a battery of short-term mutagenesis and
    carcinogenesis bioassays to the evaluation of soluble
    organics from diesel particulates. Presented at the
    International Symposium on Health Effects of Diesel Engine
    Emissions, Cincinnati, OH.

Lower, W.R., P.S. Rose, and V.K. Drobney. 1978. In situ
    mutagenic and other effects associated with lead smelting.
    Mutation Res. 54:83-93.

Ma, T.H. 1979. Micronuclei induced by X rays and chemical
    mutagens in meiotic pollen mother cells of Tradescantia -- a
    promising mutagen test system. Mutation Res. 64:307-313.

Ma, T.H., V.A. Anderson, and S.S. Sandhu. 1981. A preliminary
    study of the clastogenic effects of diesel exhaust fumes
    using Tradescantia-Micronucleus bioassay. In: Short-Term
    Bioassays in the Analysis of Complex Environmental Mixtures
    II. M.D. Waters, S.S. Sandhu, J.L. Huisingh, L. Claxton, and
    S. Nesnow, eds. Plenum Press: New York. pp. 351-358.

Ma, T.H., G.J. Kontos, Jr., and V.A. Anderson. 1980. Stage
    sensitivity and dose response of meiotic chromosomes of
    pollen mother cells of Tradescantia to X rays. Environ.
    Expt. Bot. 20:169-174.

Ma, T.H., A.H. Sparrow, L.A. Schairer, and A.F. Nauman. 1978.
    Effect of 1,2-dibromoethane (DBE) on meiotic chromosomes of
    Tradescantia. Mutation Res. 58:251-258.

Misioroski, R.L., K. Strom, and M. Schvapil. 1979. Lung
    biochemistry of rats chronically exposed to diesel
    particulate. Presented at the International Symposium on
    Health Effects of Diesel Engine Emissions, Cincinnati, OH.

Orthoefer, J.G. 1979. The strain A mouse as an inhalation
    carcinogenesis model in diesel exhaust research. Presented
    at the International Symposium on Health Effects of Diesel
    Engine Emissions, Cincinnati, OH.

Pereira, M.A., P.S. Sabharwal, C. Ross, P. Kaur, A. Choi, and
    T. Dixon. 1979a. In vivo studies on the mutagenic effects
    of inhaled diesel exhaust. Presented at the International
    Symposium on Health Effects of Diesel Engine Emissions,
    Cincinnati, OH.

Pereira, M.A., P.S. Sabharwal, and A.J. Wyrobek. 1979b. Sperm abnormality bioassay of mice exposed to diesel exhaust. Presented at the International Symposium on Health Effects of Diesel Engine Emissions, Cincinnati, OH.

Schairer, L.A., J. Van't Hof, C.C. Hayes, R.M. Burton, and F.J. deSerres. 1978. Exploratory monitoring of air pollutants for mutagenicity activity with the Tradescantia stamen hair system. Environ. Health Perspect. 27:51-60.

Schuler, R.L., and R.W. Niemeier. 1979. A study of diesel emission on Drosophila. Presented at the International Symposium on Health Effects of Diesel Engine Emissions, Cincinnati, OH.

Sparrow, A.H., L.A. Schairer, and R. Villalobos-Pretrini. 1974. Comparison of somatic mutation rates induced in Tradescantia by chemical and physical mutagens. (abstr.). Mutation Res. 26:265.

Sparrow, A.H., and L.A. Schairer. 1976. Response of somatic mutation frequency in Tradescantia to exposure time and concentrations of gaseous mutagens. (abstr.). Mutation Res. 38:405-406.

Underbrink, A.G., L.A. Schairer, and A.H. Sparrow. 1973. Tradescantia stamen hairs: a radiobiological test system applicable to chemical mutagenesis. In: Chemical Mutagens: Principles and Methods for Their Detection, Volume 3. A. Hollaender, ed. Plenum Press: New York. pp. 171-207.

SESSION 2

APPLICATION OF
BIOASSAYS TO THE
CHARACTERIZATION OF
COMPLEX MIXTURES

MICROBIAL ASSAYS IN RESEARCH AND IN THE CHARACTERIZATION OF COMPLEX MIXTURES

Herbert S. Rosenkranz[1], Elena C. McCoy[1], and
Robert Mermelstein[2]

[1]Center for the Environmental Health Sciences, School of
Medicine, Case Western Reserve University, Cleveland, Ohio
44106, and [2]Joseph C. Wilson Center for Technology, Xerox
Corporation, Rochester, New York 14644

INTRODUCTION

Because of the unusually high reliability of short-term
microbial assays for predicting the potential carcinogenicity of
environmental agents, the emphasis of recent reports has been on
standardization of experimental procedures, interlaboratory
reproducibility, and the generation of a data base for a wide
variety of chemicals (de Serres and Shelby, 1979a,b; Mattern et
al., 1978; Seiler et al., 1980). Indeed, studies in this
laboratory have dealt with experimental factors which may affect
the mutagenic and genotoxic responses of potential carcinogens
(Rosenkranz et al., 1976a,b, 1979a,b, 1980; Rosenkranz and Leifer,
1980; Rosenkranz and Poirier, 1979). All too often, however, we
tend to forget that short-term assays can be used as research
tools in the development of new concepts related to the molecular
basis of DNA-adduct formation, its repair, and alternate
mechanisms responsible for cancer induction. In addition, the
data base itself, if developed properly, may be an excellent
source of information for the development of theoretical models
which could form the basis of a more rational approach to risk
evaluation. The present report deals with new developments in
each of these aspects and how they have led to the elucidation of
the mechanism of action of a group of chemicals which originally
were detected in complex mixtures (Rosenkranz et al., 1980).

Salmonella Tester Strains Are Not Stable

The definition of a specific genotype, a priori, implies
heritable traits that are mutable at a certain, albeit low,
frequency (approximately $1/10^7$ per cell per generation in
Salmonella). This is the basic assumption of mutagenicity assay
procedures, it being stipulated further that a mutagen will
increase this frequency in a measurable dose-dependent fashion.
Recently, however, in the course of attempting to devise rapid
diagnostic procedures for verifying the phenotypes of Salmonella
tester strains, we discovered that the phenotype of these strains
was variable and that this was related mainly to the presence of
plasmid pKM101, as well as, to a smaller extent, the absence of a
functional uvrB gene product.

Since the original intent of the exercise was to devise a
rapid and sensitive procedure for ascertaining the identity of
Salmonella tester strains, we used the API 20E System, which is a
widely used procedure in clinical microbiology laboratories for
identifying Enterobacteriaceae. The system consists of a battery
of 24 biochemical tests, the results of which are used to generate
a profile number which, with a high probability, can identify the
isolate with respect to genera and species. Our intent was not to
use the system to speciate the cultures, since we knew their
identity and we had shown earlier that the introduction into the
strains of a genetic deletion to increase their sensitivity to
mutagens precluded their identification as a Salmonella species
(Speck et al., 1975). Rather, we wanted to determine whether we
could generate a unique biochemical profile number which could be
used for identification purposes.

Initially, we tested various Salmonella cultures in use at
the time for other aspects of our mutagenicity studies. The
impression was soon gained that different subcultures of the same
strain exhibited changed phenotypes which were far beyond
permissible variations of the API 20 system. This led us to
investigate systematically pedigreed subcultures and subclones of
individually-cloned cultures. However, as these studies were done
in parallel with mutagenicity assays, only those cultures that
gave the appropriate responses in control tests were used:
ampicillin-resistance for Salmonella TA98 and TA100 and crystal
violet-sensitivity for all (Ames et al., 1975). In addition, to
be included in the study, the cultures had to give the expected
response when exposed to control mutagens (sodium azide for TA1535
and TA100, 9-aminoacridine for TA1537 and 2-nitrofluorene for TA98
and TA1538).

It was thus found that while Salmonella typhimurium LT2, the progenitor of all the tester strains, reproducibly gave a unique identifiable profile number, some of the tester strains exhibited significant variations in some of the phenotypes. Strains harboring plasmid pKM101 (TA98 and TA100) exhibited even greater variations in phenotypes (Table 1). The shifts increased with the number of passages.

There are certain facts that are germane to these observations:

1) These variations increased with passage number. It must be remembered, however, that each passage involves at least 27 divisions, i.e., if the shifts occur at random in time, many of our subclones might in fact be mixtures.

2) Only certain phenotypes are affected. Thus, the inability to produce hydrogen sulfide is not affected. Obviously, since these Salmonellae do not possess the enzyme (even a cryptic one) due to a deletion, there is no mechanism for introducing this activity either by a mutation or by a shift in the regulatory mechanism.

3) Each of the phenotypes which does show variations probably involves at least five different gene products. Since, presumably, the phenotypes tested are not linked (were this the case it would, in effect, defeat the purpose of the diagnostic API 20 assay), it is probable that the variations that are seen reflect events occurring at other segments of the Salmonella genome as well. It can be estimated that of the phenotypes tested, those that are active in Salmonella approximate 4% of the total genome.

The possible basis of the phenotypic shifts is a matter of some puzzlement. In view of the fact that they appear to be conserved for several passages, they may, indeed, reflect genomic changes. Since phenotypic shifts are not seen in "wild type" strains, it may be safely concluded that they reflect the genetic manipulations that were used to make the tester strains more sensitive to mutagens: the uvrB lesion, the deep-rough character and plasmid pKM101. In this connection it should be remembered that the variations were maximal in those strains bearing pKM101, a plasmid coding for error-prone DNA repair (Goze et al., 1979; McCann et al., 1975; Monti-Bragadin et al., 1977; Todd et al., 1979; Walker, 1978).

Table 1. Phenotypic Shifts in *Salmonella typhimurium* Strains

Strain	Passage[b]	Biochemical Tests[a]																							
		1	2	3	4	5	6	7	8	9	10	11	12	13	14	15	16	17	18	19	20	21	22	23	24
TA1538	03	−	−	+	+	−	−	−	−	−	−	−	+	+	+	+	+	−	+	−	−	−	−	−	−
	05																								
	07																								
	11																	CR[c]							
	19																		CR						
TA98	03	−	−	+	+	−	−	−	−	−	−	−	−	+	+	+	−	−	−	−	−	−	−	−	+
	06							CR		CR				CR	CR										
	09									CR	CR	CR	CR			CR	CR	CR			CR		CR	CR	CR
	12														CR										
	15																								
TA100	01	+	−	+	−	−	−	−	−	+	−	−	+	+	−	+	−	−	−	+	+	−	+	−	−
	09	CR		CR						CR			CR				CR				CR		CR	CR	CR
	10	CR			CR					CR			CR		CR		CR				CR		CR	CR	CR
	12	CR	CR	CR						CR					CR		CR	CR			CR		CR	CR	CR

[a]1:o-Nitrophenyl-β-D-galactosidase; 2:Arginine dihydrolase; 3:Lysine decarboxylase; 4:Ornithine decarboxylase; 5:Citrate utilization; 6:H_2S production; 7:Urease; 8:Tryptophan deaminase; 9:Indole production; 10:Acetoin production; 11:Gelatinase; 12:Glucose fermentation; 13:Mannitol fermentation; 14:Inositol fermentation; 15:Sorbitol fermentation; 16:Rhamnose fermentation; 17:Sucrose fermentation; 18:Melibiose fermentation; 19:Amydaline fermentation; 20:Arabinose fermentation; 21:Oxidase; 22:Reduction of nitrate to nitrite; 23:Reduction of nitrate to nitrogen gas and amine; 24:Motility.

[b]Passages are not necessarily lineal.

[c]Indicates changed reaction when compared with the earliest subculture tested.

If it be assumed that the phenotypic shifts reflect genetic changes, the occurrence of which is amplified by a nonfunctioning uvrB gene product and an error-prone DNA repair, then there are four types of mutagenic events that might be responsible: spontaneous mistakes in duplication; endogenously produced mutagens (see also Sargentini and Smith, 1981) such as acetaldehyde and accumulation of formic acid (the latter could cause acid-induced depurinations, as the Salmonella tester strains are deficient in formic acid hydrogen lyase) (Speck et al., 1975); exogenous mutagens such as ultraviolet light, pyrolysis products present in nutrients and environmental agents (for this latter class of mutagens, the deep-rough character of the tester strains would facilitate penetration); and plasmid-induced chromosomal rearrangements or integration.

Because of the possible effects of these changed phenotypes on the expression of induced mutations in the tester strains, it would seem that the establishment of a large stock of early passage cultures would be a strategy to circumvent possible problems.

Genetic Drift and the Origin of Cancer

There are many theories seeking to explain the induction of cancer. For the past 20 to 25 years, it has been recognized that carcinogens share with genotoxicants an ability to alter cellular DNA. The pioneering studies of Miller and Miller (1977) and their associates on the biotransformation of procarcinogens to electrophiles provided a chemical basis to support DNA modifications as the initiating event in carcinogenesis. This, together with the extraordinary predictive ability of microbial assays (Ames, 1979) provided renewed credibility for the somatic mutation theory of cancer (Straus, 1981). However, because the somatic theory did not account for all of the experimental observations, alternate explanations involving other DNA rearrangements have been proposed, e.g., tandem duplication of src genes (Pall, 1981), the involvement of transposons (Cairns, 1981).

The serendipitous finding of genetic drift, in the absence of selective pressure, in certain bacterial strains may well have a bearing on the mechanistic theories seeking to explain the basis of chemically-induced cancers. It combines elements of several possible mechanisms: mutation (somatic mutation theory), plasmids (i.e., mutations due to either transposon insertion or uneven crossing-over due to error-prone DNA repair) and finally transposon-mediated DNA rearrangements. Because microbial strains can be manipulated readily, the present findings can provide the basis of additional experiments, e.g., the effect of the simultaneous presence of error-prone repair genes and transposons (pKM101 and Tn10), the effect of curing pKM101 on phenotypes and genotypes.

Reproducibility of the Salmonella Mutagenicity Assay

As mentioned earlier, much emphasis has been placed on developing standard procedures for maximum reproducibility (de Serres and Shelby, 1979a,b). The need for such procedures was exemplified by the quantitative results obtained in an international study in which 42 chemicals were analyzed by the procedures in use at that time in each of the participating laboratories which performed the Salmonella assay (de Serres and Ashley, 1981). On the other hand, a collaborative study involving four different laboratories revealed improved consonance of results when experimental procedures were carried out under rigorously controlled conditions (e.g., composition of the medium, physiological state of the tester microorganisms, etc.) (Dunkel, 1979; Dunkel and Simmon, 1980).

To gain an appreciation of the extent of the intralaboratory variance that is obtained when experimental conditions are rigorously controlled, a number of our findings involving a single strain (TA98) are summarized here. It was found that each strain displays a characteristic pattern of spontaneous frequencies of mutations in the absence of chemicals (Figure 1) and this is influenced to some extent by the presence of the usual solvent (DMSO, Table 2). In addition, the distribution of mutagenic signals in response to exposure to a single dose of 2-nitrofluorene (100 µg per plate; each assay done in triplicate) was unusually broad (Figure 2). The significance of this finding will be discussed later (see below). Finally, the mutagenicity observed in response to a single concentration of a chemical requiring metabolic activation (2-aminoanthracene) was also extremely broad (Table 2 and Figure 3), presumably reflecting not only the biological variability of the microorganism but also the differences in activities of the various batches of S9 preparations that were used. All of these variables must, of course, be taken into consideration when analyzing the results of mutagenicity assays and, indeed, data such as these have led to the development of computer programs which automatically and cumulatively evaluate results with respect to internal consistency as well as to confidence levels (Chu et al., 1981).

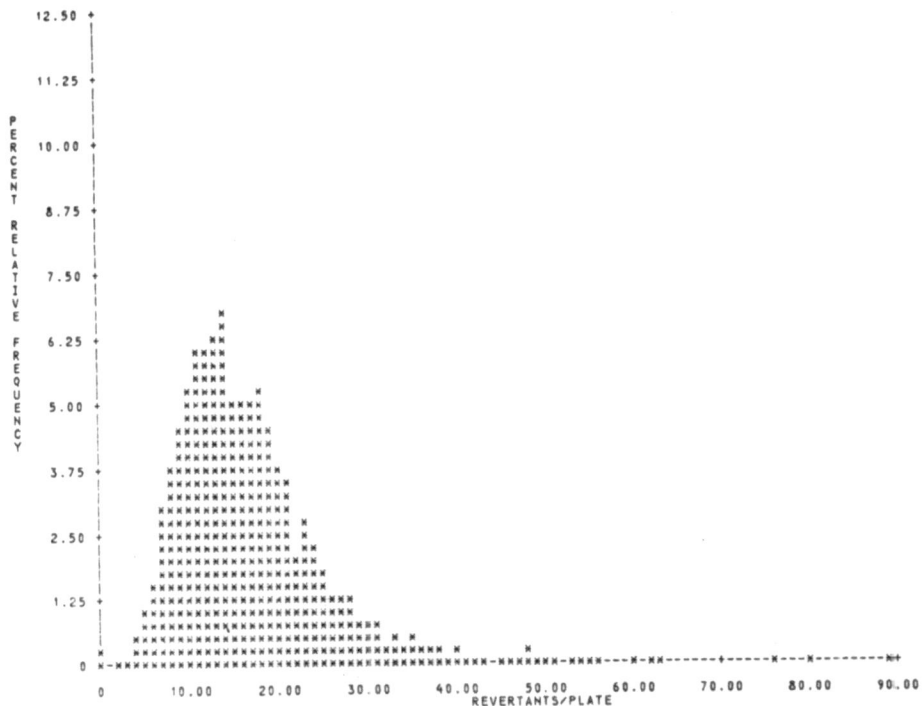

Figure 1.   Characteristics of <u>Salmonella typhimurium</u> TA98.
             Frequency of spontaneous revertants in the absenc2 of
             mutagen and of solvent.   N = 2373 determinations.

The variability of the response of 2-nitrofluorene in TA98 is
worthy of a more detailed analysis in view of our interest in the
mutagenic properties of the environmental nitroarenes, a group of
chemicals which appears to be ubiquitous in our environment
(Mermelstein et al., 1982; Rosenkranz et al., 1982).

It is, of course, recognized that single-dose determinations,
even when they are carried out in triplicate, are fraught wit˥
uncertainties and that the mutagenic potency of a chemical is best
determined from the linear (ascending) portions of dose-response
curves. Indeed, when this is done, for most chemicals the results
obtained are quite reproducible and can be used with confidence
for the determination of structure-activity relationships. When,

Table 2.   Characteristics of <u>Salmonella</u> <u>typhimurium</u> TA98

Chemical	µg/Plate	Solvent	S9	N[a]	Mean[b]	S.E.
None	0	None	-	2373	16.9	0.2
None	0	DMSO	-	1375	11.9	0.6
2-Nitrofluorene	100	DMSO	-	556	719	11.6
2-Aminoanthracene	1	DMSO	-	899	17.6	1.1
2-Aminoanthracene	1	DMSO	R[c]N[d]	402	613	20.2
None	0	DMSO	RN	150	12.5	0.5
2-Aminoanthracene	1	DMSO	RI[e]	398	452	17.5
None	0	DMSO	RI	153	14.5	0.4
2-Aminoanthracene	1	DMSO	M[c]N	375	244	10.6
None	0	DMSO	MN	144	12.9	0.4
2-Aminoanthracene	1	DMSO	MI	373	623	39
None	0	DMSO	MI	141	13.4	0.4
2-Aminoanthracene	1	DMSO	H[c]N	401	774	24
None	0	DMSO	HN	158	13.7	0.4
2-Aminoanthracene	1	DMSO	HI	404	516	17
None	0	DMSO	HI	157	13.3	0.4

[a]Number of determinations.
[b]The mean is expressed as rev/plate.
[c]Rat, mouse, and hamster livers, respectively.
[d]Non-induced.
[e]Induced.

however, such analyses are carried out for 2-nitrofluorene, the experimental variability remains (Figure 4). The result of a further set of 18 dose-response curves for 2-nitrofluorene and TA98 substantiates this variability (Table 3).

There are several features which become apparent when the results (Table 3) are scrutinized further: Specimens which were run concurrently (Experiments 12 and 13) gave results which agreed with one another. Chronologically, the earlier results tended to give higher specific mutagenicities.

In retracing these chronologies, it was found that although, initially, the protocol called for the use of growing cultures, no systematic effort was made to use early exponential growth-phase cultures. Some of the cultures might have been middle- and late-exponentially growing cultures. As our attempts to standardize the assay continued, every effort was made to use only

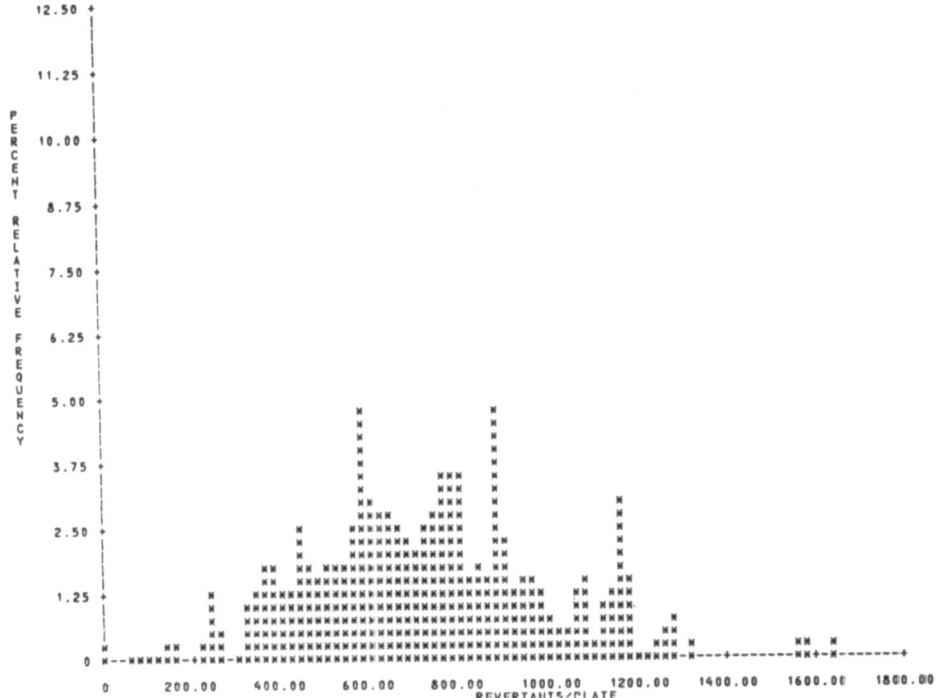

Figure 2.   Response of <u>Salmonella typhimurium</u> TA98 to
2-nitrofluorene (100 μg/plate). N = 556
determinations.

early exponential (2 x 10⁸ cell/ml) phase cultures with the result
that the mutagenicity of 2-nitrofluorene decreased and results
were more erratic. No such findings were made with non-nitrated
test chemicals.

It was only recently that these inconsistencies involving
2-nitrofluorene were clarified, largely due to recognition of the
potent mutagenicity of nitroarenes and their wide distribution in
the environment (Mermelstein et al., 1981). It appears that for
nitroarenes, early exponentially growing cells are not optimal.
Much more reproducible results are obtained when resting cultures
are used. In addition, for some nitroarenes (for example,
1,8-dinitropyrene) the mutagenicity is increased up to fourfold
when such resting cultures are used (Mermelstein et al., 1981).

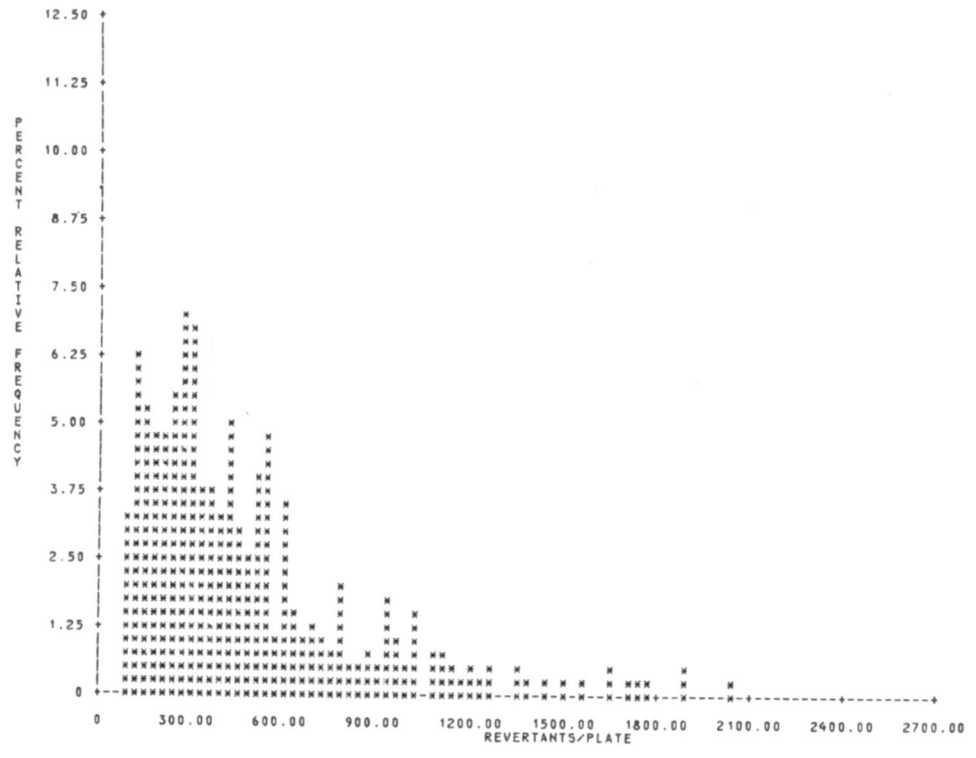

Figure 3. Response of <u>Salmonella</u> <u>typhimurium</u> TA98 to
2-aminoanthracene (1 μg/plate) in the presence of S9
from the livers of Aroclor-induced rats. N = 398
determinations.

There are several reasons which appear to be responsible for
this phenomenon (Rosenkranz et al., in press):

1) Nitroarenes must be reduced to the corresponding arylhy-
    droxylamines to express their mutagenicity. This is
    accomplished by bacterial enzymes (nitroreductases).

2) Arylhydroxylamines are notoriously oxygen-labile, being
    reoxidized to nitroarenes in the presence of oxygen.

3) Some of the bacterial nitroreductases are oxygen-
    sensitive.

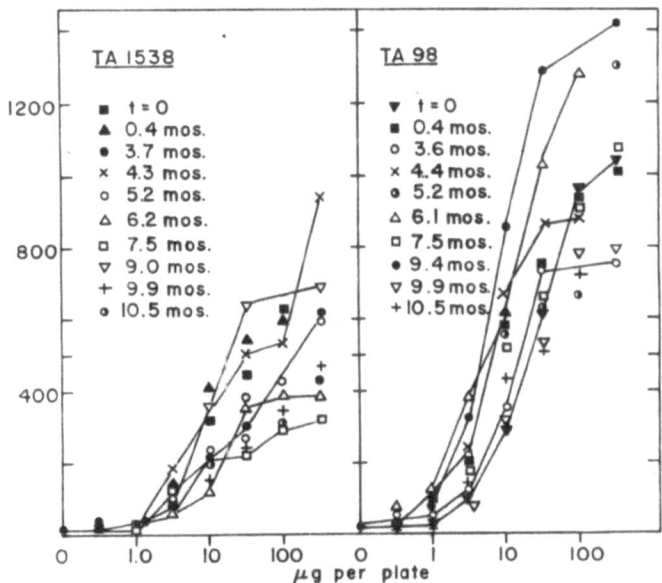

Figure 4.   Mutagenicity of 2-nitrofluorene for Salmonella
typhimurium TA98 and TA1538 (Cheli et al., 1980).

4) The mutagenicity of nitroarenes, including that of
2-nitrofluorene, is increased when the cells are
incubated anaerobically during the period of their
susceptibility to mutagenesis (Löfroth, 1981; Rosenkranz
and Poirier, 1979). It would appear that when cells
enter the resting phase, they start fermenting and, in
effect, their level of molecular oxygen decreases. In
view of the above, this increases their susceptibility to
the mutagenic actions of nitroarenes.

Obviously, the lesson to be learned is that conditions that
are optimal for one set of chemicals may not be so for another
group. This, of course, presents a problem with respect to our
goals of standardizing mutagenicity assays. It would seem that

Table 3.   Mutagenicity of 2-Nitrofluorene
for Salmonella typhimurium TA98[a]

Experiment	Date	Rev/µg	Experiment	Date	Rev/µg
1	03/13/78	79.3	10	09/25/79	61.6
2	07/11/78	60.1	11	10/29/79	85.7
3	09/08/78	45.7	12	02/26/80	43.4
4	10/19/78	79.2	13	02/26/80	47.2
5	04/02/79	65.6	14	04/08/80	44.6
6	04/12/79	26.7	15	05/16/80	49.7
7	05/22/79	34.4	16	05/26/80	22.9
8	06/18/79	45.7	17	08/19/80	40.1
9	08/13/79	43.8	18	11/17/80	24.8

[a]Mean 50.0 ± 18.6; DMSO control 11.9 ± 6.6.

since many of our environmental mixtures, especially those derived
from incomplete combustion processes, contain nitroarenes
(Rosenkranz et al., in press), these are important facts to bear
in mind.

Activity-Directed Fractionation of Complex Mixtures

Philosophically, when dealing with the toxic and potentially
carcinogenic properties of environmental mixtures, the main
concern has tended to be with the most abundant components present
therein. This had led to the adoption of benzo(a)pyrene as a
surrogate for determining the quality of the environment (e.g.,
the Clean Air Act) (U.S. Congress, 1977; U.S. Environmental
Protection Agency, 1979a,b). This, in turn, has meant that many
of our analyses of complex mixtures dealt with the benzo(a)pyrene
content; in effect, they were mass-oriented. Minor components
with potent biological activities might have been overlooked.
This was presumably the situation with the nitroarenes. We now
know that they have been present in environmental specimens,
especially those derived from diesel emissions or the ambient
atmosphere. However, their presence in mixture was probably less
than 100 ppm and, with the main emphasis being on mass, they were
missed.

We identified the nitroarenes in complex mixtures when a
mutagenic signal was detected in extracts of certain carbon blacks

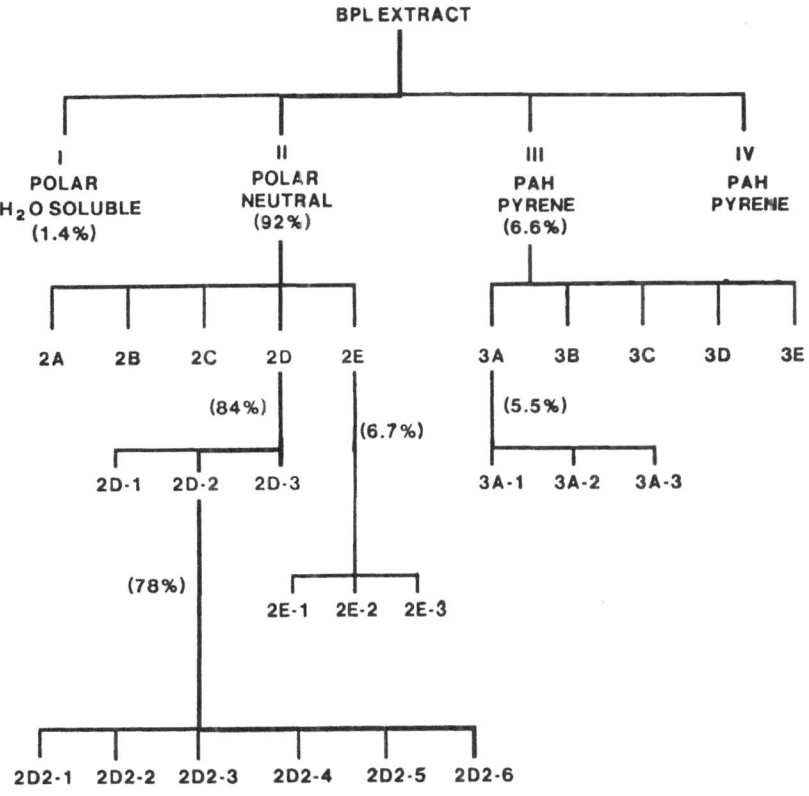

Figure 5.   Fractionation of a carbon black extract that resulted
             in the isolation and identification of nitropyrenes.

(Löfroth et al., 1980; Rosenkranz et al., 1980). An
activity-directed fractionation scheme (Figure 5) led to the
isolation of a fraction containing less than 2% of the mass and in
excess of 84% of the mutagenicity. Further fractionation resulted
in the isolation of a mixture that consisted primarily of either
dinitrofluoranthene or dinitropyrenes. The HPLC profile resembled
the nitration product of pyrene and not of fluoranthene
(Figure 6). The pattern of mutagenic responses confirmed that the
sample contained primarily dinitropyrenes. Analytical as well as
synthetic procedures revealed that we were dealing with 1-nitro-,
1,3-, 1,6-, and 1,8-dinitropyrene, 1,3,6-tri- and
1,3,6,8-tetranitropyrene (Rosenkranz et al., 1980).

     Further careful analyses revealed that the original carbon
black contained at most 100 ppm nitropyrenes (the variability
between batches was, however, substantial; some batches had a
nitropyrene content of 5 ppm), which resulted from the nitric acid

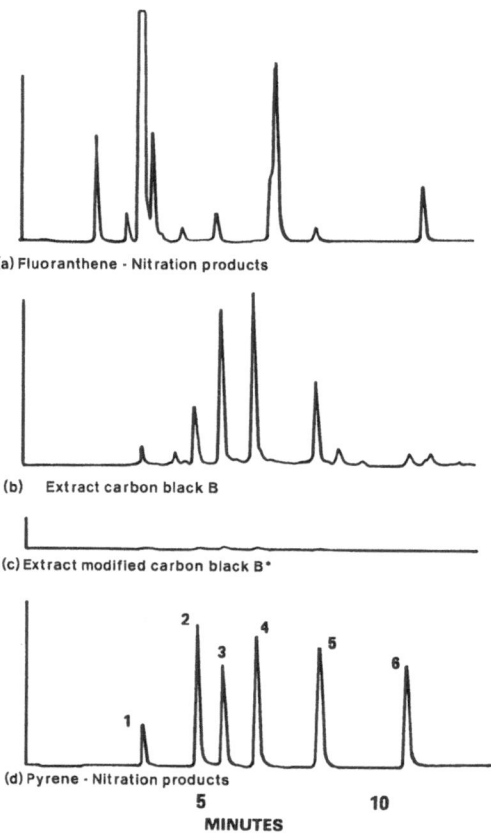

Figure 6.   HPLC chromatograms.   (A) Fluoranthene nitration
products.   (B) Carbon black B, extract from 6 mg.
(C) Carbon black B* (new carbon black prepared under
controlled conditions), extract from 6 mg.   (D) Pyrene
nitration products.   Peaks were identified as
1) 1-nitropyrene, 2) 1,3-dinitropyrene, 3) 1,6-dinitro-
pyrene, 4) 1,8-dinitropyrene, 5) 1,3,6-trinitropyrene,
and 6) 1,3,6,8-tetranitropyrene.   Data obtained on a
Zorbax CN (35°C; 2 ml/min, 5 to 60% i-propanol-hexane;
400-nm ultraviolet detector) (Rosenkranz et al., 1980).

oxidation of carbon black. The nitrated species were in no way
essential to the integrity or function of the final carbon black
and represented only a trace of impurity. This recognition
permitted modification of the manufacturing process which resulted
in a great decrease in the amount of nitroarenes present in the

final product (see Figure 6C, Carbon black B*) (Rosenkranz et al.,
1980).

Further studies of the properties of the nitroarenes revealed
that they are a group of extraordinarily active bacterial
mutagens. They have a ubiquitous distribution as they are formed
as by-products of incomplete combustion processes during the
coincidental presence in the exhaust of polycyclic aromatic
hydrocarbons, oxides of nitrogen, and traces of acid. The basis
of their biological activity is the subject of a number of studies
in various laboratories at this time (Rosenkranz and Mermelstein,
in press). We shall return to the properties of the nitroarenes
at a later time when we shall use them to illustrate certain basic
principles.

Principles of Mutagenic Specificities

In analyzing mutagenicity data there are a number of basic
species specificities in the mutagenic responses that are useful
in making preliminary classifications of the type of structure
involved in the mutagenic responses.

Although Salmonella typhimurium tester strain TA1535 responds
to base-substitution mutagens, its plasmid-containing derivative
TA100 responds not only to base-substitution mutagens but to
frameshift mutagens that form DNA adducts, as well.

Frequently Salmonella typhimurium TA1535 and E. coli WP2uvrA
are used interchangeably to detect base-substitution mutagens.
This substitution may not be valid for theoretical as well as
practical reasons:

1) There are, of course, species differences between E. coli
    and Salmonella; these may involve the structure and
    composition of their respective DNAs and their repair.

2) The basis of the reverse mutations that are scored in
    Salmonella TA1535 and E. coli WP2uvrA are also different.
    They involve an ochre mutation or its suppression in
    E. coli; the target in Salmonella is undetermined, but
    some experimental findings suggest a reversion at a GC
    base pair.

3) Salmonella TA1535 is a deep-rough strain (deficient in
    the lipopolysaccharide envelope) which increases its
    permeability to large molecules (Ames et al., 1975).

4) It is known that the mutagenicity and survival of
    microorganisms to genotoxic agents is dependent, to some
    extent, on the presence and nature of prophages and
    plasmids. Most bacterial species are known to harbor
    such episomes. The nature of the cryptic episomes
    present in Salmonella TA1535 and E. coli WP2uvrA have not
    been elucidated, but they are likely to be different.
    (This phenomenon is known to be of importance in
    unbalanced growth, e.g., thymineless death, which is one
    manifestation of toxicity due to genotoxic agents.)

5) Some chemicals presumed to induce base-substitution
    mutations do so only in one or the other of these
    species. These include nitrofurans, halogenated epoxides
    and others (Table 4).

Chemicals which induce frameshift mutations by virtue of
their ability to intercalate between DNA base pairs without
forming covalent DNA adducts will mutagenize strain TA1537, but in
the absence of S9, they will not generally affect the other tester
strains (e.g., proflavin, 9-aminoacridine, quinacrine) (Table 5).
The mutagenicity of such chemicals will not be reduced in strain
TA1977, the uvrB[+] analog of TA1537 (McCoy et al., 1981). This is
due to the fact that the uvrB[+] gene product is an enzyme that
recognizes and acts to repair covalent DNA adducts.

Table 4. Base-Substitution Mutations: Specificity

Chemical	Revertants per microgram			
	TA1535	TA100	TA98	WP2uvrA
Pivalolactone	0	1.6		0.13
NFTF[a]	0	6900		1186
Tetrachloroethylene oxide	15.6[b]			0
BMBA[c]	0	87	197	

[a]N-(4-Nitro-2-furyl) 2-thiazoyl formamide.
[b]Revertants per millimole.
[c]7-Bromomethyl-12-methylbenz(a)anthracene.

Table 5. Frameshift Mutations: Specificity[a]

Revertants per nanomole

Chemical	TA100	TA1538	TA98	TA1537	TA1977	TA1978
9-Aminoacridine	0	0	0	1.7	1.8	1.8
Quinacrine	0	0	0	3.0		
Proflavin	0	0	0	1.6	1.6	
Entozon	24	288	338	10	5	23
N2-Fluorenylhydroxamic acid (+S9)	31	22	10	0.04		
BMBA[b]	29	16	66	4.8		
1,3-Dinitropyrene[c]	8,359	15,600	28,600	13,400	0	0

[a]S9 derived from Aroclor-induced rat liver.
[b]BMBA: 7-Bromomethyl-12-methylbenz(a)anthracene.
[c]Growing cultures were used.

Chemicals which induce frameshift mutations as a result of covalent linkages with the DNA mutagenize primarily strain TA1538 and its plasmid-containing derivative, TA98. There is some "spillover" into TA1537. However, such chemicals exhibit greatly diminished activities in TA1977 and TA1978, the uvrB$^+$ analogs of TA1537 and TA1538, respectively (Table 5).

For chemicals that are capable of intercalating as well as of forming covalent adducts (e.g., Entozon, a nitroacridine), the contribution of intercalation and adduct formation can be deduced from the comparison of the activities of the chemical in the various tester strains. It was found that for Entozon, intercalation accounted for only a small percentage (1%) of the total activity (Table 5). In general, for chemicals capable of intercalating as well as of forming adducts, the latter is the major contributor to the mutagenicity.

Simple procedures to confirm the intercalating properties of frameshift mutagens involve determination of physical chemical properties of the DNA-ligand mixture: sedimentation in sucrose gradients or migration in agarose gels of closed circular DNA molecules, demonstration of spectral shifts, decreases in the temperature of the helix-to-coil transition. Such techniques have established that some chemicals do indeed induce frameshift mutations by adduct formation as well as by intercalation mechanisms (McCoy et al., 1981).

A disproportionate increase in the activity of a chemical in TA98 as compared to the activity displayed in TA1538 may indicate gross DNA structural distortions which may cause the induction of error-prone DNA-repair enzymes. This infidel DNA-repair process has been implicated in carcinogenesis (Echols, 1981; Radman, 1977, 1980; Radman et al., 1977; Sargentini and Smith, 1980). This topic will be discussed in a different context later.

Taking Advantage of Unique Microbial Metabolic Pathways

Those who criticize short-term microbial assays remind us that microorganisms are not mammalian cells and hence they cannot mimic the metabolism unique to mammals. This is, of course, true; as a matter of fact, the situation is even more complex, as there are sites on the microorganism (especially the gram-negative ones such as Salmonella) which are unique. This applies especially to the cell envelope, which is, of course, the reason for the therapeutic effectiveness of certain antibiotics. They act on the external structure which has no parallel in mammals. Metals, which may be carcinogenic for mammals, also act upon the cell surface of microorganisms, even though they may have an ability to react with a cellular DNA if the latter is in the test tube (Coward et al., 1973a,b; Coward and Rosenkranz, 1975; Rosenkranz and Carr, 1972; Rosenkranz and Rosenkranz, 1972). On the other hand, unlike cultured mammalian cells which lose their ability to metabolize xenobiotics upon continued passage, microbial cells, by their very nature, retain their synthetic capacities.

Because of the ease with which microorganisms can be manipulated genetically, the genes controlling the metabolic functions of microorganism can be altered such that a specific enzyme is deleted, altered, or even increased. Some of the enzymes susceptible to such alterations include the nitroreductases, the azoreductases, glutathione S-transferase and others. The enzymes listed are thought to be involved in the activation or detoxification of environmental agents; hence, having microorganisms that have lost this capacity could be very useful when the assays are coupled to microsomal enzymes. In addition, because some microorganisms possess enzymic machinery for oxidizing polycyclic aromatic hydrocarbons or degrading halogenated hydrocarbons, the possibility exists of introducing the genes controlling these pathways (using transposon technology) into the tester microorganisms and thereby being able to study a series of reactions of environmental impact in microorganisms which can be mutagenized in known manners.

Because of our interest in the nitroarenes, we have concentrated on the role and properties of nitroreductases at this time. Using techniques of "classical" microbial genetics, we have isolated nitroreductase-deficient Salmonella tester microorganisms and were able, thereby, to show the following:

1) the presence in microsomal preparations of enzymes capable of activating nitro-containing chemicals to mutagens (Rosenkranz and Poirier, 1975; Rosenkranz and Speck, 1975, 1976),

2) that nitroarenes are indeed reduced to hydroxylamino intermediates and that these arylhydroxylamines bypass the nitroreductase block (Table 6 and Mermelstein et al., in press), and

3) that microorganisms contain a family of nitroreductases which differ in substrate specificites (McCoy et al., 1981; Rosenkranz et al., 1982) (Table 7).

Table 6. Mutagenicity of Arylhydroxylamines for Nitroreductase-
Deficient Salmonella

Chemical	Revertants per nanomole			
	TA100	TA100NR	TA98	TA98NR
1-Nitronaphthalene	1.0	0	0.1	0
1-Hydroxylaminonaphthalene	10.6	11.8	1.9	2.9
2-Nitronaphthalene	1.3	0.1	0.01	0.02
2-Hydroxylaminonaphthalene	15.0	16.1		
2-Nitrofluorene			5.0	0.9
2-Hydroxylaminofluorene[a]			5.9	6.4
8-Nitroquinoline	0	0		
8-Hydroxylaminoquinoline[a]	2.7	0.4		

[a]Nitroarene + Zn + NH_4Cl.

Table 7. Mutagenicity of Nitro-Containing Chemicals in
Nitroreductase-Deficient Salmonella

Chemical	Revertants per nanomole		
	TA98	TA98NR	TA98/1,8-DNP$_6$
Niridazole	390	0.5	390
Nitrofurantoin	5.2	124	
4-Nitroquinoline-1-oxide	126		110
2-Nitronaphthalene	0.2	0.01	
2-Nitrofluorene	14	0.8	145
2,7-Dinitro-9-Fluorenone	1,459	184	137
1-Nitropyrene	453	35	199
1,3-Dinitropyrene	144,760	24,750	2,750
1,6-Dinitropyrene	183,570	190,900	45,890
1,8-Dinitropyrene	254,000	264,160	5,840
1,3,6-Trinitropyrene	40,700	36,630	25,640
1,3,6,8-Tetranitropyrene	15,600	10,600	14,000
Entozon	338	328	318

Thus, while the original nitroreductase-deficient strain TA98NR lacks the enzyme that is required to activate nitrofurans, nitroimidazoles, nitronaphthalenes, nitrofluorenes and 1-nitropyrene to mutagens, this enzyme (the "classical" nitroreductase) recognizes neither dinitropyrenes, nor 4-nitroquinoline-1-oxide nor the nitroacridine Entozon. Further studies led to the isolation of a bacterial strain lacking the enzyme which recognizes 1,8-dinitropyrene. This enzyme also recognizes 1,3-dinitropyrene but does not act upon other nitro-containing substrates (Table 7). 4-Nitroquinoline-1-oxide, Entozon, and trinitropyrene are not recognized by either of these enzymes, which suggests the existence of additional nitroreductases.

Results of the type described above have led to the realization that eukaryotes probably also possess families of nitroreductases. Because some mammalian cells are mutagenized by nitroarenes (Terada et al., 1981) while others are not, even while retaining their susceptibility to mutagenesis by nitrofurans and/or 4-nitroquinoline-1-oxide (Cole et al., 1982), suggests that all cells do not contain all of these nitroreductases. Possibly, some of these enzymes are lost during prolonged cultivation of the cells. Such a loss is known to occur with respect to some of the cytochrome P450-dependent microsomal enzymes.

Various nitroreductase-deficient microorganisms have been used for the presumptive identification of nitroarenes in diesel emissions, ambient atmosphere, and cigarette-smoke condensates (McCoy and Rosenkranz, 1982; Rosenkranz, 1982; Rosenkranz et al., 1981).

The recognition of the existence of a family of nitroreductases differing in substrate specificities led to studies which demonstrated that in bacteria, such enzymes may be inducible (unpublished results). This in turn resulted in two realizations, each of which might be of some significance:

1) the possibility of isolating constitutive microbial strains which produce elevated levels of specific nitroreductases and which, therefore, are hypersensitive to the mutagenic action of specific nitroarenes or groups thereof, and

2) the possibility that nitroreductases are also inducible in mammalian cells.

This provided the rationale for the prolonged incubation
of mammalian cells and the subsequent demonstration of the
mutagenicity of 1,8-dinitropyrene in mouse lymphoma cells,
when incubation was for 48 h (Cole et al., 1982). On the
other hand, only the standard incubation period (3 h) was
required for the demonstration of the mutagenicity of
2-nitrofluorene, 4-nitroquinoline-1-oxide, and nitrofurans
in these cells (Amacher et al., 1979; Cole et al., 1982);
thus implying that the nitroreductases recognizing
nitrofluorene, 4-nitroquinoline-1-oxide, and nitrofurans
were present constitutively and did not require induction.

Finally, these studies showed that although bacterial
nitroreductases have a broad substrate specificity, there are some
chemicals which are mutagenic in mammalian cells (Hsie, 1980) as
well as carcinogenic for rodents (8-nitroquinoline) and yet are
not mutagenic for Salmonella. It was established that this lack
of mutagenicity is due to an inability of the bacterial
nitroreductase to reduce 8-nitroquinoline (Karpinsky et al.,
1982). However, chemically reduced 8-nitroquinoline was mutagenic
for Salmonella (Table 6).

Demonstration of the New Type of Mutagenicity of Nitroarenes Not Dependent Upon Reduction of the Nitro Function

Studies of the properties of nitroarenes led to an
examination of the mutagenic properties of additional members of
the group. This examination revealed an unusual situation.
5-Nitroacenaphthene was a direct-acting mutagen whose mutagenicity
was reduced in nitroreductase-deficient strains (Table 8).
However, the mutagenicity of this chemical was greatly increased
in the presence of rodent S9. The most potent preparation was S9
derived from the livers of Aroclor-induced hamsters (unpublished
results). Interestingly, however, the S9-enhanced mutagenicity
was also evidenced in nitroreductase-deficient microorganisms.
This finding can be taken to indicate that the S9-induced
metabolite was not dependent upon the nitro function for its
mutagenicity. Indeed, studies (El Bayoumi and Hecht, 1981) of the
S9-induced bioconversion of 5-nitroacenaphthene have revealed the
presence of 1-hydroxy- and 2-hydroxy-5-nitroacenaphthene, both of
which are more mutagenic than the parent compound. Both of these
hydroxylated derivatives showed increased mutagenicity in the
presence of S9, suggesting that they are not the ultimate
mutagens.

Table 8. Effect of Metabolic Activation on Mutagenicity
of 5-Nitroacenaphthene

S9	Revertants per microgram				
	TA100	TA100NR	TA98	TA98NR	TA98/1,8-DNP$_6$
None	22.1	4.1	23.0	13.6	2.7
RI[a]	88.9	45.4	96.8	87.4	59.6

[a]S9 from livers of Aroclor-induced rats.

These findings suggest that maximum mutagenicity is seen when
the nitro function remains intact and the ring is hydroxylated.
Since unsubstituted acenaphthene is not mutagenic even in the
presence of S9 (unpublished results), the present results suggest
that the nitro function directs ring oxidation.

The same situation may also prevail with respect to
6-nitrobenzo(a)pyrene. This chemical is only mutagenic in the
presence of S9 (Pitts et al., 1982). Yet full S9-mediated
mutagenicity is expressed in the nitroreductase-deficient strain
TA98NR. An indirect metabolite, 3-hydroxy-6-nitrobenzo(a)pyrene
is detected during the S9-induced metabolism of
6-nitrobenzo(a)pyrene (Fu et al., 1981).

SPECIES SPECIFICITY OF METABOLIC ACTIVATION

One of the unresolved puzzles in oncology is the species
specificity of some of the carcinogens. An approach to the
elucidation of this problem involves the use of hepatic S9
preparations from induced and uninduced rats, mice, and hamsters.
The chemicals which were included in this study are identical to
those used for the National Cancer Institute Bioassay Program
(Dunkel and Simmon, 1980) and the S9 preparations were derived
from the same strains of animals. Although there were some
notable exceptions, by and large, most of the S9 preparations
exhibited a similar spectrum of activities. There was no
correlation between carcinogenicity in one species and
demonstrable S9-induced mutagenicity only in preparations derived
from the susceptible species.

Even though a chemical is rendered mutagenic by S9 preparations from different species, the extent of activation may be different. It is also conceivable that each of these preparations activates the chemical to a different intermediate or that each of the preparations converts a chemical to several chemical species but that the proportions of each vary with the individual S9 preparation. Finally, the possibility must be considered that detoxification or inactivation may be more active in one species than in the other and that it may proceed by species-specific pathways.

An approach to answering these questions would be to systematically study the metabolites produced by different S9 preparations as well as to isolate and identify the DNA adducts that are formed therefrom. Such an approach is, of course, feasible, but due to the effort and costs involved, it has been restricted in the past to the analysis of very few chemicals. An alternate approach would be an analysis of the quantitative aspects of the mutagenic response of a chemical using various S9 preparations. To determine whether such an analysis would be informative, 2-aminoanthracene was chosen. There are several reasons for such a choice:

1) The metabolites of 2-aminoanthracene are unknown but they could presumably involve several pathways, such as the oxidation of the 2-amino group to the corresponding 2-hydroxylamino moiety followed by esterification to a stronger electrophile, oxidation of the ring to yield anthraquinone and other hydroxylated products.

2) Following activation by S9, 2-aminoanthracene is mutagenic for all Salmonella tester strains.

3) Because of the multiplicity of possible metabolic intermediates and in view of the fact that Salmonella strains with different mutagenic specificities respond to 2-aminoanthracene, it is conceivable that each species could respond to the mutagenicity of a different intermediate and that different proportions of each of these could be generated by each of the S9 preparations.

Although quantitative data for all of the tester strains were obtained and S9 was derived from Aroclor-induced as well as uninduced animals, for illustrative purposes we shall be concerned only with strains TA1535, TA1537, and TA1538, and S9 preparations derived from uninduced animals. It is quite clear that the response elicited by the different S9 preparations (Table 9) varies but not in a uniform manner for all of the tester strains

Table 9. Metabolic Activation of 2-Aminoanthracene

S9	Revertants per nanomole		
	TA1538	TA1537	TA1535
RN[a]	125.3	27.8	20.7
MN[a]	43.6	5.8	15.6
HN[a]	170.8	5.0	23.2

[a]Liver S9 derived from uninduced rats, mice, and hamsters,
respectively.

(Table 10). These findings can be interpreted as indicating that
the tester strains appear to respond to different mutagenic
determinants and that the mutagenic species generated by each of
the S9 preparations are different both qualitatively and
quantitatively, as evidenced by the response of individual strains
to different S9 preparations.

It will be of interest to correlate the information obtained
from this type of analysis to animal carcinogenicity data,
especially with respect to species specificity and carcinogenic
potency. In addition, we contemplate generating such data for
procarcinogens which differ in a species-specific manner with
respect to their microsome-mediated biotransformation, especially
if the nature of the metabolites and the structure of the specific
DNA adducts are known. Conceivably, such analyses could
eventually be extended to an analysis of tissue-specific metabolic
activation.

In conclusion, we feel that the relatively simple
mutagenicity assay procedures are capable of generating
information related to the nature of the DNA adduct (mutagenic
specificities) and species specificity in metabolic activation
(relative mutagenic potencies). We intend to test such an
analysis by selecting some promutagens and analyzing both
quantitatively and qualitatively the nature of the metabolites
generated in relationship to their mutagenic specificities and
potencies.

Table 10.   Analysis of the S9-Induced Mutagenicity
            of 2-Aminoanthracene

S9	TA1538	TA1537	TA1535
Normalized to RN			
RN[a]	1.00	1.00	1.00
MN[a]	0.34	0.21	0.76
HN[a]	1.46	0.18	1.12
Normalized to TA1538			
RN	1.00	0.22	0.16
MN	1.00	0.09	0.36
HN	1.00	0.03	0.14
Normalized to TA1538 and to RN			
RN	1.00	1.00	1.00
MN	1.00	0.41	2.25
HN	1.00	0.14	0.88

[a]Liver S9 derived from uninduced rats, mice, and hamsters,
respectively.

PROSPECTS FOR THE FUTURE

     Those concerned with the potential carcinogenic hazard to man
are struck by the fact that there are literally tens of thousands
of chemicals in commerce and industry that are as yet unevaluated.
Perhaps 4,000 to 6,000 of these have been tested in one or more
short-term bioassays, and only approximately 500 of these have
been tested adequately in animal carcinogenicity studies. The
situation is even more chaotic if it is realized that there does
not appear to exist a rationale for systematically testing the
backlog of chemicals in short-term assays. Accordingly, there are
many data but a dearth of information relating to the reliability
of the various tests in predicting carcinogenicity of certain
groups of chemicals, on the relative performance of the various
assays, and finally, how they can be used to evaluate risk to man.
Clearly, a data file is needed so that some of these questions can
be answered. It is hoped that the assessment phase of the
U.S. Environmental Protection Agency-sponsored Gene-Tox Program
will address itself to this problem. Although this is a much
needed project, by itself, it will not solve the problem of the
enormous backlog of chemicals which need to be evaluated, nor will

it greatly improve our ability to make rational risk assessments
that could lead to intelligent risk versus benefit decisions.

At the Center for the Environmental Health Sciences of Case
Western Reserve University, we are approaching these problems in a
concerted manner. We are establishing our own computer-based
mutagenicity/carcinogenicity data file. The data to be entered
will include mutagenic and carcinogenic potencies and functional
groups. We are investigating approaches to correlating the
features responsible for mutagenic potencies of structurally
related classes of chemicals (e.g., nitrosamines; nitroarenes;
aminobiphenyls; polycyclic aromatic hydrocarbons). For this
endeavor we are fortunate in having the support and collaboration
of our colleagues in the Chemistry Department, who have the proven
expertise as well as computer-based resources to perform
Quantitative Structure Activity Relationships (QSAR) based upon
Linear Free Energy (LFE) features that in turn make use of
molecular orbital theory, molecular mechanics, and pattern
recognitions, as well as non-LFE (NLFE). The latter appear to be
more applicable to biological reactions involving multiple
activation and detoxification pathways (Andreozzi et al., 1980;
Petit et al., 1979; Kikuchi et al., 1979a,b).

The availability of reliable biological data for a series of
chemical analogs differing in isomerism and the extent of
substitution will permit quantitative correlations to be made
relating to the features of the molecule that are required for the
expression of biological activity.

One of the major advantages of such a collaborative effort
derives from the fact that since it combines expertise in
chemistry and biology, it would be a simple matter, if needed, to
synthesize, purify, identify, and determine the biological
properties of chemicals that will permit determination of the
relevance of alternate mechanisms.

Preliminary experience has indicated that meaningful results
can be generated using quantitative data for more than one
Salmonella tester strain. Thus, using TA1538 and its
plasmid-bearing strain TA98, it was found that data obtained with
the former permitted quantitative analyses of parameters related
to metabolic conversion to the ultimate mutagen and subsequent
formation of the DNA adduct, while data obtained with TA98,
because it involves a component of error-prone DNA repair, can be
used as a measure of the magnitude of the DNA distortion induced
by the adduct. Because it has been suggested that error-prone DNA
repair is involved in the initial carcinogenic event (Echols,
1981; Radman, 1977, 1980; Radman et al., 1977; Sargentini and

Smith, 1981), it is conceivable that a structure activity relationship based upon such data will provide a better predictive model.

In addition to using <u>Salmonella</u> mutagenicity data, we intend to perform the same set of analyses whenever quantitative data are available on structurally related chemicals for other biological systems as well (e.g., point mutations in mammalian cells, sister chromatid exchanges, carcinogenicity).

The ultimate aim of such a study is to identify potentially harmful chemicals on the basis of structural features and thence to establish priorities for action or for further testing.

Because it is not sufficient to correlate structural features of the causative agents with short-term bioassays or with animal carcinogenicity data, we are attempting, together with members of the Department of Biometry of Case Western Reserve University, to build models which will enable us to extrapolate to carcinogenic risks in man. Of course, such analyses are heavily dependent upon the availability of quantitative carcinogenicity data for different species of animals and relating the human cancer experience (both pathology and epidemiology) with animal carcinogenicity data.

Finally, the decisions relating to the risk versus benefit to society of the continued exposure to a chemical agent is, essentially, a political decision. We feel, however, that it is incumbent upon us to offer the best possible scientific advice. There are procedures based upon "trade-off" analyses (Pareto optimum concept) that are used in hierarchical analyses of complex systems, which appear to be well suited to the study of health-related effects of developing technologies. Together with our colleagues in the Center for Large Scale Systems and Policy Analysis who have international renown in performing such analyses (Haimes, 1977), we are attempting to use such methodologies for analyzing health-related effects.

ACKNOWLEDGMENTS

We are grateful to our colleagues at Case Western University School of Medicine who have made us feel welcome in our new environment and who made their expertise available to use. Special thanks are due to Dr. Edward A. Mortimer, Jr. for his warm and inspiring guidance.

REFERENCES

Amacher, D.E., S.C. Paillet, and G.N. Turner. 1979. Utility of
    the mouse lymphoma L51789/TK assay for the detection of
    chemical mutagens. In: Mammalian Cell Mutagenesis: The
    Maturation of Test Systems—Banbury Report No. 2. A.W. Hsie,
    J.P. O'Neill, and V.K. McElhony, eds. Cold Spring Harbor
    Laboratory: Cold Spring Harbor, NY. pp. 277-289.

Ames, B.N., J. McCann, and E. Yamasaki. 1975. Methods for
    detecting carcinogens and mutagens with the Salmonella/-
    mammalian-microsome mutagenicity test. Mutation Res.
    31:347-364.

Ames, B.N. 1979. Identifying environmental chemicals causing
    mutations and cancer. Science 204:587-593.

Andreozzi, P., G. Klopman, and A.J. Hopfinger. 1980. Theoretical
    study of N-nitrosamines and their presumed proximate
    carcinogens. Cancer Biochem. Biophys. 4:209.

Cairns, J. 1981. The origin of human cancers. Nature
    289:353-357.

Cheli, C., D. DeFrancesco, L.A. Petrullo, E.C. McCoy, and
    H.S. Rosenkranz. 1980. The Salmonella mutagenicity assay:
    reproducibility. Mutation Res. 74:145-150.

Chu, K.C., K.M. Patel, A.H. Lin, R.E. Tarone, M.S. Linhart, and
    V.C. Dunkel. 1981. Evaluating statistical analyses and
    reproducibility of microbial mutagenicity assays. Mutation
    Res. 85:119-132.

Cole, J., C.F. Arlett, J. Lowe, and B.A. Bridges. 1982. The
    mutagenic potency of 1,8-dinitropyrene in cultured mouse
    lymphoma cells. Mutation Res. 93:213-220.

Coward, J.E., H.S. Carr, and H.S. Rosenkranz. 1973a. Silver
    sulfadiazine: effect on the ultrastructure of Pseudomonas
    aeruginosa. Antimicrob. Ag. Chemother. 3:621-624.

Coward, J.E., H.S. Carr, and H.S. Rosenkranz. 1973b. Silver
    sulfadiazine: effect on the growth and ultrastructure of
    Staphylococci. Chemotherapy 19:348-353.

Coward, J.E., and Rosenkranz, H.S. 1975. Electron microscopic
    appearance of silver sulfadiazine-treated Enterobacter
    cloacae. Chemotherapy 21:231-235.

de Serres, F.J., and M.D. Shelby. 1979a. Recommendations on data
    production and analysis using the Salmonella/microsome
    mutagenicity assay. Mutation Res. 64:159-165.

de Serres, F.J., and M.D. Shelby. 1979b. The Salmonella
    mutagenicity assay: recommendations. Science 203:563-565.

de Serres, F.J., and J. Ashby. 1981. Evaluation of Short-Term
    Tests for Carcinogens. Elsevier/North Holland: Holland.

Dunkel, V.C. 1979. Collaborative studies on the Salmonella/-
    microsome mutagenicity assay. J. Assoc. Off. Anal. Chem.
    62:874-882.

Dunkel, V.C., and V.F. Simmon. 1980. Mutagenic activity of
    chemicals previously tested for carcinogenicity in the
    National Cancer Institute Bioassay Program. In: Molecular
    and Cellular Aspects of Carcinogenic Screening Tests.
    R. Montesano, H. Bartsch, and L. Tomatis, eds. IARC
    Scientific Publication No. 27, International Agency for
    Research on Cancer: Lyons. pp. 283-301.

Echols, H. 1981. SOS functions, cancer and inducible evolution.
    Cell 25:1-2.

El Bayoumy, K., and S.S. Hecht. 1981. Comparative metabolism of
    nitropolynuclear aromatic hydrocarbons. In: Sixth
    International Symposium Polynuclear Aromatic Hydrocarbons
    Abstracts. Battelle Laboratories: Columbus, Ohio.

Fu, P.P., M.W. Chou, L.E. Unruh, F.A. Beland, F.F. Kadlubar,
    D.A. Casciano, R.H. Heflich, and F.E. Evans. In vitro
    metabolism of 6-nitrobenzo(a)pyrene: identification and
    mutagenicity of the metabolites. In: Sixth International
    Symposium Polynuclear Aromatic Hydrocarbons Abstracts.
    Battelle Laboratories: Columbus, Ohio.

Goze, A., and R. Devoret. 1979. Repair promoted by plasmid
    pKM101 is different from SOS repair. Mutation Res.
    61:163-179.

Haimes, Y.Y. 1977. Hierarchical Analyses of Water Resources
    Systems: Modeling and Optimization of Large-Scale Systems.
    McGraw-Hill: New York.

Hsie, A.W. 1980. Quantitative mutagenesis and mutagen screening with Chinese hamster ovary cells. In: The Predictive Value of Short-Term Screening Tests in Carcinogenicity Evaluation. G.M. Williams, R. Kroes, H.W. Waaijers, and K.W. Van de Poll, eds. Elsevier/North Holland: Amsterdam. pp. 89-102.

Karpinsky, G.E., E.C. McCoy, H.S. Rosenkranz, and R. Mermelstein 1982. The chemical activation of non-mutagenic nitrated polycyclic aromatic hydrocarbons to mutagens. Mutation Res. 92:29-37.

Kikuchi, O., A.J. Hopfinger, and G. Klopman. 1979a. Electronic structure and reactivity of four stereoisomers of benzo(a)-pyrene-7,8-diol-9,10-epoxide. Cancer Biochem. Biophys. 4:1.

Kikuchi, O., A.J. Hopfinger, and G. Klopman. 1979b. A new type of semiempirical molecular orbital method for large molecules. J. Theor. Biol. 77:129.

Löfroth, G., E. Hefner, I. Alfheim, and M. Møller. 1980. Mutagenic activity in photocopies. Science 109:1037-1039.

Löfroth, G. 1981. Comparison of the mutagenic activity in carbon black particulate matter and in diesel and gasoline engine exhaust. In: Application of Short-Term Bioassays in the Analysis of Complex Environmental Mixtures II. M.D. Waters, S.S. Sandhu, J.L. Huisingh, L. Claxton, and S. Nesnow, eds. Environmental Science Research, Vol. 22. Plenum Press: New York. pp. 319-336.

Mattern, I.E., and H. Greim. 1978. Report of a workshop on bacterial in vitro mutagenicity test systems. Mutation Res. 53:369-378.

McCann, J., N.E. Spingarn, J. Kobori, and B.N. Ames. 1975. Detection of carcinogens as mutagens: bacterial tester strains with R factor plasmids. Proc. Natl. Acad. Sci. USA 72:979-983.

McCoy, E.C., H.S. Rosenkranz, and R. Mermelstein. 1981. Evidence for the existence of a family of bacterial nitroreductases capable of activating nitrated polycyclics to mutagens. Environ. Mutagen. 3:421-427.

McCoy, E.C., E.J. Rosenkranz, L.A. Petrullo, and H.S. Rosenkranz. 1981. Frameshift mutations: relative roles of simple intercalation and of adduct formation. Mutation Res. 90:21-30.

McCoy, E.C., E.J. Rosenkranz, L.A. Petrullo, H.S. Rosenkranz, and
    R. Mermelstein. 1981. Structural basis of the mutagenicity
    in bacteria of nitrated naphthalene and derivatives.
    Environ. Mutagen. 3:499-511.

McCoy, E.C., E.J. Rosenkranz, H.S. Rosenkranz, and R. Mermelstein.
    1981. Nitrated fluorene derivatives are potent frameshift
    mutagens. Mutation Res. 90:11-20.

McCoy, E.C., and H.S. Rosenkranz. 1982. Cigarette smoking may
    yield nitroarenes. Cancer Lett. 15:9-13.

Mermelstein, R., D.K. Kiriazides, M. Butler, E.C. McCoy, and
    H.S. Rosenkranz. 1981. The extraordinary mutagenicity of
    nitropyrenes in bacteria. Mutation Res. 89:187-196.

Mermelstein, R., H.S. Rosenkranz, and E.C. McCoy (in press). The
    microbial mutagenicity of nitroarenes. In: The Genotoxic
    Effects of Airborne Agents, Brookhaven National Laboratory
    Symposium. Plenum Press: New York.

Miller, J.A., and E.C. Miller. 1977. Ultimate chemical
    carcinogens as reactive mutagenic electrophiles. In:
    Origins of Human Cancer. H.H. Hiatt, J.D. Watson, and
    J.A. Winsten, eds. Cold Spring Harbor Laboratory: Cold
    Spring Harbor, NY. pp. 605-627.

Monti-Bragadin, C., S. Venturini, and P.A. Todd. 1977.
    Interaction between two error-prone DNA repair systems in
    Escherichia coli. FEMS Microbiol. Lett. 2:125-128.

Pall, M.L. 1981. Gene-amplification model of carcinogenesis.
    Proc. Natl. Acad. Sci. USA 78:2465-2468.

Petit, B., B. Potenzone, Jr., A.J. Hopfinger, G. Klopman, and
    M. Shapiro. 1979. A hierarchical QSAR molecular structure
    calculator applied to a carcinogenic nitrosamine data base.
    ACS Symposium Series No. 112 on Computer Assisted Drug
    Design. American Chemical Society: Washington, DC.

Pitts, J.N., Jr., D.M. Lokensgard, W. Harger, T.S. Fisher,
    V. Majia, J.J. Schuler, G.M. Scorziell, and Y.A. Katzenstein.
    1982. Mutagens in diesel exhaust particulate:
    identification and direct activities of 6-nitrobenzo(a)-
    pyrene, 9-nitroanthracene, 1-nitropyrene and 5H-phenanthro
    4,5-(bcd) pyran-5-one. Mutation Res. 103:241-249.

Radman, M. 1977. Inducible pathways in deoxyribonucleic acid
     repair, mutagenesis and carcinogenesis. Biochem. Soc. Trans.
     5:1194-1199.

Radman, M., G. Villani, S. Boiteux, M. Defais, P. Caillet-Fauquet,
     and S. Spadari. 1977. On the mechanism and genetic control
     of mutagenesis induced by carcinogenic mutagens. In:
     Origins of Human Cancer, Book B. H.H. Hiatt, J.D. Watson,
     and J.A. Winsten, eds. Cold Spring Harbor Laboratory: Cold
     Spring Harbor, NY. pp. 903-922.

Radman, M. 1980. Is there SOS induction in mammalian cells?
     Photochem. Photobiol. 32:823-830.

Rosenkranz, E.J., E.C. McCoy, R. Mermelstein, and H.S. Rosenkranz.
     1982. Evidence for the existence of distinct nitroreductases
     in Salmonella typhimurium: roles in mutagenesis.
     Carcinogenesis 3:121-123.

Rosenkranz, H.S., and H.S. Carr. 1972. Silver sulfadiazine:
     effect on the growth and metabolism of bacteria. Antimicrob.
     Ag. Chemother. 2:367-372.

Rosenkranz, H.S., and S. Rosenkranz. 1972. Silver sulfadiazine:
     interaction with isolated DNA. Antimicrob. Ag. Chemother.
     2:373-383.

Rosenkranz, H.S., and W.T. Speck. 1975. Mutagenicity of
     metronidazole: activation by mammalian liver microsomes.
     Biochem. Biophys. Res. Comm. 66:520-525.

Rosenkranz, H.S., B. Gutter, and W.T. Speck. 1976. Mutagenicity
     and DNA-modifying activity: a comparison of two microbial
     assays. Mutation Res. 41:61-70.

Rosenkranz, H.S., and W.T. Speck. 1976. Activation of
     nitrofurantoin to a mutagen by rat liver nitroreductase.
     Biochem. Pharmacol. 25:1555-1556.

Rosenkranz, H.S., W.T. Speck, and B. Gutter. 1976. Microbial
     assay procedures: experience with two systems. In: In
     Vitro Metabolic Activation in Mutagenesis Testing.
     F.J. de Serres, J.R. Fouts, J.R. Bend, and R.M. Phelpot, eds.
     North Holland Publishing Co.: Amsterdam. pp. 337-363.

Rosenkranz, H.S., E.C. McCoy, M. Anders, W.T. Speck, and
    D. Bickers. 1979. The use of microbial assay systems in the
    detection of environmental mutagens in complex mixtures. In:
    Application of Short-Term Bioassays in the Fractionation and
    Analysis of Complex Environmental Mixtures. M.D. Waters,
    S. Nesnow, J.L. Huisingh, S.S. Sandhu, and L. Claxton, eds.
    Plenum Press: New York. pp. 3-42.

Rosenkranz, H.S., E.C. McCoy, L. Biuso, and W.T. Speck. 1979.
    Short-term microbial assays in the assessment of carcinogenic
    risk. In: The Use of Alternatives in Drug Research.
    A.N. Rowan, and C.J. Stratman, eds. Macmillans Press:
    London.

Rosenkranz, H.S., and L.A. Poirier. 1979. An evaluation of the
    mutagenicity and DNA-modifying activity in microbial systems
    of carcinogens and non-carcinogens. J. Natl. Cancer Inst.
    62:873-892.

Rosenkranz, H.S., G. Karpinsky, and E.C. McCoy. 1980. Microbial
    assays: evaluation and application to the elucidation of the
    etiology of cancer. In: Short-Term Mutagenicity Test
    Systems for Detecting Carcinogens. K. Norpoth and
    R.C. Garner, eds. Springer-Verlag: Berlin. pp. 19-57.

Rosenkranz, H.S., and Z. Leifer. 1980. Detection of carcinogens
    and mutagens with DNA repair-deficient bacteria. In:
    Chemical Mutagens, Principles and Methods for their
    Detection. Vol. 6. F.J. de Serres, ed. Plenum Press: New
    York. pp. 109-147.

Rosenkranz, H.S., E.C. McCoy, D.R. Sanders, M. Butler,
    D.K. Kiriazides, and R. Mermelstein. 1980. Nitropyrenes:
    isolation, identification and reduction of mutagenic
    impurities in a carbon black and toners. Science
    209:1039-1043.

Rosenkranz, H.S., E.C. McCoy, R. Mermelstein, and W.T. Speck.
    1981. A cautionary note on the use of
    nitroreductase-deficient strains of Salmonella typhimurium
    for the detection of nitroarenes as mutagens in complex
    mixtures including diesel exhausts. Mutation Res.
    91:103-150.

Rosenkranz, H.S. 1982. Direct-acting mutagens in diesel
    exhausts: magnitude of the problem. Mutation Res. 101:1-10.

Rosenkranz, H.S., G.E. Karpinsky, M. Anders, E.J. Rosenkranz, L.A. Petrullo, E.C. McCoy and R. Mermelstein (in press). Adaptability of microbial mutagenicity assays to the study of problems of environmental concern. In: Induced Mutagenesis: Molecular Mechanisms and Their Implications for Environmental Protection. Plenum Press: New York.

Rosenkranz, H.S., and R. Mermelstein (in press). The nitroarenes: genotoxicity and mutagenicity. Mutation Res.

Sargentini, N.J., and K.C. Smith. 1981. Much of spontaneous mutagenesis in Escherichia coli is due to error-prone DNA repair: implications for spontaneous carcinogenesis. Carcinogenesis 2:863-872.

Seiler, J.P., I.E. Mattern, M.H.L. Green, and D. Anderson. 1980. Second European workshop on bacterial in vitro mutagenicity test systems (Ames Test Meeting 1979). Mutation Res. 74:71-75.

Speck, W.T., P.D. Ellner, and H.S. Rosenkranz. 1975. Mutagenicity testing with Salmonella typhimurium strains, I. Mutation Res. 28:27-30.

Straus, D.S. 1981. Somatic mutation, cellular differentiation, and cancer causation. J. Natl. Cancer Inst. 67:233-241.

Terada, M., M. Nakayasu, H. Sakamoto, K. Wakabayashi, M. Nagao, H.S. Rosenkranz, and T. Sugimura. 1981. Mutagenic activity of nitropyrenes and heterocyclic aromatic amines on Chinese hamster cells with diphtheria toxin as a marker. In: Third Int. Conf. Environ. Mutagen. Abstracts. p. 128.

Todd, P.A., C. Monti-Bragadin, and B.W. Glickman. 1979. MMS mutagenesis in strains of Escherichia coli carrying the R46 mutagenic enhancing plasmid: phenotypic analysis of Arg[+] revertants. Mutation Res. 62:227-237.

U.S. Congress. 1977. The Clean Air Act as amended August 1977. U.S. Government Printing Office, Serial No. 95-11, November.

U.S. Environmental Protection Agency. 1979a. Health assessment document for polycyclic organic matter. EPA-Z/102. Office of Research and Development: Washington, DC.

U.S. Environmental Protection Agency. 1979b. Preliminary
    assessment of the sources, control and population exposure to
    airborne polycyclic organic matter (POM) as indicated by
    benzo(a)pyrene (B[a]P). EPA-Z/100. Office of Air Quality
    Planning and Standards: Research Triangle Park, NC.

Walker, G.C. 1978. Isolation and characterization of mutants of
    the plasmid pKM101 deficient in their ability to enhance
    mutagenesis and repair. J. Bacteriol. 133:1203-1211.

FRACTIONATION OF AN OIL SHALE RETORT PROCESS WATER: ISOLATION OF PHOTOACTIVE GENOTOXIC COMPONENTS

Gary F. Strniste,[1] Judy M. Bingham,[1] W. Dale Spall,[2]
Joyce W. Nickols,[1] Richard T. Okinaka,[1] and David J.-C. Chen[1]

[1]Genetics and [2]Toxicology Groups, Life Sciences Division, Los
Alamos National Laboratory, Los Alamos, NM 87545

INTRODUCTION

We have previously shown that near ultraviolet light (NUV)
and natural sunlight activate a variety of complex mixtures
(products and waste streams from the retorting of oil shale),
inducing genotoxic responses in cultured mammalian cells (Strniste
and Brake, 1980; Strniste and Chen, 1981; Strniste et al., 1981;
Chen and Strniste, 1982; Strniste et al., 1982a,b). In this
report we present recent data concerning our attempts to
fractionate and to chemically identify photoactive, genotoxic
components in a particularly bioactive oil shale retort process
water. Liquid-liquid extraction methods and reverse-phase high
pressure liquid chromatography (HPLC) were employed in the
fractionation scheme, and individual samples were assessed for
potential photoinduced toxicity and mutagenicity in cultured
Chinese hamster ovary (CHO) cells. For comparison, these
fractionated water samples were also assessed for mutagenic
potential using the Salmonella histidine reversion assay with
microsomal activation (Ames et al., 1975). The photoactive
fractions of the process water have been analyzed by gas
chromatography/mass spectrometry (GC/MS) to define their chemical
composition. Several classes of chemical compounds, including
alkylated aromatic hydrocarbons, aromatic amines, aldehydes,
amides, and ketones are prevalent and most likely contribute to

the photoinduced genotoxic potential of this oil shale retort
process water.

MATERIALS AND METHODS

Primary Fractionation

The retort process water used in these experiments was
obtained from a holding tank used to separate crude shale oil from
water at the Paraho aboveground retort facility at Anvil Points,
Colorado. A chemical description of this particular water and its
genotoxic properties has been reported elsewhere (Holland and
Stafford, 1981). Initial fractionation of the water was
accomplished by the acid/base extraction scheme shown in Figure 1.
After removal of CH_2Cl_2, the residual solids from each fraction
were redissolved to the original volume of the starting material
(process water) in either dimethyl sulfoxide (DMSO) or water:DMSO
(1:1 v/v). Dilutions of these samples were used in subsequent
bioassays.

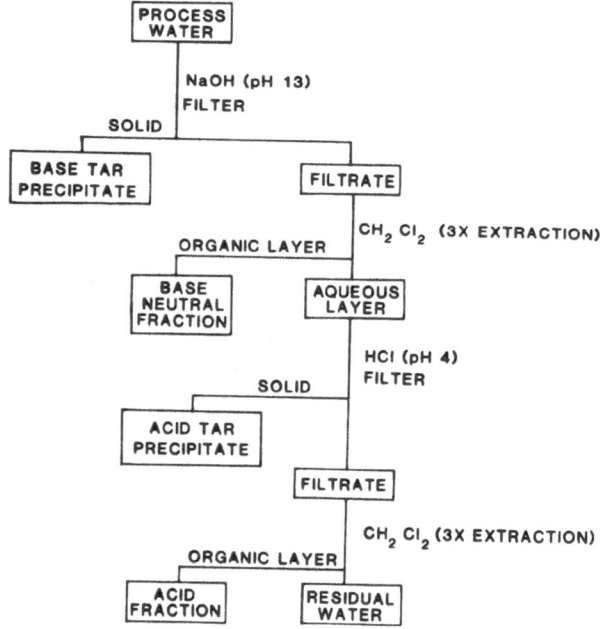

Figure 1. Acid/base extraction scheme.

Secondary Fractionation

A sample of the base/neutral fraction of the process water was dissolved in glass-distilled methanol at 20 mg/ml residual solids (RS), and 2-ml aliquots were applied to a semipreparative Bondapak C_{18}/Porasil B (Waters Associates, Inc; 37- to 75-μm particle size) column (2.3 mm [ID] x 2 m). Chromatography was achieved using a Beckman model 334 HPLC system equipped with a Hitachi variable-wavelength spectrophotometer. A gradient of 0 to 100% glass-distilled acetonitrile at a flow rate of 1 ml/min was programmed for the 150-min run. Two-ml fractions were collected and fractions from six consecutive runs were pooled. The volatiles were removed by heating at 55°C at 1 atm pressure for ~ 24 h. Pooled, fractionated samples were weighed and redissolved in DMSO or methanol.

Ames/Salmonella Bioassay

The standard plate assay as described by Ames et al. (1975) using **Salmonella** test strains TA98 and TA100 was performed with the addition of S9 fraction obtained from the livers of Aroclor 1254-induced male Sprague/Dawley rats (Okinaka et al., 1981). S9 fraction was added at 15 μl/plate (465 μg protein). Data were compiled from the results of at least two independent experiments in which the average number of his^+ revertants induced from three plates per dose for six different doses (0 to 33 μl) was determined. The slopes of the dose-response curves were determined by linear regression analysis. Experimental background mutation frequencies for TA98 and TA100 were 39 ± 11 and 184 = 13, respectively. Mutation frequencies (revertants/plate) for diagnostic mutagens for TA98 were 327 ± 10 and 395 ± 54 for 2-nitrofluorene (1.5 μg/plate) and benzo(a)pyrene (1.8 μg/plate), respectively; and for TA100 were 413 ± 40 and 348 ± 38 for methyl-N'-nitro-N-nitrosoguanidine (2.0 μg/plate) and benzo(a)pyrene (1.8 μg/plate), respectively.

CHO Bioassays

CHO cells, line AA8-4, were maintained in liquid culture as previously described (Strniste and Chen, 1981). The protocols used for measuring cytotoxicity (colony-forming assay) and mutagenicity (at the hypoxanthine-guanine phosphoribosyl transferase [HGPRT] locus) using this cell line have been reported elsewhere (Strniste and Chen, 1981; Strniste et al., 1982a,b). Plating efficiencies for untreated cells were regularly between 80 to 90%. The numbers of 6-thioguanine-resistant (6-TG^R) mutants

induced in CHO have been corrected for efficiency of plating
determined for each dose point at time of selection.

Photoactivation with NUV

In photoactivation experiments, plated CHO cells in
alpha-minimal essential medium (MEM; GIBCO) supplemented with 1%
fetal calf serum (FCS) were exposed to dilutions of various
fractionated samples for 1 h in the dark followed by exposure to
NUV (300 to 400 nm wavelength) after the removal of the treatment
medium. Irradiated cells were refed with medium plus 10% FCS, and
cytotoxic and mutagenic events were determined as previously
described (Strniste and Chen, 1981). The NUV source was two
parallel 15-W black lights (General Electric #F15T8 BLB); the
incident fluence was 7.5 $J/m^2/s$. Details of the irradiation
procedure have been described elsewhere (Strniste and Chen, 1981;
Strniste et al., 1982a).

GC/MS Methodologies

All data were collected using a Hewlett Packard model 5984A
GC/MS. A 60-m x 0.25-mm WCOT, fused silica capillary column
coated with SE-30 (J & W Scientific, Grade AA) was vented directly
into the ion source of the mass spectrometer. The mass
spectrometer was controlled to scan from 25 amu to 300 amu in
0.8 s. The gas chromatograph was temperature programmed from 30°C
to 270°C at 4°C/min. A linear flow velocity of 21.6 cm/s,
measured by the retention time of argon, was maintained during the
experiment. Spectra were collected and corrected for background
by computer subtraction of adjacent spectra. Initial compound
identifications were verified by either comparison to the EPA/NIH
or Aldermaston spectral libraries, or by comparison to authentic
compound spectra.

RESULTS AND DISCUSSION

Primary fractionation (acid/base extraction) of the retort
process water partitioned genotoxic components primarily into the
base/neutral (B/N) fraction. As shown in Table 1, > 80% of the
recovered mutagenic activity, when assessed in the Ames test with
either TA98 or TA100 test strains, resides in this B/N fraction.
Residual mutagenic activities exist in other fractions, primarily
in the base tar precipitate. Recovery of mutagenic activity in
the various fractions is essentially 100% of the activity found in

Table 1. Mutagenic Activity of Acid/Base Fractionated Oil Shale Retort Process Water in the Ames/Salmonella Test[a]

Sample	Total Dissolved Solids (mg/ml)[b]	A350 (units/ml)[c]	his+ Revertants Induced/μl	
			TA98 (+S9)	TA100 (+S9)
Process water (6 SEPT 78.005)[d]	205	72.0	103.0	11.0
Base tar precipitate	90.4	5.2	9.6	0.8
Base/neutral fraction	6.1	9.6	53.0	10.0
Acid tar precipitate	63.3	4.0	0.5	0.0
Acid fraction	5.2	3.2	1.0	0.0
Residual water	40.0	18.0	0.6	0.0

[a] Numbers of his+ revertants induced per plate determined for 6 sample doses (0 to 33 μl). Each fraction was adjusted to original volume of process water with 1:1 DMSO:H_2O.

[b] Dissolved solids obtained by heating fixed volumes of samples at 55°C until a constant weight was achieved.

[c] Absorptivity at 350-nm wavelength determined spectrophotometrically with dilutions made with glass-distilled methanol.

[d] Number used for particular sample at the Los Alamos Synfuel Repository.

the original process water when assessed with TA100; however, only
~ 60% of this activity is recovered when the fractions are
assessed with TA98.

In Figure 2 we show both photoinduced cytotoxicity and
mutagenicity in CHO cells pretreated with dilutions of the process
water or various fractions obtained by the scheme shown in
Figure 1. It is apparent that essentially all photoactive
components (toxic and mutagenic) partition into the B/N fraction.
The B/N sample was further fractionated using semipreparative
reverse-phase HPLC. A total of ~ 240 mg B/N fraction was
chromatographed in six consecutive runs. Figure 3 shows
absorption profiles (monitored at two wavelengths) and the
distribution of dissolved residual solids for the chromatographed
B/N fraction. A total of 15 samples were obtained from pooling
selected fractions. These samples were first assessed for
mutagenic activity using the standard Ames/<u>Salmonella</u> bioassay
with TA98 plus microsomal activation. The results are shown in
Figure 4 (open symbols). Two peaks of mutagenic activity are
present, accounting for about 40% of the activity present in the
B/N fraction. However, > 80% of this recovered activity resides
in Fractions 7 to 13 (HPLC retention times [t_r] of
56.1 to 132 min) which contains only 20% of the dissolved residual
solids. The peak of this activity (Fraction 9) was pooled from
fractions having t_r of 80.1 to 90 min (see Figure 3). A similar
profile for mutagenic potential was observed when each fraction
was assessed with TA100 plus S9 (data not shown).

Each pooled fraction was then assessed for photoinduced
cytotoxic and mutagenic activity in cultured CHO cells.
Appropriate dilutions were made to established minimum tolerated
doses (nontoxic when exposed in the dark). Induced 6-TG^R mutants
(HGPRT$^-$)/min NUV exposure were determined from the linear portions
of the dose-response curves. The photoinduced mutagenic response
in CHO cells for all fractions tested is also shown in Figure 4
(closed symbols). Recovery of photoinduced mutagenic activity in
all fractions was calculated to be ~ 35% of that observed for the
B/N fraction itself. The distribution of the photoactive
fractions of the chromatographed B/N sample is qualitatively
similar to the distribution seen for mutagenic activity when
assessed by the Ames test with microsomal activation. However,
about 50% of the photoinduced activity resides in Fraction 3
(HPLC t_r of 20.1 to 30 min) with the remaining photoactivity
spread among several pooled fractions (Fractions 7 to 13) of t_r of
56.1 to 132 min. Fraction 3 is also greatest in photoinduced
toxicity in CHO cells of the 15 pooled HPLC fractions tested (data
not shown).

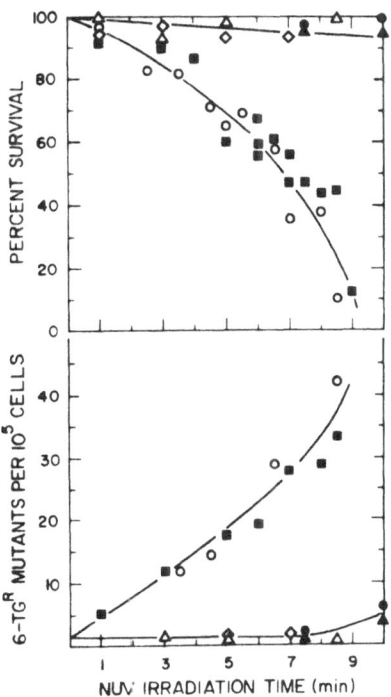

Figure 2.   Photoinduced cytotoxicity and mutagenicity in CHO cells
            pretreated with 1:300 dilutions (v/v) of oil shale
            retort process water (■) or samples fractionated from
            it: (O), base/neutral; (●), base precipitate;
            (△), acid; (▲), acid precipitate; and (◇),
            residual water.

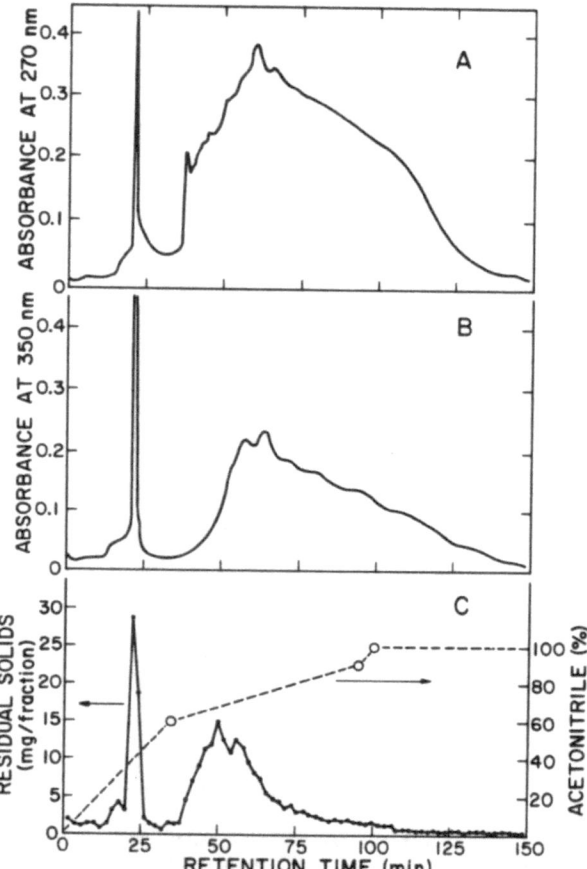

Figure 3.   HPLC of a base/neutral fraction of an oil shale retort
             process water.   (a) Absorbance (units/ml) at 270 nm for
             1:100 dilution of each fraction; (b) Absorbance
             (units/ml) at 350 nm for 1:10 dilution of each
             fraction; (c) Dissolved residual solids in each
             fraction (pooled from six consecutive runs); total
             240 mg applied, total 237 mg recovered.   Broken line
             represents the programmed acetonitrile gradient.

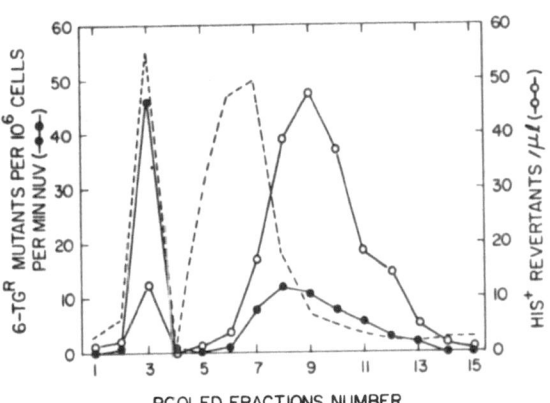

Figure 4.   Distribution of mutagenic activity in <u>Salmonella</u> (Ames
            test) and in CHO cells (NUV-induced) for pooled
            fractions from HPLC eluent of a base/neutral fraction
            of an oil shale retort process water.   In the Ames
            test, <u>S. typhimurium</u> strain TA98 with microsomal
            activation was used.   In the CHO/HGPRT assay,
            activation was achieved with 300 to 400 nm light.   The
            broken line represents the relative mass distribution
            of the dissolved residual solids recovered in each
            pooled sample.   Open symbols (o) represent Ames
            bacterial mutagenicity results and closed symbols (●)
            represent CHO mutagenicity results.

       Several pooled HPLC fractions showing photoinduced activity
in CHO, including Fractions 3, 7, and 9, were analyzed by GC/MS
methods to identify the chemical species present.   Resulting gas
chromatographs are shown in Figure 5.   Due to the chemical
complexity of these fractions, only predominant chemical classes
will be discussed.   A more detailed analysis of the chemical
composition of these fractions will be published elsewhere.
Compound classes present in the B/N fraction include polycyclic

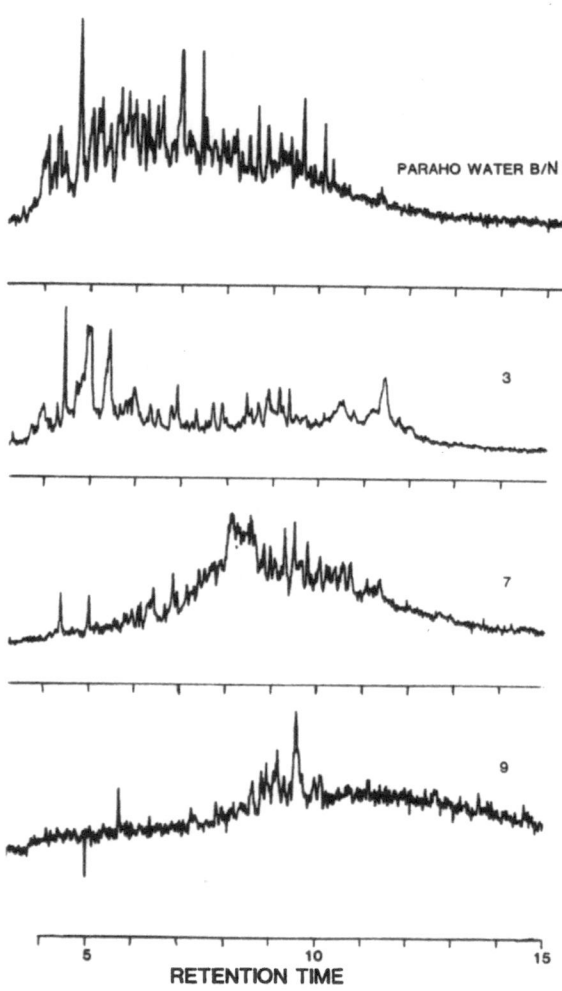

Figure 5.   From top to bottom.   Gas chromatographs of the original
            B/N sample and of HPLC-pooled Fractions 3, 7, and 9.

aromatic hydrocarbons (PAHs) and alkylated PAHs up to four rings, aromatic and alkyl aromatic amines up to three rings, pyrazine and alkylated pyrazines, pyridine and alkylated pyridines, alkanes up to fourteen carbon atoms, aliphatic nitriles, alkylated phenols, a variety of furan-based compounds, methyl ketones, and several cyclic ketones.

As one progresses from Fractions 1 to 15, the polarity of the compounds present decreases. Fraction 3 contains a wide variety of compounds, principally pyrazines, pyridines, furans, furfurals, and ketones. The first four groups all have the parent compound and a variety of alkylated members of each class. Methyl, ethyl, propyl, and isopropyl alkyl substituents are present with the concentrations of the alkylated compounds decreasing as the chain length increases. Up to seven carbons may be substituted onto the rings. The predominant ketones present are methyl ethyl ketone, methyl vinyl ketone, and cyclopentanone.

Fraction 7, which corresponds to the maximum of the residual solids mass peak, is extremely complex. The majority of the species present are alkyl aromatic hydrocarbons containing two and three fused rings. The positional isomers from multiple alkylated ring systems cannot be easily identified by GC/MS. In general, alkylated naphthalene, acenaphthalene, fluorene, anthracene and phenanthrene with alkylation of up to six carbon atoms are the predominant species. Methyl-alkyl ketones containing seven to ten carbon atoms are also present. Several highly alkylated phenolic compounds as well as phenols containing two and three rings are also found in this fraction.

Fraction 9, which corresponds to the major peak of Ames/Salmonella biological activity, is also quite complex. Major species are alkylated three- and four-fused-ring aromatic compounds, and several quinolines. Some aromatic amines, aldehydes, and amides are also present. The quinolines include quinoline, isoquinoline, and two benzoquinolines. Methylated quinolines are present at ~ 1/50th the concentrations of the parent compounds. The major aromatic amines are 1- and 2-naphthylamine, and 1- and 2-anthramine and phenanthramine. Concentrations of these compounds are less than 10 ppm (of the original water) with the three-ring compounds appearing at less than 5 ppm concentrations. Several aldehydes are present, including n-butyraldehyde, benzaldehyde, and several alkylated benzaldehydes. Amides present include propioamide and butyramide. Other amides may be present, but volatility limitations would prevent their detection.

CONCLUSIONS

Acid/base extraction of an oil shale retort process water
resulted in the partitioning of > 90% of the photoactive
components into the base/neutral fraction (when assessed in the
CHO/HGPRT/NUV-activation assay). Subsequent reverse-phase HPLC
fractionation of the B/N fraction resulted in a bimodal
distribution of photoinduced mutagenic activity. A similar
qualitative distribution of mutagenic activity was observed when
the HPLC eluent was assessed in the Ames/Salmonella test with
metabolic activation. Two photoactive fractions (nos. 3 and 9)
contain a variety of chemical species according to GC/MS analysis.
Classes of chemical compounds including the more hydrophilic
furans, furfurals, pyrazines, pyridines, and ketones, and the more
hydrophobic alkylated three- and four-fused-ring aromatics
probably contribute to the photoinduced genotoxicity of this
process water. Additional fractionation schemes will be necessary
to better chemically resolve these photoactive, genotoxic
components.

In addition to providing data for the assessment of potential
health and environmental risks resulting from phototransformation
of industrial effluents, the process of photoactivation should be
useful in assessing genotoxic potential of complex mixtures in in
vitro cell assays where metabolic activation processes are lacking
or are affected by the complex mixtures themselves.

ACKNOWLEDGMENTS

This work was funded by the U.S. Department of Energy and the
U.S. Environmental Protection Agency.

REFERENCES

Ames, B.N., J. McCann, and E. Yamasaki. 1975. Method for
    detecting mutagens with the Salmonella/mammalian-microsome
    mutagenicity test. Mutation Res. 31:347-364.

Chen, D.J., and G.F. Strniste. 1982. Cytotoxic and mutagenic
    properties of shale oil byproducts, II: comparison of
    mutagenic effects at five genetic markers induced by retort
    process water plus near ultraviolet light in CHO cells.
    Environ. Mutagen.

Holland, L.M., and C.G. Stafford, eds. 1981. The Los Alamos
    Integrated Oil Shale Health and Environmental Program: A
    Status Report. LA-8665-SR. Los Alamos National Laboratory:
    Los Alamos. 102 pp.

Okinaka, R.T., B.J. Barnhart, and D.J. Chen. 1981. Comparison
    between sister chromatid exchange and mutagenicity following
    exogenous metabolic activation of promutagens. Mutation Res.
    91:57-61.

Strniste, G.F., and R.J. Brake. 1980. Toxicity and mutagenicity
    of shale oil retort product waters photoactivated by near
    ultraviolet light. Environ. Mutagen. 2:268.

Strniste, G.F., and D.J. Chen. 1981. Cytotoxic and mutagenic
    properties of shale oil byproducts, I: activation of retort
    process waters with near ultraviolet light. Environ.
    Mutagen. 3:221-231.

Strniste, G.F., D.J. Chen, and R.T. Okinaka. 1982a. Genotoxic
    effects of sunlight-activated waste waters in cultured
    mammalian cells. J. Natl. Cancer Inst.

Strniste, G.F., D.J. Chen, and R.T. Okinaka. 1982b. Sunlight
    activation of shale oil byproducts as measured by genotoxic
    effects in cultured Chinese hamster cells. In: Polycyclic
    Aromatic Hydrocarbons, Volume 6. M. Cooke and A.J. Dennis,
    eds. Battelle Press: Columbus, OH.

Strniste, G.F., R.T. Okinaka, and D.J. Chen. 1981.
    Photoactivation of shale oil byproducts. Environ. Mutagen.
    3:308.

THE INTEGRATION OF BIOASSAY AND PHYSIOCHEMICAL INFORMATION FOR

COMPLEX MIXTURES

Larry D. Claxton

Genetic Toxicology Division, Health Effects Research
Laboratory, U.S. Environmental Protection Agency, Research
Triangle Park, North Carolina 27711

INTRODUCTION

To understand and estimate human environmental risk,
investigators need to identify, quantitate, and evaluate the
natural and anthropogenic pollutants in our environment.
According to Eugene Sawicki, "The main barriers to estimating
human environmental risks stem from a lack of knowledge of the
chemical composition of our environment, the failure to use the
information we have, and our indecision as to what to measure"
(Sawicki, 1978). This is especially evident in the areas of
genetic toxicology, analytical chemistry, and environmental
engineering, where ongoing research has been somewhat isolated.
Only recently, with the development of new analytical tools and
sensitive short-term tests, has interdisciplinary research into
environmental pollutants by engineers, chemists, and genetic
toxicologists expanded. This interdisciplinary research involving
complex environmental mixtures has created problems not only in
the research methods used, but also in the integrating and
summarizing of information that can be disseminated and
intelligently evaluated. The purpose of this paper, therefore, is
to illustrate how physiochemical and biological information can be
integrated to provide a more complete evaluation of a complex
mixture. Although any or all complex mixtures could be used as
illustrations, this paper will concentrate on air pollution data.

Nitrogen (N_2), oxygen (O_2), and the rare gases comprise more than 99.9% of the molecules within the earth's atmosphere, yet these substances are of secondary importance to researchers. Indeed, researchers are most interested in reactive organic gases and respirable particles. The combined concentration of all organic reactive gases seldom totals more than 10 ppm (Graedel, 1978), while the Total Suspended Particulate concentration (TSP) averages approximately 65 $\mu g/m^3$ in the United States (Council on Environmental Quality, 1980). Ambient air and emissions from ambient-air sources, therefore, provide excellent examples of complex environmental mixtures that are difficult to collect, analyze, and evaluate. This paper will focus on four areas of concern: creating proper research goals, approaches to research planning, evaluating different research strategies, and data analysis methods.

CREATING PROPER RESEARCH GOALS

In my opinion, the most common reason for difficulty in data integration and evaluation is the lack of a clear and definable goal for a project and for intermediate steps within a project. Although various physiochemical and biological procedures and protocols may have only minor modifications, the selection of procedures and the integration of information varies greatly according to the ultimate goal of the project. Some examples follow.

Estimating Health Risks

When estimating health risks is the ultimate goal, one is more thorough and cautious in selecting and performing bioassays. For example, one would not use only a bacterial assay, but instead would choose a series of tests whose end points indicate different types of toxic and genotoxic effects. In vivo mammalian tests are used for "more realistic" exposure and toxic-effect evaluation; studies estimating exposure to some compounds, group of compounds or particles, or other relevant material must also be completed. When the data is examined, one should find good correlations between the various bioassays and physiochemical results.

Improving Industrial Technology

Integrating physiochemical and biological information can help improve industrial technology in areas other than health effects. For example, if one wanted to produce an industrial product without nitroarenes, one might include microbial mutagenicity testing with specialized strains of bacteria in the quality control procedures. This use of a single bioassay would provide a rapid and inexpensive method for detecting nitroarene contaminants.

Evaluating, Characterizing, and Improving Bioassays

Since complex mixtures, by definition, usually include hundreds of chemicals, investigators have begun to explore in a systematic fashion the response of the various bioassays to different chemical classes, the potential of synergism and similar effects, and the reliability and cross-correlation of different bioassays.

Supplying Basic Research Knowledge

Complex-mixture work has aided the development of basic research techniques and knowledge. For example, identifying nitroaromatic mutagens in various complex mixtures with microbial assays has helped in the development of better physiochemical methods for identifying and quantifying nitroarenes. Developing bioassay techniques has helped in the characterization of physiological extraction of organic material from various types of airborne particles. With this type of basic research, only a few techniques may be employed; however, these techniques may have to be merged together in a unique fashion.

APPROACHES TO RESEARCH PLANNING

After developing proper goals, one must ask basically two questions. First, how much information is needed to satisfy the goal? Second, what is the most efficient manner by which my co-workers and I can attain that information in the allotted time and with limited resources? Six possible approaches have been reviewed in a prior publication (Claxton, in press): 1) chemical/ biological predictions, 2) pairing method, 3) chemical search method, 4) mutagen search method, 5) bioassay identification method, and 6) multi-integration.

Briefly, the generalized chemical/biological comparison approach is a crude extrapolation of thought about one complex mixture based upon what is known about other complex mixtures. The pairing method relies upon already available knowledge. In this method, literature searches done to identify compounds

associated with a complex mixture are followed by literature
searches to identify the biological activity of these compounds.
In the chemical search method, a list of known toxicants is
assembled and the complex mixture is assayed (or chemically
analyzed) for these chemicals. With the mutagen search method,
one bioassays those chemicals identified with the complex mixture.
The decision to use either the chemical search or mutagen search
process may depend on whether one is working in a primarily
chemical or bioassay laboratory. The bioassay identification
method is generally reserved for special situations in which
specialized strains of an organism give clues as to the specific
genotoxicants present. The last approach listed,
multi-integration, is obviously the integration of all previous
approaches. This method also includes bioassay-directed
fractionation.

EVALUATION OF DIFFERENT RESEARCH STRATEGIES

Each approach presented has certain advantages and
disadvantages; each approach can be performed to a different
extent. This section demonstrates the type of information that
can be gathered with each approach and the usefulness of that
information.

Chemical/Biological Predictions

This is essentially the educated-guess approach for which
there can be different levels of knowledge and guessing involved.
For example, if one had thoroughly reviewed and understood the
chemistry and toxicology of cigarette smoke and roofing tar
emissions, one might be willing to speculate about the toxicology
of diesel emissions for an academic discussion. But one would be
less likely to make managerial or political decisions based on
this knowledge. If other information was included, however, one
might be willing to make more critical decisions. For example, if
a scientific manager knew the results of a physiochemical screen
of both residential home heaters and wood combustors along with
the toxicology and chemistry of cigarette smoke, roofing tar
emissions, exhausts of gasoline and diesel vehicles, and coke-oven
emissions, he or she could decide whether to concentrate research
efforts on oil heaters or on wood stoves. This approach is rapid
and inexpensive, since it relies upon a generalized knowledge that
can be derived from review articles.

Pairing Method

The pairing method helps to remove much of the guesswork from the decision process and concentrate upon knowledge of the mixture itself. For example, a search of the open literature prior to 1979 identified 184 compounds associated with diesel exhaust. Forty-four of these compounds were listed in published mutagenicity reports; however, only 39 of these compounds had published data that could be readily evaluated. Twenty-one were positive in one or more mutagenicity assays. Of these 184 compounds, I could find 7 reported as carcinogens. In addition to that information, other toxicity data are available for 84 of the compounds. Altogether, there is toxicity data published on 108 of these compounds. This type of analysis sounds definitive until one realizes that these compounds contribute less than 5% of the total organic mass that is exhausted from diesel vehicles, and many of these organic exhaust compounds condense onto carbonaceous soot particles. This type of analysis gives one a clearer indication of what toxic activity may be associated with the complex mixture; however, it still leaves many unsolved problems. For example, what is the actual human exposure? Is there atmospheric transformation of these compounds? Will binding to soot particles decrease or increase target-site exposure? The questions one might ask are endless. The process, however, is moderately rapid and inexpensive.

Chemical Search and Mutagen Search Methods

The pairing method could be made simpler if one had to examine for only a few specific chemicals. This is the advantge of the chemical search and mutagen search methods. If one examines EPA's List of Priority Pollutants, one finds that 13 of 129 priority pollutants are known to be emitted by diesel engines and 14 of 129 are known to be emitted by gasoline engines. In many cases, however, additional qualitative and quantitative chemistries will have to be done to know which pollutants of interest are associated with the complex mixture. The reverse type of process is the mutagen search method. The major drawback to this process is one may overlook the many genotoxicants not represented or listed.

Bioassay Identification Process

By using a variety of indicator strains and assay conditions, bioassays can give clues as to the type of genotoxicants present in a bioassay. For example, testing gasoline- and diesel-exhaust

extracts with the normal indicator strains of <u>Salmonella</u>
<u>typhimurium</u> demonstrates that the genotoxicants are mainly
frameshift mutagens. When results with and without exogenous
activation are compared, diesel organics appear to have mostly
direct-acting activity in contrast to gasoline-vehicle exhaust
organics that show both direct and indirect activity. Both of
these complex mixtures are relatively nontoxic to the indicator
strains. Also, specialized tester strains indicate that
nitroarenes play a larger role in diesel-exhaust mutagenicity than
in gasoline-exhaust mutagenicity. One learns from this comparison
that genotoxicant identification for gasoline-exhaust organics
should concentrate initially on polynuclear aromatic hydrocarbons
(PNAs), while efforts with diesel-exhaust organics should
emphasize nitroaromatics and other non-PNA compounds.

Multi-Integration Processes

There are almost innumerable ways in which to integrate each
of these processes. One manner of direct integration is
bioassay-directed fractionation. With this procedure,
physiochemical fractions of a complex mixture are bioassayed and
those fractions contributing the largest proportion of toxic
activity are subfractionated and bioassayed. This procedure
continues until the individual genotoxicants are identifed and
quantitated. This method is the most efficient and most reliable
means available for identifying genotoxicants associated with a
complex mixture; however, there are still major drawbacks. First,
this process is slow, costly, and requires an extensive
coordinating process. Second, large representative quantities of
the complex mixture are usually needed. This is especially
difficult with ambient-air samples, for example. Next, the
bioassay problems or difficulties change with fractionation. For
example, a neat sample may be very toxic to the indicator
organism, masking genotoxic activity, but present no other testing
problems (e.g., solubility). Upon fractionation, some fractions
may no longer exhibit this masking effect, allowing the detection
of genotoxic effects. Because of the concentrating effect of
fractionation, however, the different fractions now will exhibit
different solubility, uptake, and toxicity problems; therefore,
each fraction is likely to be handled in a slightly different
manner.

Even with these difficulties, most fractions can be tested;
diesel exhausts provide a clear example. With a typical
light-duty diesel sample, the acid, insoluble-tar, polar-neutral,
and PNA fractions are generally positive. If one combines tester
strain information with this initial information (see Table 1),

Table 1.   Information for Multi-Integration of S. typhimurium Data
for Diesel Exhaust Work

Category	Response of Diesel Exhaust Samples
Bioassay of chemical fractions	
Acid	Positive
Base	Negative or questionable
Insoluble Tar	Positive
Polar Neutral	Positive
PNA	Positive
Nonpolar neutral	Negative
Response by strain	
TA98, TA100, TA1538, TA1537	Positive
TA1535	Negative
Effect of activation	Same or similar to nonactivation (direct-acting)
Response with nitroreductase-deficient strains	Decreased from parental strain
Identified compounds	B(a)P, pyrene, fluorenone, etc.

then one learns the following: active fractions contain primarily
frameshift mutagens (comparing TA98 and TA1535); the active
compounds are primarily direct acting; there is little need for
continued interest in the genotoxicity of the nonpolar-neutral
fraction, since it also is not very toxic to the indicator strain;
and, there may be spillover of non-PNA compounds with this
fractionation scheme (as seen by activity in TA98 without S9), or
some atypical PNA compounds may be present. Coupling this
information with information from the use of
nitroreductase-deficient strains, the percent of the total
mutagenicity each fraction contributes, a list of previously
identified compounds associated with diesel exhaust, and a little
educated guessing, a researcher could easily conclude that more
research needs to be applied toward the genotoxicity,
identification, and environmental control of 3-, 4-, and 5-ring
nitroarenes.

DATA SUMMARY METHODS

In our discussion of these approaches, we have moved from
qualitative information to quantitative information. Again, each
has its value; however, each has some major drawbacks as well. As
we move toward more quantitative information, it is important to
understand where the field of genetic toxicology stands in
relation to data management and statistical analysis. Again, my
comments are limited primarily to the Ames bacterial assay.

When the Ames test first became established as a bioassay,
researchers disagreed about which guidelines should be followed to
judge a compound as either negative or positive in the assay. For
a number of years, there was a general consensus that if a
substance produced a regularly increasing dose-response curve and
at one or more doses the revertants per plate were double the
spontaneous count, the substance was positive in the bioassay.
For most substances, that rule was quite adequate; however, for
some substances that also exhibited considerable toxicity, this
guideline provided definite problems. This guideline also
promoted reporting bacterial mutagenicity data as either positive
or negative without supporting data or an indication of the level
of genotoxicity within that bioassay.

In the last five years, various methods for analyzing
S. typhimurium plate incorporation test data have been published.
In the last year, three modeling (or semi-modeling) methods for
the Ames plate incorporation tests have appeared in peer-reviewed
journals. These three statistical methods (Stead et al., 1981;
Myers et al., 1981; Margolin et al., 1981) provide three different
approaches. The method of Myers et al. provides for the easiest
mathematical computations but may not model the data as well;
curves generated by the methods of Stead et al. appear to fit most
data extremely well but are based upon a Poisson distribution.
The work of Margolin et al. provides for extensive analysis of
data based on negative binomial models but is computationally very
complex, even with extensive computerization. Recent work of
Hughes et al. (personal communication, 1982, Research Triangle
Institute, Research Triangle Park, NC) compared two of these
models for six vapor-phase compounds (see Table 2). These two
methods gave very similar results for these data. This may be a
trend that will be seen with all three models or may just be
fortuitous. The point is that we only now have the tools to do
these comparisons. It would be premature to eliminate any one
method or to accept only one method.

Table 2. Initial Mutagenicity Slopes for Six Vapor-Phase Compounds by Two Statistical Methods

Compound	Initial Mutagenicity Slope (S) (rev/μg)		Mutagenic Response from Model	
	Myers et al. (1981)	Stead et al. (1981)	Myers et al. (1981)	Stead et al. (1981)
Ethylene dichloride	0.0000	0.0000	−	−
Ethylene oxide	0.1453	0.1464	+	+
Styrene oxide	1.5374	1.5261	+	+
Ethylene dibromide	1.0103	1.0498	+	+
Propylene oxide	0.6502	0.6366	+	+
Butylene oxide	0.4846	0.4768	+	+

CONCLUSIONS

Integrating bioassay and physiochemical information depends on the question to be answered and the degree of confidence wanted in the answer. As one moves from academic exploration toward the modeling of human risks, the quantitative evaluation of complex mixtures becomes more extensive, providing a maze of data to integrate. The analysis of individual chemicals within a complex mixture must be consistent with and supportive of the bioassay results of the complex mixture and components of the mixture. The method of bioassay-directed fractionation provides for this final type of integrated analysis of a complex mixture. Quantitative integration of physiochemical and biological results also depends on the the level of confidence and variability of each assay. Systematic comparison of statistical methods for bioassay evaluation has only recently begun; however, these recent models show promise.

REFERENCES

Claxton, L.D. (in press). Review of fractionation and bioassay characterization techniques for the evaluation of organics associated with ambient air particles. In: Genotoxic Effects of Airborne Agents. R. Tice, ed. Plenum: New York. pp. 19-34.

Council on Environmental Quality. 1980. Environmental Quality-1980: The Eleventh Annual Report of the Council on Environmental Quality. p. 168.

Graedel, T.E. 1978. Chemical Compounds in the Atmosphere. Academic Press: New York. p. 1.

Margolin, B., N. Kaplan, and E. Zeiger. 1981. Statistical analysis of the Ames Salmonella/microsome test. Proc. Natl. Acad. Sci. USA 78:3779-3783.

Myers, L.E., N.H. Sexton, L.I. Southerland, and T.J. Wolff. 1981. Regression analysis of Ames test data. Environ. Mutagen. 3(5):575-586.

Sawicki, E. 1978. Atmospheric genotoxicants--what numbers do we collect? In: Application of Short-Term Bioassays in the Fractionation and Analysis of Complex Environmental Mixtures. M.D. Waters, S. Nesnow, J.L. Huisingh, S.S. Sandhu, and L. Claxton, eds. Plenum Press: New York. pp. 173-194.

Stead, A.G., V. Hasselblad, J.P. Creason, and L. Claxton. 1981. Modeling the Ames test. Mutation Res. 85:13-27.

SESSION 3

DEVELOPMENT OF SHORT-TERM BIOASSAYS: MUTAGENICITY AND CYTOTOXICITY

ISOLATION OF MUTAGENS FROM DRINKING WATER: SOMETHING OLD,

SOMETHING NEW

John C. Loper[1,2] and M. Wilson Tabor[2]

[1]Department of Microbiology and Molecular Genetics, and
[2]Department of Environmental Health, University of Cincinnati
College of Medicine, Cincinnati, OH 45267

INTRODUCTION

In our previous presentation in this Symposium series, we
described the development of a coupled <u>Salmonella</u> mutagenesis/
reverse-phase high performance liquid chromatography (HPLC)
procedure for the subfractionation of mutagens from complex
mixtures of drinking water organics. Using the 20-year-old
drinking water residue mixture available to us at that time, we
demonstrated the reproducibility of this system. A group of
potent, microsomal-activation-dependent, strain TA100 mutagens was
detected and partially purified. For one of the repeat HPLC
subfractions, further purification by gas chromatography (GC) of
multiple aliquots on an SE 30 column yielded 10-μg amounts of
purified compound (Tabor and Loper, 1980; Loper and Tabor, 1981).
This substance is the "something old" alluded to in our title.
Part of the "something new" is that we have determined this
substance to be a previously unrecognized mutagen; we will have
more to say about it later (see Discussion).

Also "new" are some results of our tests of this <u>Salmonella</u>/
HPLC subfraction procedure using residues of recent drinking water
samples. For this work we instituted an XAD resin extraction
procedure suitable for processing 50- to 1500-l drinking water
samples. The extraction procedure has been applied to repeat
samples from two cities that draw upon different types of surface

water sources. A report of this work and our conclusions from it
form the major part of this presentation.

MATERIALS AND METHODS

Bacterial Strains and Bioassays

Salmonella typhimurium strains TA98 and TA100 were obtained
from B.N. Ames. Verification of their properties and their use
with standard positive and negative controls was as described
earlier (Loper et al., 1978). Microsomal activation was conducted
using PCB (Aroclor 1254)-induced rat liver microsomes from Litton
Bionetics (Kensington, MD). Detection of mutagenesis in
experimental samples was based upon a dose-dependent response
which exceeded the control (zero dose) value by at least twofold
(total revertant colonies per plate/control colonies per
plate > 2). Where the amount of sample was limited, as often
occurred with subfractions, semiquantitative determinations of
mutagenesis were based upon as little as 2 plates/tester strain,
with 1/3 and 2/3 aliquot/plate, the size of the aliquot
nevertheless being chosen to yield at least the twofold response.
Fractions and subfractions were not taken to dryness so residue
was not determined by weight; all recoveries of bioactivity were
based upon expressions of mutagenesis per liter equivalent,
representative of the original water sample.

Chemicals

Type I water for HPLC experiments was generated in our
laboratory using a Continental water conditioning system as
previously described (Tabor and Loper, 1980). This water, and
CH_3OH, CH_2Cl_2, and CH_3CN obtained distilled-in-glass from Burdick
and Jackson Laboratories, Inc. (Muskegon, MI), were degassed by
sonication under reduced pressure for 15 min immediately prior to
use. Hexane and acetone (HPLC grade) were obtained from Matheson
Coleman and Bell (Cincinnati, OH). All other chemicals were of
reagent grade and were used as obtained.

HPLC and Isolation of Mutagenic Subfractions

HPLC separations were performed on a Waters Associates
(Milford, MA) ALC/GPC 204 system fitted with a 254-nm absorbance
detector and Waters Associates columns as described below.
Analytical-scale HPLC separations employed a 3.9-mm x 30-cm
prepacked column of 10-μm silica particles bonded with

octadecylsilane and protected with a guard column as previously described (Tabor and Loper, 1980; Loper and Tabor, 1981). Preparative separations were obtained using a Waters radial compression module (RCM) fitted with an 8-mm x 10-cm column packed with 10-μm silica particles bonded with octadecylsilane. Final HPLC separations in normal phase employed an 8-mm x 10-cm column of 10-μm silica particles (PORASIL^R). Coupled bioassay/reverse-phase HPLC applied our previously described method (Tabor et al., 1980; Tabor and Loper, 1980). Analytical-scale HPLC elutions were performed using a water-to-CH$_3$CN linear gradient; preparative-scale elutions were accompanied by combinations of gradient and isocratic steps involving the same solvents. Fractions were processed for bioassay using the SEPPAK^R procedure previously described (Tabor et al., 1980; Tabor and Loper, 1980). Normal-phase HPLC was accomplished isocratically using CH$_2$Cl$_2$, or with a starting solvent of 50% hexane:50% CH$_2$Cl$_2$ followed by a linear gradient to 100% CH$_2$Cl$_2$.

Mass Spectrometry

High-resolution mass spectrometry (MS) was performed on a Kratos (Manchester, U.K.) MS-80 by direct probe inlet of the sample using perfluorokerosene as the internal mass standard. Data were continually collected using a Data General Nova 4X-based DS 55 computer.

RESULTS

Method Development -- Organic Residue Extraction

Nonvolatile organics were extracted from drinking water using XAD-2 and XAD-7 columns in series. This approach was based upon results of multiple column experiments by others (Leenheer and Huffman, 1976; vanRossum and Webb, 1978) and upon our own preliminary experiments using a drinking water known to contain mutagenically active nonvolatile organics (unpublished). Resins were obtained from the Applied Science Division of the Milton Roy Company (State College, PA), prepared and evaluated according to U.S. Environmental Protection Agency criteria (EPA-600/7-78-201). Resins were stored until use in amber bottles under CH$_3$OH, and columns were formed by gravity from slurries in the same reagent. Column configuration and bed volumes were designed to accommodate 1500 l of water containing total organic carbon levels typical of finished drinking water from surface sources (1 to 10 ppm), based upon theoretical performance properties of XAD resins (Leenheer

and Huffman, 1976). A diagram of the gauge, valve, and stainless
steel columns used in the system is presented in Figure 1.

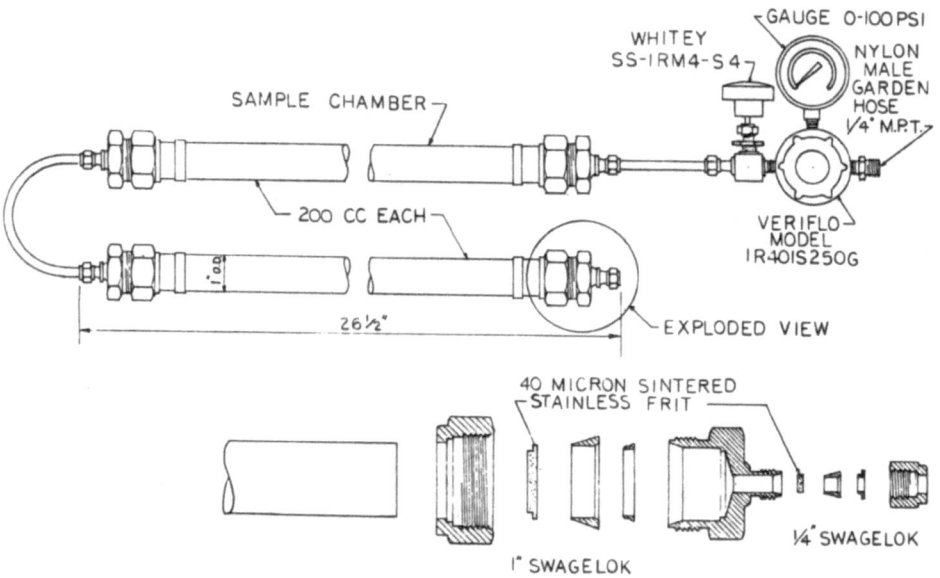

Figure 1.   Schematic of the column components of the collection
            unit.

Drinking water was obtained from two widely separated U.S.
cities that draw upon surface waters of distinctly different
environmental types. Since several days could intervene between
the initiation of collection and the elution of residues from the
columns, extensive prefiltration was included in the system to
minimize microbial contamination. The prefiltration consisted of
one of the empty steel columns packed with silanized glass wool as
a prefilter, followed by a Millipore (Bedford, MA) filter assembly
containing an AP40 fiberglass filter plus a Durapore 0.45-μm
filter. Samples of from 70 to 500 l of water at 30 to 55 psi were
processed, with filters changed as necessary to maintain flow at
100 to 60 ml/min; sample size was determined as volume of effluent
collected in a suitable container. Each column was eluted

separately using 2.5 column volumes of an 85:15 mixture (v/v) of
HPLC-grade hexane:acetone (LeBel et al., 1979). Eluates were
concentrated by rotary evaporation under reduced pressure at 30°C
to 10^{-4} times the original water volume. Concentrates were stored
at 4°C under nitrogen in Teflon-sealed, amber glass vials.
Aliquots for bioassay were dried and then dissolved in dimethyl
sulfoxide (DMSO) or more generally, were combined with DMSO and
the original solvents then removed under a gentle stream of
nitrogen gas. Samples for HPLC were concentrated by the latter
method after the addition of a small volume of CH_3CN. This
procedure has been employed for all routine extraction of water
from the two cities, which is being sampled on a seasonal basis.
Based upon results given in the next section, additional
experimental samples have been extracted without the use of XAD-7.

Direct-Acting Mutagenesis in XAD-Extracted Water Residues

With a few exceptions, all of the mutagenic activity for
either TA98 or TA100 was found in the residues extracted by the
XAD-2 columns. This mutagenesis was independent of, and partially
inhibited by, the presence of a microsome activation system, a
pattern typical of many other drinking water samples (reviewed in
Loper, 1980). The data in Figure 2a, although obtained as part of
a slightly different experiment, are representative of these
general results.

For the experiment that yielded the results in Figure 2, a
typical drinking water sample was filtered and then split, with
one-half being extracted by each resin. Although the compounds
extracted by XAD-7 are expected to include some too polar to be
retained by XAD-2, the mixtures extracted by either resin alone
showed the same patterns of mutagenic activity (compare Figures 2a
for XAD-2 and 2b for XAD-7). On one occasion, however, a raw
water sample processed by XAD-2 and XAD-7 in tandem showed
mutagenic activity residing only in the XAD-7-extracted residue
(unpublished).

Mutagenic Isolation

The data of mutagenic responses plotted in Figure 2 allow
some calculation of the sample size needed to undertake the
isolation of unknown compounds based upon their bioactivity. At
two plates/dose, assay at these doses using one strain, both plus
and minus microsomal activation, requires 5-1 equivalents/assay.
Plating only minus activation and at the minimum levels described
in Methods reduces this to more nearly 0.5-1 equivalents/assay.

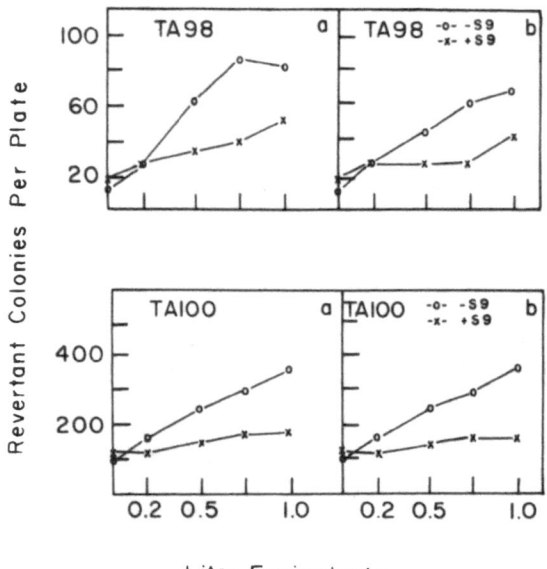

Figure 2.   Mutagenic activity of residues extracted by XAD-2 or
             XAD-7.

Our strategy has been to begin isolation on residues from 300- to
500-1 samples.   Individual preparative-scale separations involve
20- to 100-1 equivalents/HPLC.

     For residues from either city, mutagenic activity was found
predominantly in relatively nonpolar fractions.   Differences among
residues from separate cities, with respect to relative elution
patterns of mutagenic activity for TA98 versus TA100, and with
respect to the interactive effects of repooling subfractions upon
assayable mutagenic activity, have been documented elsewhere
(Loper and Tabor, 1982).   To get a further indication of the
mutagenic complexity of a sample, an HPLC elution of City 1
residue was collected into the fractions depicted in Figure 3.   A
composite representation of the mutagenic activity of these
fractions, assayed up to a dose of 10-1 equivalents/plate appears

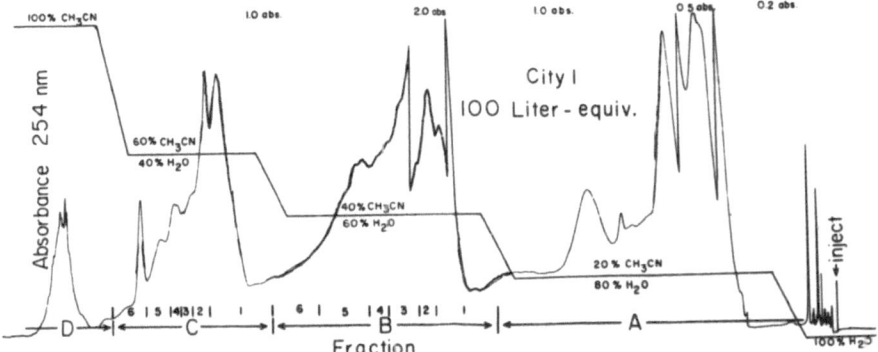

Figure 3.   Absorbance profile of City 1 residue
separated by reverse-phase HPLC.

in Figure 4. Clearly, among a multitude of peaks absorbing at
254 nm, there are several that show differential levels of TA98
and TA100 mutagenesis. Clearly, too, the purification and
identification of some of these by our present techniques would
require a quite large initial sample.

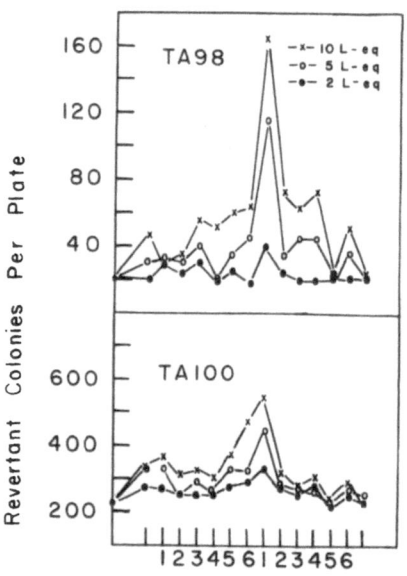

Fraction Plated

Figure 4. Direct-acting mutagenesis of TA98 and TA100 by
fractions indicated in Figure 3.

For the attempted purification of a mutagen, we chose a large
sample residue from City 2 that showed a high level of
direct-acting TA98 mutagenesis (see Figures 5c and 5d). This
residue was fractioned by the step elution depicted in Figure 3 to

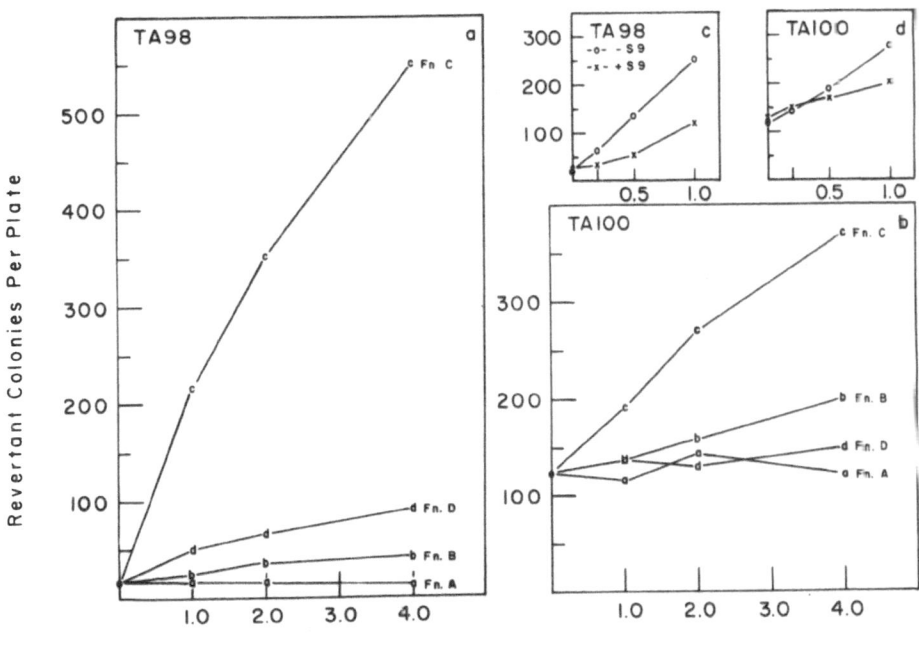

Figure 5.   Mutagenic activities of HPLC subfractions (A,B) and of
            parent residue (C,D) from City 2.

yield Fractions A, B, C, and D; the mutagenic activities of these
fractions are presented in Figure 5a for TA98 and in Figure 5b for
TA100. Fraction B containing the predominant activity for TA98
was rechromatographed by reverse-phase HPLC and the resultant
mutagenic subfraction was partitioned further by normal-phase HPLC
on PORASIL[R]. An isocratic elution of that column using CH_2Cl_2
produced the 254-nm absorption profile shown in Figure 6a; all the
mutagenic activity was contained in fraction 3. This activity was
specific for TA98. (An amount that gave a doubling of total TA98
revertants over the zero dose control count gave < 0.025 doublings
of TA100.) Efforts at GC revealed only that the subfraction
contained material of low volatility. Temperature-programmed GC

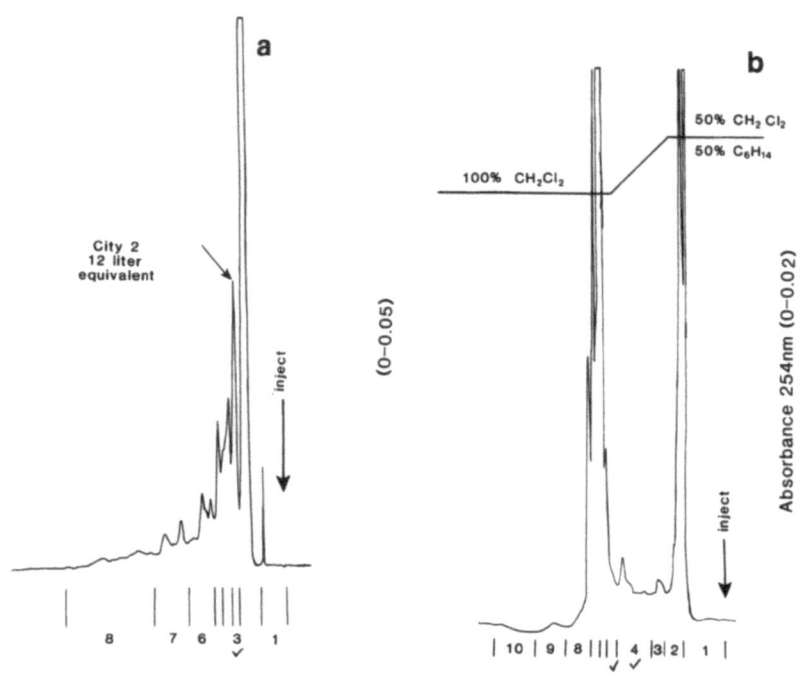

Figure 6.   Normal-phase HPLC separations of mutagenic subfractions
             on PORASIL[R].

runs on 3% OV17 yielded only apparent decomposition products
between 260° and 320°C. The material in fraction 3 (Figure 6a)
was examined further using high-resolution MS. These results are
summarized in Figure 7. Analysis of the total ion chromatogram
and the m/e ion 327 chromatogram presented in Figure 7a revealed
the electron impact spectrum shown in Figure 7b.

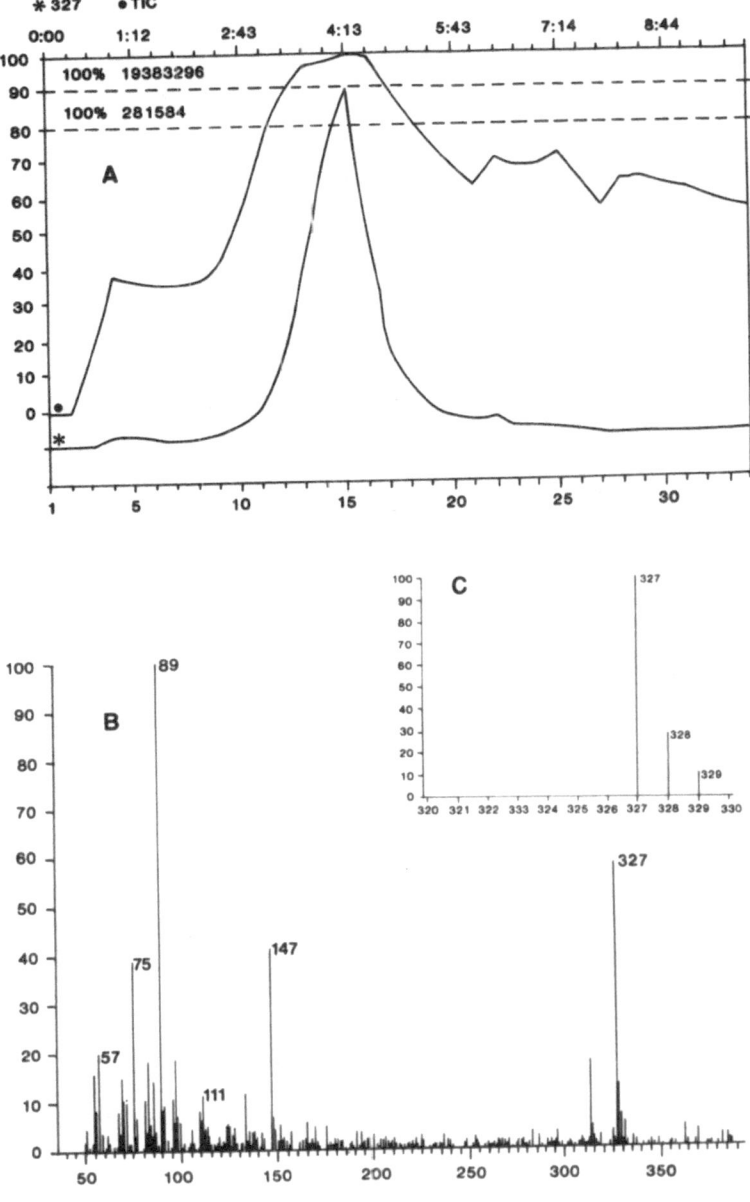

Figure 7.   Total ion current chromatogram and high-resolution
electron impact MS for the mutagenesis fraction in
Figure 6a.

For a repeat isolation, fraction B material was again
rechromatographed by reverse phase and then by normal phase. This
time, a gradient elution of the PORASIL^R column was used as
described in Methods; the results are plotted in Figure 6b.
Contents of each of these tubes were assayed for TA98 mutagenesis,
and were examined independently by high-resolution MS in a
specific ion monitoring search (SIMS) for the ions m/e 57, 89,
147, and 327. For scans with positive matches on all four ions, a
SIMS was conducted for the six m/e near 327 (327.3094, 328.3115,
329.3110, 329.9871, 331.9814, 332.9962). Verification of a match
was also based upon the percent relative intensity ratio of these
peaks to 327.3094 of 100, 30.6, 6.14, 4.6, 12.9, and 7.5%,
respectively. This specific ion pattern is shown in Figure 7c.

The result was that the compound that constituted the major
component of fraction 3 (Figure 6a) was shown to be present in
fractions 4 and 5 of this second preparation (Figure 6b). These
two fractions also contained all the characteristic TA98
mutagenesis in this elution.

The mutagenic fractions identified in Figure 6 are not
entirely homogenous, and given the wide range of mutagenic potency
among known mutagens (McCann and Ames, 1977), we cannot say that
the fractions' major component with this distinguishing mass
spectrum is the TA98 mutagen. Further characterization will
require fractionation of additional residue.

The mass spectrum in Figure 7 does not match any in the
NIH/EPA library. However, we recently identified its presence in
residues of two previous samples we examined from this city. This
proved to be a relatively rapid procedure. By following the
elution protocols that we had established for the mutagen during
the previous bulk isolations, residues proportional to only 12 l
of sample water were chromatographed two times by reverse-phase
HPLC; suspect fractions were found positive for the compound upon
direct probe inlet high-resolution MS.

DISCUSSION

Considering all the problems associated with the assessment
of adverse health effects from dilute organics in drinking water,
we have concluded the area of our greatest contribution might be
the development of a feasible method for separating and
identifying major mutagens from residue mixtures. It was always
presumed that previously unknown mutagens might be involved, so
that no ideal concentration procedure could be predicted (Kopfler,
1981), and that a definitive comparison of alternate concentration

methods could be conducted once the properties of a set of the major nonvolatile drinking water mutagens was established. The XAD-2/XAD-7 tandem resin approach to residue extraction was chosen for its versatility in recovery of many known compounds, plus its portability for field use in extracting large water samples at different locations.

Among short-term bioassays, the relative merits of the Salmonella mutagenesis assay in the isolation of possible mutagens/carcinogens from residues of this type have been discussed (Loper, 1980a; Loper, 1980b). Our current findings support those earlier conclusions. As a practical matter, our assays to monitor progress in mutagen subfractionation do not include all the standard Salmonella tester strains; indeed we reduce the dose-response plating to a minimum with only one strain whenever possible.

These procedures have permitted the isolation of two distinctly different chemical mutagens to purity or near purity from separate initial residues. It appears that both of these will prove to be previously unknown drinking water constituents. In the case of the "something old," sufficient residue was available to permit repeat isolation and rigorous characterization of one of the mutagenic isomers. The compound separated homogeneously by GC with a variety of different columns and temperature programs, and we can link its bioactivity directly with its m/e characterization. Nevertheless, establishing the molecular structure of this substance has been a challenging problem, and its properties have yet to be proven by synthesis. Based upon data obtained by several analytical techniques, we have identified the compound as a polychlorinated unsaturated aliphatic ether (Tabor, MS). It is a potent mutagen for TA100 and for TA1535, but has little or no activity for the other standard Salmonella tester strains (Tabor and Loper, 1980; Loper et al., MS). Interestingly, it is not active with strain hisG46, the prototype strain for those named above (Distlerath et al., in preparation). All this mutagenesis is entirely dependent upon microsomal activation, using liver S9 preparations from specifically induced rats or mice. Phenobarbital, PCB (Aroclor 1254), and 1,2,4-trichlorobenzene were good inducers, but preparations induced with 3-methylcholanthrene or uninduced S9 were essentially unable to activate this ether (Distlerath et al., 1982). The promutagen thus shows many mutagenic similarities to the S-chloroallyl thiocarbamate herbicides diallate and triallate (Schuphan and Casida, 1979; Distlerath et al., 1982; Loper et al., MS).

The mutagen highly purified from current drinking water residue is predominantly, if not entirely, specific for TA98, not TA100, and is direct acting. It has repeatedly been isolated to near homogeneity based upon its 254-nm absorption in chromatography profiles. It also shows physical properties of temperature stability and UV absorbance spectra (unpublished) which appear to be compatible with the major constituent found when these isolated fractions are subjected to MS analysis. At present, the compound represented by the m/e data in Figure 7 is our only candidate for the basis of the isolated mutagenic activity, but this relationship is not proven.

Even so, the relative ease with which we could identify these mass spectra among previously extracted residues is quite encouraging. It suggests that once specific water mutagens are characterized, both as to their chromatographic separation properties and their mass spectra, this information could be applied readily to research their ubiquity and concentration among drinking waters. Theoretically, if present in water at the same level as for the samples we have studied, the compound represented in Figure 7 could be detected from a dozen liters of water within two working days of its receipt in our laboratory.

Based upon our previous studies, we have suggested that drinking water residues contain a great number of nonmutagenic compounds, some of low mutagenicity, and a few that are highly mutagenic (Tabor et al., 1979; Loper and Tabor, 1981). Work from several laboratories indicates that most of this mutagenic activity is direct acting (Loper, 1980). Recently, Cheh and co-workers have described the inactivation of such electrophilic mutagenesis by sulfite treatment (Cheh et al., 1980a,b) or with 4-nitrothiophenol (NTP) (see Cheh et al., 1981). Using this approach, Cheh and also Baird and co-workers (Jacks et al., in press) have presented evidence for a number of compounds with different levels of reactivity for NTP among mutagenic residue mixtures from chlorinated high TOC water.

We have described a general procedure for mutagens in drinking water which extends from the extraction of a complex mixture of nonvolatile residues to the isolation of mutagenic subfractions and their characterization by high-resolution MS. Our results show that major mutagens of distinctly different mutagenic and chemical properties are amenable to isolation by this procedure. These results also suggest that once such mutagens are defined in terms of their HPLC fractionation and their mass spectra, it should be possible to detect their presence and approximate concentration in drinking water, based upon a few days' study of 10- to 20-l water samples.

ACKNOWLEDGMENTS

   Our appreciation to Dr. F.C. Kopfler for advice in the
application of XAD resins, to Mr. Roy E. Jones for unstinting
support in the collection of water samples, and to
Mrs. S.K. MacDonald for expert laboratory assistance. This
research was supported by research grants from the
U.S. Environmental Protection Agency.

REFERENCES

Cheh, A., J. Skochdopole, P. Koski, and L. Cole. 1980a.
    Nonvolatile mutagens in drinking water: production by
    chlorination and destruction by sulfite. Science 207:90-92.

Cheh, A.M., J. Skochdopole, and C. Heilig. 1980b. Destruction of
    direct-acting mutagens in drinking water by nucleophiles:
    implications for mutagen identification and mutagen
    elimination from drinking water. In: Water Chlorination:
    Environmental Impact and Health Effects, Volume 3.
    R.L. Jolley, W.A. Brungs, and R.B. Cumming, eds. Ann Arbor
    Science: Ann Arbor. pp. 803-815.

Cheh, A.M., and R.E. Carlson. 1981. Determination of potentially
    mutagenic and carcinogenic electrophiles in environmental
    samples. Anal. Chem. 53:1001-1008.

Distlerath, L., J.C. Loper, and M.W. Tabor. 1982. Microsomal
    activation of promutagenic chlorinated propenyl ethers.
    Program Abstract Ab-1. Presented at the 13th Annual Meeting
    of the Environmental Mutagen Society, Boston, MA.

Jacks, C.A., J.T. Gute, L.B. Neisess, R.J. Van Sluis, and
    R.B. Baird. (in press). Health effects of water reuse:
    characterization of mutagenic residues isolated from
    reclaimed surface, and groundwater supplies. In: Water
    Chlorination: Environmental Impact and Health Effects,
    Volume 4. R.L. Jolley, W.A. Brungs, and R.B. Cumming, eds.
    Ann Arbor Science: Ann Arbor.

Kopfler, F.C. 1981. Alternative strategies and methods for
    concentrating chemicals from water. In: Application of
    Short-Term Bioassays in the Fractionation and Analysis of
    Complex Environmental Mixtures, Volume 2. M.D. Waters,
    S.S. Sandhu, J.L. Huisingh, L. Claxton, and S. Nesnow, eds.
    Plenum Press: New York. pp. 141-153.

LeBel, G.L., D.T. Williams, G. Griffith, and F.M. Benoit. 1979.
Isolation and concentration of organophosphorous pesticides
from drinking water at the ng/l level using macroreticular
resin. J. Assoc. Off. Anal. Chem. 62:241-249.

Leenheer, J.A., and E.W.D. Huffman, Jr. 1976. Classification of
organic solutes in water using macroreticular resin. J. Res.
U.S. Geol. Survey 4:737-751.

Loper, J.C., D.R. Lang, R.S. Schoeny, B.B. Richmond,
P.M. Gallagher, and C.C. Smith. 1978. Residue organic
mixtures from drinking water show in vitro mutagenic and
transforming activity. J. Toxicol. Environ. Health 4:919-938.

Loper, J.C. 1980a. Overview of the use of short-term biological
tests in the assessment of the health effects of water
chlorination. In: Water Chlorination: Environmental Impact
and Health Effects, Volume 3. R.L. Jolley, W.A. Brungs, and
R.B. Cumming, eds. Ann Arbor Science: Ann Arbor.
pp. 937-945.

Loper, J.C. 1980b. Mutagenic effects of organic compounds in
drinking water. Mutation Res. 76:241-268.

Loper, J.C., and M.W. Tabor. 1981. Detection of organic mutagens
in water residues. In: Application of Short-Term Bioassays
in the Fractionation and Analysis of Complex Environmental
Mixtures, Volume 2. M.D. Waters, S.S. Sandhu, J.L. Huisingh,
L. Claxton, and S. Nesnow, eds. Plenum Press: New York.
pp. 155-165.

Loper, J.C., M.W. Tabor, and L.M. Distlerath. (MS). Chloroallyl
ethers: new promutagens isolated from an old drinking water
residue.

Loper, J.C., M.W. Tabor, and S.K. MacDonald. (in press).
Mutagenic subfractions from non-volatile organics in drinking
water. In: Water Chlorination: Environmental Impact and
Health Effects, Volume 4. R.L. Jolley, W.A. Brungs, and
R.B. Cumming, eds. Ann Arbor Science: Ann Arbor.

McCann, J., and B.N. Ames. 1977. The Salmonella/microsome
mutagenicity test: predictive value for animal
carcinogenicity. In: Origins of Human Cancer, Book C.
H.H. Hiatt, J.D. Watson, and J.A. Winsten, eds. Cold Spring
Harbor Laboratory: Cold Spring Harbor, NY. pp. 1431-1450.

Schuphan, I. and J.E. Casida. 1979. S-chloroallyl thiocarbamate herbicides: chemical and biological formation and rearrangement of diallate and triallate sulfoxides. J. Agric. Food Chem. 27:1060–1067.

Tabor, M.W. (MS). Structure elucidation of 3-(2-chloroethoxy)-1,2-dichloropropene, a new promutagen from an old drinking water residue.

Tabor, M.W., and J.C. Loper. 1980. Separation of mutagens from drinking water using coupled bioassay/analytical fractionation. Int. J. Environ. Anal. Chem. 8:197–215.

Tabor, M.W., J.C. Loper, and K. Barone. 1980. Analytical procedures for fractionating non-volatile mutagenic components from drinking water concentrates. In: Water Chlorination: Environmental Impact and Health Effects, Volume 3. R.L. Jolley, W.A. Brungs, R.B. Cumming, eds. Ann Arbor Science: Ann Arbor. pp. 899–910.

vanRossum, P., and R.G. Webb. 1978. Isolation of organic water pollutants by XAD resins and carbon. J. Chromatog. 150:381–392.

MUTAGENICITY TESTING OF COMPLEX ENVIRONMENTAL MIXTURES WITH

CHINESE HAMSTER OVARY CELLS

A.P. Li, A.L. Brooks, C.R. Clark, R.W. Shimizu, R.L. Hanson, and J.S. Dutcher

Inhalation Toxicology Research Institute, Lovelace Biomedical and Environmental Research Institute, P.O. Box 5890 Albuquerque, NM 87185

INTRODUCTION

Because of the positive correlation between mutagenicity and carcinogenicity, short-term tests for mutagenicity are used widely as initial screening tests for substances of potential carcinogenicity. The Ames test (Ames et al., 1975) quantitating the reversion of histidine-requiring Salmonella typhimurium auxotrophs to histidine-nonrequiring prototrophs, has been used in different laboratories to study chemical mixtures obtained through extraction of environmental pollutants. A wide data base obtained from the Ames test includes studies of urban air particles (Pitts et al., 1977), automobile exhaust (Huisingh et al., 1978; Ohnishi et al., 1980; Clark et al., 1981), and coal combustion products (Fisher et al., 1979; Clark and Hobbs, 1980). Ames test-directed chemical fractionation has been used as an approach to identify mutagens in complex environmental mixtures (Dutcher et al., 1981).

Concern, however, has been raised on the proper interpretation of the Ames test data. There are diverse differences between bacteria and mammalian cells including membrane structure, metabolic pathways, DNA repair systems, and genome organization. Such differences may affect their response to mutagens. Results from mammalian mutation assays, therefore, should help in the interpretation of the wide data base obtained using the Ames test, allowing one to place research emphasis on

183

findings that have greater implications for estimating human risks.

We report here our application of a well-defined mutation system, the Chinese hamster ovary cell/hypoxanthine-guanine phosphoribosyl trnsferase (CHO/HGPRT) mutation system developed by Hsie and co-workers (Hsie et al., 1979) to study the mutagenicity of complex environmental mixtures. Modifications were made to increase the efficiency and sensitivity of the system. The complex environmental mixtures studied include extracts of organic chemicals from diesel and spark-ignition automobile exhaust particles, coal-combustion fly ash, coke-oven emissions, roofing tar, and cigarette-smoke condensate.

MATERIALS AND METHODS

Source of Complex Environmental Mixtures

Diesel (1980 Audi 5000) and spark-ignition (1980 Chevrolet Citation, oxidation catalyst-equipped) automobile exhaust particles were collected in a dilution tunnel during vehicle chassis dynomometer tests (EPA Federal Test Procedure/hot start, 75°F) at the Bartlesville Energy Technology Center, Bartlesville, OK. Coal-combustion fly ash was collected from the bag filter of an experimental fluidized-bed combustor in Morgantown, WV. The automobile-exhaust particles and coal-combustion fly ash were extracted using dichloromethane in our laboratory as previously described (Clark and Hobbs, 1980; Li, 1981a; Clark et al., 1981). Extracts prepared from coke-oven mains, roofing tar, and cigarette-smoke condensate were provided by Dr. Joellen Lewtas of the Health Effects Research Laboratory, U.S. Environmental Protection Agency (EPA), Research Triangle Park, NC. The collection of EPA samples has been previously reported (Huisingh et al., 1980).

Mutation Assays with CHO Cells

CHO cells cloned K_1-BH_4 were originally obtained from Dr. A.W. Hsie of Oak Ridge National Laboratory, Oak Ridge, TN. The cells were routinely maintained as reported (Li, 1981a, 1981b). The growth of CHO cells as unattached culture (Li, 1980) and application of the procedure to the CHO/HGPRT mutation assay (Hsie et al., 1979; Li, 1981b) are as previously reported. Advantages of the unattached cultures over those of conventional monolayer and suspension cultures are as follows:

1) Trypsinization not required for cell detachment.
   Decrease in time, effort and cost in subculturing,
   and lower chances of bacterial and fungal
   contamination.

2) Use of nontissue (bacteriological) culture plates
   with a cost of 10 to 20% of similar tissue culture
   plates.

3) Same incubator space requirements as monolayer
   cultures (much less than that of suspension
   cultures).

4) Easily adaptable to laboratories established for
   monolayer cultures.

The sensitivity of the CHO/HGPRT assay mutation assay was
increased through increasing the sampling size from 10^6 cells to
5×10^6 cells. Detailed description of this modification will be
reported elsewhere (Li and Shimizu, MS). Briefly, 2.5×10^6 cells
were plated 24 h before treatment in 75-cm^2 flasks. The cell
number on the day of treatment was ~ 5×10^6 cells/flask. After
treatment, 5×10^6 cells/sample were subcultured as unattached
cultures. Selection was performed by placing 10^6 cells/100-mm
plates in a medium of 0.33% agar (Noble) in selective medium
(Ham's F12 medium with 5% dialyzed newborn-calf serum) and
10 μM 6 thioguanine (TG), on a pre-poured bottom layer of 1% agar
in selective medium. There were 5×10^6 cells (5 selection
plates) used/sample. To determine cloning efficiency, 200 cells
were plated in similar medium with agar without TG. Data were
expressed as mutants/10^6 survivors, calculated as the ratio of the
number of mutant colonies observed/10^6 cells plated to the cloning
efficiency. Dose of mutagens used was expressed as micrograms cf
mutagens per milliliter of treatment medium (μg/ml).

Other Mutation Assays

Mutation at the Na^+-K^+-dependent ATPase gene locus
(Na^+-K^+-ATP) in CHO cells was quantitated as reported (Li, in
press) on day 3 after treatment, using 0.5 mM ouabain as selective
agent. Sister chromatid exchange in CHO cells was performed
according to the procedures of Takehisa and Wolff (1977). Ames
test was performed according to the procedures of Ames et al.
(1975). S. typhimurium (TA98) was used.

Synergism between Diesel Exhaust Particle Extract (DEPE) and
Benzo(a)pyrene (B[a]P)

CHO cells were treated with either DEPE (60 µg/ml) alone, or
B(a)P (1.0 µg/ml) alone, or a combination of DEPE and B(a)P.
Genotoxicity induced was measured using three different end
points: induction of mutation at the HGPRT gene locus, mutation
at the Na^+-K^+-ATP, and sister chromatid exchange as described
above. The expected genotoxicity (genotoxicity [DEPE only] +
genotoxicity [B(a)P only]) was compared to that observed for the
combined treatment (DEPE + B[a]P). Synergistic interaction would
be suggested by a higher-than-expected genotoxicity.

Materials

Ham's F12 medium and newborn-calf serum was purchased from
Flow Lab. Ham's F12 medium without hypoxanthine was purchased
from K.C. Biological, Inc. Ethyl methane sulfonate (EMS), TG, and
ouabain were obtained from Sigma Chemical Co. B(a)P was obtained
from Aldrich Chemical Co. Aroclor 1254-induced rat-liver S9 was
obtained from Litton Bionetics.

RESULTS

A comparison of the data from the modified CHO/HGPRT assay
with 5X increase in sampling size and selection in soft agar to
that from the original procedure is shown in Figure 1. No
dose-response relationship was obtained with the original
procedure for EMS at doses of 0 to 50 µg/ml (Figure 1A). Using
our modified soft-agar assay, a definite dose response was
obtained (Figure 1B). At higher concentrations of EMS
(50 to 400 µg/ml), no difference in mutagenicity between the two
assays was observed (data not shown).

Mutagenicity of different complex environmental mixtures in
CHO cells at the HGPRT gene locus are shown in Table 1. The
extracts tested were obtained from different inhalable pollutants
including coke-oven soot, roofing tar, coal-combustion fly ash,
automobile exhaust, and cigarette smoke. The extracts were tested
at several concentrations with S9 and had a range of potencies
from ~ 0.05% (cigarette-smoke condensate) to ~ 5% (roofing tar) of
that of the mutagenicity of B(a)P.

A comparison of the mutagenicity of a coal-combustion fly ash
extract determined by the Ames test and CHO/HGPRT assay is shown

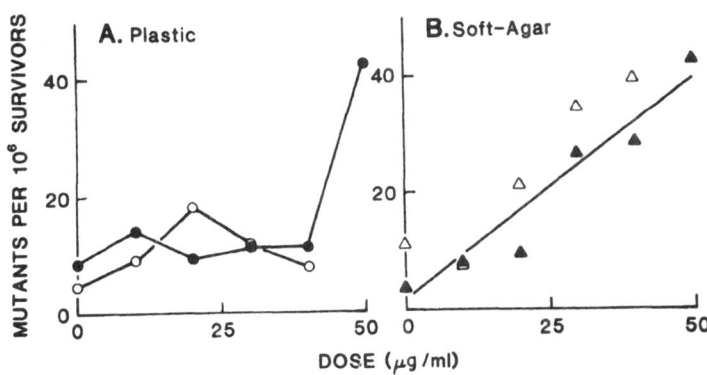

Figure 1.   Mutagenicity of low doses (0 to 50 µg/ml) of EMS as
as determined using the original CHO/HGPRT assay of
Hsie et al. (A) and our modified assay with increased
sampling size (B). Results of two independent
experiments (open and closed symbols) are shown.

in Figure 2. The extract was a direct-acting mutagen in
S. typhimurium TA98, with the activity slightly decreased by
exogenous Aroclor 1254-induced rat-liver S9 (Figure 2A).   In CHO
cells, the extract was cytotoxic. The cytotoxicity was decreased
when liver S9 was added (Figure 2B). Mutagenicity of the extract
in CHO cells, unlike that observed in the Ames test, was detected
only in the presence of liver S9 (Figure 2B).

    Mutagenicity of extracts from exhaust particles from a diesel
car and a spark-ignition car are shown in Figure 3.   The
mutagenicity of the extracts from the exhaust particles from the
spark-ignition car operated on three fuels (indolene, indolene
with 10% ethanol, and indolene with 10% methanol) were also
compared.   The three extracts from the spark-ignition exhaust
particles showed similar mutagenicity which was slightly higher
than the diesel exhaust particle extracts (Figure 3A).   For a
comparison of the emission rates of mammalian mutagens of the
diesel and spark-ignition cars, the data were replotted with
mutagenicity versus miles driven to yield the equivalent amount of
extract (weight of extract ÷ percent of extractable chemicals from
exhaust particles ÷ particulate emission rate) (Figure 3B).   Our

Table 1. Mutagenic Activity of Different Complex Environmental
Mixtures at the HGPRT Gene Locus in CHO Cells[a]

Source	Mutagenicity[b]	% of B(a)P
B(a)P	127	100
Roofing tar	6.0	4.7
Coke-oven mains	5.8	4.6
Coal-combustion fly ash	3.8	3.0
Automobile-exhaust particles		
Spark-ignition vehicle		
Indolene	0.69	0.54
Indolene with 10% ethanol	0.53	0.42
Indolene with 10% methanol	0.52	0.41
Diesel vehicle	0.22	0.17
Cigarette-smoke condensate	0.063	0.05

[a]With exogenous activation (Aroclor 1254-induced rat-liver S9).
[b]Slope of dose-response curves after linear regression of the data
(mutants/10^6 survivors/μg extract/ml treatment medium).

results showed a 100-fold higher emission rate of mutagenic
combustion product associated with the particulate exhaust from
the diesel car than the spark-ignition car. The major portion of
the difference is attributed to the higher particle emission rate
of the diesel-powered vehicle (0.37 g/mi) compared to the gasoline
spark-ignition vehicle (0.04 g/mi).

We observed a synergistic interaction between DEPE and B(a)P.
CHO cells were treated with DEPE alone, B(a)P alone, or a
combination of B(a)P and DEPE. The observed genotoxicity of the
combined treatment was higher than that expected from the
individual treatment (genotoxicity expected = genotoxicity [DEPE
only] + genotoxicity [B(a)P only]). This observation was made for
three end points of genotoxicity: mutation at HGPRT gene locus
(Figure 4A), mutation at Na^+-K^+-ATP gene locus (Figure 4B), and
sister-chromatid-exchange induction (Figure 4C).

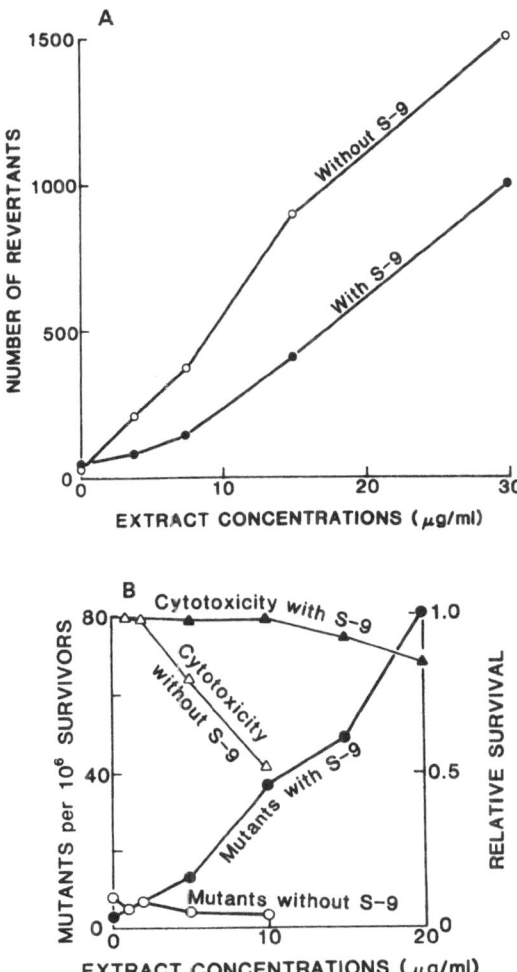

Figure 2. Mutagenicity of coal-combustion fly ash extract in
S. typhimurium TA98 (A) and CHO cells (B). Assays
were performed both with and without Aroclor 1254-
induced rat-liver S9 as indicated.

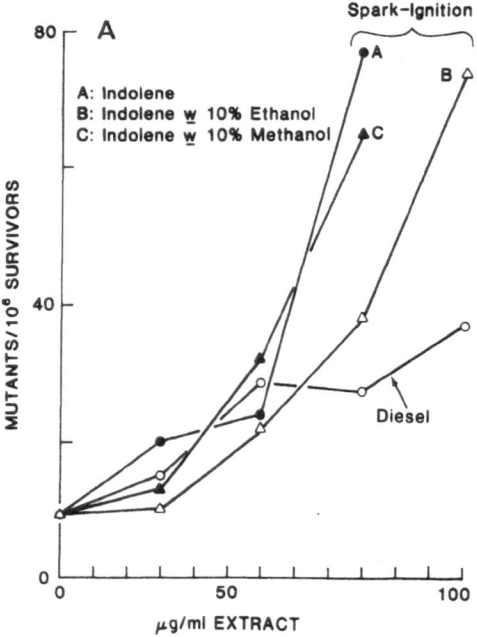

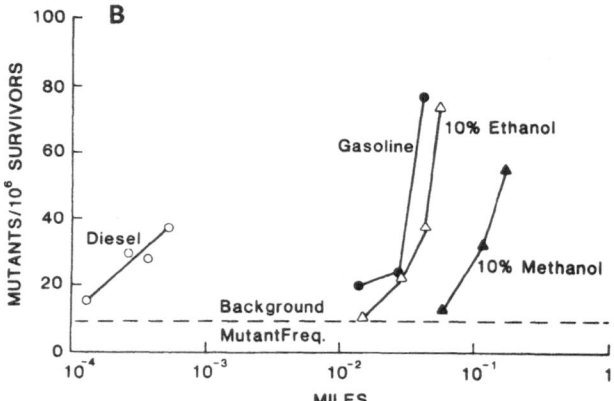

Figure 3.   A comparison between diesel and spark-ignition
automobiles in the mutagenicity of the exhaust particle
extracts (A) and the emission rates of mutagens (B).
The emission rate is the number of miles the vehicle
was driven to yield an extract with the measured
potency as shown.   The spark-ignition automobile was
operated using three different fuel blends as
indicated.

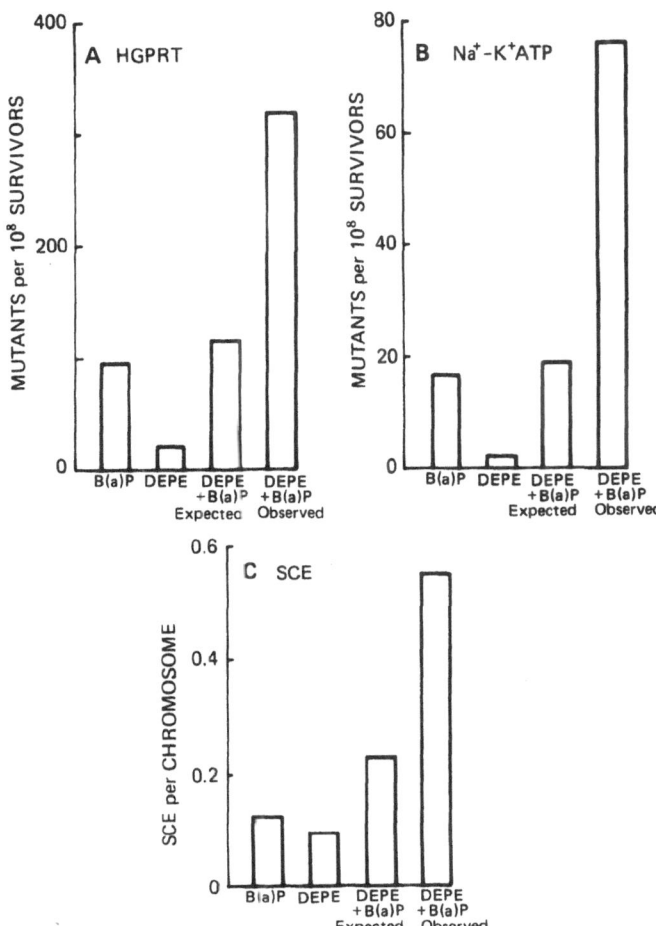

Figure 4.   Synergistic interaction between DEPE and B(a)P in
            genotoxicity as measured by induction of mutations at
            the HGPRT gene locus (A), Na^+-K^+-ATP gene locus (B),
            and sister chromatic exchanges (C). The concentrations
            of DEPE and B(a)P used are 60 μg/ml and 1.0 μg/ml,
            respectively. Assays were performed with exogenous
            activation (Aroclor 1254-induced rat-liver S9).

DISCUSSION

We have modified the CHO/HGPRT mutation assay developed by Hsie et al. (1979). The modifications decreased the cost- and time-commitment of the assay, and increased its ability to detect low mutagenic responses. Our modifications will be useful for routine testing of environmental samples. These modifications are especially useful in the assaying of complex environmental mixtures which, because of either solubility or cytotoxicity, can be tested only at concentrations that can induce only a weak mutagenic response in CHO cells.

The difference between Ames test and CHO/HGPRT assay for the coal-combustion fly ash extract suggests that the two assays may be sensitive to different mutagens. Experimental evidence demonstrating that nitroaromatics may be responsible for the differences are as follows. Nitroaromatics have been found to be extremely potent mutagens in bacteria (Rosenkranz et al., 1980), and tentatively identified in the coal-combustion fly ash extract that we used (Hanson et al., 1982). The same extract had a lower mutagenic potency when tested in nitroreductase-deficient strains of $\underline{S. \text{ typhimurium}}$ TA98 (Li et al., MS). Since nitroaromatics are only weakly mutagenic in CHO cells, the mutagenicity of the extract in CHO cells was probably due to the presence of other mutagens, for instance, polynuclear aromatic hydrocarbons that are promutagens. Our observations with the coal-combustion fly ash extract may be extended to extracts from other pollutants containing nitroaromatics. Because of their mammalian origin, data obtained using the CHO cells may be more relevant for extrapolation to human health risk than data obtained from bacteria. Further interpretation of our data, however, awaits a better understanding of the carcinogenicity of nitroaromatics and the relative importance of metabolism of the chemicals by animal-gut bacteria.

Using the exhaust particle extracts from diesel and spark-ignition automobiles, we showed that conclusions cannot be drawn based solely on the mutagenic potencies of the extracts. For instance, the exhaust particle extracts from the spark-ignition automobile were more mutagenic than the DEPE, suggesting that a higher health risk may be imposed by the spark-ignition automobile exhaust. However, when considering the distance driven, we found that the emission rate of mutagens per unit distance traveled was ~ 100X higher for the diesel automobile than the spark-ignition automobile. Our data, therefore, emphasize the need to consider physical and chemical emission data in addition to the mutagenic activities of the extracts. It should be noted that our analysis was not a definitive comparison of all diesel to spark-ignition cars, since we studied only one diesel and one spark-ignition automobile and we studied only the mutagens associated with the exhaust particles. Research with more automobiles, studying the mutagenicity of chemicals associated with both the gaseous and particulate exhaust, is required for a complete evaluation of the health risk associated with the exhaust emissions of the two types of automobile.

Our finding that the genotoxicity observed was higher than expected when CHO cells were treated with a combination of DEPE and B(a)P suggests a synergistic interaction between the two agents. Such synergistic interaction between chemicals associated with DEPE and B(a)P may raise an important question. We are living in an environment being contaminated by a large variety of pollutants from different sources. If synergistic interaction between pollutants actually occurs, the laboratory studies addressing the exposure of one pollutant would tend to underestimate the actual risk. Our observation made with the DEPE have been confirmed with extracts from diesel exhaust particles of cars of different manufacturers (Li and Royer, 1982), with different end points of genotoxicity as reported here, as well as with a different mutagen (N-methyl, N'-nitro, N-nitrosoguanidine) (Li and Royer, 1982). It is to be noted, however, that our observation on the synergistic interaction was made using high doses of diesel exhaust particle-associated chemicals and B(a)P that human cells would never encounter under normal routes of exposure to these environmental pollutants. The synergism observed here will be a significant health risk only if the phenomenon occurs at realistic doses of both agents. Understanding the dose-response relationship of the synergistic interaction at realistic dose levels, the mechanism of the synergism (e.g., competition for detoxifying enzymes, inhibition of DNA repair, alterations in membrane permeability), as well as investigation of the occurrence of this phenomenon in whole animals, should be performed for the elucidation of the implication of our observation to human health risk.

ACKNOWLEDGMENT

The authors would like to thank Drs. Roger McClellan, Jim Sun, Ron Wolff, and T.R. Henderson of our Institute for their critical review of our paper, Millie Deland, Dianne Mead, Amy Federman, and Sylvia Crain for their technical assistance, and Kathleen Epler for editorial assistance. We are especially grateful to Drs. William Marshall and Ted Namen of Bartlesville Energy Technology Center for providing the automobile exhaust particles, and Dr. Joellen Lewtas of the U.S. Environmental

Protection Agency (EPA) for providing the coke-oven mains, roofing
tar, and cigarette-smoke condensate samples. Research was
supported by the U.S. Department of Energy Contract
no. DE-AC04-76EV01013 and the EPA via interagency agreement
no. EPA-IAG-05-561.

REFERENCES

Ames, B.N., J. McCann, and E. Yamasaki. 1975. Methods for
    detecting carcinogens and mutagens with the Salmonella/-
    mammalian-microsome mutagenicity test. Mutation Res.
    31:347-364.

Clark, C.R. and C.H. Hobbs. 1980. Mutagenicity of effluents from
    an experimental fluidized bed coal combustor. Environ.
    Mutagen. 2:101-109.

Clark, C.R., R.E. Royer, A.L. Brooks, R.O. McClellan,
    W.F. Marshall, T.M. Naman, and D.E. Seizinger. 1981.
    Mutagenicity of diesel exhaust particle extracts: influence
    of car type. Fund. Appl. Toxicol. 1:260-265.

Dutcher, J.S., R.E. Royer, J.O. Hill, C.E. Mitchell, and
    C.R. Clark. 1981. (in press). Identification of components
    of biologically active fractions of low BTU gasifier coal
    tars. Presented at the Chemical Characterization of
    Hazardous Substance in Synfuels Symposium, Seattle, WA.

Fisher, G.L., C.E. Chrisp, and O.G. Raabe. 1979. Physical
    factors affecting the mutagenicity of fly ash from a
    coal-fired power plant. Science 204:879-881.

Guerin, M.R., C.H. Ho, T.K. Rao, B.R. Clark, and J.L. Epler.
    1980. Polycyclic aromatic primary amines as determinant
    chemical mutagens in petroleum substitutes. Environ. Res.
    23:42-53.

Hanson, R.L., T.R. Henderson, R.L. Carpenter, C.R. Clark, and
    C.H. Hobbs, T.M. Harvey, and D.F. Hunt. (MS).
    Identification of mutagens in coal combustion fly ash
    extracts with gas chromatograph-mass spectrometry (GC/MS) and
    MS/MS.

Hsie, A.W., D.B. Couch, J.P. O'Neill, J.R. San Sebastion, P.A. Brimer, R. Machanoff, J.C. Riddle, A.P. Li, J.C. Fuscoe, N. Forbes, and M.H. Hsie. 1979. Utilization of a quantitative mammalian cell mutation system CHO/HGPRT, in experimental mutagenesis and genetic toxicology. In: Strategies for Short-Term Testing for Mutagens/Carcinogens. B.E. Butterwork, ed. CRC Press: West Palm Beach. pp. 39-54.

Huisingh, J., R. Bradow, R. Jungers, L. Claxton, R. Zweidinger, S. Tejada, J. Bumgarner, F. Duffield, M. Waters, V.F. Simmon, C. Hare, C. Rodriguez, and L. Snow. 1978. Application of bioassay to the characterization of diesel particle emissions. In: Application of Short-Term Bioassays in the Fractionation and Analysis of Complex Environmental Mixtures. M. Waters, S. Nesnow, J. Huisingh, S. Sandhu, and L. Claxton, eds. EPA 600/9-78-027. U.S. Environmental Protection Agency, Research Triangle Park, NC. pp. 1-32.

Huisingh, J.L., R.L. Bradow, R.H. Jungers, B.D. Harris, R.B. Zweidinger, K.M. Cushing, B.E. Gill, and R.E. Albert. 1980. Mutagenic and carcinogenic potency of extracts of diesel and related environmental emissions: Study design, sample generation, collection and preparation. In: Health Effects of Diesel Engine Emissions: Proceedings of an International Symposium, Volume 2. W.E. Pepelko, R.M. Danner, and N.A. Clarke, eds. EPA-600/9-80-057b. pp. 788-800.

Li, A.P. 1980. Growth of mammalian cells as unattached cultures on nontissue culture plates. J. Tissue Culture Methods 6:71-73.

Li, A.P. 1981a. Antagonistic effects of animal sera, lung and liver cytosols, and sulfhydryl compounds on the cytotoxicity of diesel exhaust particle extracts. Toxicol. Appl. Pharmacol. 57:55-62.

Li, A.P. 1981b. Simplification of the CHO/HGPRT mutation assay through the growth of Chinese hamster ovary cells as unattached cultures. Mutation Res. 85:165-175.

Li, A.P. (in press). Quantitation of mutation at the hypoxanthine-guanine phosphoribosyl transferase and Na^+-K^+-ATPase gene loci in Chinese hamster ovary cells. J. Tissue Culture Methods.

Li, A.P., C.R. Clark, R.L. Hanson, T.R. Henderson, and C.H. Hobbs. (MS). Comparative mutagenicity of a coal-combustion fly ash extract in Salmonella typhimurium and Chinese hamster ovary cells.

Li, A.P. and R.E. Royer. 1982. Diesel exhaust particle extract enhancement of chemical-induced mutagenesis in cultured Chinese hamster ovary cells: possible interaction of diesel exhaust with environmental carcinogens. Mutation Res. 103:349-355.

Li, A.P. and R. Shimizu. (MS). Increasing the sensitivity of CHO/HGPRT mutation assay through increasing sampling size.

Ohnishi, Y., K. Kachi, K. Sato, I. Tohara, H. Takeyoshi, and H. Tokiwa. 1980. Detection of mutagenicity activity in automobile exhaust. Mutation Res. 77:229-246.

Pitts, J.N., D. Grosjean, T.M. Mischke, V.F. Simmon, and D. Poole. 1977. Mutagenic activity of airborne particulate organic pollutants. Toxicol. Lett. 1:65-70.

Rosenkranz, H.S., E.C. McCoy, D.R. Sanders, M. Butler, D.K. Kiriazides, and R. Mermelstein. 1980. Nitropyrenes: isolation, identification and reduction of mutagenic impurities in a carbon black in toners. Science 209:1039-1043.

Takehisa, S. and S. Wolff. 1977. Induction of sister chromatid exchanges in Chinese hamster cells by carcinogenic mutagens requiring metabolic activation. Mutation Res. 45:263-270.

BIOASSAYS OF OIL SHALE PROCESS WATERS IN PARAMECIUM AND SALMONELLA

Joan Smith-Sonneborn, Elizabeth A. McCann, and
Ronald A. Palizzi

Department of Zoology and Physiology, University of Wyoming
Laramie, WY 82071

INTRODUCTION

Oil shale deposits located in the western United States represent an abundant resource for recoverable synthetic crude oil (Yen and Chilingarin, 1976). Water coproduced with shale oil and decanted from it is called oil shale process water. This water originates primarily from combustion, dehydration of minerals, and groundwater (Farrier et al., 1977). For every 100,000 barrels of shale oil produced per day, a process water production of 1.6 million gallons is estimated (Hendrickson, 1975; Shih et al., 1979; Strniste and Chen, 1981). Technologies must be developed for the safe handling, containment, and disposal of such process waters.

Evaluation of cytotoxicity and genotoxicity of effluents produced in association with wastes from oil shale energy source technologies was carried out using a two-tiered approach: the prokaryotic standard histidine reversion Ames and eukaryotic Paramecium bioassays.

The established mutagenesis assay in the ciliated protozoan Paramecium tetraurelia (Kimball, 1950, 1963, 1965, 1969; Kimball and Gaither, 1951; Kimball and Perdue, 1962, 1967; Kimball et al., 1959) has been used as a prescreen for environmental hazards (Smith-Sonneborn, 1979; Smith-Sonneborn et al., 1981a,b). The detection of induced dominant or recessive lethals by potential

environmental hazards can be accomplished since cells of these
organisms undergo the self-fertilization process of autogamy
resulting in homozygosity. Thus, parent cells are exposed to a
test agent, the survivors of the treatment multiply and are
starved to induce autogamy, and the induced damage in the parental
genome is expressed in the next generation of progeny cells
(Sonneborn, 1970, 1974). An increased fraction of dead or slow
growers among the progeny of agent-treated parents versus controls
reflects the induction of lethal and detrimental mutations in the
parental "germ-line" micronucleus.

The effects of the oil shale process waters on the
prokaryotic Salmonella system are compared with those obtained
with the eukaryotic Paramecium assay. Since the chromosomal
organization of eukaryotes differs from that of prokaryotes, some
DNA damage peculiar to the organization of the chromosome would be
expected to be detected in eukaryote protozoans with chromosome
structure akin to mammalian cells.

The cumulative data indicate that qualitative and
quantitative differences exist in the sensitivities of the two
bioassays. In Paramecium, the S9 microsomal fraction for
activation of promutagens to their mutagenic form has been found
to be unpredictable. Therefore, at present, the Ames assay is
more reliable for detection of promutagens. The Paramecium
bioassay, on the other hand, appears to be more sensitive to
direct-acting mutagens and detects genotoxic activity in complex
environmental mixtures or extracts not significantly mutagenic in
the Ames assay (Smith-Sonneborn et al., 1981a,b).

In the present study, genotoxic and cytotoxic effects of
dilutions of oil shale process waters from three different
retorting processes have been evaluated using the eukaryotic
protozoan Paramecium tetraurelia and the prokaryotic bacterial
Salmonella assays. Process waters from above-ground, true in
situ, and modified in situ retorting processes coproduced with
shale from deposits in the western United States were analyzed.
Significant mutagenicity has been detected in waste water from
both the above-ground and true in situ retorting processes in both
the Salmonella and Paramecium bioassays. The Salmonella assay
required the addition of the rat-liver S9 fraction to provide
enzymatic conversion of promutagens to their active form; the
Paramecium bioassay was more sensitive to genotoxic effects from
the oil shale retort waters without the addition of the rat-liver
fraction. Mutagenicity of the dilutions tested from the modified
in situ retort process waters was detected only in the Paramecium
system. The above-ground retort process water was the most
cytotoxic sample.

MATERIALS AND METHODS

Oil Shale Process Waters

Samples were kindly supplied by the Laramie Energy Technology
Center and included process waters from: an above-ground surface
retort, 79-(Paraho-77/78)-COW-OO-C (PRHO); a simulated in situ
retort experiment, 79-(150T-SISR-R-17)-OOW-OO-U01 (SISR); a
vertical modified in situ retort condenser water,
79-(Oxy-6-Condensate)-OOW-OO-U01 (Oxy 6 C); a vertical modified in
situ retort water, 79-(Oxy-6)-02W-OO-C (Oxy 6 R); a true in situ
retort, 79-(Geokinetics-17)-OOW-OO-U (Gekn); and a true in situ
oil shale combustion experiment, 76-(Omega-9)-OOW-OO-F (Om 9).

The positive controls for the Ames and Paramecium bioassays
included benzo(a)pyrene (B[a]P) and 4-nitroquinoline-N-oxide
(Sigma Chemical Corp., St. Louis, MO).

Bacterial Strains and Ames Mutagenicity Testing

The Salmonella typhimurium tester strains TA98 and TA100 were
received from Bruce N. Ames (McCann et al., 1975; Ames et al.,
1975). The strains were stored, grown, and used for testing as
described in Ames et al. (1975). The S9 microsomal fraction was
received from Litton Bionetics (Kensington, MD) or was prepared
from livers of male rats that had received a single
intraperitoneal injection of Aroclor 1254 (PCB, Analabs, North
Haven, CT). The values presented represent the mean of a minimum
of two experiments with triplicate plates in each trial, ± the
standard error of the mean (SEM) unless otherwise indicated.

Paramecium Cultivation

P. tetraurelia, stock 51, temperature-insensitive, was used.
The culture medium inoculated 24 h before use with Klebsiella
aerogenes was adjusted to pH 6.7. The cells were carried in daily
isolation lines (Figure 1) to provide cells of known age. Since
death after autogamy increases with age (Rodermel and Smith-
Sonneborn, 1977), we presume repair decreases with age.
Therefore, clones 50 to 80 fissions old were used. Older clones
can show too high baseline mutagenesis, which can complicate
analyses.

Autogamous Cells -- Daily Isolations

Clone	A	B	C	D	E	F	G	H	- Sublines (A-H)
Day 1	0	0	0	0	0	0	0	0	
Day 2	0	0	0	0	0	0	0	0	
Day 3	0	0	0	0	0	0	0	0	
Day 4	0	0	0	0	0	0	0	0	
	↓	↓	↓	↓	↓	↓	↓	↓	
Day 40	0	0	0	0	0	0	0	0	0 = Depression

Figure 1.   Daily isolation procedure. The isolation line is
initiated when single isolates of fertilized cells are
placed in each depression on day 1. On day 2, the
number of cells from each single cell is counted and
one cell is transferred to the next depression. This
procedure is repeated daily and provides the daily
fission rate and the clonal age of the cell (the sum of
the number of fissions/day since day 1). Thus, cells
of known clonal age are available.

Cells can be aged while sister cells (in back depressions)
can be used as a source of cells in autogamy. Parent cells can be
maintained while progeny cells are obtained for determination of
characteristics. The parent cells are available for further
analysis.

This daily isolation procedure was used both to provide a
source of known aged cells for experiments and to maintain the
parent cells for the mutagenesis experiment described below.

Paramecium Toxicity Bioassay

Individual cells were isolated with a micropipette under a
dissecting microscope to separate depressions containing test
dilutions of the agent. The fraction of 16 cells which did not
survive 24 h of exposure was observed. The LC_{50} value was the
concentration at which 30 to 50% of the cells did not survive.

Paramecium Mutagenicity Assay

The main steps are the following: treatment of synchronized
cells; isolation of single treated cells and growth of each into a

clone; induction of autogamy in the treated clones; isolation of
16 cells in autogamy, i.e., 16 randomly-chosen haploid-chromosome
sets from each clone; and determination of how many of these
isolates die or grow poorly, i.e., contain lethal or poor growth
mutations made homozygous by autogamy. Using a micropipette under
a dissecting microscope, synchronized cells were obtained by
removing cells with the typical morphology of dividers from mass
cultures of cells of known clonal age. The dividers were
incubated at 27°C for 2 h to provide cells which were at a stage
refractory to DNA repair (Kimball and Perdue, 1962).

Synchronized cells, 2 h after the previous cell division,
were incubated for 2 h in the following: food only; dilutions of
the retort water in food; dilutions of the retort water plus
induced S9; positive control with S9; positive control without S9;
and food plus S9. Forty cells were placed in each treatment or
control group for 2 h. Then, the cells were washed by three
repeated transfers to agent-free medium. The cells were placed in
daily isolation lines as described above. Cells not transferred
(in back depressions) served as a source of starved parent cells
in autogamy, since starvation induces this self-fertilization
process.

When 90 to 100% of an observed sample was found to be in
autogamy (Smith-Sonneborn, 1974), 16 autogamous progeny were
dispensed, one cell per depression, into food. Normally
20 parents for each experimental group with 16 progeny each
(320 offspring) were used. Duplicates of each group were made.
Since P. tetraurelia has two diploid micronuclei, or four genes at
each locus, the chance of missing a particular mutation in any
autogamous cell is $3/4^{16}$.

The isolated cells were allowed to multiply, and autogamous
survival was scored after 3 days, as follows: viable, cells which
have cleared the medium of bacterial food; detrimentals or slow
growers, cells which have not cleared the medium in 3 days; and
lethals, not surviving or a few moribund cells.

RESULTS

Paramecium

PRHO was 5 to 15 times more toxic than the other waters
(Table 1).

Table 1. Cytotoxicity of Oil Shale Process Water in __Paramecium__

Process Water	LC_{50}[a]
PRHO	1:250
Oxy 6 C	1:50
Om 9	1:25
SISR	1:20
Oxy 6 R	1:20
Gekn	1:15

[a]LC_{50} is the dilution at which 30 to 50% of the treated cells did not survive 24 h of exposure to the retort water.

The genotoxicity of the oil shale retort water was assayed at the LC_{50} dilution, as the maximal concentration of the agent tested. Due to variation in clonal sensitivity to the waters, some parent cells did not survive these concentrations and the autogamous progeny bioassay could not be carried out.

The results of the combined triplicate experiments using six process waters are seen in Table 2. Om 9 showed significantly increased ($p < 0.05$) lethals and detrimentals in individual and combined experiments; PRHO and Oxy 6 C showed significantly increased ($p < 0.05$) progeny lethality and detrimentals in one of three trials and two of three trials, respectively, in individual experiments (Cochran, 1977; Mendenhall, 1975).

In __Paramecium__, the inclusion of induced S9 did not result in a significant increase in progeny lethality, although Gekn, 1:20 plus induced S9, produced marginally significant damage ($p < 0.10$) in one trial. Rather, the presence of induced S9 protected the parent cells from cytotoxicity and the induction of genotoxic effects with PRHO, Om 9, Oxy 6 C, and Oxy 6 R. The absence of induction of mutagenesis with S9, however, must be considered relative to the effect of the positive promutagen control, B(a)P. Previously, B(a)P plus the induced rat-liver S9 fraction produced significant mutagenesis in __Paramecium__, although some sporadic results were noted (Smith-Sonneborn et al., 1981a,b). More recently, the significant positive mutagenesis ($p < 0.05$) with B(a)P plus S9 was found in only 7 of 32 trials, although the B(a)P plus S9 value was higher than food-only controls in 26 of 32 trials. The probability that the positive promutagen value would be higher than controls in 81% of the trials, by chance

Table 2. Genotoxicity of Oil Shale Process Waters in **Paramecium**

Process Water[a]	% Lethals and Detrimentals					
	PRHO	Gekn	Om 9	Oxy 6 C	Oxy 6 R	SISR
Food control	2.2	1.8	1.7	4.5	2.5	2.9
Water + S9 (A)[b]	1.1	1.6	1.0	5.1	2.4	2.9
Water + S9 (B)[b]	1.3	2.6	3.4	4.4	3.6	1.5
Water only (A)	3.3	0.8	6.0[d]	5.5	2.7	1.3
Water only (B)	3.6[c]	2.2	7.6[d]	6.7[c]	3.8	2.0

[a]The process waters were tested in three replicate experiments with a minimum of 100 treated parent cells and 1600 progeny at each dilution used, (A) and (B).

[b]The dilutions tested, (A) and (B), respectively, for the process waters were the following: PRHO, 1:250 and 1:500; Gekn, 1:15 and 1:20; Om 9, 1:25 and 1:50; Oxy 6 C, 1:75 and 1:100; Oxy 6 R, 1:20 and 1:25; SISR, 1:30 and 1:50.

[c]Although this value is not significantly different (p > 0.05) from controls when the three replicate experiments are pooled, in one of three individual experiments (PRHO) and two of three individual experiments (Oxy 6 C), a significant increase (p < 0.05) in the percent lethals and detrimentals among the autogamous progeny of water-treated parent versus control cells was detected.

[d]These values represent a significant increase (p < 0.05) in the percent lethals and detrimentals in the progeny of parents treated with Om 9 water versus controls (Cochran, 1977; Mendenhall, 1975).

alone, is only 0.02%, if in fact the test promutagen did not differ significantly from controls. Thus, although the B(a)P plus S9 had a significant positive genotoxic effect when considered over all experiments, the lack of frequently reproducible significant results (p < 0.05) in individual experiments casts doubt on the validity of the bioassay for detection of genotoxic agents which require activation to their mutagenic form.

However, it should be noted that the values at given concentrations of retort waters were higher in the absence of S9 in all tests except Gekn. On the other hand, the positive

direct-acting mutagen used, 4-nitroquinoline-N-oxide, was
significantly mutagenic and was active when the process waters,
Oxy 6 R, SISR, and Gekn, did not induce significant direct
genotoxic damage in the treated parent cells. The mean
4-nitroquinoline-N-oxide value is 7.7% for 13 trials at 1.9 µg/ml.

Salmonella Ames Assay

The effect of oil shale process waters on reversion mutations
in TA98 and TA100 indicated that the above-ground retort water,
PRHO, in the presence of induced S9, resulted in a 9- to 10-fold
increase in TA98 revertants (Table 3). No other process water
induced significant increases in revertants with or without the
addition of induced S9. Since Gekn was the only other retort
water which produced a higher number of revertants than S9 only,
this water was examined for effects at higher concentrations.
When increased volumes of Gekn process water were used, increased
numbers of revertants were observed (Table 4).

Comparison of Paramecium and Ames Assays

When the two bioassays are used to evaluate the genotoxic
effects of the process waters, similarities and differences are
noted (Table 5).

DISCUSSION

The Ames and **Paramecium** bioassays indicate that certain oil
shale process waters contain genotoxic and cytotoxic constituents.
The PRHO above-ground retort water was 5 to 10 times more .toxic
than any other to **Paramecium**, and toxicity in decreasing order was
PRHO, Oxy 6 C, SISR, Om 9, Oxy 6 R, and Gekn. Similar results
with the same waters were obtained using daphnia, fathead minnows,
and rainbow trout (Bergman, 1980). The toxicity of these waters
is correlated with ammonia concentration: PRHO, 31.0 g/l;
Oxy 6 C, 9.4 g/l; Om 9, 3.5 g/l; Gekn 2.0 g/l; and Oxy 6 R,
1.0 g/l (Bergman, 1980). PRHO also contains higher concentrations
of arsenic and zinc, unique to this process water, and suggests
that these elements may be characteristic of the shale rock
composition.

Table 3. Mutagenicity of Oil Shale Process Waters with the Ames Salmonella Assay

Process Water (100 µl)	Mean No. of Revertants ± SEM[a]			
	With S9		Without S9	
	TA98[b]	TA100[c]	TA98	TA100
PRHO	384 ± 25	300 ± 53[d]	15 ± 1	99 ± 34
Gekn	49 ± 5	188 ± 14	12 ± 4	153 ± 17
Om 9	38 ± 4	182 ± 9	12 ± 1	166 ± 8
Oxy 6 C	26 ± 4	155 ± 21	25 ± 4	184 ± 10
Oxy 6 R	36 ± 8	224 ± 39	22 ± 3	154 ± 12
SISR	37 ± 5	177 ± 10	10 ± 4	164 ± 14

[a]The above data represent a minimum of two separate experiments with triplicate plates in each trial. The mean number of colonies ± SEM are given.

[b]The TA98 control with no treatment yielded 16 ± 4 colonies/plate. The positive control mutagen included was 4.5 µg/ml B(a)P. The mean values were B(a)P only, 16 ± 3; S9 only, 43 ± 2; B(a)P plus S9, 185 ± 10.

[c]The TA100 control with no treatment yielded 176 ± 6 colonies; the positive control mutagen included 4.5 µg/ml B(a)P. B(a)P only, 170 ± 4; S9 only, 181 ± 6; B(a)P plus S9, 857 ± 80.

[d]Marginal activity was seen with TA100.

Comparison of the known chemical composition of the process waters which induced genotoxic damage with the nongenotoxic Oxy 6 R process water (Fox et al., 1978; Bergman, 1980) did not reveal any qualitative differences in chemical composition; however, quantitative differences were seen.

The Gekn sample, which tests marginal mutagenicity (p < 0.10) in Paramecium and in Ames tests at 200 µl, showed a large number of peaks eluting after naphthalene in reverse-phase high pressure liquid chromatography (HPLC) chromatograms suggesting the presence of polynuclear aromatics not found with PRHO, Om 9, and Oxy 6 C. The combustion dynamics of the vertical modification retort seems to favor the presence of these organics (Bergman, 1980). Also of concern is whether the process water was a filtered sample;

Table 4.   Mutagenicity of Geokinetics Process Water in Ames Assay[a]

	Mean No. of Revertants ± SEM	
Volume (μl)	TA98[b]	TA100[c]
With S9		
300[de]	81 ± 6	252 ± 9
200[e]	80 ± 10	188 ± 14
100	49 ± 5	226 ± 11
50	48 ± 7	226 ± 2
25	37 ± 2	--
Without S9		
300[e]	1 ± 2	--
200[e]	12 ± 4	198 ± 4
100	21 ± 5	153 ± 17
50	12 ± 3	235 ± 12
25	11 ± 4	155 ± 7

[a]The data represent a minimum of three plates at each dilution tested and are expressed as the mean number of colonies/plate ± SEM.

[b]For TA98, the no-treatment controls yielded 9 ± 2 colonies/plate. The positive mutagen control was B(a)P. The mean numbers of colonies per plate were B(a)P only, 13 ± 2; S9 only, 27 ± 3; B(a)P plus S9, 124 ± 13.

[c]For TA100, the control was 130 ± 20, the positive mutagen control was B(a)P. The mean numbers of colonies per plate were B(a)P only, 124 ± 26; S9 only, 160 ± 18; B(a)P plus S9, 484 ± 85.

[d]At 200 and 300 μl, the Geokinetics process water had greater than twice the mean number of colonies seen in plates containing S9 only (27 ± 3) in this experiment and previous studies (Table 3).

[e]The process water may be cytotoxic at this concentration.

polynuclear aromatic hydrocarbons (PAHs) were removed when the retort waters were filtered (Hurtubise et al., 1978a,b). The Om 9 sample was filtered when this true in situ combustion water effluent was acquired and processed (Farrier et al., 1977) and was the only filtered water tested in this study.

Table 5. Mutagenicity of Oil Shale Process Waters in __Paramecium__
          and Ames Assays

| | Bioassay | |
Process Water	__Paramecium__	Ames
PRHO	-,+	+
Gekn	-	-,+
Oxy 6 C	-,+	-
Om 9	+	-
Oxy 6 R	-	-
SISR	-	-

-,+ In the __Paramecium__ bioassay, the sample induced significant
    mutagenicity (p < 0.05) in one of three trials. In the Ames
    bioassay, the sample induced < 2X reproducible increase in
    revertants in the treated versus S9 control value.
 + Significant (p < 0.05) in triplicate trials.
 - Not significantly different from controls in any trial.

        Detection of promutagens like PAHs in the oil shale process
waters then can be dependent on whether the sample has been
filtered and the bioassay used. The fact that the source of
variation in the __Paramecium__ bioassay, when the agent requires
metabolic activation from the promutagen to mutagenic form is not
yet known, restricts the utility of this bioassay to direct-acting
mutagens.

        The mutagenicity of process waters from above-ground, true in
situ, and modified vertical technologies has been tested in
Chinese hamster ovary cells (CHO) and the Ames assay and these
samples were found to be mutagenic with light activation in the
CHO and metabolic activation in the Ames assay (Strniste and Chen,
1981). Since the quality and quantity of contaminants in the
process waters can vary due to sampling and processing variations
and to storage procedures, the results of our bioassays and those
of Strniste and Chen cannot be directly compared. However, three
bioassays, CHO mammalian cells, __Paramecium__, and Ames, indicate
genotoxicity of oil shale process waters.

Therefore, continued study of the nature of the contaminants and caution in handling, containment, and storage of the oil shale process waters are indicated.

ACKNOWLEDGMENTS

This study was supported by Department of Energy Contract no. DE AC02 77EV04477.

REFERENCES

Ames, B.N., J. McCann, and E. Yamasaki. 1975. Methods for detecting carcinogens and mutagens with Salmonella/mammalian microsome mutagenicity test. Mutation Res. 31:347-364.

Bergman, H.L. 1980. Effects of Aqueous Effluents from In Situ Fossil Fuel Processing Technologies on Aquatic Systems. DOE/LETC/10058-T1. Department of Energy: Washington, DC.

Cochran, E.G. 1977. Sampling Techniques. Wiley and Sons: New York.

Farrier, D.S., R.E. Poulson, Q.D. Skinner, J.C. Adams, and J.P. Bower. 1977. Acquisition, processing and storage for environmental research of aqueous effluents from in situ oil shale processing. Proceedings of the 2nd Pacific Chemical Engineering Congress, Denver, CO, Volume II. p. 1031.

Fox, J.P., D.S. Farrier, and R.E. Poulson. 1978. Chemical Characterization and Analytical Considerations for an In Situ Oil Shale Process Water. LETC/RI-78/7. Department of Energy: Washington, DC.

Hendrickson, T.A., ed. 1975. Synthetic Fuels Data Handbook. Cameron Engineers, Inc.: Denver. pp. 3-112.

Hurtubise, R.J., G.T. Skar, and R.E. Poulson. 1978. Determination of benzo(a)pyrene in shale oil by solid surface fluorescence. Anal. Chimica Acta 97:13-19.

Hurtubise, R.J., J.D. Phillip, and G.T. Skar. 1978. Determination of benzo(a)pyrene in a filtered retort water sample of solid-surface fluorescence. Anal. Chimica Acta 101:333-338.

Kimball, R.F. 1950. The effect of radiation on genetic mechanisms of Paramecium aurelia. J. Cell Comp. Physiol. 35(suppl.):157-169.

Kimball, R.F. 1963. X-ray dose rate and dose fractionation studies on mutations in Paramecium. Genetics 48:581-595.

Kimball, R.F. 1965. The induction of repairable premutational damage in Paramecium aurelia by the alkylating agent triethylene melamine. Mutation Res. 2:414-425.

Kimball, R.F. 1969. Studies on mutations induced by ultraviolet irradiation in Paramecium aurelia with special emphasis on photoreversal. Mutation Res. 8:79-89.

Kimball, R.F. and N. Gaither. 1951. The influence of light upon the action of UV on Paramecium aurelia. J. Cell Comp. Physiol. 37:211-233.

Kimball, R.F. and S.W. Perdue. 1962. Studies on the refractory period for the induction of recessive lethal mutations by X rays in Paramecium. Genetics 49:1595-1607.

Kimball, R.F. and S.W. Perdue. 1967. Comparison of mutagenicity of X rays and triethylene melamine in Paramecium with emphasis on the role of mitosis. Mutation Res. 4:37-50.

Kimball, R.F., N. Gaither, and S. Wilson. 1959. Reduction of mutations by postirradiation treatment after ultraviolet and various kinds of ionizing radiation. Radiat. Res. 10:490-497.

McCann, J., N.D. Spingarn, J. Kobari, and B.N. Ames. 1975. Detection of carcinogens: bacterial strains with R factor plasmids. Proc. Natl. Acad. Sci. USA 72:979-983.

Mendenhall, W. 1975. Introduction to Probability and Statistics, 4th ed. Duxbury Press: Belmont. p. 186.

Rodermel, S. and J. Smith-Sonneborn. 1977. Age-correlated changes in expression of micronuclear damage and repair in Paramecium tetraurelia. Genetics 87:259-274.

Shih, C.C., J.E. Cotter, C.H. Prien, and T.D. Nevens. 1979. Technological Overview Reports for Eight Shale Oil Recovery Processes. EPA-600/7-79-075. U.S. Environmental Protection Agency: Cincinnati, OH.

Smith-Sonneborn, J. 1974. Acridine orange fluorescence: a temporary stain for paramecia. Stain Technol. 49:77-80.

Smith-Sonneborn, J. 1979. Use of a ciliated protozoan as a t system to detect toxic and carcinogenic agents. In: In-Vitro Toxicity Testing of Environmental Agents: Current Future Possibilities. Monte Carlo.

Smith-Sonneborn, J., R.A. Palizzi, C. Herr, and G.L. Fisher. 1981a. Mutagenicity of fly ash particles in Paramecium. Science 211:180-182.

Smith-Sonneborn, J., G.L. Fisher, R.A. Palizzi, and C. Herr. 1981b. Mutagenicity of coal fly ash: a new bioassay for mutagenic potential in a particle feeding ciliate. Envir Mutagen. 3:239-252.

Sonneborn, T.M. 1970. Methods in Paramecium research. In: Methods in Cell Physiology. D.M. Prescott, ed. Academic Press: New York. pp. 241-541.

Sonneborn, T.M. 1974. Paramecium aurelia. In: Handbook of Genetics. R.C. King, ed. Plenum Press: New York. pp. 469-595.

Strniste, G.F. and D.J. Chen. 1981. Cytotoxic and mutagenic properties of shale oil byproducts, I: activation of ret process waters with near ultraviolet light. Environ. Mutagen. 3:221-231.

Yen, T.F. and G.V. Chilingarin. 1976. Introduction to oil shales. In: Oil Shale. T.F. Yen and G.V. Chilingarin, Elsevier Scientific Publishing Co.: Amsterdam. pp. 1-2.

A SEARCH FOR THE IDENTITY OF GENOTOXIC AGENTS IN THE AMBIENT AIR

USING THE TRADESCANTIA BIOASSAY

Lloyd A. Schairer, Richard C. Sautkulis, and Neal R. Tempel

Biology Department, Brookhaven National Laboratory, Upton
New York 11973

INTRODUCTION

The theme of the conference on "Application of Short-Term Bioassays in the Analysis of Complex Environmental Mixtures" concerned the state of the art of bioassay systems from cultured microbes to laboratory animals and their application to the assessment of human health effects of airborne environmental contaminants. The major emphasis for short-term bioassays has been placed on bacterial and mammalian cell lines. However, for increased perspective on the state of the art of specific in vitro assays, it is important to consider the environmental impact on whole organisms by reviewing the contributions made by in vivo assays. The more classical nonmammallian in vivo systems such as Drosophila, Zea mays, and Tradescantia are characterized by well-defined genetic bases, versatility in mode of treatment, relatively low cost, short term, and/or high sensitivity to both physical (radiation) and chemical mutagens (Underbrink et al., 1973; Vogel and Sobels, 1976; Nix and Brewen, 1978; Plewa, 1978; Schairer et al., 1978a). This paper will deal exclusively with somatic mutation in the Tradescantia stamen hair, describing the system briefly, demonstrating its relevance to environmental mutagen assessment, and discussing its adaptation for in situ ambient atmosphere monitoring.

MATERIALS AND METHODS

The Tradescantia Stamen Hair System

     The stamen hair system has been described in detail elsewhere
(Underbrink et al., 1973; Sparrow et al., 1974; Schairer et al.,
1978a; Schairer et al., 1978b; Thompson and Nawrocky, 1980), so
only certain features will be reviewed here. The plant used
exclusively in the field studies described here is clone 4430, a
diploid interspecific hybrid (T. subacaulis x T. hirsutiflora)
produced at Brookhaven. This clone is a hybrid between pink- and
blue-flowering parents with blue being dominant over pink. The
visible marker used in this test system is the phenotypic change
in pigmentation from blue to pink in mature flowers. The
pigmentation change (hereafter called mutational or pink event) is
induced in young developing floral tissue and is expressed 5 to
18 days later as isolated pink cells or groups of pink cells in
the stamen hairs of mature flowers. The pink events are
essentially nonlethal; large mutant sectors indicate genetic
alteration early in the development of that tissue.

     The stock plants are easily maintained by vegetative
propagation and flower continuously throughout the year in
controlled-environment growth chambers. The plant material
treated consists of unrooted, fresh cuttings, each with a young
inflorescence containing flower buds in a range of developmental
stages. Following exposure to either chemical or physical
mutagens, the cuttings are grown in aerated Hoagland's nutrient
solution under standard conditions (Underbrink et al., 1973), and
the flowers are analyzed each day as they bloom for approximately
two weeks after treatment. Induced pink-event rates are expressed
as the mean of the rates for several consecutive peak response
days, usually days 11 to 15 for acute X rays and 7 to 12 for acute
chemical exposures. Detailed descriptions of laboratory
techniques for radiation and chemical exposures and methods for
calculating mutation rates are given elsewhere (Underbrink et al.,
1973; Sparrow et al., 1974; Schairer et al., 1978b).

Chemical Exposures under Laboratory Conditions

     The techniques developed for vapor-phase chemical exposures
in the laboratory have been described in detail elsewhere (Sparrow
et al., 1974; Nauman et al., 1979), so comments here will be
limited to a few basic observations.

     Young inflorescences of Tradescantia clone 4430 were exposed
to gaseous 1,2-dibromoethane (DBE), an alkylating agent used as a

standard chemical mutagen in our studies. The mutation frequency
increased linearly with both increasing chemical concentration
(0.1 to 100 ppm) and duration of exposure (2 to 144 h). These
data may be expressed in terms of total dose by plotting induced
mutation frequency against the product of concentration (ppm) and
duration of exposure (hours) (Figure 1). For purposes of
comparison, a standard curve for x-ray effect is shown in rads.
Slope and shape of the curve for DBE induction of color change
resemble those for radiation injury.

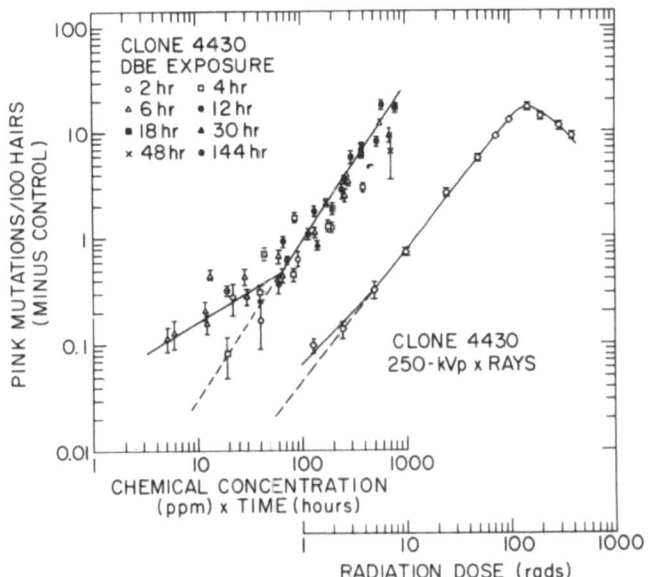

Figure 1.   DBE-induced mutation response plotted against total
            dose (ppm X hours of exposure) results in a linear
            response curve. The standard x-ray curve is shown for
            comparison. Broken lines represent extrapolation of
            high dose-response curves.

Of particular interest are the shapes of the dose-response curves which show a significant decrease in slope in the low dose range. Differences in mechanisms responsible for mutation at high and low doses are unknown for chemicals, but, as with radiation results, the response is greater at low doses (Figure 1). Extrapolation from high to low levels of mutagen exposures would lead to an underestimate of the hazard (Nauman et al., 1977; Nauman and Sparrow, 1978).

Chemical Exposures under Field Conditions (Ambient Air)

Laboratory studies with chemicals indicated that the stamen hair system was highly sensitive to gaseous chemicals (Sparrow et al., 1974; Nauman et al., 1976; Nauman et al., 1979) and that the system should respond to in situ exposures to industrial pollution at ambient levels (Schairer et al., 1978a). The criteria for monitoring air pollution for mutagenicity in the field include a roadworthy vehicle to house test organisms during exposure, exposure of the test organisms under suitable culture conditions, a constant flow of untempered ambient air, and a chronic exposure capability simulating natural exposures of plants and animals. The mobile monitoring vehicle (MMV) shown in Figure 2 has three controlled environment growth chambers on board: one of the chambers serves as a clean-air control, the second is used for ambient-air exposures, and the third is a backup for either function. Ambient air is drawn into the exposure chamber through a 4-in glass duct at a continuous flow rate of about 18 ft^3/min, a maximum of 1 air change every 2 min. An air filter train is used to scrub the air continually in the chamber serving as the concurrent control.

One of the difficulties in monitoring ambient air at a fixed site is that wind direction varies and, as a result, no point source of pollution can be analyzed on a predictable schedule. In the interest of complete characterization of the ambient air (both as to mutagenicity and chemical composition), point source definition, and minimizing the number of Tenax tubes required for continuous monitoring, the exposure chamber on the MMV was modified in the following manner: the wind monitoring instruments were connected to a system of electrical relays such that the ambient air intake blower is activated only when the wind is in the desired direction. The same relay activates the Tenax tube sampling pump so that simultaneous ambient air fumigation and chemical sampling are accomplished. A solenoid valve seals the Tenax tube whenever it is not activated for monitoring. Hence, both mutagenicity and chemical concentration can be correlated with winds from any desired direction.

Figure 2.   View of the front of the Mobile Monitoring Vehicle
            (MMV) which shows remote mounting of the air
            conditioning and filter train. Battelle massive air
            sampler is shown on trailer to the right of the MMV;
            Hi-vol sampler for measuring total suspended
            particulates is on a stand to the left of the MMV.

     The primary function of the research group at Brookhaven
National Laboratory is to perform the mutagenicity testing of the
ambient air using the Tradescantia stamen hair system. However,
an integral part of that study is the identification and
quantification of the airborne chemical pollutants present during
exposure. EPA-designated instrumentation on board the MMV is used
to continuously monitor the levels of SO_2, NO_2, NO, NO_x, O_3, CO,
and THC. Organic vapors are collected on Tenax-GC cartridges and
subsequently identified and quantified by gas chromatography/mass
spectrophotometry at Research Triangle Institute, Research
Triangle Park, NC. Total suspended particulates are measured with

a Hi-vol sampler (on stand behind MMV in Figure 2), while Battelle massive air samplers (Figure 2, right) were used to collect particulates in 3.5 to 20, 1.7 to 3.5, and < 1.7 micron fractions for subsequent bioassessment (Hughes et al., 1980).

Field exposures of the plant samples are accomplished in the following manner: fresh cuttings of Tradescantia clone 4430 are made from stock plants grown in controlled environment chambers at Brookhaven National Laboratory; they are hand carried to the test site by car or airplane; cuttings are placed in the chambers in glass containers filled with Hoagland's nutrient solution; and the cuttings are exposed continuously for a 10-day period. At the end of the exposure, the cuttings are taken back to Brookhaven National Laboratory for posttreatment analysis of the flowers as they bloom each day. Exposures of a few hours to several weeks could be made but a 10-day period was chosen for the Tradescantia plants because it was long enough to maximize the sensitivity of the system and to simulate chronic exposures in the workplace, yet short enough to permit analysis of a sufficient number of flowers over the peak response period back at Brookhaven. The peak mutation response period following a 10-day exposure is 11 to 17 days after the start of the exposure. The mean of the mutation rates for a 5- to 7-day scoring period results in an observed rate for a given test site based on an average stamen hair population between 200,000 and 300,000. A population of 300 cuttings in each ambient-air and control chamber will yield enough data to resolve as small as a 10% increase in pink events over the background frequency.

RESULTS

Results of In Situ Exposures to Ambient Air

The first field trials for the MMV were conducted in the summer of 1976 in Elizabeth, NJ. Over the next four years a total of 17 additional sites throughout the United States were monitored in a preliminary study (Houston and El Paso each had two different sites.). The results are summarized in Figure 3. This study had two objectives: to demonstrate the adaptability of the stamen hair system to ambient air monitoring, and, if mutagenicity were observed, to look for causative agents common to positive sites. These sites were selected because of known high levels of human cancer incidence or the presence of high levels of suspect compounds in the atmosphere. Two exceptions were the sites at Grand Canyon, AZ, and Pittsboro, NC. It was deemed essential to conduct a "clean ambient air" exposure to verify the efficiency of the filter on the concurrent control chamber and to eliminate the

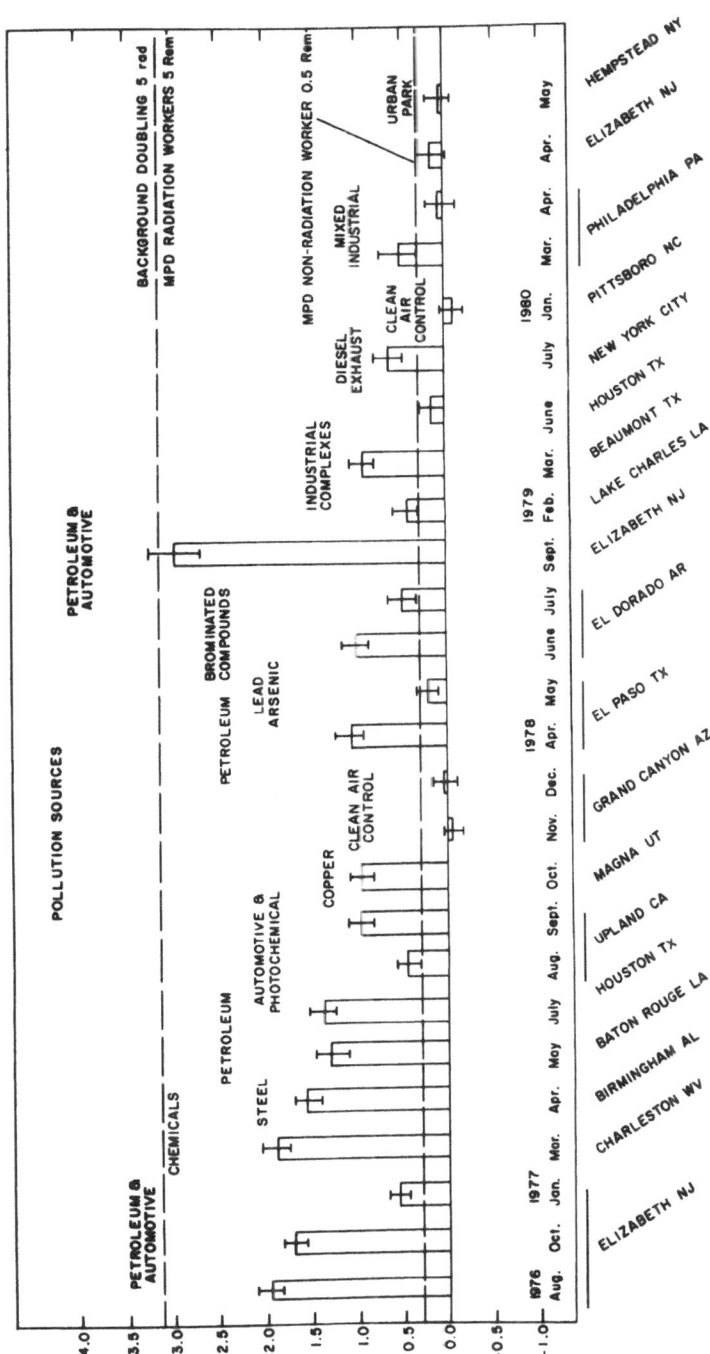

Figure 3. The mutagenicity of ambient air as measured by Tradescantia in the MMV is summarized for eighteen test sites visited. Horizontal broken lines represent mutation response of stamen hairs to established maximum permissible dosages (MPD) of radiation for comparison with observed chemical mutagen responses.

possibility of an artifact generated in the ambient air chamber.
Two exposures at Grand Canyon produced similar results with no
significant difference in background mutation frequency between
the control and ambient air samples. Since the Grand Canyon test
represented a good baseline exposure to clean air in situ and
included all of the stress of shipping plant material, field
handling, etc., the weighted mean for four replicated ambient air
samples ($3.35 \pm 0.09/10^3$ hairs) was determined and this mutation
frequency was established as the standard for comparison for all
other field sites monitored. The results of mutagenicity
monitoring at all sites are shown in Figure 3 as the net induced
mutation frequency following subtraction of the Grand Canyon rate.
Clearly significant increases in pink-celled mutation frequency
were observed at many industrial sites. The most consistent
mutagenic response was that associated with petroleum refining.
Mutation rate varied not only from site to site but also with
repeated exposures at the same site. The induced mutation
frequency ranged from 16.6% above control at Elizabeth, NJ, in
January to nearly a background doubling of 90.6% in September.
Some of this variation in response was undoubtedly due to
seasonable change in effluent production and varying wind
direction; prevailing winds in the summer and fall are SW, while
winter winds are NW. The fixed location of the Elizabeth site
placed petroleum refining operations directly upwind for normal
prevailing summer winds. It is important to note that atmospheric
monitoring at fixed sites is very dependent upon wind direction
and speed. Although trailer locations were selected downwind from
the desired source of pollution, a false negative may result from
unpredicted wind changes, extended periods of rain, or other
environmental factors.

All of the field sites monitored by Tradescantia plants
showed generally similar spectra of organic compounds by the Tenax
sampling procedure. Based on data from these preliminary surveys,
no compound or class of compounds was associated with sites
demonstrating biologically active atmospheres. The only
conclusion to be drawn was that all sites downwind of petroleum
processing industries showed some degree of mutagenicity in the
test plants. No specific compounds could be identified as the
responsible agent in this preliminary survey.

Quantitative analysis of organic vapors were made for
20 selected compounds from samples from each site. The
distribution of these compounds is shown in Figure 4. In this
figure the test site results are plotted in order of decreasing
somatic mutation frequency from left to right and the observed
chemical concentration for each site is represented by the solid
bars numerically from left to right. Hence, any of the

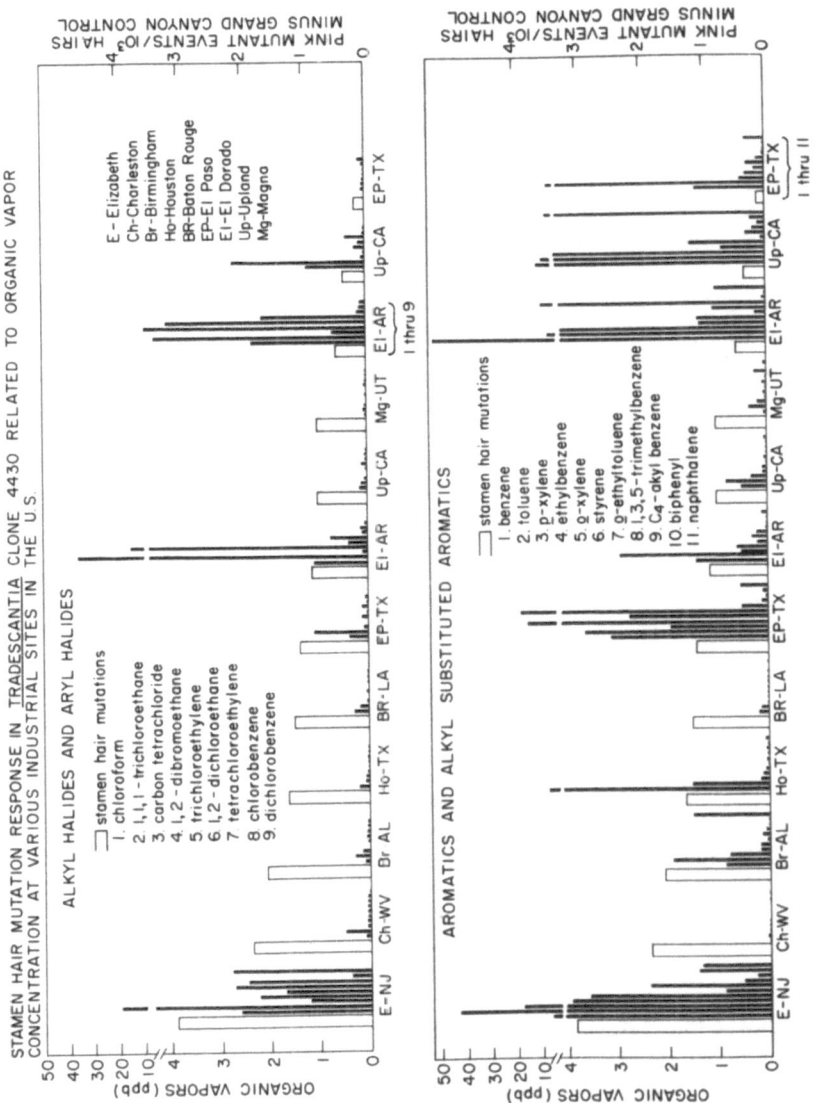

Figure 4. Stamen hair mutation response in *Tradescantia* clone 4430 related to organic vapor concentration at various industrial sites in the U.S. Solid bars represent chemicals listed numerically from left to right.

20 compounds quantified associated with mutation frequency would be expected to decrease as plotted from left to right. Clearly, Elizabeth had the highest levels of most compounds, but Charleston, Birmingham, Houston, and Baton Rouge ranked next in mutation frequency without observed high levels of organic vapors. The El Paso (petroleum site) and El Dorado atmospheres were intermediate to low in mutation response but had generally high levels of most compounds. The brominated compounds at El Dorado stressed the plants enough to reduce flowering and perhaps to reduce apparent mutation frequency because of the toxic effect. Benzene and toluene were essentially ubiquitous and, except for possible synergistic reactions, are not likely to be the primary suspect compounds. These data suggested that a more detailed study would be required to identify specific mutagenic compounds in the ambient atmosphere.

As a first approach to a more comprehensive study, Elizabeth, NJ, was selected because of known high levels of pollution, demonstrable mutation response in Tradescantia, and proximity to Brookhaven facilities. Several exposures of Tradescantia plants had been made during different seasons of the year and generally high mutation responses were obtained when winds were from the SW or perhaps some combination of SW and N. Figure 5 summarizes data for exposures at Elizabeth. Each shaded sector represents the cumulative number of hours of winds from the directions indicated. It is evident that, although the summer winds are prevailing SW, a great deal of variability was experienced. The 25% increase above control for the September 1980 exposure resulted from exposure to only the SW winds (32 h). The success of this SW exposure led to a series of experiments in which the results of exposure to SW winds were compared with the response of exposure to all other winds. These results are summarized in Figure 6. The mutagenic response for the October, November, June, and August Elizabeth exposures were essentially negative; the prevailing SW winds did not materialize, but instead clear NW Canadian winds persisted. In September 1981, the MMV was moved to a site in Staten Island, NY, where NW winds would carry petroleum refining effluent to the test site. The results of September and October exposures (Figure 6, last two bars) were negative; the winds again were not optimal since they were primarily SW and W.

Results of Individual Compound Testing Under Laboratory Conditions

A comparison of quantitative analyses for nine compounds from both positive and negative exposures at Elizabeth is made in Figure 7. Only one chemical, 1,1,1-trichloroethane, was among those which were mutagenic at low concentrations (see Table 1).

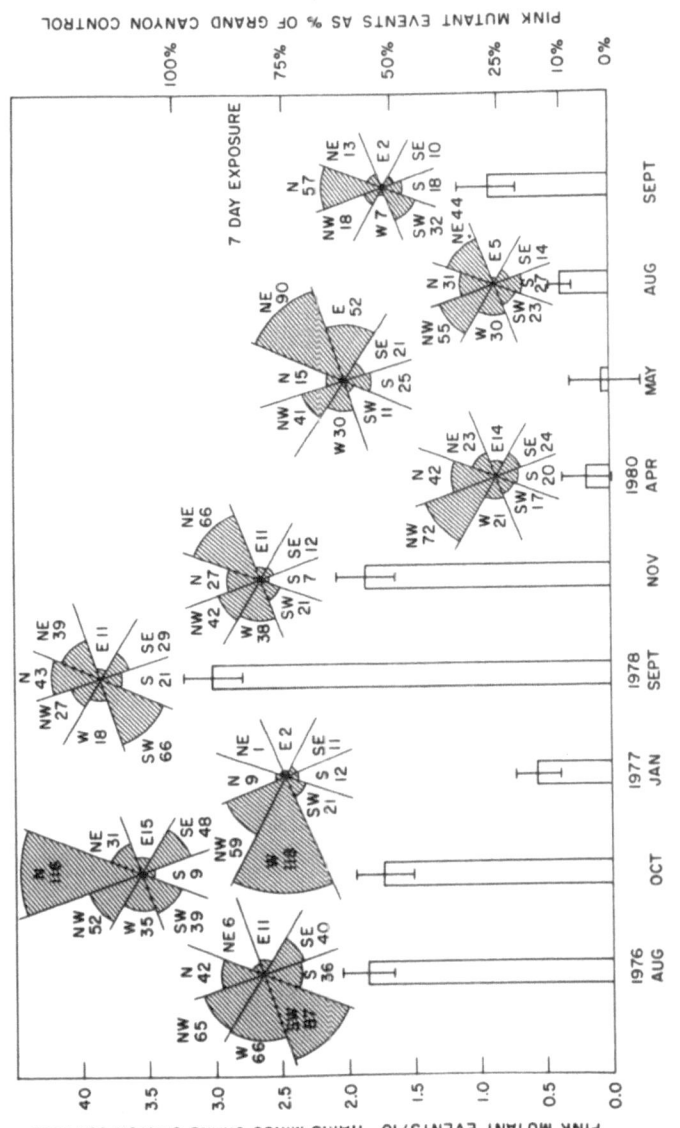

Figure 5. Mutagenicity of ambient air at Elizabeth, NJ, as measured by *Tradescantia* in the MMV (10-day exposures). Shaded sectors show cumulative number of hours in which winds were in the designated directions in each exposure.

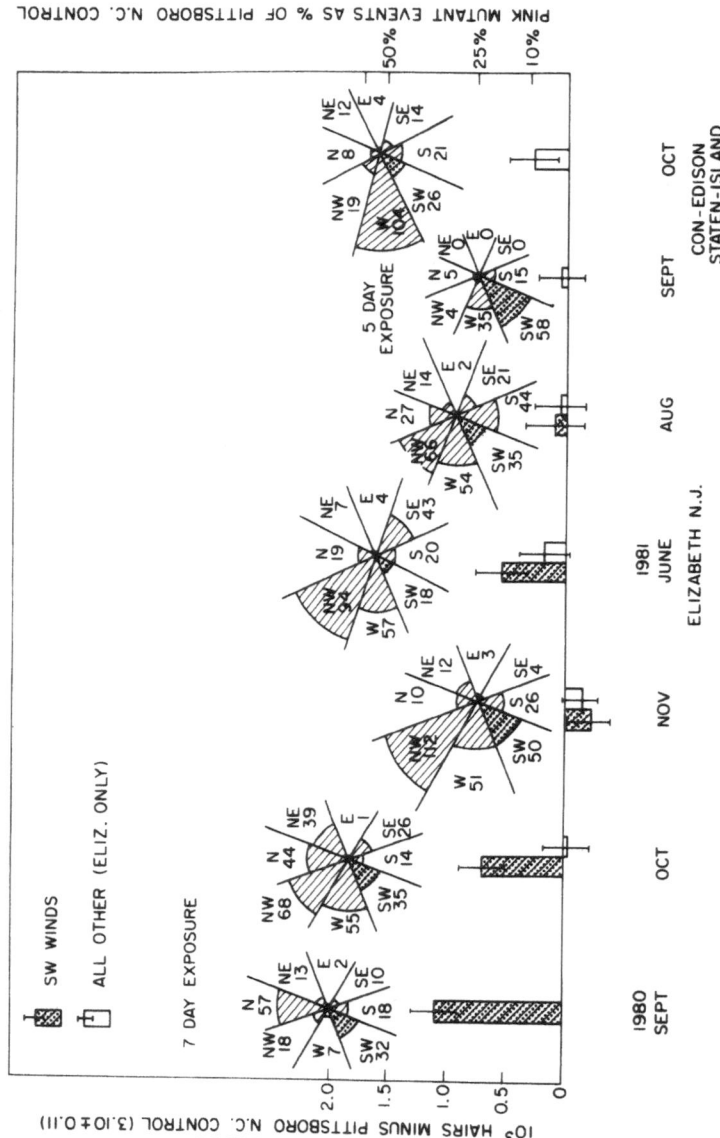

Figure 6. Mutagenicity of ambient air at Elizabeth, NJ, and at the Con Edison plant on Staten Island, NY. Shaded sectors show cumulative number of hours in which winds were in the designated directions during exposure.

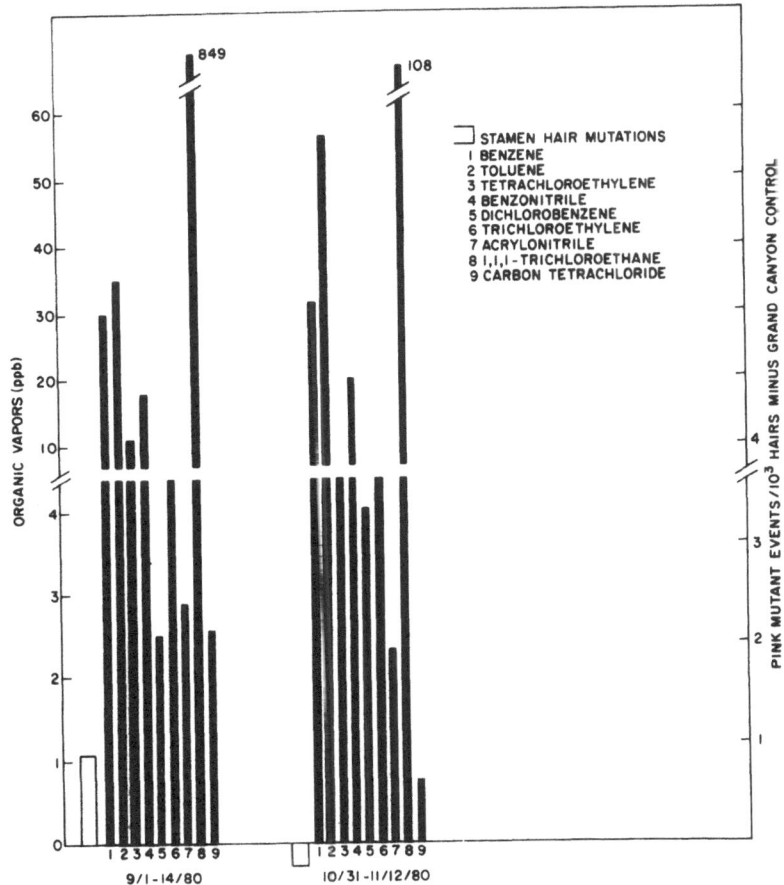

Figure 7.   Stamen hair mutation response in <u>Tradescantia</u>
clone 4430 related to organic vapor concentration at
Elizabeth, NJ.

        The reliability of the recovery potential of Tenax was tested
with the following procedure. Using a gas chromatograph
calibrated against standards to verify concentrations of
1,1,1-trichloroethane and trichloroethylene, Tenax tubes were
exposed to 13.8 1 of vapors of these compounds at concentrations
ranging from about 0.3 to 3 ppm. At 21°C this 13.8 1 of gas
should have been less than the calculated breakthrough volume for
these compounds. The results of these recovery tests are shown in

Table 1.   Mutagenicity of Compounds Identified in Ambient Air in
           Elizabeth, NJ - Tradescantia Stamen Hair Assay

Compound	Lowest Conc. Used Giving + Effect		Mutagenic Response % Control
	M (liquid)	ppm (gas)	
Benzonitrile	10^{-5}		178
Acrylonitrile	10^{-5}		163
1,1,1-Trichloroethane		0.2	32
Trichloroethylene		1.0	29
Methylene chloride		127	32
Dichlorobenzene		120	26
Toluene		1725	34
Benzene		3850	74
Tetrachloroethylene		0.5-1035	0
Carbon tetrachloride		1-1000	0

Table 2.   The recovery per tube ranged from about 1/2 to 1/10 of
the calculated exposure. This discrepancy was further complicated
by an apparent increasing loss of chemical as the exposure
increased. These results indicate that breakthrough of compounds
in this class occurs sooner than anticipated.

DISCUSSION

The most unique application of this plant system is its
ability to respond to low levels of airborne compounds. Without
the constraints of sterile culture and complicated experimental
procedures required to collect, concentrate, and subsequently
elute chemical pollutants, the Tradescantia plant is very
adaptable to in situ monitoring at suspect industrial sites.
Collection of data on the mutagenicity of ambient air pollution is
an area of testing not amenable to mammalian or in vitro systems
and hence an area in which Tradescantia can make a unique
contribution.

The high sensitivity and versatility of the Tradescantia
stamen hair system were put to good use by employing the plant as

Table 2. Reproducibility of Tenax Recovery for Known Exposure
Levels of Two Chemicals

Level of Exposure		Recovery	
(ppm)	(ng/tube)	(ng/tube)	(%)
1,1,1-Trichloroethane			
0.3	22,500	10,000	44.4
0.75	56,600	18,000	31.8
1.1	82,700	24,000	29.0
2.8	210,000	21,000	10.0
2.8	210,000	40,000	19.0
Trichloroethylene			
0.2	14,800	7,900	53.4
0.7	51,800	8,900	17.2
0.75	55,700	11,000	19.7
1.6	118,000	16,000	13.6
1.6	118,000	16,000	13.6

an in situ monitor for mutagens in ambient air in polluted
industrial sites. Preliminary results from many sites showed a
significant increase in mutation rate with one (Elizabeth, NJ)
approaching a background doubling. The environment most
consistently mutagenic was that downwind from petroleum
refineries. No specific compounds or groups of compounds have as
yet been correlated with the positive sites, although several have
been shown to be mutagens. Studies involving the fractionation
and analysis of complex mixtures are continuing with more emphasis
on possible synergistic reactions between organic and/or inorganic
compounds. The data presented above demonstrate the need for a
larger "window" for identification and quantification of airborne
organic compounds than is presently available with Tenax.

The relevance of the mutation increase attributed to exposure
to ambient air pollution can be demonstrated by comparison with
the radiation levels required to produce similar effects. Over
25 years of intensive radiobiological studies have generated
health hazard exposure limits for industrial radiation workers of

5 rem per year and for nonradiation workers 0.5 rem per year. Figure 3 shows the pollution-induced mutation increment for the various sites monitored, and the broken horizontal lines indicate the mutation responses in the stamen hair system for 5 and 0.5 rem, respectively. It is clear that one urban site (Elizabeth, NJ) showed a pollution response equal to that for the maximum permissible dose for a radiation worker and most sites had a mutation response exceeding that permissible for nonradiation workers. The inference to be drawn from these data is that mutation frequency induced by environmental contamination from airborne chemicals alone is already of a magnitude greater than that postulated as excessive for radiation exposure. Based on the responses of the Tradescantia stamen hair bioassay to in situ exposures to ambient air, many industrial sites warrant extensive investigation with the more expensive, long-term bioassays to determine the degree of human health hazard present.

Greater use of basic radiobiological data should be encouraged. There is a very large background of radiation data describing dose-response curve patterns (Nauman and Sparrow, 1978), dose-rate effects (Nauman et al., 1975), and nuclear factors influencing sensitivity (Sparrow et al., 1968). Dose-response curve patterns have been shown to be similar in Tradescantia for both physical and chemical agents. One particularly important example is the experimental observation that dose-response curves for both radiation and chemical mutagens show a reduced slope at very low doses. The consequence of this observation is that extrapolation of high dose curves to low dose effects would lead to an underestimate of the potential mutagenic response and health hazard (Nauman et al., 1977; Nauman and Sparrow, 1978). Other phenomena established by radiation studies should be explored for chemicals. Predictive potential for effects of environmental mutagens is of equal importance to earlier studies related to nuclear fallout and its impact on ecology in general but food chain in particular.

ACKNOWLEDGMENTS

This work was supported jointly by the U.S. Department of Energy, National Institute of Environmental Health Sciences, and U.S. Environmental Protection Agency. The authors acknowledge with thanks the special efforts of Dr. E. Pellizzari and staff of the Research Triangle Institute for organic vapor analysis. The many hours of flower analysis by Mr. E.E. Klug, Ms. A.F. Nauman, Ms. M.M. Nawrocky, Ms. V. Pond, Ms. R.C. Sparrow, and Dr. M. Emmerling-Thompson are also gratefully acknowledged.

REFERENCES

Hughes, T.J., E. Pellizzari, L. Little, C. Sparacino, and A. Kolber. 1980. Ambient air pollution: collection, chemical characterization and mutagenicity testing. Mutation Res. 76:51-83.

Nauman, C.H., P.J. Klotz, and L.A. Schairer. 1979. Uptake of tritiated 1,2-dibromoethane by *Tradescantia* floral tissues: relation to induced mutation frequency in stamen hair cells. Environ. Exp. Bot. 19:201-215.

Nauman, C.H., and A.H. Sparrow. 1978. Problems of extrapolation from high dose to low dose in *Tradescantia* mutation studies. Environ. Health Perspect. 22:161-162.

Nauman, C.H., A.H. Sparrow, and L.A. Schairer. 1976. Comparative effects of ionizing radiation and two gaseous chemical mutagens on somatic mutation induction in one mutable and two non-mutable clones of *Tradescantia*. Mutation Res. 38:53-70.

Nauman, C.H., A.H. Sparrow, A.G. Underbrink, and L.A. Schairer. 1977. Low-dose mutation response relationships in *Tradescantia*: principles and comparison to mutagenesis following low-dose gaseous chemical mutagen exposure. In: Radiobiological Protection, First European Symposium on Rad-Equivalence. Commission of the European Community: Luxemburg. pp. 13-23.

Nauman, C.H., A.G. Underbrink, and A.H. Sparrow. 1975. Influence of radiation dose rate on somatic mutation induction in *Tradescantia* stamen hairs. Radiat. Res. 62:79-96.

Nix, C.E., and B. Brewen. 1978. The role of *Drosophila* in chemical mutagenesis testing. In: Application of Short-Term Bioassays in the Fractionation and Analysis of Complex Environmental Mixtures. M.D. Waters, S. Nesnow, J.L. Huisingh, S.S. Sandhu, and L. Claxton, eds. Plenum Press: New York.

Plewa, M.J. 1978. Activation of chemicals into mutagens by green plants: a preliminary discussion. Environ. Health Perspect. 27:45-50.

Schairer, L.A., J. Van't Hof, C.G. Hayes, R.M. Burton, and F.J. deSerres. 1978a. Exploratory monitoring of air pollutants for mutagenicity activity with *Tradescantia* stamen hair system. Environ. Health Perspect. 27:51-60.

Schairer, L.A., J. Van't Hof, C.G. Hayes, R.M. Burton, and
    F.J. deSerres. 1978b. Measurement of biological activity of
    ambient air mixtures using a mobile laboratory for in situ
    exposures: Preliminary results from the Tradescantia plant
    test system. In: Application of Short-Term Bioassays in the
    Fractionation and Analysis of Complex Environmental Mixtures.
    M.D. Waters, S. Nesnow, J.L. Huisingh, S.S. Sandhu, and
    L. Claxton, eds. Plenum Press: New York. pp. 419-440.

Sparrow, A.H., L.A. Schairer, and R. Villalobos-Pietrini. 1974.
    Comparison of somatic mutation rates induced in Tradescantia
    by chemical and physical mutagens. Mutation Res. 26:265-276.

Thompson, M. Emmerling-, and M.M. Nawrocky. 1980. Genetic basis
    for using Tradescantia clone 4430 as an environmental monitor
    of mutagens. J. Heredity 71:261-265.

Underbrink, A.G., L.A. Schairer, and A.H. Sparrow. 1973.
    Tradescantia stamen hairs: a radiobiological test system
    applicable to chemical mutagenesis. In: Chemical Mutagens:
    Principles and Methods for Their Detection, Volume 3.
    A. Hollaender, ed. Plenum Press: New York. pp. 171-207.

Vogel, E., and F.H. Sobels. 1976. The function of Drosophila in
    genetic toxicity testing. In: Chemical Mutagens:
    Principles and Methods for Their Detection, Volume 4.
    A. Hollaender, ed. Plenum Press: New York. pp. 93-142.

RELATIONSHIP(S) BETWEEN MUTATION AND CYTOTOXICITY INDUCED <u>IN VITRO</u>

J.H. Carver,[1] A.D. Mitchell,[2] and M.D. Waters[3]

[1]Chevron Environmental Health Center, Richmond, CA 94802
[2]Biochemical Genetics Department, SRI International, Menlo
Park, CA 94025; [3]Genetic Toxicology Division, Health Effects
Research Laboratory, U.S. Environmental Protection Agency
Research Triangle Park, NC 27711

INTRODUCTION

Rapid, short-term tests are currently used to identify
compounds mutagenic <u>in vitro</u>. <u>In vivo</u> tests are then used to
confirm a compound's activity in the intact animal and to assess
oncogenic and genetic risk. Carver et al. (1979) presented the
hypothesis that cytotoxicity induced by an agent <u>in vitro</u> might
estimate potential mutagenicity <u>in vivo</u>. They compared the
induced mutation frequency per unit exposure dose (M) of 22 known
mutagens with the unit decrease in survival; a relative increase
in cytotoxicity (defined as the failure of cultured cells to
undergo continued cell division) was usually accompanied by a
proportional increase in mutagenicity. The correlation between
cytotoxicity and mutagenicity does not necessarily imply that
lethal and mutagenic events arise from the same type of lesion;
factors such as (de)toxification, biochemical reactions, and
repair mechanisms may affect both end points. The relationship
implies, however, that the maximum potential mutagenic potency of
agents <u>in vitro</u> might be estimated from measurement of cytotoxic
potency. Some compounds highly toxic <u>in vitro</u> are not
significantly mutagenic <u>in vitro</u>, but agents rarely induce
mutation in the complete absence of a dose-related increase in
toxicity.

We examine here an expanded literature base for this hypothesis, primarily data from the Chinese hamster (V79 and CHO) and mouse lymphoma (L5178Y) test systems. Observed relationships between in vitro mutation and cytotoxicity are described for 62 known mutagens and toxins, 17 selected pesticides, and 9 complex mixtures derived from environmental emissions. The use of such comparisons to determine mutagenic/cytotoxic potency in vitro and to set priority for other testing is discussed. Correlations between in vitro and in vivo assays are presented.

METHODS

The induced mutation frequency per viable cell (IMF) per unit exposure dose (M or μg/ml) was calculated by estimating from the authors' data the slope of IMF as a function of exposure dose. The data used included ⩾ 2 dose points in the most linear range, but none for survival < 10% was used. The D_{37} values (concentration in M or μg/ml required to kill 63% of the initial cell population) were estimated from the authors' survival data. Killing was defined as the inability of a cell to undergo continued cell division in vitro resulting in a viable colony. All D_{37} values were predicted from survival ranging from 100% to ⩽ 35% except acephate, ICR-171, and simazine, where the values are only approximate. Experimental conditions varied from system to system and among laboratories using a single system. When used, metabolic activation (MA) was Aroclor-induced rat-liver S9, S15, or the microsomal fraction.

The equation for linear regression in Figures 1, 2, and 3 is log (IMF/M or IMF/μg/ml) = a(-log D_{37}) -b. The lower detection limit of the assays (shaded) was estimated assuming that at IMF of approximately 10^{-6}, the Poisson error is large, and determinations are not statistically significant; and the maximum feasible exposure dose is usually ⩽ 10-fold higher than the D_{37}. The upper shaded area shows that no agents have exceeded a proposed biological limit of approximately 1 forward mutation induced at a given locus/100 cells (Carver et al., 1979). This assumes that the loci tested are representative of the entire genome and refers to compounds whose cytotoxicity is due solely to mutational events.

RESULTS

The 79 compounds evaluated are shown in Table 1. In Figure 1, the induced mutation frequency per viable cell per unit exposure dose is shown for 62 chemicals as a function of the

Table 1.　Compounds Evaluated for Relationship(s) Between Mutagenicity and Cytotoxicity

Compound Name	Abbr.	References
Chemicals		
2-Acetylaminofluorene	2A	Amacher et al., 1981; Clive et al., 1979.
4-Acetylaminofluorene	4A	Clive et al., 1979.
Adriamycin	A	Suter et al., 1980.
Benzo(c)phenanthrene diol epoxide 2	B	Wood et al., 1980.
N-Butyl-N-nitrosourea	BN	Couch and Hsie, 1978.
Cyclophosphamide	CP	Clive et al., 1979.
Diethylnitrosamine	DE	Clive et al., 1979.
Diethylstilbestrol	DS	Clive et al., 1979.
Diethylsulfate	ES	Couch et al., 1978.
Dimethylnitrosamine	DM	Clive et al., 1979; Frantz and Malling, 1976; Kuroki et al., 1977.
Dimethylsulfate	MS	Couch et al., 1978.
Diphenylnitrosamine	DP	Clive et al., 1979.
Ethylene dibromide	EB	Clive et al., 1979.
Ethyl methanesulfonate	EM	Adair et al., 1980; Adair and Carver, MS; Carver et al., 1980; Clive et al., 1979; Cole and Arlett, 1976; Couch et al., 1978; Hsie et al., 1975; O'Neill and Hsie, 1977; O'Neill and Hsie, 1979; Peterson et al., 1979; Suter et al., 1980; Van Zeeland and Simons, 1976.
N-Ethyl-N'-nitro-N-nitroso-guanidine	EG	Couch and Hsie, 1978; Peterson et al., 1979.

(continued)

Table 1. (continued)

Compound Name	Abbr.	References
N-Ethyl-N-nitrosourea	EN	Couch and Hsie, 1978; Suter et al., 1980.
Furylfuramide (AF-2)	AF	Clive et al., 1979; Wild, 1975.
Hydrazines (4)	H	Rogers and Back, 1981.
ICR compounds (19)	I	Adair and Carver, MS; Fuscoe et al., 1979; O'Neill et al., 1978.
Isopropyl methanesulfonate	IM	Couch et al., 1978.
Methyl iodide	MI	Clive et al., 1979.
Methyl methanesulfonate	MM	Adair and Carver, MS; Clive et al., 1979; Cole and Arlett, 1978; Couch et al., 1978; Peterson et al., 1979; Suter et al., 1980.
Mitomycin C	MC	Adair and Carver, MS; Clive et al., 1979.
N-Methyl-n'-nitro-N-nitroso-guanidine	MG	Bhuyan et al., 1976; Carver et al., 1980; Clive et al., 1979; O'Neill et al., 1977; Peterson et al., 1979; Suter et al., 1980.
N-Methyl-N-nitrosourea	MN	Bradley et al., 1980; Couch and Hsie, 1978; Suter et al., 1980.
Nitrogen mustard	HN	Suter et al., 1980.
Nitrosoureas, other (6)	N	Bradley et al., 1980.
Platinum (II) chloroamines (4)	Pt	Johnson et al., 1980; O'Neill et al., 1977.
β-Propiolactone	P	Clive et al., 1979.
Streptozotocin	SZ	Bhuyan et al., 1976; Bradley et al., 1980.

Compound Name	Abbr.	References
Uracil mustard	U	Clive et al., 1979.
Vinblastine	VB	Suter et al., 1980.
Vincristine	VC	Suter et al., 1980.
Pesticides		
Acephate	A	Jones et al., MS.
Azinphos-methyl	AM	Jones et al., MS.
Benomyl	B	Jones et al., MS.
Bromacil	Br	Jones et al., MS.
Cacodylic acid	CA	Jones et al., MS.
Captan	C	Jones et al., MS.
Crotoxyphos	Cr	Jones et al., MS.
Demeton	D	Jones et al., MS.
Diallate	Da	Jones et al., MS.
Disulfoton	Ds	Jones et al., MS.
Folpet	F	Jones et al., MS.
Monocrotophos	Mc	Jones et al., MS.
Monuron	M	Jones et al., MS.
Parathion-methyl	P	Jones et al., MS.
Simazine	S	Jones et al., MS.
Triallate	Ta	Jones et al., MS.
Trichlorfon	Tc	Jones et al., MS.

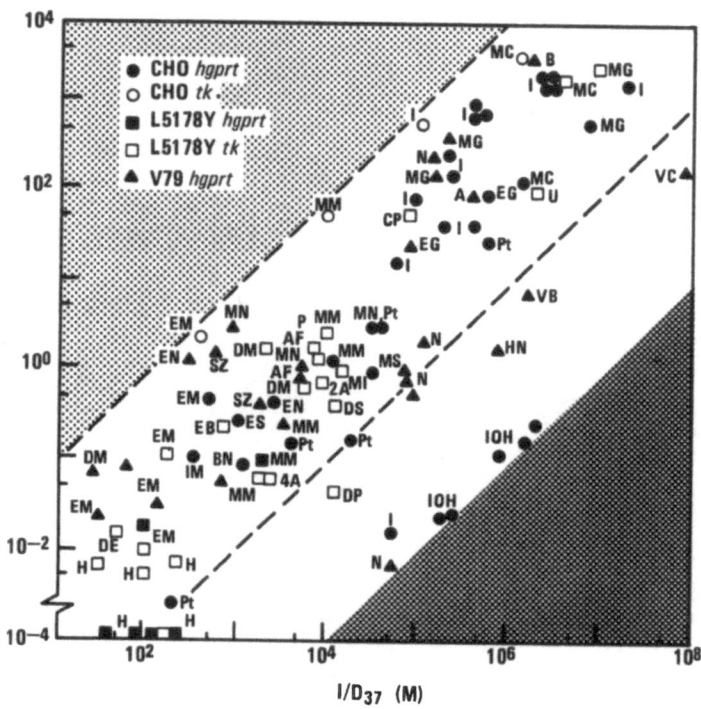

Figure 1. Relationship(s) between mutagenicity and cytoxicity.
Regression parameters pertaining to Figure 1 are in
Table 2. Test systems include (1) CHO cells (hgprt,
n = 39; tk, n = 5); (2) Chinese hamster V79 cells
(hgprt, n = 27); (3) mouse lymphoma L5178Y cells
(hgprt, n = 6; tk, n = 24). The following compounds
are considered as cytotoxic agents rather than as
mutagens: ICR-1910H, ICR-1700H, ICR-3400H, ICR-3720H,
ICR-2920H, ICR-283, DP, HN, VB, VC, selected N and H
(Bradley et al. 1980; Rogers and Back, 1981). They are
not included in the statistical analyses. Dotted lines
are the 95% confidence bands for a single y; r^2 is
coefficient of determination.

Table 2.   Regression Parameters

Cell Lines	Locus	a	b	r^2	n
CHO, V79, L5178Y	hgprt	0.94	3.31	0.80	58
L5178Y, CHO	tk	1.10	3.90	0.83	27
Combined cell lines		0.99	3.51	0.82	85
± 1 S.E.		± 0.05	± 0.23		

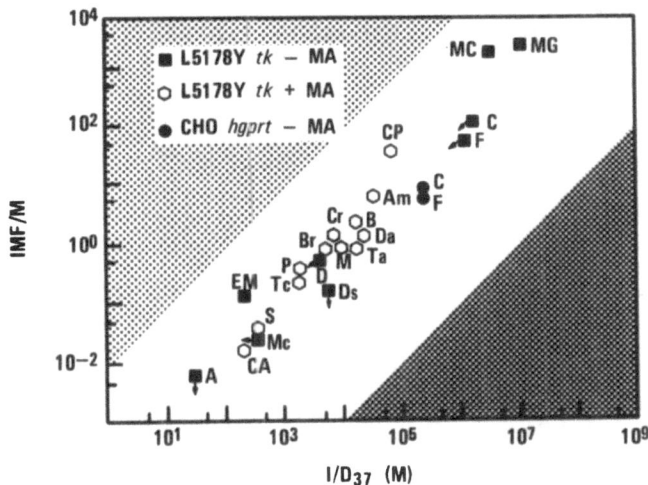

Figure 2.   Relationship(s) between mutagenicity and cytotoxicity for 17 pesticides.

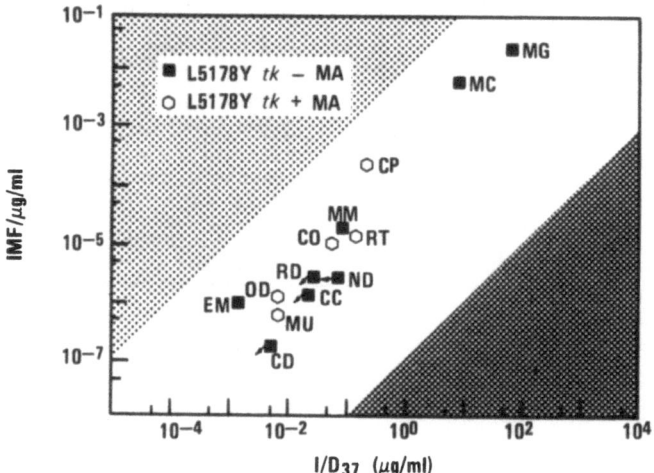

Figure 3.   Relationship(s) between mutagenicity and cytotoxicity
            for environmental emissions.

reciprocal of D_{37} (M). In Figure 2, data from the L5178Y system
for 17 pesticides (Jones et al., 1982; Waters et al., 1982) are
shown. Tested with and without MA, results are shown for maximal
mutagenicity:cytotoxicity. For data without MA (6/17), arrows
indicate the effect of added MA. Captan and folpet have also been
tested in CHO without MA (O'Neill et al., 1981). Because captan
(and presumably folpet) react with protein thiols (Bridges, 1975),
in vitro results may depend upon the levels of media serum
proteins and MA proteins, if any. For the pesticides tested, the
relationship between IMF and cytotoxicity shows a relative
increase in mutagenicity. The slope, a, of the regression
relationship is 0.85 ± 0.06 (b = 3.36 ± 0.23, r^2 = 0.93, n = 19).
Positive controls (EM, CP, MC, MG) are shown for reference, but
are not included in the regression analysis. If the 17 compounds
in Figure 2 are combined with the 47 chemicals analyzed in
Figure 1, the parameters for the regression analysis are:
a = 0.98 ± 0.05, b = 3.53 ± 0.20, r^2 = 0.82, n =104.

Evaluation of complex mixtures requires an alternative method of calculating dose. Using μg/ml as the unit of exposure dose and expressing cytotoxicity as the reciprocal of D_{37} in μg/ml, we analyzed data from the L5178Y system for environmental emissions (Figure 3, Mitchell et al., 1980) and from the CHO system for a coal liquified crude oil (Hsie et al., 1980). All were conducted with and without MA, and maximal results are shown. In the presence of MA, coke oven (CO), roofing tar (RT), and Mustang gasoline engine emission (MU) samples were more cytotoxic and more mutagenic; the Oldsmobile diesel sample (OD) was less cytotoxic but more mutagenic. Cigarette-smoke condensate (CC), Caterpillar diesel emissions (CD), and VW rabbit diesel emission samples (RD) were more toxic and more mutagenic without MA. The Nissan diesel emission sample (ND) was more toxic without, but mutagenicity was similar with and without MA. The acetone subfraction of the crude oil sample (CR) required MA for significant mutagenic activity, but MA reduced the observed cytotoxicity. Positive controls calculated for IMF per μg/ml and D_{37} (μg/ml) are not included in the regression analysis. The relationship for these complex mixtures (n = 9) is a = 1.10, b = 3.39, r^2 = 0.84.

DISCUSSION

In vitro results for 79 chemicals and 9 complex mixture samples tested at two gene loci in three rodent cell systems support the theory that a quantitative estimate of potential in vitro mutagenicity can be made from the cytotoxicity induced in vitro. Exceptions are noted for 15 compounds where mutagenicity and cytotoxicity are not as tightly coupled as would be predicted, e.g., the ICR-OH series, HN, VB, VC, DP. The OD and ND emission samples also demonstrated some independence of cytotoxic and mutagenic responses in the absence and presence of metabolic activation. The structure-activity studies of Clive (1979) with structural analogs of hycanthone methanesulfonate clearly show similar dissociations of toxic and mutagenic effects. Other reports have documented that cytotoxicity may not be due solely to induction of lethal mutation (Baker et al., 1979; Bradley and Erickson, 1981; Hsu et al., 1979). A recent study (Raffeto et al., 1979) suggested that only selected types of toxic lesions are capable of increasing sister chromatid exchange. Peterson and co-workers (1979) have further suggested that mutagenic and alkali-labile lesions are associated, but not identical, and that neither lesion produces cytotoxicity. However, mutagenicity, as measured by forward mutations that result in inactive gene products, might be expected to correlate with a minimal cytotoxic response. Thus, mutagenic agents should confer a certain amount of obligatory cytotoxicity. If further cytotoxicity is induced by independent pathways, the maximal mutagenic potential predicted by the observed cytotoxicity would not be observed.

Admittedly we are, as discussed by Clive et al. (1979), mainly correlating doses biologically active in vitro. However, the fact that the slope is 0.98 ± for 64 compounds (104 data points) strengthens our previous conclusion that empirically, in vitro cytotoxic potency is closely correlated with in vitro mutagenic potency. Not only do mutagenicity and cytotoxicity increase with increasing dose; for a large number of agents, they do so proportionally with dose. Thus, as a first approximation, the proposed relationship appears credible for a variety of known toxic and mutagenic agents. It can be used to discuss potency defined as the relative effects of in vitro mutagenicity and cytotoxicity. For example, a compound might be identified by coordinates (log x, log y) to reflect its position on the plot in Figure 1. In the L5178Y TK$^{+/-}$ system, the in vitro potency of MG (7.0, 3.5) is greater than tht of MM (4.0, 0.3), which is greater than EM (2.2, -1.0). In both CHO and V7.9, MM (4.0, 0.1) > EM (2.6, -0.3) and MM (3.2, -0.8) > EM (1.7, -1.5), respectively. Note this is on a molar basis, which is the critical parameter in considering relative potential hazards of environmental exposure. Some inter- and intra-system variability is noted, e.g., DM (3.7, -0.2; 3.3, 0.3) (1.5, -1.1) and AF (3.8, 0.2) (3.6, -0.2) in L5178Y and V79, respectively. Data showing complex activation parameters for DM (Haag and Sipes, 1980) and AF (Murthy and Najaria, 1980) suggest that such variability may be more closely related to differences in experimental conditions than to absolute differences in cell sensitivity.

The in vitro cytotoxic response may also have utility for setting priority for in vitro mutagenicity testing. Agents toxic at very low concentrations would be presumed to have more potential for inducing significant mutagenicity and would be tested before those agents toxic at higher dose levels. Then, agents demonstrating a higher mutagenic/cytotoxic potency (or highly toxic) in vitro would be given priority for testing in in vivo bioassays. The observed relationship may also be applicable to determining whether an agent's toxic effects are related to DNA (genotoxic) damage or other mechanisms, e.g., deleterious effects on membranes, RNA, or protein. We suggest that compounds whose ratio of mutagenicity to cytotoxicity is high show the closest coupling between genetic injury and lethality. When induced cytotoxicity related to non-DNA damage predominates, the mutagenic activity expected from the cytotoxicity is not seen. We previously acknowledged this class of compounds (Carver et al., 1979). Highly toxic chemicals like VC (8.0, 2.5), I170-OH (6.2,

-0.3), or HN (5.9, 0.4) are clearly identified as inherently cytotoxic by their x coordinate.

Potency in short-term tests for mutagenicity does not always correlate with potency in carcinogenicity bioassays predicting risk to the intact animal. DP and 4A, usually considered weakly or noncarcinogenic in vivo (Clive et al., 1979) are relatively weak mutagens but moderately toxic in vitro (Figure 1). The potency of EM and MM varies under in vitro (and in vivo) experimental conditions, perhaps due to chemical lability. In vitro, MM is more potent than EM (Figure 1; McCann and Ames, 1976) and the reverse is noted in vivo (Loveless reference in Clive et al., 1979). However, other in vivo data for MM indicate that the tumorigenic potential of MM may have been underestimated (Magee and Swan reference in Clive et al., 1979).

Lassiter (1977) has classified occupational carcinogens from data in the IARC series; potency in vivo is estimated to be MS > MG > EN > P > H > SZ, compared to our in vitro estimation of MG > MS > P > SZ > EN > H. However, Lassiter's criteria were based on considerations of route of administration, tumor site, latency period, and incidence. They did not consider dose levels in vivo. Also, variability within the in vitro systems may be related to differences in experimental conditions rather than to absolute differences in cell sensitivity. The in vitro ranking of the complex environmental emissions (Figure 3) is approximately RT > CO > ND > RD > CC > OD > MU > CD. In the skin carcinogenesis assay with SENCAR mice, the tumorigenic activity potency in vivo based on either tumor multiplicity or tumor incidence is approximately CO > ND > RT > RD,MU,OD > CD (Stephen Nesnow, U.S. Environmental Protection Agency, Research Triangle Park, NC, personal communication).

Another consideration in correlating in vitro and in vivo activities is metabolic activation. The data for pesticides A, C, F, D, MC (Figure 2) and environmental emissions CC, CD, ND, RD (Figure 3) show the effect of adding MA to direct mutagens in vitro. In vitro results which decrease significantly with added MA may, in fact, be more predictive of in vivo results. For example, C is inefficient in vivo, perhaps due to the compound's low water solubility and avidity for thiols. In contrast to the in vitro data without MA (Figure 2); Tezuka et al., 1980), adding MA protein to C reduces the observed mutagenicity and toxicity (Jones et al., 1982).

The assessment of in vitro mutagenic/cytotoxic potency described here is clearly not a definitive predictor of carcinogenic potency in the intact animal. However, cytotoxicity

in vitro may estimate maximum mutagenic potential in vitro. More
specific end points can then be used to determine the observed
mutagenic potency. These relationship(s) in vitro may be useful
in determining priority levels for in vivo testing.

ACKNOWLEDGMENTS

We thank Debbie L. Newlin and Dr. E.R. Jones for assistance
with statistical analyses and Vickie L. Laughlin for manuscript
preparation. Contributed data obtained in part from EPA contract
68-02-2947.

REFERENCES

Adair, G.M., J.H. Carver, and D.L. Wandres. 1980. Mutagenicity
    testing in mammalian cells, I: derivation of a Chinese
    hamster ovary cell line heterozygous for the adenine
    phosphoribosyltransferase and thymidine kinase loci.
    Mutation Res. 72:187-205.

Adair, G.M. and J.H. Carver. (MS). Mutagenicity testing in
    mammalian cells, III: induction and expression of mutations
    at multiple drug-resistance marker loci in Chinese hamster
    ovary cells.

Amacher, D.E., S.C. Paillet, and J.A. Elliott. 1981. The
    metabolism of N-acetyl-2-aminofluorene to a mutagen in
    L5178Y/TK$^{+/-}$ mouse lymphoma cells. Mutation Res. 89:311-320.

Baker, R.M., W.C. Van Voorhis, and L.A. Spencer. 1979. HeLa cell
    variants that differ in sensitivity to monofunctional
    alkylating agents, with independence of cytotoxic and
    mutagenic responses. Proc. Natl. Acad. Sci. USA
    76:5249-5253.

Bhuyan, B.K., A.R. Peterson, and C. Heidelberger. 1976.
    Cytotoxicity, mutations and DNA damage produced in Chinese
    hamster cells treated with streptozotocin, its analogs, and
    N-methyl-N'-nitro-N-nitrosoguanidine. Chem.-Biol. Interact.
    13:173-179.

Bradley, M.O. and L.C. Erickson. 1981. Comparison of the effects
    of hydrogen peroxide and X-ray irradiation on toxicity,
    mutation, and DNA damage/repair in mammalian cells (V-79).
    Biochim. Biophys. Acta 654:135-141.

Bradley, M.O., N.A. Sharkey, K.W. Kohn, and M.W. Layard. 1980.
    Mutagenicity and cytotoxicity of various nitrosoureas in V-79
    Chinese hamster cells. Cancer Res. 40:2719-2725.

Bridges, B.A. 1975. The mutagenicity of captan and related
    fungicides. Mutation Res. 32:3-34.

Carver, J.H., G.M. Adair, and D.L. Wandres. 1980. Mutagenicity
    testing in mammalian cells, II: validation of multiple
    drug-resistance markers having practical application for
    screening potential mutagens. Mutation Res. 72:207-230.

Carver, J.H., F.T. Hatch, and E.W. Branscomb. 1979. Estimating
    maximum limits to mutagenic potency from cytotoxic potency.
    Nature 279:154-156.

Clive, D., K.D. Johnson, J.F.S. Spector, A.G. Batson, and
    M.M.M. Brown. 1979. Validation and characterization of the
    L5178Y/TK$^{+/-}$ mouse lymphoma mutagen assay system. Mutation
    Res. 59:61-108.

Cole, J. and C.F. Arlett. 1976. Ethyl methanesulfonate
    mutagenesis with L5178Y mouse lymphoma cells: a comparison
    of ouabain, thioguanine and excess thymidine resistance.
    Mutation Res. 34:507-526.

Cole, J. and C.F. Arlett. 1978. Methyl methanesulfonate
    mutagenesis in L5178Y mouse lymphoma cells. Mutation Res.
    50:111-120.

Couch, D.B. and A.W. Hsie. 1978. Mutagenicity and cytotoxicity
    of congeners of two classes of nitroso compounds in Chinese
    hamster ovary cells. Mutation Res. 57:209-216.

Couch, D.B., N.L. Forbes, and A.W. Hsie. 1978. Comparative
    mutagenicity of alkylsulfate and alkanesulfonate derivatives
    in Chinese hamster ovary cells. Mutation Res. 57:217-224.

Frantz, C.N. and H.V. Malling. 1976. Bromodeoxyuridine
    resistance induced in mouse lymphoma cells by microsomal
    activation of dimethylnitrosamine. J. Toxicol. Environ.
    Health 2:179-187.

Fuscoe, J.C., J.P. O'Neill, R.M. Peck, and A.W. Hsie. 1979.
    Mutagenicity and cytotoxicity of nineteen heterocyclic
    mustards (ICR compounds) in cultured mammalian cells. Cancer
    Res. 39:4875-4881.

Haag, S.M. and I.G. Sipes. 1980. Differential effects of acetone
    or Aroclor 1254 pretreatment on the microsomal activation of
    dimethylnitrosamine to a mutagen. Mutation Res. 74:431-438.

Hsie, A.W., P.A. Brimer, T.J. Mitchell, and D.G. Gosslee. 1975.
    The dose-response relationship for ethyl methanesulfonate-
    induced mutations at the hypoxanthine-guanine phosphoribosyl
    transferase locus in Chinese hamster ovary cells. Somatic
    Cell Genet. 1:247-261.

Hsie, A.W., P.A. Brimer, J.P. O'Neill, J.L. Epler, M.R. Guerin,
    and M.H. Hsie. 1980. Mutagenicity of alkaline constituents
    of a coal-liquified crude oil in mammalian cells. Mutation
    Res. 78:79-84.

Hsu, I., G.T. Bowden, and C.C. Harris. 1979. A comparison of
    cytotoxicity, ouabain-resistant mutation, sister-chromatid
    exchanges, and nascent DNA synthesis in Chinese hamster cells
    treated with dihydrodiol epoxide derivatives of
    benzo(a)pyrene. Mutation Res. 63:351-359.

Lassiter, D.V. 1977. Occupational Carcinogenesis. In: Advances
    in Modern Toxicology, Volume 3: Environmental Cancer.
    H.F. Kraybill and M.A. Mehlman, eds. John Wiley and Sons:
    New York. pp. 63-86.

McCann, J. and B.N. Ames. 1976. Detection of carcinogens as
    mutagens in the Salmonella/microsome test: assay of
    300 chemicals, discussion. Proc. Natl. Acad. Sci. USA
    73:950-954.

Johnson, N.P., J.D. Hoeschele, R.O. Rahn, J.P. O'Neill, and
    A.W. Hsie. 1980. Mutagenicity, cytotoxicity, and DNA
    binding of platinum (II)-chloroamines in Chinese hamster
    ovary cells. Cancer Res. 40:1463-1468.

Jones, D.C.L., V.F. Simmon, K.E. Mortelmans, A.D. Mitchell,
    E.L. Evans, M.M. Jotz, E.S. Riccio, D.E. Robinson, and
    B.A. Kirkhart. (MS). In vitro mutagenicity studies of
    environmental chemicals. SRI Project LSU-7558; contract
    No. 68-02-2947. U.S. Environmental Protection Agency.

Kuroki, T., C. Drevon, and R. Montesano. 1977. Microsome-
    mediated mutagenesis in V79 Chinese hamster cells by various
    nitrosamines. Cancer Res. 37:1044-1050.

Mitchell, A.D., E.L. Evans, E.S. Riccio, K.E. Mortelmans, and
    V.F. Simmon. 1980. Mutagenic and carcinogenic potency of
    extracts of diesel and related environmental emissions: In
    vitro mutagenesis and DNA damage. In: Health Effects of
    Diesel Engine Emissions. W.E. Pepelko, R.M. Danner, and
    N.A. Clarke, eds. EPA-600/9-80-057b. U.S. Environmental
    Protection Agency: Cincinnati, OH. pp. 810-840.

Murthy, M.S.S. and K.B. Najaria. 1980. Deactivation of furyl
    furamide (AF-2) by rat-liver microsomes and its implication
    in short-term tests for mutagenicity/carcinogenicity.
    Mutation Res. 77:127-134.

O'Neill, J.P., P.A. Brimer, R. Machanoff, G.P. Hirsch, and
    A.W. Hsie. 1977. A quantitative assay of mutation induction
    at the hypoxanthine-guanine phosphoribosyl transferase locus
    in Chinese hamster ovary cells (CHO/HGPRT system):
    development and definition of the system. Mutation Res.
    45:91-101.

O'Neill, J.P., D.B. Couch, R. Machanoff, J.R. San Sebastian,
    P.A. Brimer, and A.W. Hsie. 1977. A quantitative assay of
    mutation induction at the hypoxanthine-guanine phosphoribosyl
    transferase locus in Chinese hamster ovary cells (CHO/HGPRT
    system): utilization with a variety of mutagenic agents.
    Mutation Res. 45:103-109.

O'Neill, J.P. and A.W. Hsie. 1977. Chemical mutagenesis of
    mammalian cells can be quantified. Nature 269:815-817.

O'Neill, J.P., J.C. Fuscoe, and A.W. Hsie. 1978. Mutagenicity of
    heterocyclic nitrogen mustards (ICR compounds) in cultured
    mammalian cells. Cancer Res. 38:506-509.

O'Neill, J.P. and A.W. Hsie. 1979. Phenotypic expression time of
    mutagen-induced 6-thioguanine resistance in Chinese hamster
    ovary cells (CHO/HGPRT system). Mutation Res. 59:109-118.

O'Neill, J.P., N.L. Forbes, and A.W. Hsie. 1981. Cytotoxicity
    and mutagenicity of the fungicides captan and folpet in
    cultured mammalian cells (CHO/HGPRT system). Environ.
    Mutagen. 3:233-237.

Peterson, A.R., H. Peterson, and C. Heidelberger. 1979.
    Oncogenesis, mutagenesis, DNA damage, and cytotoxicity in
    cultured mammalian cells treated with alkylating agents.
    Cancer Res. 39:131-138.

Raffetto, G., S. Parodi, P. Faggin, and A. Maconi. 1974.
Relationship between cytotoxicity and induction of sister-
chromatid exchanges in mouse foetal cells exposed to several
doses of carcinogenic and non-carcinogenic chemicals.
Mutation Res. 63:335-343.

Rogers, A.M. and K.C. Back. 1981. Comparative mutagenicity of
hydrazine and 3 methylated derivatives in L5178Y mouse
lymphoma cells. Mutation Res. 89:321-328.

Suter, W., J. Brennand, S. McMillan, and M. Fox. 1980. Relative
mutagenicity of antineoplastic drugs and other alkylating
agents in V79 Chinese hamster cells: independence of
cytotoxic and mutagenic responses. Mutation Res. 73:171-181.

Tezuka, H., N. Ando, R. Suzuki, M. Terahata, M. Moriya, and
Y. Shirasu. 1980. Sister-chromatid exchanges and
chromosomal aberrations in cultured Chinese hamster cells
treated with pesticides positive in microbial reversion
assays. Mutation Res. 78:177-191.

Van Zeeland, A.A. and J.W.I.M. Simons. 1976. Linear dose-
response relationships after prolonged expression times in
V79 Chinese hamster cells. Mutation Res. 35:129-138.

Waters, M.D., S.S. Sandhu, V.F. Simmon, K.E. Mortelmans,
A.D. Mitchell, T.A. Jorgenson, D.C.L. Jones, R. Valencia, and
N.E. Garrett. (In Press). Study of pesticide genotoxicity.
In: Genetic Toxicology: An Agricultural Perspective.
R. Fleck and A. Hollaender, eds. New York: Plenum Press.

Wild, D. 1975. Mutagenicity of the food additive AF-2, a
nitrofuran, in Escherichia coli and Chinese hamster cells in
culture. Mutation Res. 31:197-199.

Wood, A.W., R.L. Chang, W. Levin, D.E. Ryan, P.E. Thomas,
M. Croisy-Delcey, Y. Ittah, H. Yagi, D.M. Jerina, and
A.H. Conney. 1980. Mutagenicity of the dihydrodiols and
bay-region diol-epoxides of benzo(c)phenanthrene in bacterial
and mammalian cells. Cancer Res. 40:2876-2883.

THE MOUSE OOCYTE TOXICITY ASSAY

James S. Felton and R. Lowry Dobson

Biomedical Sciences Division, Lawrence Livermore National Laboratory, University of California, P.O. Box 5507 Livermore, CA 94550

INTRODUCTION

Certain chemicals, including energy-related pollutants such as polycyclic aromatic hydrocarbons (PAHs) are highly destructive to mouse female germ cells (Felton et al., 1980). Like ionizing radiation, these chemicals readily kill oocytes, and the gametic cell loss can be quantified to provide dose-response curves and _in vivo_ measurements of chemical cytotoxicity.

Although twenty-five years have passed since it was recognized that chemicals can destroy oocytes in mice (Marchant, 1957; Mody, 1960; Kuwahara, 1967; Krarup, 1969), only recently has the oocyte system been used for detecting low-exposure effects and quantitatively evaluating hazards. Studies of exposures to tritium and gamma rays (Dobson and Cooper, 1974; Dobson and Kwan, 1976; Dobson et al., 1976; Dobson, 1976; Dobson and Kwan, 1977) were the model for our work on PAHs, in which differences in genetic control of PAH-metabolizing activity were shown to influence benzo(a)pyrene (B[a]P) and 3-methylcholanthrene (MC) killing of oocytes (Felton et al., 1978). Using MC as a representative test compound, it was found that chemical vulnerability of mouse oocytes varies systematically with the animal's age (Dobson et al., 1978a), and that the variation closely follows the pattern of sensitivity to radiation (Dobson et al., 1978b). While oocyte killing by PAHs in adult mice has been reported to be correlated with induction of aryl hydrocarbon hydroxylase activity in the target ovary (Mattison and Thorgeirsson, 1977), more complicated relationships apparently operate in the maternal-fetal system where in utero exposure occurring transplacentally causes large oocyte depletion in the developing fetus (Felton et al., 1978).

Here we describe a "standard" toxicity assay based on the mouse oocyte system, and we show its application to toxicity determinations of complex mixtures as well as of pure compounds. For the standard assay we use the juvenile mouse because its oocytes are especially sensitive, and we use exposure via i.p. injection because this route allows ready control of dosage. To facilitate comparisons of oocytotoxicity among chemical agents, we have developed the Oocyte Toxicity Index (OTI), the ratio of whole-animal toxic dose to oocyte LD_{50}. Measured OTIs are found to range from less than 1 (for benzene) to almost 40 (for 7,12 dimethylbenz[a]anthracene [DMBA]), and even higher for certain compounds (e.g., more than 50 for ethyl nitrosourea).

In addition, this intact-mammal system has important versatility, allowing the evaluation of various other exposure routes, including ingestion and transplacental exposure. Transplacental exposure is of particular interest: PAHs administered orally to the pregnant animal have been found to cause even greater oocyte deficiency in offspring than in the mother (MS. in preparation).

MATERIALS AND METHODS

Chemicals

The chemicals used were of the highest purity readily available commercially and include the following (with their sources): bleomycin, mitomycin-C, and MC (Sigma); urethane, methylmethane sulfonate, and B(a)P (Aldrich); DMBA and dibromochloropropane (Chemical Procurement Laboratories); N-ethyl-N-nitrosourea (Fluka); procarbazine (Hoffman-LaRoche); and decarbamoyl mitomycin-C (kind gift of Dr. M. Sasaki).

Animals

All mice were of the C57BL/6 strain and were purchased from Simonsen Laboratory (Gilroy, CA). They were housed either 6 per cage or as single litters per cage and allowed free access to food and water. A 12-h light-dark cycle was maintained.

Chemical Exposures of Mice

For determination of oocyte LD$_{50}$ and OTI by the standard assay, chemicals were dissolved in corn oil, dimethyl sulfoxide (DMSO), or saline and given by i.p. injection. Dilutions were prepared so that animals received the same volume of corn oil (20 µl/g body weight), DMSO (5 µl/g body weight), or saline (25 µl/g body weight) for all doses. Control mice received only corn oil, DMSO, or saline. Injections were given 12 days after birth, and sacrifice was 14 days later.

For transplacental exposure-route studies, maternal exposure by ingestion was fractionated into three doses given by gavage on days 14, 15, and 16 of gestation. Sacrifice of offspring was at specified times after birth.

Whole-Animal Toxic Dose

As used here for OTI determinations, whole-animal toxic dose is defined as that dose which kills at least two but not more than three of five test animals.

Histological Preparation and Oocyte Enumeration

Briefly, ovaries were removed at sacrifice, fixed in Zenker-formol (Helly's solution), serially sectioned in paraffin at 5 µm, and stained with Delafield's hematoxylin and eosin. Primordial oocytes (stages 1, 2, and 3a of Pederson, 1972) were enumerated at 1000X magnification in every 20th section. Total counts were then compared with controls to give oocyte survival.

RESULTS

Dose-Response Curves

Figure 1 shows representative dose-response curves for three PAHs. Responses are sigmoidal, with measurable oocyte loss occurring at impressively low PAH doses. And the oocyte LD$_{50}$ for DMBA is 20 times lower than that for B(a)P. Oocyte-killing effectiveness among these three PAHs is related in the same way as their known mutagenicity/carcinogenicity (DMBA > MC > B[a]P).

We have tested 77 chemicals in this manner, each with a minimum of 6 doses and 5 animals per dose. Forty-two of the 77 compounds tested (55%) are ones that are classified as

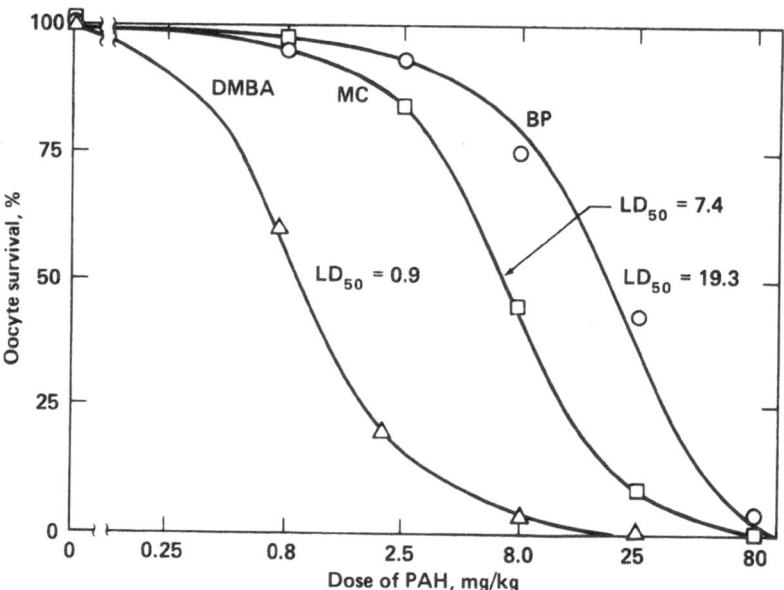

Figure 1.   Comparative dose-response curves for DMBA, MC, and
             B(a)P. Each data point represents oocyte survival
             (oocyte count expressed as percentage of control) for a
             minimum of five animals. SEM/M was uniformly less than
             15%.

mutagen/carcinogens (McCann et al., 1975), and of these, 21 were
positive in the oocyte-killing assay. (Most of the
mutagen/carcinogens that proved negative were either metals or
nitrogen-containing heterocyclic compounds and aromatic amines
extensively metabolized in the liver.) It is of interest that no
nonmutagen/noncarcinogen has been found to kill oocytes.

Among compounds showing positive results in this assay,
quantitative comparison can be made by means of oocyte LD_{50} values
or, more tellingly, by OTIs. Measured oocyte LD_{50}s (Table 1)
range more than a thousand-fold from a low of 0.9 μmol/kg for
bleomycin to a high of 2,500 μmol/kg for urethane. Of special

Table 1. Comparison of Whole-Animal Toxic Dose, Oocyte LD_{50}, and
          Oocyte Toxicity Index (OTI) for Selected Compounds

Compound	Whole-Animal Toxic Dose[a]	Oocyte LD_{50}[a]	OTI[b]
Bleomycin	110	0.94	117
N-Ethyl-N-nitrosourea	1,400	26	54
7,12-Dimethylbenz(a)anthracene	96	2.7	36
Urethane	13,500	2,500	5.4
Methyl methanesulfonate	1,400	350	4.0
Mitomycin-C	7.5	3.0	2.5
Dibromochloropropane	1,300	690	1.9
Procarbazine	1,200	810	1.5
Decarbamoyl mitomycin-C	100	160	0.7

[a] μmol/kg.
[b] Ratio of whole-animal toxic dose (see text) to the oocyte LD_{50}.

interest, however, are those compounds having high OTIs, for they
represent chemicals that cause germ-cell destruction out of
proportion to their general toxicity to the animal. And this
toxicologically important feature of a given compound, otherwise
easily overlooked, is readily identified and measured by the
oocyte assay.

    It may be noted that there is no necessary relationship
between low oocyte LD_{50} and high OTI. For example, mitomycin-C, a
compound that kills oocytes at a very lose dose
(LD_{50} = 3 μmol/kg), also kills 2 to 3 out of 5 mice at
7.5 μmol/kg, a dose only 2.5 times higher. In contrast, bleomycin
and N-ethyl-N-nitrosourea, both potent in the oocyte system,
require substantially higher relative doses to kill the whole
animal. It may be noted too that OTIs even less than unity can be
determined. Decarbamoyl mitomycin-C, for example, gives an OTI
of 0.7 (see Table 1). This measurement is possible because
significant oocyte killing is measured at doses below 100 μmol/kg,
the whole-animal killing dose, so that a regression line can be
fit to the data and an oocyte LD_{50} calculated. Thus, an OTI less
than 1 is measurable, and this is because of the reproducible
sensitivity of the oocyte system in the low-effect region of the
dose-response curve.

We have used the oocyte system also to test in vivo
cytotoxicity of complex mixtures, for example crude and
hydrotreated shale-oil samples. Crude shale oil injected into
juvenile mice at a particular dose level (5 mg/kg) killed 41% of
the oocytes, while equivalent doses of hydrotreated shale oil
produced no significant oocyte loss, the data correlating well
with Salmonella and CHO-cell mutagenesis assays done on the same
materials and with cytotoxicity in CHO cells in vitro (Timourian
et al., 1981).

Maternal-Fetus Exposure

We have shown (Felton et al., 1978) that oocyte killing in
the fetus following exposure via MC injection into the mother is
greater (at least up to three days after MC administration) than
in the newborn when it is itself injected. Also, we have explored
the material-fetal exposure system further (see Figure 2) by
measuring oocyte loss in offspring after material oral ingestion
of MC and of B(a)P during pregnancy. Not only was there massive
oocyte loss in the female offspring, but the degree of cell
killing was higher than in the mother herself (MS. in
preparation). In these studies, a fractionated dose of 150 μg/g
of MC or B(a)P was administered by gavage to the mother over
days 14, 15, and 16 of gestation; and it is noteworthy that the
same dose when given intraperitoneally on day 15, although toxic
to oocytes, causes significantly less cell loss (Felton et al.,
1978) than when given by the more natural oral route.

DISCUSSION

Data from dose-response curves can be used for determining
detection limits, thus allowing comparisons between the oocyte
system and other toxicity assays. Such comparisons, using MC as a
standard test compound, are shown in Table 2. While exposure
levels as low as 0.5 mg/kg have been reported to cause tumors in
mice (at very low incidence, and after 6 or more months of
latency) and to produce histidine revertants in certain bacterial
strains, mouse-oocyte loss is also found at very low exposure
levels, orders of magnitude lower than those detected by
dominant-lethal and micronuclei tests. The mouse oocyte toxicity
assay measures the destruction of crucially important,
irreplaceable mammalian cells in vivo, and results can be
available within three weeks. Furthermore, its flexibility and
ease in accommodating various exposure routes, including ingestion
and transplacental passage, are particularly valuable in
whole-animal risk evaluation. The mouse oocyte toxicity assay

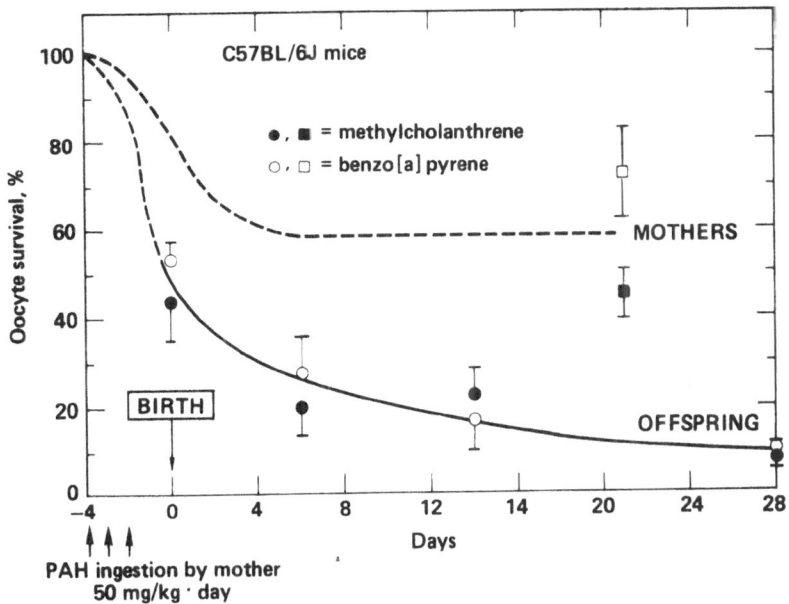

Figure 2.   Numbers of surviving primordial oocytes (expressed as
            percentage of control) counted in ovaries of offspring
            and mothers at various times after maternal exposure to
            MC or B(a)P. Shown are mean values ± SEM.
            Interpolated portions of curves (broken lines) are
            shaped to conform with available time-response data.

should be viewed not as a replacement for other tests, but as an
important addition that makes use of the completely intact
mammalian physiology and metabolizing machinery in the
quantification of toxic effects.

ACKNOWLEDGMENTS

     Work performed under the auspices of the U.S. Department of
Energy by the Lawrence Livermore National Laboratory under
contract number W-7405-ENG-48 with support from EPA under
interagency agreement EPA-IAG-D5-E681-AQ.

Table 2. Approximate Lower Detection Limits for
3-Methylcholanthrene-Induced Effects

Test System	Approximate Detection Level $(\mu g/g)$[a]
Mouse oocyte killing (juvenile)	5
Mouse sperm-head abnormalities (adult)[b]	20
Mouse sarcoma (6-mo latency)[c]	0.5
<u>Salmonella</u> mutations (strain TA1538)[d]	5
<u>Salmonella</u> mutations (strain TA98)[e]	0.5
Dominant lethals in mice[f]	> 100
Micronuclei in mice[g]	> 500

[a]Estimated "concentration" of MC in mice and in bacterial plates.
[b]Wyrobek and Bruce, 1976.
[c]Bryan and Shimkin, 1943; O'Gara et al., 1965.
[d]Felton and Nebert, 1975.
[e]McCann et al., 1975.
[f]Epstein and Bateman, 1971.
[g]Heddle and Bruce, 1977.

REFERENCES

Bryan, W.R., and M.B. Shimkin. 1943. Quantitative analysis of
    dose-response data obtained with three carcinogenic
    hydrocarbons. J. Natl. Cancer Inst. 3:505-531.

Dobson, R.L., and M.F. Cooper. 1974. Tritium toxicity: effect
    of low-level ^{3}HOH exposure on developing female germ cells in
    the mouse. Radiat. Res. 58:91-100.

Dobson, R.L., J.A. Arrington, and T.C. Kwan. 1976. Tritium
    toxicity: increased relative biological effectiveness of
    HOH^3 with protraction of exposure. In: Radioecology and
    Energy Resources, Volume 2. C.E. Cushing, Jr., ed. Dowden,
    Hutchinson, and Ross: Stroudsburg. pp. 349-353.

Dobson, R.L., and T.C. Kwan. 1976. The RBE of tritium radiation
    measured in mouse oocytes: increase at low exposure levels.
    Radiat. Res. 66:615-625.

Dobson, R.L.   1976.   Low-level, chronic exposure to tritium: an
    improved basis for hazard evaluation.   In: Biological and
    Environmental Effects of Low-Level Radiation, Volume 2.
    IAEA: Vienna.   pp. 57-63.

Dobson, R.L., and T.C. Kwan.   1977.   The tritium RBE at low-level
    exposure: variation with dose, dose rate, and exposure
    duration.   Curr. Top. Radiat. Res. Q. 12:44-62.

Dobson, R.L., J.S. Felton, and T.C. Kwan.   1978a.   Destruction of
    female germ cells in the mouse by 3-methylcholanthrene:
    effect of animal's age.   Mutation Res. 53:100-101.

Dobson, R.L., C.G. Koehler, J.S. Felton, T.C. Kwan, B.J. Wuebbles,
    and D.C.L. Jones.   1978b.   Vulnerability of female germ cells
    in developing mice and monkeys to tritium, gamma rays, and
    polycyclic aromatic hydrocarbons.   In: Developmental
    Toxicology of Energy-Related Pollutants.   D.D. Mahlum,
    M.R. Sikov, P.L. Hackett, and F.D. Andrew, eds.
    U.S. Department of Energy: Washington, DC.   pp. 1-14.

Epstein, A.J., and S.S. Bateman.   1971.   Detection of chemical
    mutagens by dominant lethal assay in the mouse.   In:
    Chemical Mutagens, Volume 2.   A. Hollaender, ed.   Plenum
    Press: New York.   pp. 541-568.

Felton, J.S., and D.W. Nebert.   1975.   Mutagenesis of certain
    activated carcinogens in vitro associated with genetically
    mediated increases in monooxygenase activity and cytochrome
    P_1-450.   J. Biol. Chem. 250:6769-6778.

Felton, J.S., T.C. Kwan, B.J. Wuebbles, and R.L. Dobson.   1978.
    Genetic differences in polycyclic-aromatic-hydrocarbon
    metabolism and their effects on oocyte killing in developing
    mice.   In: Developmental Toxicology of Energy-Related
    Pollutants.   D.D. Mahlum, M.R. Sikov, P.L. Hackett, and
    F.D. Andrew, eds.   U.S. Department of Energy: Washington,
    DC.   pp. 15-26.

Felton, J.S., D.S. Orwig, L.E. Olsen, G.S. Knize, G.J. Ryan,
    T.C. Kwan, and R.L. Dobson.   1980.   Oocyte killing in mice as
    an in vivo gametic cytotoxicity assay: results for
    45 chemicals.   Environ. Mutagen. 2:213.

Heddle, J.A., and W.R. Bruce. 1977. Comparison of tests for mutagenicity or carcinogenicity using assay of sperm abnormalities, formation of micronuclei and mutations in Salmonella. In: Origins of Human Cancer. H.H. Hyatt, J.D. Watson, and J.A. Winsten, eds. Cold Spring Harbor Laboratory: New York. pp. 1549-1557.

Krarup, T. 1969. Oocyte destruction and ovarian tumorigenesis after direct application of a chemical carcinogen (9,10-dimethyl-1,2-benzanthrene) to the mouse ovary. Int. J. Cancer 4:61-75.

Kuwahara, I. 1967. Experimental induction of ovarian tumors in mice treated with single administrations of 7,12-dimethyl-benz(a)anthracene, and its histopathological observation. Gann 58:253-266.

Marchant, J. 1957. The chemical induction of ovarian tumors in mice. Br. J. Cancer 11:452-464.

Mattison, D.R., and S.S. Thorgeirsson. 1977. Genetic differences in mouse ovarian metabolism of benzo(a)pyrene and oocyte toxicity of 3-MC. Biochem. Pharmacol. 26:909-912.

McCann, J., E. Choi, E. Yamasaki, and B.N. Ames. 1975. Detection of carcinogens as mutagens in the Salmonella/microsome test: assay of 300 chemicals. Proc. Natl. Acad. Sci. USA 72:5135-5139.

Mody, J.K. 1960. The action of four carcinogenic hydrocarbons on the ovaries of IF mice and the histogenesis of induced tumors. Br. J. Cancer 14:256-266.

O'Gara, R.W., M.G. Kelly, J. Brown, and N. Mantel. 1965. Induction of tumors in mice given a minute single dose of dibenz(a,h)anthracene or 3-methylcholanthrene as newborns: a dose-response study. J. Natl. Cancer Inst. 35:1027-1042.

Pedersen, T. 1972. Follicle growth in the mouse ovary. In: Oogenesis. J.D. Biggers and S.W. Schuetz, eds. University Park Press: Baltimore. pp. 361-376.

Timourian, H., A. Carrano, J. Carver, J.S. Felton, F.T. Hatch, D.S. Stuermer, and L.H. Thompson. 1981. Comparative mammalian genetic toxicology of shale oil products assayed _in vitro_ and _in vivo_. In: Proc. Symp. Health Effects Investigation of Oil Shale Development. W.H. Griest, M.R. Guerin, and D.L. Coffin, eds. Ann Arbor Science: Ann Arbor. pp. 173-188.

Wyrobek, A.J., and W.R. Bruce. 1975. Chemical induction of sperm abnormalities in mice. Proc. Natl. Acad. Sci. USA 72:4425-4429.

APPLICATION OF BOVINE MACROPHAGE BIOASSAYS IN THE ANALYSIS OF TOXIC AGENTS IN COMPLEX ENVIRONMENTAL MIXTURES

Gerald L. Fisher, Karen L. McNeill, and Charles J. Democko

Health and Environmental Sciences, Battelle Columbus Laboratories, 505 King Avenue, Columbus, OH 43201

INTRODUCTION

The pulmonary alveolar macrophage (PAM) plays a primary role in the defense of the lung against inhaled particulate matter. Fine particles deposited in the nonciliated regions of the lung are rapidly phagocytized and, if of biological origin, are rapidly killed and metabolized. In recent years, concern has focused on the effects of anthropogenic particulate matter with an emphasis on the PAM toxicity of trace metals and combustion products. Generally, however, previous efforts have described the utility of PAM derived from standard laboratory-animal models. In this report, we describe the methodology and application of the bovine PAM in the in vitro evaluation of metal and particulate toxicity.

Our efforts have been directed to the application of the bovine model because large quantities of relatively homogenous populations of PAM are readily harvested from the bovine lung. Our previous investigations have indicated that the PAM obtained from the lungs of cattle at slaughter phagocytize test particles in vitro with kinetics similar to that of human PAM lavaged from normal, healthy donors. Furthermore, the bovine PAM demonstrates enhanced phagocytic capacity over time in tissue culture in a manner similar to that of human cells (Fisher et al., 1982). In contrast to the bovine PAM, cells from rats or mice require much longer culture times to demonstrate similar enhancement in

phagocytosis (Fisher and McNeill, 1979). Cell yields from lavage
of a single bovine lung lobe often exceed 100 x 10^6 cells, while
lavage of a whole murine lung produces approximately
1 x 10^5 cells. Furthermore, because the cells are obtained from
the same donor under a unique set of lavage conditions, a genetic
and historic homogeneity is present that is lacking when PAM from
many donors are pooled. We have initiated characterization of
lysosomal enzyme activity by studies of bovine PAM elastase and
elastase inhibitors (Valentine et al., 1981). Clearly, with the
widespread availability of bovine-derived tissue culture
components and biochemical supplies, a long-standing precedent
exists for the use of bovine-derived materials.

MATERIALS AND METHODS

Cell Collection

Pulmonary alveolar macrophages were harvested from the
axillary or cardiac lung lobe of a healthy adult steer. The lung
samples were obtained from a local slaughterhouse. Procurement of
the lung lobe involved initial visual examination of the entire
lung for gross abnormality. If the lung was free of visual or
palpable lesions, the bronchus leading to the lobe was isolated by
clamping it with a pair of hemostats and the lobe was excised
approximately 4 cm from the clamp. The lobe was placed in a
plastic bag and covered with crushed ice. Within the hour, the
lobe was taken to the laboratory, where a blunt 15-ga needle was
secured in the bronchus and the lobe was lavaged. Fifty-ml
aliquots of 2 to 4°C Ca-Mg-free, phosphate-buffered saline (PBS)
were introduced into the lobe using a sterile 60-cc syringe until
distention of the lobe. Lavage fluid was withdrawn and stored on
ice in 250-ml sterile centrifuge tubes. A total of 500 ml of
fluid was used for each lobe. Lavage fluid was centrifuged at
150 x g for 30 min. The supernatant was decanted and the cells
were resuspended and pooled using Minimal Essential Medium (MEM;
Gibco, Grand Island, NY). A 1:10 dilution of a portion of the
cell pool with trypan blue was counted in a hemacytometer chamber
to determine cell concentration and viability. Smears of the cell
suspension were stained with Wright-Giemsa and cellular
differentials were determined. The specimen was discarded if it
contained more than 10% neutrophils, lymphocytes, and other
contaminating cell types. Cells were resuspended and then diluted
to maintain appropriate test particle mass:cell ratios.

Functional Assays

Macrophages (2×10^5 viable cells) were added to Leighton tubes containing complete medium. The complete medium was composed of 79% MEM containing 1% antibiotic-antimycotic, 1% l-glutamine, and 20% heat-inactivated fetal bovine serum. Macrophages were attached to glass coverslips in Leighton tubes to obtain a homogeneous, adherent cell population. The cells were then exposed to the control or test particles in the medium and incubated for a period of 20 h. At the end of the incubation, viability was evaluated for two coverslips from each group and cells were challenged with inert, carbonized latex microspheres at a 20:1 particle:cell ratio, and reincubated for 30 min (Fisher et al., 1978a). At the end of 30 min, coverslips were rinsed with PBS, air-dried, and stained with Wright-Giemsa. Cells that lost adherence were collected and counted from the individual tubes after the initial 20-h exposure period and the 30-min challenge with microspheres. Coverslips were then stained with Wright-Giemsa, mounted on slides, and scored, 200 cells per coverslip (50/quadrant) for phagocytic uptake of the inert microspheres.

Dose Preparation

Glass beads (1 to 4 µm, Particle Information Services, Grants Pass, OR) were used as negative control (biologically inert) particles and α-quartz (216 min-u-sil, Whittaker, Clark, and Daniels, Inc., South Plainfield, NJ) was used for the positive control. Both materials were treated in the same manner as the fly ash samples. After weighing, particles (1 to 10 mg) were wet with 20 µl of approximately a 5% solution of Tween-80 in Hanks buffered salt solution (Gibco, Grand Island, NY). The particle suspension was then diluted in MEM and sonicated for 20 min. Appropriate serial dilutions were made from the stock suspension after vortexing. To avoid settling, the particle suspensions were vortexed approximately 10 s each time material was pipetted into a culture tube. To obtain soluble and particulate fractions of the oil fly ash, particles were weighed and suspended in media as previously described and incubated for a period of 20 h. At the end of the incubation, the suspensions were centrifuged at 500 x g for 20 min. The supernatant (soluble fraction) was pipetted into a sterile container and the remaining particles were washed with 0.5 ml of media. The supernatant from each wash was added to the soluble fraction and both the soluble material and the particles were stored at -70°C for subsequent studies. A second soluble fraction was obtained in a similar manner with the following exception. After initial centrifugation, the particle pellet was

equally divided. One-half was washed as described above for the
initial soluble particle fraction and the other half was brought
to the proper concentration with fresh media and reincubated for
20 h.

Test Materials

A cyclonic test particle sampler was designed and fabricated
by Dr. Andrew McFarland, Department of Civil Engineering, Texas
A & M University, for classification of oil fly ash from the
smokestack of a power plant into fine (respirable) and coarse
fractions. The sampler was mounted in the stack of a 485-MW,
oil-fired power plant burning Indonesian crude oil of low-sulfur
(0.16%) and low-ash (< 0.01%) content. Because of sample
limitation, particle sizing was performed by light microscopy by
estimating the diameter of the projected circular area. Particles
greater than 0.625 μm were counted and tallied into one of six
size classes, the last class containing particles from 6.25 to
12.5 μm. The count distribution was converted to a volume
distribution by assuming all particles were spheres and
multiplying the count by the cube of the geometric mean of the
class boundaries. The cumulative distribution of volume was
calculated and fit to a lognormal distribution using the logarithm
of the diameter and conversion of the volume distribution to a
standard z function. After linearizing both parameters, a
standard linear regression was performed.

Volume medium diameter (VMD) of the fine fraction as
determined from the regression analysis was 5.6 μm and the
geometric standard deviation (σg) was 2.3. Correlation
coefficient (r) for the fit was 0.9997. The coarse fraction had a
VMD of 8.9 μm, a σg of 2.5, and an r of 0.9993. VMDs for the
other particulate samples were determined by Coulter analysis at
Texas A & M University. The VMDs of four size-classified,
stack-collected coal fly ash samples were 20, 6.3, 3.2, and
2.2 μm, each with a σg of 1.8 (McFarland, et al., 1977); VMDs of
α-quartz and glass beads were 2.0 and 2.1, with σg's of 1.8 and
2.0, respectively.

Trace Elements

Metal dose-response and interaction studies were conducted
using soluble forms of nickel and vanadium. Stock vanadium and
nickel solutions were atomic absorption standards (Fisher) in the
form of V_2O_5 dissolved in dilute HCl and Ni $(NO_3)_2 \cdot 6H_2O$
dissolved in water.

Statistical Approach

Results are presented as the percentage of control values in Tables 1 and 2 to combiné a series of experiments performed at different times. Statistics were done using data from individual experiments comparing dosage level to control values. Treated and control comparisons were performed using a two-tailed Student t-test with an α value of 0.025.

RESULTS

Four size-classified fractions of western U.S. coal fly ash, glass beads (1 to 4 µm), or silica were incubated with bovine macrophages for 20 h. The physical and chemical properties of the coal fly ash have been previously described (Fisher et al., 1978b; Hansen and Fisher, 1980). Particle size data for the test particles are presented in Table 1. Test particle concentrations were 0.3, 0.1, 0.03, and 0.01 mg/ml for all test substances except quartz. Based on previous studies, quartz was used as a positive control at one concentration, 0.1 mg/ml. Adherence was affected only at the highest concentrations of the finest coal fly ash sample ($p < 0.01$) and glass beads ($p < 0.05$). Extensive cell lysis occurred at this concentration and made further interpretation of the adherence data difficult.

The phagocytic activity (Table 1) of the macrophages was significantly inhibited ($p < 0.001$) at the 0.3-mg/ml dose with the control particles (glass beads). Similar inhibition was observed with all other materials tested at this concentration. The 3.2-µm fraction of coal fly ash had the greatest effect on the cells, causing a significant depression of activity at all 4 doses. The least active fly ash fractions were cuts 1 and 2, the fractions with the largest particle size. The positive control, quartz, inhibited phagocytosis to 22% of control activity at 0.1 mg/ml. Changes in viability were not seen in any of the coal fly ash fractions and only slightly lower viability was observed at the 0.3-mg/ml dose of glass beads and the 0.1-mg/ml dose of silica.

Oil fly ash from the coarse sample was found to have significant ($p < 0.001$) effects on the viability and phagocytosis at all tested concentrations. Adherence was not affected. The fine samples of oil fly ash also did not affect adherence, but viability was reduced for all but the lowest dose. Phagocytosis was again significantly depressed ($p < 0.001$) for the three higher doses of fly ash; but the lowest dose of the fine ash sample did not depress phagocytosis to as great a degree as the coarse

Table 1. Macrophage Function After Exposure to Fly Ash, Glass Beads, or Silica

Phagocytosis (% Control)

Test Particle	0.3 (mg/ml) x̄ (SD)	n	0.1 (mg/ml) x̄ (SD)	n	0.03 (mg/ml) x̄ (SD)	n	0.01 (mg/ml) x̄ (SD)	n
SiO$_2$, VMD 2.0 μm	--		22[a] (32)	12	--		--	
Glass beads, VMD 2.1 μm	55[a] (15)	16	87 (14)	16	94 (9)	4	100 (7)	4
Coal fly ash, VMD								
Cut 1, 20 μm	54[a] (11)	4	77[a] (7)	4	88 (11)	4	94 (16)	4
Cut 2, 6.3 μm	75[a] (7)	4	98 (4)	4	101 (8)	4	98 (12)	4
Cut 3, 3.2 μm	58[a] (16)	4	61[a] (1)	4	85[a] (8)	4	83[a] (4)	4
Cut 4, 2.2 μm	67[a] (3)	4	77[a] (9)	4	71[a] (16)	4	89 (10)	4
Oil fly ash, VMD								
Coarse, 8.9 μm	0[a] (0)	8	0[a] (0)	8	40[a] (23)	8	47[a] (9)	8
Fine, 5.6 μm	2[a] (1)	4	7[a] (3)	4	44[a] (11)	4	75[a] (9)	4

[a] significantly different from control values, $p < 0.05$.

Table 2. Function and Viability of Bovine Macrophages Exposed to Soluble and Particle Fractions of Fine Oil Fly Ash.

| | Phagocytosis (% Control) | | | | | | | | Viability (% Control) | | | |
| | 0.3 mg/ml | | 0.1 mg/ml | | 0.03 mg/ml | | 0.01 mg/ml | | 0.3 mg/ml | 0.1 mg/ml | 0.03 mg/ml | 0.01 mg/ml |
	$\bar{x}$	(SD)	$\bar{x}$	(SD)	$\bar{x}$	(SD)	$\bar{x}$	(SD)	$\bar{x}$	$\bar{x}$	$\bar{x}$	$\bar{x}$
Soluble fraction												
#1	0[a]	(−)	9[a]	(9)	96	(5)	88	(9)	12	36	85	92
#2	0[a]	(−)	2[a]	(2)	98	(6)	96	(7)	0	50	94	85
Particle fraction												
#1	0[a]	(−)	16[a]	(9)	79	(16)	88	(6)	12	36	85	92
#2	20[a]	(5)	91	(22)	82	(26)	93	(9)	34	85	93	96

[a]significantly different from control values, $p < 0.001$.

sample. Glass beads showed a depression of phagocytic activity at
0.3 mg/ml and 0.1 mg/ml (p < 0.02).

Fine oil fly ash and glass beads were incubated in MEM for
20 h to evaluate the relative contributions of soluble and
particulate components to the observed biological activity
(Figure 1). The glass bead soluble fraction had no effect on
adherence, phagocytosis, or viability at any dose level. The
particulate fraction of glass beads demonstrated a significant
depression of phagocytosis and viability only at the highest
concentration (0.3 mg/ml).

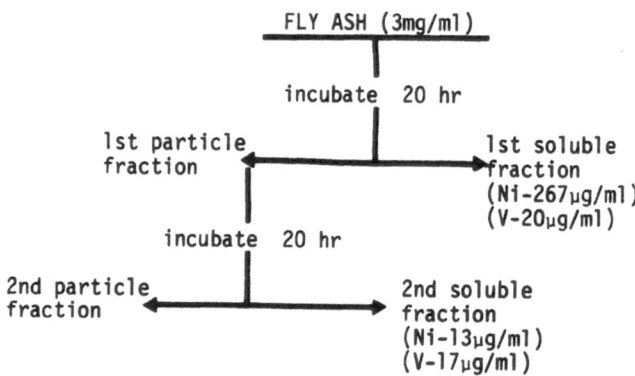

Figure 1.   Sequential separation of soluble and particle
             components in oil fly ash.

Because toxicity of the particle fraction was observed after
the first 20-h extraction, a second sequential 20-h incubation was
performed. Thus, two soluble and two particle fractions were
sequentially separated from the same sample as previously
described (Figure 1). Table 2 shows that viability was decreased
in the two highest concentrations for the first and second
extracts (soluble fractions) and the first remaining particle
fraction; viability in the second particle fraction was depressed
only at the highest concentration. Phagocytosis follows the same
pattern (p < 0.01). The second extraction resulted in fly ash
particles that were less inhibitory of phagocytosis than either

the parent material or particles that were only extracted one time.

Elemental analyses of nickel (Ni) and vanadium (V) in the two soluble fractions (Figure 1) of oil fly ash (3 mg/ml) were performed by atomic absorption spectroscopy. The Ni and V concentrations in the first soluble fraction were 267 and 20 µg/ml, respectively; the second soluble fraction contained 13 µg/ml of Ni and 17 µg/ml of V. In contrast to the oil fly ash, previous studies of the finest coal fly ash sample indicated dissolution of no detectable V or Ni after shaking 0.5-g samples in 50 ml water at room temperature for 24 h, i.e., solution concentrations of less than 0.02 and 0.2 µg/ml for V and Ni, respectively.

Because Ni and V appeared to be readily soluble from the oil fly ash and are generally considered biologically active, further studies were performed to evaluate macrophage response to the individual metals. Macrophages were extremely sensitive to V; the EC_{50} (concentration inhibiting 50% of the phagocytic activity) was ~ 0.4 µg/ml with an associated decrease in viability to 60% of control values. Ni was less toxic to macrophages; the EC_{50} was ~ 10 µg/ml without an effect on viability.

DISCUSSION

In these studies, the bovine PAM was utilized to evaluate the relative toxicity of trace metals and combustion products. Of the four size-classified fractions of coal fly ash, the finest, most respirable particles were the most toxic. Interestingly, it appeared that within the respirable mode, the 3.2-µm fraction demonstrated a significant inhibition of phagocytosis at a concentration lower than that of the finest fraction (2.2 µm). Although these results need further verification, they are consistent with our previous studies of fly ash mutagenicity (Chrisp et al., 1978; Fisher et al., 1979; and Sonneborne et al., 1980). These earlier studies also demonstrated that the finest fractions were the most biologically active and that the mutagenicity of the 3.2-µm fraction was greater than that of the 2.2-µm fraction. It may well be that similar chemical processes occurred in the formation of these fly ash particles resulting in the apparent correlation of particle size and biological activity. In this regard, Fisher and Natusch (1979) have reviewed the size dependency of the physical and chemical properties of coal fly ash.

Oil fly ash was found to be more cytotoxic than coal fly ash. Comparison of equivalent particle sizes (i.e., cut 3, coal fly ash and fine oil fly ash) indicates that the specific cytotoxicity of oil fly ash is approximately 30 times that of coal fly ash. Furthermore, the biologically active component(s) of the oil ash are readily extracted under tissue culture conditions.

As previously demonstrated in studies with rabbit PAM, vanadium was found to be a potent macrophage toxicant. Waters et al. (1974) demonstrated a 50% reduction in phagocytosis by rabbit PAM at 6 µg V/ml; the equivalent EC_{50} for bovine PAM was 0.4 µg/ml. It is not known whether these differences reflect a greater sensitivity of bovine cells or differences in methodology. Furthermore, the sensitivity of human PAM to vanadium has not been reported. In a subsequent study, Waters et al. (1975) reported that on the basis of viability, vanadium was approximately 40 times more toxic than nickel. Our studies, which indicate a 25-fold differential for vanadium compared to nickel, are basically in agreement, although the bovine macrophage is more sensitive than the rabbit PAM.

Garrett et al. (1981a) reported an enhanced sensitivity of the rabbit macrophage to particulate and chemical insults when cells were exposed in serum-free medium. They reported that silica exposure in serum-free medium at 0.1 mg/ml resulted in 16% of control survival and 12% of control ATP content; these results are in agreement with our study indicating that 0.1 mg/ml silica resulted in 22% of control phagocytosis. Thus, it is possible that bovine macrophage response in media containing serum is similar to the rabbit PAM response in serum-free media. Garrett et al. (1981b) also compared the macrophage toxicity of particles from coal gasification, fluidized-bed combustion, and conventional coal combustion. Conventional coal fly ash was found to be most toxic. The EC_{50} for rabbit macrophages without serum was 0.2 mg/ml for both viability and ATP. These results are in general agreement with our studies of fine fly ash at 0.3 mg/ml, again indicating a possible similarity in response to the bovine macrophage in complete media compared to the rabbit macrophage in serum-free media.

Our studies indicate that the active component of the oil ash may be vanadium, and clearly demonstrate the enhanced toxicity of these oil fly ash samples compared to other particulate matter. Further efforts are necessary to specifically demonstrate vanadium as the active component of oil fly ash. Most importantly, other oil fly ash samples should be evaluated to assess the representativeness of our findings.

ACKNOWLEDGMENTS

The authors would like to thank Dr. Andrew McFarland for Coulter information on particle size and Mr. Bruce A. Prentice for the light microscopy particle size analysis. This work was supported by the Electric Power Research Institute (Contract No. RP-1639.2) and the California Air Resources Board.

REFERENCES

Chrisp, C.E., G.L. Fisher, and J. Lammert. 1978. Mutagenicity of respirable coal fly ash. Science 199:73-75.

Fisher, G.L., C.E. Chrisp, and O.G. Raabe. 1979. Physical factors affecting the mutagenicity of fly ash from a coal-fired power plant. Science 205:879-881.

Fisher, G.L., and K.L. McNeill. 1979. Effects of sera on the in vitro function of pulmonary alveolar macrophages. UCD 472-125. National Technical Information Service: Springfield, VA. pp. 56-58.

Fisher, G.L., K.L. McNeill, G.L. Finch, F.D. Wilson, and D.W. Golde. (MS) Functional evaluation of lung macrophages from cigarette smokers and nonsmokers.

Fisher, G.L., K.L. McNeill, C.B. Whaley, and J. Fong. 1978a. Attachment and phagocytosis studies with murine pulmonary alveolar macrophages. J. Reticuloendothel. Soc. 24:243-252.

Fisher, G.L., and D.F.S. Natusch. 1979. Size-dependence of the physical and chemical properties of fly ash. Analytical Methods for Coal and Coal Products, Volume III. C. Karr, ed. Academic Press: New York. pp. 489-541.

Fisher, G.L., B.A. Prentice, D. Silberman, J.M. Ondov, A.H. Biermann, R.C. Ragaini, and A.R. McFarland. 1978b. Physical and morphological studies of size-classified coal fly ash. Environ. Sci. Technol. 12:477-451.

Garrett, N.E., J.A. Campbell, H.F. Stack, M.D. Waters, and J. Lewtas. 1981a. The utilization of the rabbit alveolar macrophage and Chinese hamster ovary cell for evaluation of the toxicity of particulate materials, I. Environ. Res. 24:345-365.

Garrett, N.E., J.A. Campbell, H.F. Stack, M.D. Waters, and
   J. Lewtas. 1981b. The utilization of the rabbit alveolar
   macrophage and Chinese hamster ovary cell for evaluation of
   the toxicity of particulate materials, II. Environ. Res.
   24:366-376.

Hansen, L.D., and G.L. Fisher. 1980. Elemental distribution in
   coal fly ash particles. Environ. Sci. Technol. 14:1111-1117.

McFarland, A.R., R.W. Bertch, G.L. Fisher, and B.A. Prentice.
   1977. A fractionator for size-classification of aerosolized
   solid particulate matter. Environ. Sci. Technol. 11:781-784.

McFarland, A.R., and D.C. Russell. 1981. Design and calibration
   of a stack sampler for oil fly ash. Texas A & M Air Quality
   Laboratory publication no. 4126-01-07-81-ARM: College
   Station, TX.

Sonneborne, J., R.A. Palizzi, C. Herr, and G.L. Fisher. 1981.
   Mutagenicity of fly ash particulates in paramecium. Science
   211:180-182.

Valentine, R., W. Goettlich-Riemann, G.L. Fisher, and R.B. Rucker.
   1981. An elastase inhibitor from isolated bovine pulmonary
   macrophages. Proc. Soc. Exp. Biol. Med. 168:238-244.

Waters, M.D., M.W. Gardner, C. Aranyi, and D.L. Coffin. 1975.
   Metal toxicity for rabbit alveolar macrophages in vitro.
   Environ. Res. 9:32-47.

Waters, M.D., D.E. Gardner, and D.L. Coffin. 1974. Cytotoxic
   effects of vanadium on rabbit alveolar macrophages in vitro.
   Toxicol. Appl. Pharmacol. 28:253-263.

DETERMINATION OF DIRECT-ACTING MUTAGENS AND CLASTOGENS IN OIL
SHALE RETORT PROCESS WATER

D.J. Chen,[1] L.L. Deaven,[2] J. Meyne,[2] R.T. Okinaka,[1] and
G.F. Strniste[1]

[1]Genetics and [2]Experimental Pathology Groups, Life Sciences
Division, Los Alamos National Laboratory, Los Alamos, NM
87545

INTRODUCTION

Shale oil products contain various metabolically active and
photoactive genotoxic components (Strniste and Chen, 1981; Okinaka
et al., 1981). In addition, our preliminary observations
indicated that retort process water contains direct-acting
mutagens that cause significant increases in 6-thioguanine-
resistant (6-TG^R) mutants in Chinese hamster ovary (CHO) cells
(Chen et al., 1981). However, we have been unable to demonstrate
the occurrence of direct-acting mutagens in these process waters
when tested in the standard Ames/Salmonella assay (Nickols and
Strniste, 1981). In this report, we present results concerning
the dose response of direct-acting mutagenicity in an aboveground
retort process (ARP) water and various fractions from it. Since
many mutagenic agents are also clastogenic, we have compared the
cytogenetic and mutagenic effects of this process water under the
same experimental conditions.

MATERIALS AND METHODS

Test Materials

The shale oil process (ARP) water used in this study was
obtained from a holding tank at an aboveground retort facility

that used oil shale deposits located in the western U.S. in the
Green River formation. The product water was filtered before use
as described elsewhere (Strniste and Chen, 1981). ARP water was
further fractionated into base/neutral (B/N), base tar precipitate
(BTP), acid (A), acid tar precipitate (ATP), and residual water
(RW) fractions according to the acid-base extraction procedure
described by Strniste et al. (1982) in these proceedings.

Cytotoxicity and Mutagenicity

CHO cells, line AA8-4, were maintained in suspension culture
as previously described (Strniste and Chen, 1981). Between
2×10^5 and 1×10^6 AA8-4 cells were plated onto 100-mm dishes
containing 12 ml alpha-MEM medium with 10% fetal calf serum (FCS)
and incubated at 37°C for 18 h before treatment. The plated cells
were exposed to various dilutions of the ARP water in medium for
48 h. For cytotoxicity measurements, the cells were rinsed twice
with serum-free alpha-MEM medium, and 200 to 2000 cells from each
treatment dose were plated onto 60-mm dishes with alpha-MEM medium
containing 10% FCS. Cell colonies developed after incubating
dishes for 6 days, and the colonies were fixed in ethanol, stained
with 10% solution of crystal violet, and counted. In addition to
cytotoxicity measurements, treated cells were replated in 100-mm
dishes with 12 ml alpha-MEM medium containing 10% FCS and
subcultured every 2 days for an expression period of 8 days.
Selection protocols used for measuring mutagenicity at the
hypoxanthine-guanine phosphoribosyl transferase (HGPRT) locus
using this cell line have been reported elsewhere (Strniste and
Chen, 1981; Chen and Strniste, 1982). Plating efficiencies for
nontreated cells were between 80 and 90%.

Chromosome Preparation and Analysis

Cultures treated as indicated above were rinsed twice with
serum-free alpha-MEM, refed with fresh medium containing 10% FCS,
and incubated at 37°C for 24 h. Two hours before harvest,
colcemid (0.1 µg/ml) was added to each culture. The cells were
removed with trypsin, treated with 0.075M KCl for 10 min, and
fixed in 3:1 methanol:acetic acid. Slides were prepared by
flame-drying and stained in 2% Giemsa for 10 min. The percentage
of cells with at least one break or exchange per 100 metaphase
cells was determined. Achromatic lesions (gaps) were not scored.
The results are the mean of two experiments.

RESULTS AND DISCUSSION

To avoid cell density effects, cytotoxicity was determined by replating treated cells as described in Materials and Methods. Cytotoxicity induced in CHO cells after exposure to ARP water or various fractions from it for 48 h in the dark (37°C) is shown in Figure 1. The ARP water is extremely toxic to CHO cells under the described treatment conditions. The concentration of ARP water necessary for inactivation to 37% surviving fraction CHO was 0.7% (v/v). The various acid-base extracted fractions and resulting precipitates were diluted to the equivalent volume of the original process water. The most toxic of these various fractions was the BTP followed in order of decreasing toxicity by B/N, ATP, A, and RW fractions. However, the magnitude of the toxic effect for each individual fraction was less than the original process water.

In Figure 2, we show mutagenicity induced in CHO cells by dilutions of the ARP water or various fractions from it. The data indicate significant mutagenic activity for the ARP water when treated in CHO cells without metabolic activation. A tenfold increase in 6-TG^R mutants was seen for cells exposed to a 0.8% solution of the process water for 48 h in the dark. Due to the extreme cytotoxic effects (Figure 1), increased ARP concentrations resulted in a decrease in mutation frequency. When various fractions of the ARP water were tested for direct-acting mutagenic potential, negative results were obtained.

Table 1 summarizes the data on the cytogenetic effects observed in cells treated with ARP water. There was a significant, dose-related increase in the percentage of cells with chromosomal aberrations. The aberrations were primarily chromatid breaks and exchanges. At concentrations of 0.3 to 1.0% ARP water, the frequency of cells demonstrating tetraploidy and endoreduplication (diplochromosomes) ranged from 11 to 13% (see Figure 3). Tetraploid cells are often the result of endoreduplication during a previous cell division (Sutou and Tokuyama, 1974). Therefore, the apparent dose-related increases of endoreduplicated cells and the concurrent decreases in tetraploid cells are probably due to proliferative delay at the higher concentrations of the ARP water.

The cytogenetic analysis of the various fractions at concentrations of 0.1 to 3.0% indicated that only the B/N fraction contained components that induce chromosomal aberrations and endoreduplication. The 3% concentration of the B/N fraction induced chromosomal aberrations in 16% of the cells, tetraploidy in 11% of the cells, and 28.5% of the metaphases were

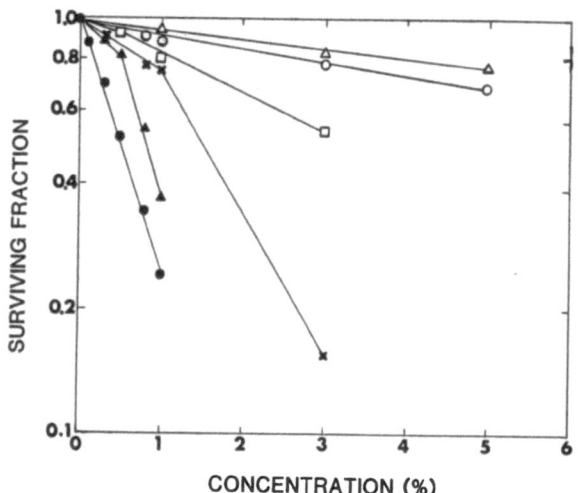

Figure 1.   Cytotoxicity in CHO cells treated with oil shale retort
            process water and various fractions from it for 48 h at
            37°C in dark. Samples used are: (●) ARP water;
            (▲) base tar precipitate; (x) base/neutral fraction;
            (□) acid tar precipitate; (o) acid fraction; and
            (Δ) residual water.

endoreduplicated. The 0.1 to 1.0% concentrations of this fraction
did not induce increased frequencies of any of these parameters.

     The lack of mutagenic potential in the various ARP fractions
indicates that components that were responsible for the mutagenic
activity may have been selectively lost during the extraction
procedure. Alternatively, direct-acting mutagenic components may
be equally distributed among the various fractions diluting their
activity (approximately fivefold) to nondetectable levels in this
particular assay. Comparing cytotoxic and mutagenic activities of
the ARP water and various fractions from it indicates that
measurements of cytotoxicity are not necessarily indicators of
mutagenic potential for these complex mixtures.

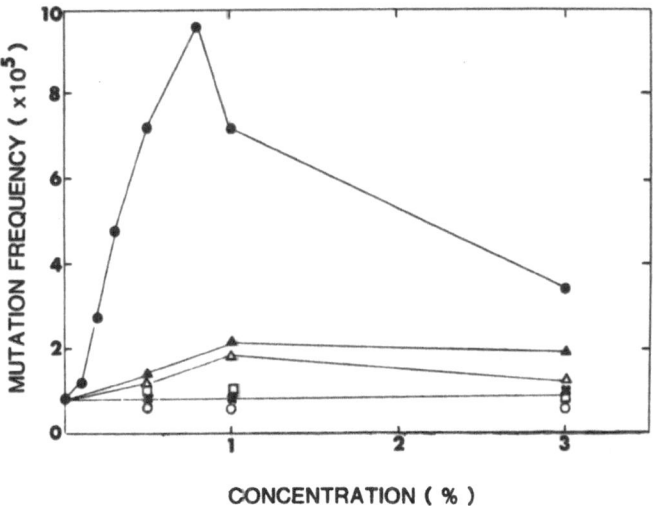

Figure 2. Mutagenicity in CHO cells treated with ARP water and various fractions from it for 48 h at 37°C in dark. Samples used are: (●) ARP water; (▲) base tar precipitate; (x) base/neutral fraction; (□) acid tar precipitate; (o) acid fraction; and (△) residual water.

The induction of endoreduplication in CHO cells following treatment with shale-derived oils has been previously reported by Deaven and Nock (1979). In addition to mutagenicity, the induction of chromosomal breaks, endoreduplication, and tetraploidy by ARP water indicates that shale-derived by-products contain direct-acting genotoxins. Furthermore, the results of this study suggest that employment of mammalian assays in a testing strategy should eliminate false negative responses observed in the Ames test for some of these direct-acting components in complex mixtures.

Table 1. Cytogenetic Effects of an Oil Shale Retort Process Water
          in CHO Cells

ARP Water %	% Cells with Chromosomal Aberrations	% Tetraploid Cells	% Endoreduplicated Cells
0	0.5	0.2	0
0.1	2.5	5.5	0
0.3	6.0	11.5	1.0
0.5	13.0	5.0	6.0
0.8	26.0	4.0	9.5
1.0	39.5	1.0	12.0

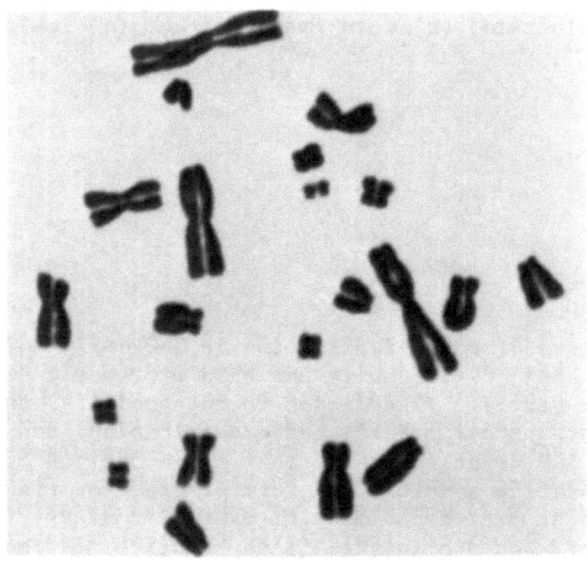

Figure 3. An endoreduplicated CHO metaphase cell observed after
           oil shale retort process water treatment.

ACKNOWLEDGMENTS

    This work was funded by the Department of Energy and the
U.S. Environmental Protection Agency. We thank Ms. E. Wilmoth for
technical assistance.

REFERENCES

Chen, D.J., R.T. Okinaka, and G.F. Strniste. 1981. Determination
    of direct-acting mutagens in shale oil retort process water.
    In: The Los Alamos Integrated Oil Shale Health and
    Environmental Program: A Status Report. LA-8665-SR. Los
    Alamos National Laboratory: Los Alamos, NM. pp. 59-60.

Chen, D.J., and G.F. Strniste. 1982. Cytotoxic and mutagenic
    properties of shale oil byproducts, II: comparison of
    mutagenic effects at five genetic markers induced by retort
    process water plus near ultraviolet light in CHO cells.
    Environ. Mutagen.

Deaven, L.L., and A. Nock. 1979. The induction of
    endoreduplication in Chinese hamster cells following
    treatment with shale-derived oils. Mamm. Chromosomes
    Newsletter 20:10.

Nickols, J., and G.F. Strniste. 1981. Ames/Salmonella mutagen
    assay of oil shale process waters. In: The Los Alamos
    Integrated Oil Shale Health and Environmental Program: A
    Status Report. LA-8665-SR. Los Alamos National Laboratory:
    Los Alamos, NM. pp. 54-57.

Okinaka, R.T., and G.F. Strniste. 1981. Exogenous metabolic
    activation of process waters in mammalian cell cultures. In:
    The Los Alamos Integrated Oil Shale Health and Environmental
    Program: A Status Report. LA-8665-SR. Los Alamos National
    Laboratory: Los Alamos, NM. pp. 57-59.

Strniste, G.F., and D.J. Chen. 1981. Cytotoxic and mutagenic
    properties of shale oil byproducts, I: activation of retort
    process waters with near ultraviolet light. Environ.
    Mutagen. 3:221-231.

Sutou, S., and F. Tokuyama. 1974. Induction of endoreduplication
    in cultured mammalian cells by some chemical mutagens.
    Cancer Res. 34:2615-2623.

UNSCHEDULED DNA SYNTHESIS IN HAMSTER TRACHEAL EPITHELIUM EXPOSED

IN VITRO TO CHEMICAL CARCINOGENS AND ENVIRONMENTAL POLLUTANTS

Leonard J. Schiff,[1] Susan F. Elliott,[1] Steven J. Moore,[1]
Mary S. Urcan,[1] and Judith A. Graham[2]

[1]Life Sciences Division, IIT Research Institute, 10 West 35th
Street, Chicago, IL 60616, and [2]Inhalation Toxicology
Division, Health Effects Research Laboratory
U.S. Environmental Protection Agency, Research Triangle Park
NC 27711

INTRODUCTION

The Syrian golden hamster has been used extensively for
studies of respiratory carcinogenesis (Saffiotti, 1969;
Nettesheim, 1972; Becci et al., 1978). Using organ cultures of
the target tissue (i.e., respiratory epithelium) obtained from the
hamster, a middle ground has been established between animal
experiments and the use of cell cultures for carcinogenesis
research (Saffiotti and Harris, 1979). Metabolic studies on
environmental chemical carcinogens have been conducted with
tracheas of hamsters (Harris et al., 1973; Kaufman et al., 1974).
Some carcinogens can be metabolically activated in the tracheal
epithelium, and their interaction with cellular macromolecules can
be localized at the molecular level by biochemical techniques or
at the cellular level by autoradiography. Studies in cultured
human bronchus have shown that the metabolic pathway leading to
the major benzo(a)pyrene (B[a]P)-DNA adducts is similar to that
found in hamsters (Harris et al., 1978). Sixty-five percent of
the B(a)P bound to DNA in human cells was removed after 10 days in
culture. Since a variety of mammalian cells, including human, can
repair damage to DNA caused by chemical carcinogens, it was
speculated that DNA repair was responsible for this observation.
Thus, by using hamster tracheas in organ culture maintaining
specific metabolic enzymes, we may obtain a system relating to
human respiratory carcinogenesis that is useful for screening

environmental chemicals and complex mixtures capable of damaging DNA.

In this paper, we report on the ability of carcinogenic and noncarcinogenic polycyclic hydrocarbons and complex environmental pollutants to elicit unscheduled DNA synthesis (UDS) in hamster tracheal epithelial cells exposed in organ culture.

MATERIALS AND METHODS

Chemicals

Pyrene, benzo(e)pyrene (B[e]P), and B(a)P were obtained from Aldrich Chemical Co. (Milwaukee, WI) and dimethylbenz(a)anthracene (DMBA) from Fisher Scientific Co. (Pittsburgh, PA). The complex environmental mixtures, coal-fired fly ash, diesel fuel exhaust extract, and cigarette smoke condensate were supplied by the U.S. Environmental Protection Agency. Coal-fired fly ash was characterized by Oak Ridge National Laboratory (Department of Energy and Electric Power Research Institute, 1979) and recently described (Schiff et al., 1981). The collection, characterization, and description of the diesel fuel exhaust extract and cigarette smoke condensate were described by Huisingh et al. (1980). The test chemicals and mixtures were dissolved in reagent-grade dimethyl sulfoxide (DMSO; Fisher Scientific Co.) and diluted in serum-free CMRL Medium 1066. The final concentration of DMSO in culture was 0.05%.

Tracheal Organ Cultures

Tracheal organ cultures were obtained from 4- to 6-week-old male Syrian golden hamsters of the Ela:ENG(SYR) strain (Engle Laboratory Animals, Inc., Farmersburg, IN). The hamsters were anesthetized with sodium methohexital (Brevital Sodium, Eli Lilly, Indianapolis, IN) by injecting intraperitoneally, 4 mg/animal. Their tracheas were removed by sterile techniques and cut between the first two cartilage rings just below the larynx and immediately above the bifurcation. Each trachea was opened from the larynx to the carina along the membranous dorsal wall and placed into a 60- x 15-mm culture dish containing approximately 2 ml of a serum-free medium consisting of CMRL Medium 1066 with crystalline bovine insulin, 1.0 µg/ml; hydrocortisone hemisuccinate, 0.1 µg/ml; glutamine, 2 mM; penicillin, 100 units/ml; and streptomycin, 100 µg/ml. The culture medium, glutamine, and antibiotics were obtained from GIBCO, and the hormones (insulin and hydrocortisone) were purchased from Sigma

Chemical Co. (St. Louis, MO). The cultures were placed in a controlled atmosphere chamber (Bellco Glass, Vineland, NJ) and gassed for 5 min at 15 l/min with a water-saturated atmosphere of 50% O_2, 45% N_2, and 5% CO_2 before incubation at 35 to 36°C. The chamber containing the culture dishes was placed on a platform rocker (Bellco Glass) and rocked approximately 10 times per minute to allow the tracheas contact with both gas and culture medium.

Autoradiographic Assay for Unscheduled DNA Synthesis

Eighteen to 24 h after initiation of tracheal cultures, the culture medium was replaced with serum-free CMRL Medium 1066 containing 2.0 mM hydroxyurea (Sigma Chemical Co.) and the explants were incubated for 2 h in the 50% O_2 mixture. This medium was then removed and replaced with serum-free CMRL Medium 1066 containing 2.0 mM hydroxyurea, 5 μCi [methyl-^{3}H]-thymidine/ml (specific activity 20 Ci/mmol; Research Products International Corp., Elk Grove, IL), and either 0.1 or 1.0 μg pyrene, B(e)P, B(a)P, or DMBA, or 1.0 or 10.0 μg coal-fired fly ash, diesel fuel exhaust extract or cigarette smoke condensate. Control cultures were incubated in similar medium containing no carcinogen or environmental pollutants. Cultures were then incubated for an additional 5 h in 50% O_2, 45% N_2, and 5% CO_2.

After incubation, the cultures were washed several times with cold Hanks balanced salt solution and fixed with 10% neutral buffered formalin. Each trachea was cut transversely into quarters of 3 to 4 rings/quarter. The four portions were embedded in paraffin as close together as possible in the middle of the block, making sure they were level and standing so that transverse sections could be microtomed. Five-micron sections were deparaffinized in xylene and dipped in Kodak NTB2 nuclear track emulsion (1:1 distilled water) at 42°C. Slides were placed in a light-tight box and exposed at 4°C for 1 week. The autoradiographs were developed and stained with hematoxylin and eosin.

The mean grain count per nucleus was made under a x100 objective lens in a Zeiss microscope. Grains were counted on 400 to 800 epithelial cells from the mucosa of each of the four sections per trachea. Background grain counts were made from an area adjacent to the nucleus and subtracted from the nuclear grain count.

A compound was considered to have induced a positive UDS response when both concentrations tested produced nuclear grain counts that exceeded those of the DMSO control by at least two standard deviations of the control value.

RESULTS

Preliminary studies were performed to determine if treatment
with hydroxyurea would inhibit the DNA repair process in tracheal
organ culture. As shown in Table 1, hydroxyurea did not have a
toxic effect and did not interfere with DNA repair. In addition,
hydroxyurea prevented epithelial cells from entering the S phase
of the cell cycle since heavily labeled nuclei were not present.

Table 1. Influence of 2 mM Hydroxyurea
on Unscheduled DNA Synthesis

Treatment	DNA Repair Synthesis[a] (grains/nucleus)
0.05% DMSO	0
0.01 μg B(a)P/ml	1.9 ± 0.6[b]
0.1 μg B(a)P/ml	5.0 ± 2.2
1.0 μg B(a)P/ml	8.0 ± 4.1
0.05% DMSO + hydroxyurea	0
0.01 μg B(a)P/ml + hydroxyurea	2.3 ± 0.5
0.1 μg B(a)P/ml + hydroxyurea	6.3 ± 2.0
1.0 μg B(a)P/ml + hydroxyurea	8.3 ± 3.0

[a]Number of grains per cell are the average of eight tracheal rings
from two different animals. Grain counts were made on 200 to
400 cells of the epithelial layer of each tracheal ring.
[b]Mean ± standard deviation.

Table 2 shows the UDS response of tracheal explants to
nontoxic doses of carcinogens and environmental pollutants. The
two concentrations of polycyclic hydrocarbons and complex
environmental pollutants tested were chosen because they did not
produce cytotoxicity in the tracheal epithelium. Cytotoxicity was
indicated when grossly degenerated exfoliated cells were observed
in the lumen of tracheal rings. Concentration-dependent
stimulation of UDS was induced by B(a)P and DMBA, as well as
diesel fuel exhaust extract and cigarette smoke condensate.
Similar treatment with B(e)P, pyrene, and coal-fired fly ash
failed to educe demonstrable UDS in hamster tracheal cultures.
Although UDS was observed in only 3 to 5% of the tracheal

Table 2.   Unscheduled DNA Synthesis in Tracheal Epithelium Treated
with Chemical Carcinogens and Particulate Air Pollutants
in Organ Culture[a]

Compound	Concentration (μg/ml)	% of Cells Showing UDS	UDS (grains/nucleus)[b]
DMSO (control)		0.5	0.7 ± 0.3
B(a)P	0.1	44	6.5 ± 2.1
	1.0	67	8.0 ± 3.4
B(e)P	0.1	5	4.8 ± 2.8
	1.0	5	6.2 ± 1.9
Pyrene	0.1	3	5.5 ± 3.1
	1.0	5	6.1 ± 2.7
DMBA	0.1	52	9.2 ± 4.4
	1.0	61	16.1 ± 5.5
Coal-fired fly ash	1.0	2	2.8 ± 1.4
	10.0	5	3.1 ± 1.5
Diesel fuel exhaust extract	1.0	48	8.6 ± 3.3
	10.0	63	14.5 ± 5.0
Cigarette smoke condensate	1.0	69	14.0 ± 6.2
	10.0	75	18.0 ± 5.5

[a]12 sections; 3 whole tracheas divided into quarters of 3 to
4 rings/quarter.
[b]Mean ± standard deviation.

epithelial cells exposed to the noncarcinogen, pyrene, the mean
grain count/nucleus was similar to the grain count of the
carcinogen, B(a)P.

In tracheal explants treated with B(a)P, DMBA, diesel fuel
exhaust extract, or cigarette smoke condensate, UDS was
demonstrated in 44 to 75% of the mucosal cells. The superficial
cells of the mucosa, consisting of both ciliated and mucous cells
showed the majority of labeling.

DISCUSSION

An autoradiographic assay for measuring carcinogen-induced DNA damage in tracheal organ cultures was examined to establish a practical test that is responsive to environmental carcinogens. Our results indicate that epithelial cells from tracheal organ cultures are capable of repairing carcinogen and complex environmental mixture-induced DNA damage as evidenced by UDS. The use of organ cultures of hamster tracheal mucosa obtained from the in vivo animal model of respiratory carcinogenesis (Saffiotti, 1972) is an important factor in establishing an assay system for environmental carcinogens. Procarcinogens can be metabolically activated in the tracheal epithelial explants, and their interaction with cellular macromolecules can be localized at the cellular level by autoradiography (Harris et al., 1973).

The results with coal-fired fly ash particles indicate that using DMSO as a solvent may not be an efficient method for leaching chemical carcinogens from airborne particulates that damage DNA. Recently, we have shown that fly ash from a coal-fired power plant can induce histological alterations in hamster tracheal epithelium similar to those of carcinogenic hydrocarbons (Schiff et al., 1981). However, to induce such an effect, incorporation of the particles into the tracheobronchial epithelium was necessary.

The hamster tracheal organ culture system described here is a relatively simple procedure that can detect direct-acting and activation-dependent carcinogens, as well as evaluate complex environmental mixtures for carcinogenic potential. The application of this UDS procedure for detecting DNA repair needs further validation with known respiratory carcinogens and complex environmental mixtures.

ACKNOWLEDGMENTS

This work was supported by U.S. Environmental Protection Agency grant R807354010.

REFERENCES

Becci, P.J., E.M. McDowell, and B.F. Trump. 1978. The respiratory epithelium, IV: histogenesis of epidermoid metaplasia and carcinoma in situ in the hamster. J. Natl. Cancer Inst. 61:577-586.

Department of Energy and Electric Power Research Institute. 1979. Identification and quantification of polynuclear organic matter (POM) on particulates from a coal-fired power plant. DOE RTS 77-58 and EPRI-1092. Electric Power Research Institute: Palo Alto, CA.

Harris, C.C., D.G. Kaufman, M.B. Sporn, H. Boren, F. Jackson, J.M. Smith, J. Pauley, P. Dedick, and U. Saffiotti. 1973. Localization of benzo[a]pyrene-^{3}H and alterations in nuclear chromatin caused by benzo[a]pyrene-ferric oxide in the hamster respiratory epithelium. Cancer Res. 33:2842-2848.

Harris, C.C., H. Autrup, and G. Stoner. 1978. Metabolism of benzo[a]pyrene in cultured human tissues and cells. In: Polycyclic Hydrocarbons and Cancer: Chemistry, Molecular Biology and Environment, Volume 2. P.O.P. Ts'o and H.V. Gelboin, eds. Academic Press: New York. pp. 331-342.

Huisingh, J.L., R.L. Bradow, R.H. Jungers, B.D. Harris, R.B. Zweidinger, K.M. Cushing, B.E. Gill, and R.E. Albert. 1980. Mutagenic and carcinogenic potency of extracts of diesel and related environmental emissions: study design, sample generation, collection, and preparation. Presented at U.S. Environmental Protection Agency Symposium on the Health Effects of Diesel Engine Emissions, Cincinnati, OH.

Kaufman, D.G., V.M. Genta, and C.C. Harris. 1974. Studies on carcinogen binding in vitro in isolated hamster trachea. In: Experimental Lung Cancer. E. Karbe and J.F. Park, eds. Springer-Verlag: Berlin, Heidelberg, New York. pp. 564-574.

Nettesheim, P. 1972. Respiratory carcinogenesis studies with the Syrian golden hamster: a review. Prog. Exp. Tumor Res. 16:185-200.

Saffiotti, U. 1969. Experimental respiratory tract carcinogenesis. In: Progress in Experimental Tumor Research, Volume 11. F. Homburger, ed. S. Karger: Basel, New York. pp. 302-333.

Saffiotti, U. 1972. The laboratory approach to the identification of environmental carcinogens. In: Proceedings of the Ninth Canadian Cancer Research Conference. P.J. Scholenfield, ed. University of Toronto Press: Toronto. pp. 23-36.

Saffiotti, U., and C.C. Harris. 1979. Carcinogenesis studies on
    organ cultures of animal and human respiratory tissues. In:
    Carcinogenesis: Identification and Mechanisms of Action.
    A.C. Griffin and C.R. Shaw, eds. Raven Press: New York.
    pp. 65-80.

Schiff, L.J., M.M. Byrne, and J.A. Graham. 1981. Fly ash-induced
    changes in hamster tracheal epithelium in vivo and in vitro.
    J. Toxicol. Environ. Health 8:431-448.

IN VITRO MUTAGENICITY TESTING OF OHIO COAL-DERIVED MATERIALS

Rita Schoeny and David Warshawsky

Department of Environmental Health, Kettering Laboratory,
University of Cincinnati Medical Center, Cincinnati, OH 45267

INTRODUCTION

In the effort to meet United States energy needs from
domestic sources, coals of many origins are being studied for
their feasibility in industrial and home heating use. The Ohio
Coal Research Laboratory Association, a consortium of
12 university laboratories, has as its objective the study of
methods by which Ohio coals may be used productively. Of
particular interest is the application of technologies to the
efficient use of high-sulfur coals by this region's basic
industries. Our concern has been with health effects of increased
coal use. It is likely that more extensive coal combustion,
particularly high-sulfur coal, would add to the air pollution
burden. Processing coal to cleaner-burning fuels, however, is
also not without risks. For example, it is known that workmen
employed in the carbonization of coal for coke or for generation
of gas have higher than expected risks of cancer at various sites
(Lloyd, 1971). Likewise, processing of coal to liquid fuels has
been associated with increased cancer risk, and carcinogenicity of
coal-derived liquids has been demonstrated experimentally (Weil
and Condra, 1960; Ketcham and Norton, 1960; Battelle, 1979).
Considering the potential biohazards of using coal in whatever
form, the necessity for evaluating risks at an early stage in
technological development becomes obvious.

Mutagenicity is a reasonable end point to use in a program of hazard assessment. The integrity of the human gene pool is compromised by increases in the mutation frequency of the population. One must also consider the possibility of mutational damage to ecological balances maintained by plant and animal species. Mutagenesis can serve as a useful indicator of other deleterious bioeffects; e.g., carcinogenicity and teratogenicity. The Salmonella/microsome assay for mutagenicity, which this laboratory has been using, has been demonstrated to have a high correlation of positive responses with known carcinogens (McCann and Ames, 1976; Rosenkranz and Poirier, 1979). This assay has been used successfully in fossil-fuel research in the recognition of potentially hazardous materials (Pelroy and Petersen, 1979; Schoeny et al., 1981; Epler et al., 1978); in the evaluation of fractionation, collection, and other study techniques (Guerin et al., 1981; Pelroy et al., 1981); and in the identification of mixture components responsible for bioeffects (Guerin et al., 1980).

MATERIALS AND METHODS

Mutagenicity Assays

The Salmonella/microsome reverse mutation assays (Ames test) have been done according to established procedures (Ames et al., 1975). Hepatic extracts for activation (S9) were prepared from male Sprague-Dawley rats treated IP with 500 mg/kg body weight Aroclor 1254 in corn oil.

Samples for Assay

Dr. William Kneller of the University of Toledo supplied a set of eight powdered coals collected from four Ohio mines. Dr. Robert L. Savage of Ohio University contributed a set of samples collected in an experimental flash carbonization unit. In addition to the Ohio coals and coal-derived materials, the following were obtained from the U.S. Environmental Protection Agency-Department of Energy (EPA-DOE) Fossil Fuels Research Materials Facility: CRM 1.13, Coal Oil A; CRM 2.11, Crude Shale Oil A; CRM 3.13, Petroleum Crude A; 1106.09, Syncrude (4-100A); 1202.07, Synthoil (13-100A). These are materials for which mutagenicity data exist in the published literature or which are similar to those studied in our laboratory. They were obtained to serve as reference standards.

Preparation, Testing, and Fractionation of Samples

All materials were stored at 5°C in the dark. Samples were prepared for testing by weighing a small amount, 20 to 70 mg, and adding dimethyl sulfoxide (DMSO) so as to obtain a presumptive concentration of 10 mg/ml. In cases where the material did not dissolve, the amount of insoluble sample was subtracted from the total to give the adjusted concentrations used in calculating the mutagenic rates. Except in those cases where the volume was very small, the sample solutions were filter sterilized prior to testing. Dilutions were made in DMSO. In some instances where there was no detectable mutagenicity, the concentration of the starting material was increased until limited by bacterial toxicity or other pragmatic considerations. Samples were applied in 0.1-ml volumes for reverse mutation assays and as 0.01-ml aliquots for forward mutation assays. At least two strains (TA98 and TA100) were used for all reverse mutation assays, both with and without S9. Whenever possible, strains TA1535, TA1538, and TA1537 were also used. Duplicate assays were performed on at least two sample preparations.

Fractionation was done by serial extraction using solvents of increasing polarity; namely, hexane, toluene, methylene chloride, and acetonitrile. The organic extracts were prepared by shaking the sample with an amount of solvent equal to 5 times the sample weight for 2 h. After centrifugation to settle particulates, the solvent was removed and an equal volume of fresh solvent added. These two extracts were pooled and evaporated. The extracted materials were prepared and tested as described in the foregoing section, using strains TA98 and TA100 only.

RESULTS AND DISCUSSION

Mutagenicity of Solutions Prepared from Ohio Coals

Results of all mutagenicity assays of the eight unfractionated coal samples, as well as their organic solvent fractions, are summarized in Table 1. In general, relatively little material from powdered coal proved to be soluble in the DMSO. To date, no other solvents have been tried on these particular samples, as past experience has indicated DMSO to be the most efficient solvent for coal-related materials. None of the samples contained DMSO-soluble material that was mutagenic for strain TA1535 (a missense mutagen detector strain) or for any other strain in the absence of exogenous metabolic activation systems, that is, in the absence of rat liver homogenate (S9). We have defined positive responses by either of two sets of criteria.

Table 1. Comparison of Coal Fraction with Unfractionated Coal
          Mutagenicity

| | Revertant colonies per milligram | | | |
| | Whole Sample[a] | | Sum of Fractions[a,b] | |
Coal Sample	TA98	TA100	TA98	TA100
1	304	–	205	5
2	393	722	191	432
3	–	–	156	1
4	–	–	< 1	6
5	120	–	220	4037
6	–	–	1	5
7	177	–	130	1
8	135	–	1	1
Kentucky Coal No. 11 A			17	40
Kentucky Coal No. 11 B			–	28

[a]Calculated from linear portion of dose-response curves done on
same occasion as fraction assays; 8 plates/dose point; – = no
linear dose response.
[b]Numbers are sums of weighted mutagenicity estimates of all
fractions.

An increase in the number of revertant colonies equal to or
greater than 2 times the number of revertants observed to have
occurred spontaneously is interpreted as a positive response. By
this standard, coal 4 appeared to be weakly mutagenic for TA1538,
while TA98 was reverted by DMSO solutions from coals 2, 5, 7,
and 8 (data not shown). The second criterion for a mutagenic
response is a sample dose-dependent increase in numbers of
revertants. Table 1 presents estimates of activity in revertants
per microgram sample derived from regression analyses of data
points that appeared to contribute to a linear dose response. All
these activities were rather low, as compared, for instance, with
that of the carcinogen/mutagen benzo(a)pyrene. In many cases,
there appeared to be a dose-dependent increase in activity that
did not reach the level of 2 times the number of spontaneous
revertants; e.g., coal 1 for TA98 and TA100, coal 4 for TA98
response (data not shown).

As a means of investigating further the mutagenic potential
of coals, we undertook a fractionation procedure based on removal
of increasingly polar materials, as described in Materials and
Methods. In all cases, the organic solvent fractions comprised
very small amounts of material; 95 to 99% of the coal mass
remained as the nonextractable residue at the end of the serial
extraction process. This residue (that is, the DMSO-soluble
components thereof) was mutagenic for coals 1, 3, and 7, and
probably active for coals 2 and 5. The methylene chloride
fraction was the only one which proved to be mutagenic for all
eight samples. For all but sample 3, the acetonitrile extract was
active in three instances (coals 1, 4, and 5), and was 10 times as
mutagenic as any other raw coal materials tested. As these
components are present in very small amounts, their overall
contribution to the mutagenicity of whole coal was rather small.

The mutagenicity of whole coal is not necessarily equal to
the sum of fraction mutagenicities. For coals 3 and 5, there was
an apparent increase in mutagenicity upon fractionation. This
could be due to artifacts of sample preparation and testing. Our
fractionation technique, however, was selected because it was
believed to be less likely to generate oxygenated or otherwise
modified components than would processes involving the use of
strong base or acid. It should be noted in this context that no
direct-acting mutagens (that is, mutagenic without S9) were found
in any fraction. It may be that some combinations of coal
materials act in an antagonistic fashion or are in some way
inactive or toxic until separated into fractions. Sample 8
appeared to lose activity upon fractionation. Again there may be
a trivial explanation, such as the loss of components during the
extraction process. However, for this sample, there was a change
of only 1% in the total sample weight after fractionation. The
fractionation scheme may instead have separated components which
acted as co-mutagens or in some other synergistic manner.

Included in Table 1 are data from the assay of two samples of
Kentucky Coal No. 11 mined from the same seam. The ash content of
sample A was 6.4%, while that of B was 18.0% (data supplied by
DOE, Pittsburgh Energy Technology Center). It appears that the
majority of Ohio coals tested have a greater degree of mutagenic
activity for Salmonella than do these Kentucky coals. It is
likely that as the concentration of whole coal materials and
organic solvent fractions tested were low and the mutagenic
response slight, the calculated numbers of revertants per
milligram may be overestimates for some of them. That certain
organic fractions had increased mutagenicities over the parent
coals has implications for coal processing. Our solvent
extraction was done at room temperature. The use of solvent

refining under conditions of increased temperature and pressure
will in all probability result in the production or concentration
of mutagenic materials. This has been the case for numerous coal
processes analyzed to date (e.g., Battelle, 1979; Guerin et al.,
1981).

Mutagenicity of Products of Flash Carbonization

A set of processed Ohio coal materials was generated in a
laboratory scale flash carbonization unit developed by
Dr. R.L. Savage of Ohio University. Flash carbonization is being
studied as an alternative to total coal gasification, in which
about half the coal energy is lost. The flash carbonization
process being studied consists of reacting powdered coal for short
periods of time under conditions of high pressure and temperatures
(2125°F to 2150°F for the samples described herein). In this
process, the gas produced (which contains a large percentage of
the coal sulfur) is separated immediately from the coal residue or
char. This hot material is quenched with steam to retain a
sufficient content of volatile to permit its use as a fuel. Thus,
there are two fuel products: a gas mixture of CO and H_2 which
could be further converted to methane or hydrogen; and a reduced
sulfur char which can be burned directly (Savage, 1980).

The samples provided for mutagenesis assay were three sets of
chars from the main collection drum or from the cyclone used to
separate the gas from the remaining residue. Also included in
this set was the Southern Ohio coal used in the flash
carbonization unit.

The char samples were soluble in DMSO to about the same
extent as were material from the Ohio coals; that is, at about
0.75 to 4 mg/ml. None proved to be mutagenic in the absence of
S9. Table 2 summarizes mutagenicity data for five Salmonella
strains. The Southern Ohio coal sample was somewhat mutagenic for
TA98, probably slightly mutagenic for TA100, and did not revert
TA1538. It was thus similar to the previously described set of
eight coals on the basis of whole coal mutagenicity. The chars
recovered from the cyclone separator were without exception more
mutagenic than those from the drum. Proximate analyses of these
materials (supplied by N.G. Patke and R.L. Savage of Ohio
University) indicated that the only major difference between
cyclone and drum samples was in the greater moisture content of
the former; i.e., 1.87 to 3.70 weight percent for drum samples,
and 12.85 to 21.25% for the cyclone material. The moisture
content of the samples does not appear to be related to their DMSO
solubility, nor are the more soluble samples more mutagenic. Of

Table 2. Mutagenicity of Flash Carbonization Materials

Sample[a]	Mutagenicity - colonies per microgram[b]				
	TA1535	TA1537	TA1538	TA98	TA100
Southern Ohio coal	0.05	0.08	0	0.14	0.19
Drum samples					
E	_0.06	0.14	0.74	0.17	0.5_
G	0	0.17	0.32	0.92	3.5_
I	0	0.16	0.48	0.38	0.80
Cyclone samples					
F	0	1.18	0.75	1.52	4.24
H	0.11	0.54	0.28	1.86	4.26
J	0.08	0.68	0.23	1.11	2.82

[a]Conditions of sample generation were the following: E and F, 2125°F, 3.5-h run; G and H, 2125°F, 3.0 h; I and J, 2150°F, 2.6 h.
[b]Colonies per microgram sample as determined by regression analysis of all dose-response data; 0 = no apparent dose response. Each plate contained 50 µl/plate S9 in standard mix (Ames, 1975).

the drum chars, one is unequivocally mutagenic (G), and the other two are slightly or marginally mutagenic. Char G, which was generated at a lower temperature, has a greater percentage of ash and sulfur than its less mutagenic counterparts. How these differences in sample composition contribute to mutagenicity is uncertain.

The carbonization samples differ from other coal-related materials we have assayed. Thus far, TA98 has been most responsive to the mutagenic action of coal gasification and liquefaction products we have assayed (Schoeny et al., 1981). For the unequivocally mutagenic char samples (F, G, H, and J), however, TA100 is the most sensitive strain. This is also true of the slightly mutagenic sample I. TA98 and TA100 respond to different spectra of mutagenic agents, albeit with some overlap. It is possible that the composition of these carbonization chars is unlike that of previously tested materials. Another novel

feature of these samples is the relatively high response with
TA1537, in some instances equal to or greater than that of the
TA1538, TA98 pair.

Comparative Mutagenicity Assays of Fossil Fuel Materials

Our purpose in assaying samples from the EPA/DOE Fossil Fuels
Research Material Facility was to obtain data that could be
compared with published results. As assay and sample preparation
techniques vary from laboratory to laboratory, such comparative
data are useful in assessing the applicability of one's
methodology.

Coal Oil A (CRM 1.13) was a dark brown liquid, and Petroleum
Crude A (CRM 3.13) and Crude Shale Oil A (CRM 2.11) were viscous
dark brown liquids. Synthoil 13-100 A (1202.07) was supplied to
the repository from an experimental run at the Pittsburgh Energy
Technology Center. Syncrude 4-100 A (1106.09), provided by the
FMC Corporation, was a COED process product which had undergone
some hydrotreating (Guerin et al., 1980). For this study, all of
these materials were tested as DMSO solutions of the
unfractionated samples. All of these were far more soluble in
DMSO than any of the other materials assayed in the study. Coal
Oil A and Syncrude were completely miscible with DMSO at 10 and
20 mg/ml. Variations in the DMSO solubility can be attributed to
the use of relatively small aliquots of these heterogeneous
mixtures.

All samples were tested for five <u>Salmonella</u> strains as
summarized in Table 3. TA1535 responded marginally to Coal Oil A
and to the Synthoil and not at all to the other samples. As with
the majority of the other fossil fuel materials we have assayed,
TA98 proved to be the most sensitive to their mutagenic activity,
followed by TA1538 and TA100. Based primarily on their reversion
of TA98, the samples can be ranked as follows: Synthoil = Coal
Oil A > Crude Shale Oil A > Syncrude > Petroleum Crude A.

These results are in reasonable agreement with those
published by Guerin et al. (1981). These workers reported that
testing of unfractionated fossil fuel materials resulted in
toxicity and questionable results. Their estimates of whole
sample activity, therefore, are sums of weighted fractional
activities, such as we report in Table 1 for Ohio coals. They
admit the possibility of synergism among sample constituents and
so claim their estimates to be accurate within ± 100%. A general
conclusion reached by this group is that unrefined coal oils and
high-boiling coal-derived residues are likely to be more mutagenic

Table 3.   Comparative Mutagenicity Testing of Fossil Fuel
Materials

Sample	Mutagenicity – colonies per microgram				
	TA1535	TA1537	TA1538	TA98	TA100
Petroleum Crude A CRM 3.13[a]	0	0.02	0	0.27	0
Crude Shale Oil A CRM 2.13[a]	0	0.15	0.61	0.88	0.75
Syncrude 1106.09[a]	0	0.10	0.03	0.15	0.81
Coal Oil A CRM 1.13[a]	0.26	0.35	0.92	2.36	1.91
Coal Oil A 1106[b]				0.53	
Synthoil 1202.07[a]	0.06	0.44	1.35	3.88	1.01
Synthoil 1202[b]				4.20	
Vehicle Oil A[c]	0		30.10	18.50	6.90

[a]Colonies per microgram determined by regression analyses of
dose-response data; 0 = no linear response.
[b]Data from Guerin et al. (1981).
[c]Data from Schoeny et al. (1981).

than petroleum and shale oils. We have not tested all the
material reported in Guerin (1980), nor can we be certain that our
samples were identical to theirs. Our estimate of
Syncrude 1106.09 mutagenicity of 0.15 rev/µg is somewhat lower
than, but within a reasonable range of, that reported by Guerin et
al. (1980). Synthoil 1202.07 has nearly identical activity
(3.88 rev/µg) to their figure of 4.2 rev/µg. It should be noted
that this activity is an order of magnitude less than that we
reported for a similar vehicle oil supplied to us by DOE (see
Table 3). These materials from the EPA-DOE repository were all
somewhat less mutagenic than various products from a coal
hydrogenation process we assayed for DOE (Schoeny et al., 1981).

This study demonstrates some applications of short-term *in
vitro* mutagenesis assays in the evaluation of potential biohazards
from the use of coal. Unequivocal evidence of mutagenicity has
been obtained from some coal-related materials (e.g., the
carbonization cyclone chars), while for other materials, raw coals
in particular, the data has been more difficult to evaluate.
While positive responses in such assays are ·no guarantee of a
human health risk, they do indicate the necessity of further
study. *In vitro* mutagenesis tests can be performed early in the

evolution of coal technologies, e.g., on experimental and pilot
plant products, as was done in our studies. Properly evaluated
mutagenicity data can then be of great use in the planning and
engineering stages. In this way, information on bioactivity can
be used to identify needs for further research in technology and
further toxicological testing. This type of approach gives
direction to the development of alternate energy sources with the
minimization of human health risks as part of the overall
development of coal technologies.

ACKNOWLEDGMENTS

This work was supported by the Ohio Coal Research Laboratory
Association and Department of Energy #ET-00222-8.

REFERENCES

Ames, B.N., J. McCann, and E. Yamasaki. 1975. Methods for
    detecting carcinogens and mutagens with the Salmonella/
    mammalian microsome mutagenicity test. Mutation Res.
    31:347-364.

Battelle Pactific Northwest Laboratories. 1979. Biomedical
    Studies of Solvent Refined Coal (SRC-II) Liquefaction
    Materials: A Status Report. PNL-3189. Battelle Pacific
    Northwest Laboratory: Richland, WA. 54 pp.

Epler, J.L., J.A. Young, A.A. Hardigee, T.K. Rao, M.R. Guerin,
    I.B. Rubin, C.H. Ho, and B.R. Clark. 1978. Analytical and
    biological analyses of test materials from the synthetic fuel
    technologies, I: mutagenicity of crude oils determined by
    the Salmonella typhimurium microsomal activation system.
    Mutation Res. 57:265-276.

Guerin, M.R., C.H. Ho, T.K. Rao, B.R. Clark, and J.L. Epler.
    1980. Polycyclic aromatic primary amines as determinant
    chemical mutagens in petroleum substitutes. Environ. Res.
    23:42-53.

Guerin, M.R., I.B. Rubin, T.K. Rao, B.R. Clark, and J.L. Epler.
    1981. Distribution of mutagenic activity in petroleum
    substitutes. Fuel 60:282-288.

Ketcham, N.H., and R.W. Norton. 1960. The hazards to health in
    the hydrogenation of coal, III: the industrial hygiene
    studies. Arch. Environ. Health 1:194-207.

Lloyd, J.W. 1971. Long-term mortality study of steelworkers, Volume 5: respiratory cancer in coke plant workers. J. Occup. Med. 13:53-67.

McCann, J., and B.N. Ames. 1976. Detection of carcinogens as mutagens in the Salmonella/microsome test: assay of 300 chemicals: discussion. Proc. Natl. Acad. Sci. USA 73:950-954.

Pelroy, R.A., and M.R. Petersen. 1979. Use of Ames test in evaluation of shale oil fractions. Environ. Health Perspect. 30:191-203.

Pelroy, R.A., D.S. Sklarew, and S.P. Downey. 1981. Comparison of the mutagenicities of fossil fuels. Mutation Res. 90:233-245.

Rosenkranz, H.S., and L.A. Poirer. 1979. Evaluation of the mutagenicity and DNA-modifying activity of carcinogens and noncarcinogens in microbial systems. J. Natl. Cancer Inst. 62:873-892.

Savage, R.L. 1980. Design and operation of a second generation flash carbonization unit for the concurrent production of low-sulfur, low-volatile char and synthesis gas (hydrogen and carbon monoxide) with fast response FT-IR analysis of the solids and gas products. UCL project proposal No. IV to Ohio Coal Research Laboratories Association.

Schoeny, R., D. Warshawsky, L. Hollingsworth, M. Hund, and G. Moore. 1981. Mutagenicity of products from coal gasification and liquefaction in the Salmonella/microsomal assay. Environ. Mutagen. 3:181-195.

Weil, C.S., and N.I. Condra. 1960. The hazards to health in the hydrogenation of coal, II: carcinogenic effect of materials on the skin of mice. Arch. Environ. Health 1:187-193.

PRESENCE OF VARIOUS TYPES OF MUTAGENIC IMPURITIES IN CARBON

BLACK DETECTED BY THE SALMONELLA ASSAY

Eva Agurell and Göran Löfroth

Department of Radiobiology, University of Stockholm
S-106 91 Stockholm, Sweden

INTRODUCTION

Aside from their use as rubber additives, carbon blacks are used as colorants in many contexts, such as photocopying, typewriting, carbon copying, and printing. The ultimate products are commonly in close proximity to man and are often worn or turned into waste, further enhancing the dissemination of their ingredients.

Carbon blacks may contain a number of impurities related to polycylic aromatic hydrocarbons (Fitch et al., 1978; Fitch and Smith, 1979). It has been shown that photocopying toners may contain mutagens detectable by the Salmonella assay (Lofroth et al., 1980) and one manufacturer has reported that their particular problem has been remedied by changing the production of the colorant carbon black (Rosenkranz et al., 1980). It has also been reported that typewriter ribbons and carbon paper likewise may contain mutagenic impurities (Alfheim et al., 1981).

Although analyses with respect to biologically active impurities can be made on readily available products, it is more rational to study single product ingredients in order to have a higher detection level and a better chance of finding the

potentially offensive material. This report presents results of
the Salmonella/microsome assays concerning the presence of
mutagenic impurities in extracts of some carbon blacks that are
used or have a potential use as colorants.

MATERIALS AND METHODS

Sample preparation

    A weighed amount, usually 1 g, of the carbon black was
Soxhlet-extracted with 200 ml benzene or acetone for 16 hours with
3-4 extractions cycles per hour and about 100 ml per extraction
cycle. The thimble consisted of a short and a long glass cylinder
between whose leveled ends a 15 cm^2 glass fiber filter was
inserted and kept fastened by a tightened stainless steel wire.
Benzene extracts were evaporated to dryness and acetone extracts
almost to dryness before dissolving the residue in dimethyl
sulfoxide (DMSO). Samples were stored frozen at -20°C prior to
and between mutagenicity assays.

HPLC-fractionation

    A concentrated extract was separated on a Spherisorb S5-ODS
column (250 mm by 4.6 mm i.d.) with aqueous methanol as elutant.
Fractions of 1 ml (1 min) were collected and each fraction was
assayed for mutagenicity in the Salmonella strains TA98 or TA98 NR
with 50, 100, and 200 µl per plate.

Mutagenicity assay

    Mutagenicity was determined by the Salmonella plate
incorporation method using bacterial cultures fully grown
overnight as described by Ames et al. (1975) and applied to
extracts of particulate matter (Lofroth, 1980; 1981). Assays were
performed with the tester strains TA98 and TA100 obtained from
Dr. B.N. Ames (Berkeley, CA) and the nitroreductase deficient
strains TA98 NR and TA98/1,8 DNP$_6$ obtained from Dr. H.S.
Rosenkranz (Cleveland, OH). The microsome-containing liver
supernatant (S9) was prepared from Aroclor 1254-induced male
Sprague-Dawley rats and was used with the necessary cofactors
(S9-mix). The amount of S9 used was 20 or 50 µl per plate.

Each sample has been assayed with one plate per dose level at three or more dose levels in at least two independent tests. The mutagenic response, expressed as revertants per mg carbon black, was calculated from the linear or almost linear part of the dose-response curve. The doses employed would, in the absence of interfering factors, permit a detection of about 1 revertant per mg carbon black. The absence of a detectable mutagenic response is thus given as less than 1 revertant per mg.

Samples having a high or the highest mutagenic response in the absence of S9 were further assayed in simultaneous tests with TA98, TA98 NR, and TA98/1,8DNP$_6$. Quercetin (Sigma Chemical Co.) was then used as concurrent positive control to check that the three strains had the same response to a non-nitro-dependent mutagen.

Samples having a low or no detectable response were assayed together with known mutagens to discover if the apparent response was due to a true low response or due to presence of compounds interfering with the mutagenicity assay. 1-Nitroyrene (Kock-Light Ltd.) and benzo(a)pyrene (Sigma Chemical Co.) were used respectively in the absence and presence of S9.

RESULTS

The carbon black gave extracts with mutagenic activities that varied widely between different samples (Table 1). Some samples did not give a detectable mutagenic response. These extracts did not decrease the response of either 1-nitropyrene for benzo(a)pyrene when added in the mutagenicity assay, showing that the low response of the sample is real. Inclusion of assays in the presence of known mutagens is essential to avoid false negative results. Extracts of several carbon black containing products have, without being bacteriotoxic, interfered in the bioassay by decreasing or abolishing the response of the added known mutagens, indicating that the extracts may contain mutagens that are not detected.

Several samples had relatively low mutagenic activity in the absence of s9 and a high enhancement in the presence of S9. Typical examples are N-330, known to be used in newspaper printing ink, Raven 1020 and 1200, Printex 140U, and Regal SRF, of which the latter presumably is used in carbon paper. Carbon black impurities that require mammalian activation in order to elicit a mutagenic response in the Salmonella assay are a number of polycyclic aromatic hydrocarbons (PAH) and heterocyclic PAH. A difference with respect to the relative response in TA98 and TA100

Table 1. Mutagenic Response of Benzene Extracts of Carbon Blacks

	Revertants per mg Carbon Black					
MANUFACTURER Sample	TA98 -S9	TA98 +S9[a]	TA98 NR -S9	TA98/ 1,8DNP6 -S9	TA100 -S9	TA100 +S9[a]
NORDIC PHILBACK						
N-330	7	90	-[b]	-	7	80
COLUMBIAN						
Raven 1020	1	15	-	-	< 1	75
Raven 1200	1	45	-	-	1	100
Raven 1255	220	↓[c]	90	50	85	↓
DEGUSSA						
Special Black 5	360	↓	300	60	115	↓
Special Black 10	310,000	↓	180,000	30,000	-	-
Colour Black FW 200	390	↓	270	40	180	↓
Printex 90	< 1	< 1	-	-	< 1	< 1
Printex 140U	2	15	-	-	1	250
CABOT						
Black Pearls, L, old	30,000	↓	24,000	3,300	-	-
Black Pearls, L, new	40	↓	25	5	5	↓
Mogul L, old	55,000	↓	45,000	9,000	8,000	↓
Mogul L, new	1	< 1	-	-	< 1	< 1
SL 7325	< 1	< 1			< 1	< 1
Regal 400R, old	60,000	↓	40,000	5,000	-	-
Regal 400R, new	6	↓	-	-	< 1	2
SL 7326	< 1	< 1	-	-	< 1	< 1
Regal SRF	10	90	-	-	4	110
Sterling V	1	9	-	-	1	19
Vulcan XG72	< 1	< 1	-	-	< 1	< 1

[a]20 or 50 µl per plate, whichever, gave the highest enhancement.
[b]Not determined.
[c]Response decreases in the presence of S9.

implies a pronounced difference in the composition of the
compounds causing the mutagenic activity. This means, for
example, that the PAH profile of Printex 140U is different from
the profiles of N-330 and Regal SRF, whereas the PAH profiles of
the latter two samples may be similar or dissimilar.

Several samples were mutagenic in the absence of S9 and
showed a decreased activity after the addition of S9. Among these
were samples of Black Pearls L, Mogul L, and Regal 400R produced
before manufacturing changes that resulted in a drastically
reduced level of dinitropyrenes (Rosenkranz et al., 1980). All
samples having the highest mutagenic response in the absence of S9
are after-treated carbon blacks. Special Black 5 and 10 and
Colour Black FW 200 are presumably nitric acid oxidized whereas
Raven 1255 has been treated with ozone. Although a number of
polycyclic aromatic hydrocarbon derivatives, such as arene oxides,
arene lactones, and arene dicarboxylic anhydrides, can be
mutagenic in the absence of S9, the different responses of the
three nitroreductase proficient and deficient TA98 strains reveal
that the major mutagenic components are nitro or nitroso
derivatives.

There is a distinct difference between Raven 1255 and most of
the other samples. An HPLC-fractionation of Raven 1255 shows that
the mutagenicity is caused by several components having different
activities in the strains TA98 and TA98 NR (Figure 1). The peak
coinciding with 1-nitropyrene (fraction 17) has an activity ratio
between TA98 and TA98 NR agreeing with that of mutagenic
1-nitropyrene (Greibrokk et al., 1982). The activity of this peak
is about 20% of the total mutagenic response of the sample in
TA98. Assuming that the peak is 1-nitropyrene, it can be
calculated that the carbon black sample contains about 25 ppm of
this mutagen. The samples does not seem to contain significant
amounts of dinitropyrenes (fractions 14-16).

Black Pearls L is a pelletized form of Mogul L. The
differences between the samples in the mutagenic activity indicate
possible variations between different production batches. A
similar wide variation has been observed for Raven 1255 with 110,
220, and 1,000 revertants per mg for three different samples.
Analysis of the same batch generally results in a lesser variation
being no more than ±20%.

Samples SL 7325 and 7326 are carbon blacks prior to the
oxidative after-treatment resulting in the types Mogul L and
Regal 400R. The absence of a detectable mutagenic activity in
these samples strongly suggests that is possible to produce carbon
blacks with a very low level of mutagenic impurities.

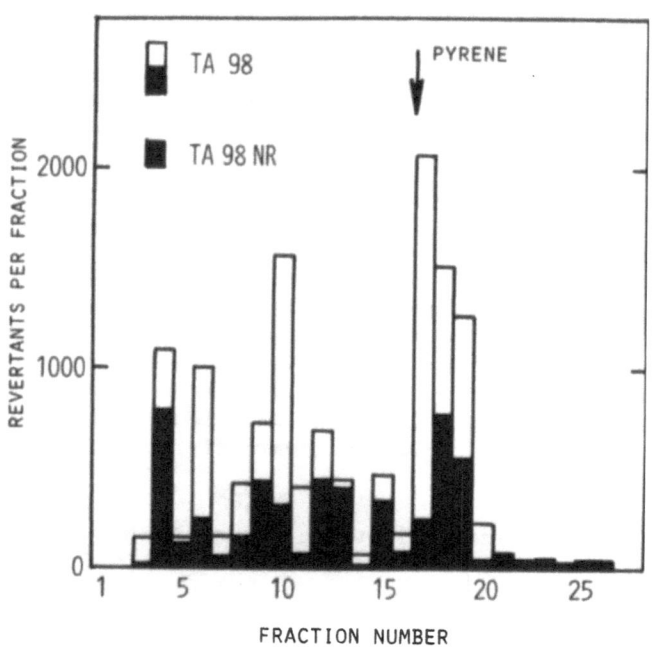

Figure 1.   The distribution of the mutagenic response in TA98 and
TA98 NR of HPLC-fractions of a benzene extract of the
carbon black Raven 1255.

Acetone has been tested as extraction solvent for some
samples. This solvent may have an efficiency similar to that of
benzene for certain types of carbon blacks and certain types of
compounds (Figure 2) but the main result is that benzene usually
is better than acetone for extracting mutagenic impurities in
carbon blacks. Toluene has also been tried but been abandoned due
to formation of a bacteriotoxic residue. The toxic residue is
already formed in the extraction procedure in the absence of both
thimble and sample. The use of toluene may require a complete
oxygen-free extraction procedure.

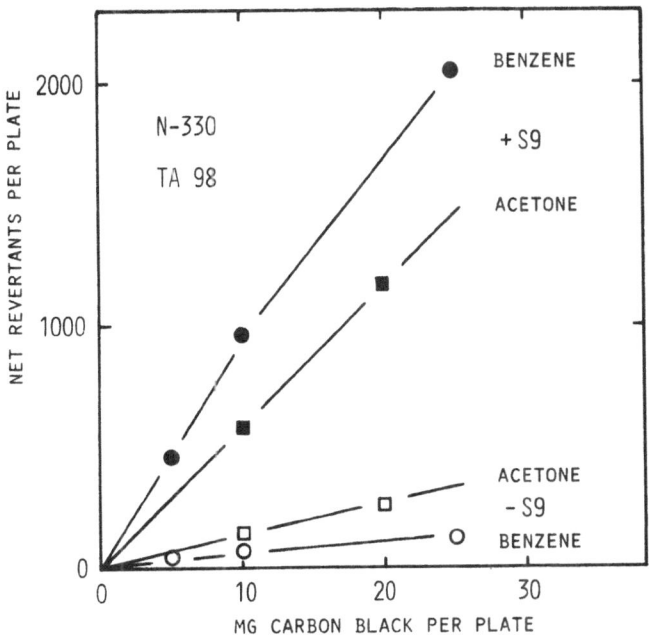

Figure 2. Mutagenic response in TA98 of benzene and acetone
extracts of carbon black N-330.

      Extraction of carbon black can be optimized with respect to
particular components (Taylor et al., 1980; Giammarise et al.,
1981). Alsberg et al. (1981) have evaluated a number of
procedures for extracting PAH and mutagenic compounds from a
single sample of carbon black. They found that benzene extracts
more high molecular weight PAH than dichloromethane and acetone,
whereas the mutagenic activity detected in TA98 in the presence of
S9 was about the same for all three solvents, about 2.3
revertants per mg carbon black. The same sample has been
extracted, using 0.5 sample sizes, and assayed within the present

study. Acetone extraction gave 8 and benzene extration 18 revertants per mg carbon black with TA98 and an optimal amount of S9. The discrepancies may be due to differences in the extraction time (16 versus 24 h), sample size (0.5 versus 1.7 g), solvent for mutagenicity assay (DMSO versus acetone) and amount of S9 per plate (20 versus 50 μl).

DISCUSSION

Kirwin et al. (1981) have screened a carbon black sample, N-339, related to N-330, in several genetic test systems by applying a DMSO suspension of the particles to the incubtion media. The lack of any detectable effects is ascribed to the fact that impurities are strongly adsorbed to the carbon black particles and may not be released during the incubation periods. Locati et al. (1979) also suggest that the strong binding of known carcinogens to carbon black annuls their potential carcinogenic risk. There seeem to be some reasons for this view provided that the exposure is to neat, unmixed carbon black as it for example occurs in carbon black manufacturing plants (Robertson and Ingalls, 1980).

Carbon blacks are, however, used in daily life in products mixed with polymers, resins, oils, and other organic chemicals. The possibilities of displacing the mutagenic impurities from the particles may thus increase. The release of carbon black impurities from new and aged products containing carbon blacks is an important subject for future studies.

The magnitude of the problem can be exemplified. The carbon black N-330 is used as a colorant in the printing ink used in a Swedish newspaper printing office. In printing its daily morning and evening papers about 5 metric tons of ink with 14% carbon black is used per day. The emission of mutagens in the carbon black from the printing office into the hands of readers corresponds to a mutagenic activity of about 60 billion revertants per day. Some of this is undoutedly transferred to the skin. The larger part is, however, ultimately turned into waste, but the eventual fate of the compounds is largely unknown.

It can be calculated, by using the specific emission of 200 revertants per g gasoline (Lofroth 1980), that a similar mutagenic activity, 60 billion revertants, may be released into the environment in the exhaust of a gasoline-powered car driven about 4 million km, i.e. about 5 km per newspaper copy.

Potential environmental problems due to the presence of mutagenic/carcinogenic compounds in carbon blacks can be substantially decreased by production and use of carbon blacks with a minimum of mutagenic impurities. Methods and results presented in this report may be a contribution in reaching this goal.

ACKNOWLEDGEMENT

This study has been supported by a grant from the Swedish Work Environment Fund.

REFERENCES

Alfheim, I., M. Møller, G. Löfroth, and L. Nilsson. 1981. Mutagenic activity from chemicals in typewriter ribbons and carbon paper. Environ. Mut. 3:392-393.

Alsberg, T., U. Rannug, U. Stenberg, and A. Sundvall. 1981. Evaluation of extraction methods for carbon black: POM analysis and mutagenicity assay. In: Sixth International Symposium on Polynuclear Aromatic Hydrocarbons. Battelle-Columbus Laboratories: Columbus, OH.

Ames, B.N., J. McCann, and E. Yamasaki, 1975. Methods for detecting carcinogens and mutagens with the Salmonella/ mammalian-microsome mutagenicity test. Mut. Res. 31:347-364.

Fitch, W.L., and D.H. Smith. 1979. Analysis of adsorption properties and adsorbed species on commercial polymeric carbons. Environ. Sci. Technol. 13:341-346.

Fitch, W.L., E.T. Everhart, and D.H. Smith. 1978. Characterization of carbon black adsorbates and artifacts formed during extraction. Anal. Chem. 50:2122-2126.

Giammarise, A.T., D.L. Evans, M.A. Butler, C.B. Murphy, D.K. Kiriazides, D. Marsh, and R. Mermelstein. 1981. Improved methodology for carbon black extraction. In: Sixth International Symposium on Polynuclear Aromatic Hydrocarbons. Battelle-Columbus Laboratories: Columbus, OH.

Greibrokk, T., G. Lofroth, L. Nilsson, and R. Toftgard. In press.
Nitroarenes: Mutagenicity in the Ames Salmonella/microsome
assay and affinity to the TCDD receptor protein. In:
Toxicity of Nitroaromatic Compounds. D.E. Rickert, ed.
Hemisphere Publishing Corp.: New York.

Kirwin, C.J., J.V. LeBlanc, W.C. Thomas, S.R. Haworth, P.E.
Kirby, A. Thilager, J.T. Bowman, and D.J. Brusick. 1981.
Evaluation of the genetic activity of industrially produced
carbon black. J. Toxicol. Environ. Health 7:973-989.

Locati, G., A. Fantuzzi, G. Consonni, I.L. Gotti, and G. Bonomi.
1979. Identification of polycyclic aromatic hydrocarbons in
carbon black with reference to cancerogenic risk in tire
production. Am. Ind. Hyg. Assoc. J. 40:644-652.

Löfroth, G. 1980. Salmonella/microsome mutagenicity assays of
exhaust from diesel and gasoline powered motor vehicles. In:
Health Effects of Diesel Engine Emissions. W.E. Pepelko,
R.M. Danner, and N.A. Clarke, eds. U.S Environmental
Protection Agency: Cincinnati, OH. EPA-600/9-80-057a.
pp. 327-342.

Löfroth, G. 1981. Comparison of the mutagenic activity in carbon
particulate matter and in diesel and gasoline engine exhaust.
In: Short-Term Bioassays in the Analysis of Complex
Environmental Mixtures II. M.D. Waters, S.S. Sandhu, J.L.
Huisingh, L. Claxton, and S. Nesnow, eds. Plenum Press: New
York. pp. 362-380.

Löfroth, G., E. Hefner, I. Alfheim, and M. Moller. 1980.
Mutagenic activity in photocopies. Science 209:1037-1039.

Robertson, J. McD., and T.H. Ingalls. 1980. A mortality study of
carbon black workers in the United States from 1935 to 1974.
Arch. Environ. Health 35:181-186.

Rosenkranz, H.S., E.C. McCoy, D.R. Sanders, M. Butler,
D.K. Kiriazides, and R. Mermelstein. 1980. Nitropyrenes:
Isolation, identification, and reduction of mutagenic
impurities in carbon black and toners. Science
209:1039-1043.

Taylor, G.T., T.E. Redington, M.J. Bailey, F. Buddingh, and C.A.
Nau. 1980. Solvent extracts of carbon black - Determination
of total extractables and analysis for benzo(a)pyrene. Am.
Ind. Hyg. Assoc. J. 41:819-825.

SESSION 4

DEVELOPMENT OF
SHORT-TERM BIOASSAYS:
CARCINOGENICITY

CRITERION DEVELOPMENT FOR THE APPLICATION OF BALB/c-3T3 CELLS
TO ROUTINE TESTING FOR CHEMICAL CARCINOGENIC POTENTIAL

J.O. Rundell, M. Guntakatta, and E.J. Matthews

Department of Molecular Toxicology, Litton Bionetics, Inc.
Kensington, MD 20895

INTRODUCTION

Short-term in vitro cell transformation assays have emerged
as potentially powerful tools for the evaluation of pure chemicals
and complex mixtures for their carcinogenic potential. Among the
more widely applied of these systems are the BALB/c-3T3 (Kakunaga,
1973), C3H/10T1/2 (Reznikoff et al., 1973), and Syrian hamster
embryo (Berwald and Sachs, 1965; DiPaolo et al., 1971) cell
transformation assays, although a variety of other related systems
have been described (Casto et al., 1974; Evans and DiPaolo, 1975;
Freeman et al., 1970; Kakunaga, 1978; Milo and DiPaolo, 1978;
Traul et al., 1979; and see Heidelberger, 1981 for a recent
review). Each of these systems measures chemical induction of
alterations in cell colony or focus morphology and in each case,
these alterations have been shown to be correlated with the
acquisition of cellular malignant properties (DiPaolo et al.,
1971; Kakunaga, 1978; Reznikoff et al., 1973). Thus, the strength
of in vitro cell transformation assays, in the context of chemical
carcinogen screening, lies in their apparent phenomenological
relation to the process of carcinogenesis per se. At the same
time, these assay systems exhibit significant technical as well as
theoretical difficulties in their conduct and evaluation. For
example, the results of cell transformation assays are generally
evaluated by tacitly assuming that the coincidence between the
morphologically transformed phenotype and the expression of

cellular malignant properties is unity. Further, no agreement
exists among investigators on the methodologies employed for dose
selection and quantitative analysis and, therefore, comparisons of
the published results of model compound analyses among
laboratories are primarily qualitative.

To address these issues as they relate to the subclone of
BALB/c-3T3 cells used in our laboratory, we have studied the
sensitivity and response characteristics of this system towards a
variety of model pure chemicals and analyzed the coincidence
between morphological transformation and malignant transformation
for a set of isolated transformants. The results of these studies
are the subject of this report.

MATERIALS AND METHODS

Cell Culture

BALB/c-3T3 clone 1-13 cells were the generous gift of
T. Kakunaga (NCI, Bethesda, MD). A subclone, C14, was established
and used for these experiments because of its relatively high
cloning efficiency and low frequency of spontaneous
transformation. 3T3 cells were cultured in Eagle's Minimal
Essential Medium (EMEM) supplemented with L-glutamine,
penicillin-streptomycin, and 3.5 to 10% selected heat-inactivated
fetal bovine serum (FBS). Cell incubations were performed at 37°C
in a water-saturated atmosphere of 95% air, 5% CO_2.

Clonal Survival Assay

3T3 cell survivals after chemical treatments were estimated
using a clonal growth assay. Briefly, approximately 200 3T3 cells
were plated in 60-mm culture dishes, incubated for 18 to 24 h to
allow for cell attachment and recovery, and then treated with
various concentrations of test chemical for the time periods
described in Results. After a total of 7 to 8 days' incubation,
the resultant colonies were washed with Hanks' Balanced Salt
Solution (HBSS), fixed in methanol, stained with 10% aqueous
Giemsa, and counted. The relative survivals to chemical
treatments were calculated as the ratio of the number of colonies
observed for each treatment condition to the number of colonies in
the untreated controls, and the results were utilized for dose
selection for the transformation assays.

Transformation Assay

Test chemical induction of 3T3 cell transformation was evaluated according to a modification of the method of Kakunaga (1973). Briefly, 1×10^4 3T3 cells were plated per 60-mm dish (20 to 40 dishes per condition) and incubated for 18 to 24 h. The plating medium was then removed and replaced with medium containing preselected doses of the test chemical, and incubation was continued for 24 to 72 h (see Results). Positive and negative control cultures were treated similarly. After treatment, the cultures were washed twice with HBSS, refed with EMEM supplemented with reduced serum concentrations (~ 3.5 to 5% v/v), and incubation was continued with refeeding twice weekly. After 28 to 31 days of incubation, the dishes were washed in HBSS, fixed and stained as previously described, and the number of foci of transformed cells was determined for each experimental condition. Foci of transformed cells were identified using published criteria (Kakunaga, 1973). Briefly, the scored foci exhibited three distinctive phenotypic qualities: marked cellular disorientation, loss of contact inhibition as reflected in dense cellular piling up, and invasiveness into the contiguous monolayer of normal cells. A normal distribution of transformed foci/dish was obtained after adding one to each frequency and then converting the number of foci to their logarithmic (base 10) equivalent value. Bailey's modification of Student's t-test (Bailey, 1959) was then used to determine whether the results for each treatment condition were significantly different from the concomitant negative control (i.e., $p \leqslant 0.05$ or 0.01). Responses at one or more treatment levels which attained the 99% confidence level were interpreted as evidence for transforming activity. Responses attaining the 95% confidence level and exhibiting evidence of dose-dependency were interpreted similarly.

Phenotypic Confirmation Studies

Transformed 3T3 cell foci were scored by phase-contrast microscopy, using the same criteria as were employed for the scoring of stained dishes, and were isolated as described by Cahn et al. (1967). Isolated foci were cultured in EMEM + 5 to 10% FBS and subcultured to obtain a total of 3 to 4 x 10^7 cells (usually 3 to 4 serial passages), and tested for their expression of the anchorage independent (ar^+) and tumorigenic phenotypes. The ar^+ phenotype (growth in soft agar) was determined by the method of MacPherson (1969) using 0.35% Noble Agar over a 0.7% Noble Agar basal layer in EMEM + 10% FBS. The frequency of the appearance of ar^+ variants for each tested isolate was defined as the number of agar colonies greater than 0.1 mm diameter per plated cell.

Frequencies greater than 1 x 10^4 were evaluated as evidence of a
significant increase in the expression of the ar^+ phenotype, since
untreated 3T3 cell populations yielded approximately
5 x 10^{-6} colonies/plated cell. Transformed focal isolates were
tested for their transplantability by subcutaneous (s.c.)
injection of 1 x 10^7 cells into x-irradiated (425 to 450 rads
whole-body—Westinghouse Quadrocondex 250 Kev, 15 ma-filtration
with 1.5 mm Cu and 1.0 mm Al) female, weanling BALB/c mice
(Charles River Labs, Wilmington, MA). The administered dose of
whole-body irradiation was selected on the basis of a preliminary
x-ray survival study in which the X ray $LD_{50/30}$ was found to be
approximately 525 rads. Animals were observed for approximately
90 to 120 days for the development of progressively growing
tumors. Selected tumor-bearing animals were necropsied and their
tumor masses submitted for cytopathological analysis (Dept. of
Pathology, Litton Bionetics, Inc., Kensington, MD).

RESULTS

3T3 Cell Transformation Studies -- Model Compound Activities

To establish test chemical dose-selection criteria and
dose-response characteristics, we evaluated the cytotoxicity and
morphological transforming activities of a panel of 27 model
compounds. These compounds were selected from several chemical
classes, including aromatic amines, nitrogen mustards,
nitrosamides, polycyclic aromatics, mycotoxins, and steroids. Of
the 27 model compounds tested, at least 15 were known rodent or
human carcinogens. The remaining compounds were suspect
carcinogens (e.g., formaldehyde, acetoxy-2-acetylaminofluorene
[AAAF], 3-methylcholanthrene [3-MCA], Arabinofuranosylcytosine
[Ara-c]) or were known or presumed noncarcinogens, including the
solvents commonly employed in the delivery of test chemicals (data
on rodent or human carcinogenicity taken from IARC Monographs and
NCI Technical reports; see IARC Monographs, 1971-1981; IARC
Working Group, 1980; NTP and NCI Technical Report Series,
1976-1981). Prior to their application in the transformation
assay, the cytotoxicity of each of the 27 model compounds was
determined by use of a clonal growth assay (see Materials and
Methods). Using this method, complete dose-response profiles were
obtained for 3T3 cell relative survivals ranging from 100% to
about 0% for each of the tested materials. These survival
profiles were used to select the concentrations of each model
compound for application in the transformation assay. The maximum
concentration chosen for each test compound was the LC_{80-90} (the
concentration required to obtain ~ 10 to 20% survival). The
LC_{80-90} concentrations for the tested compounds ranged from

0.025 µg/ml for mitomycin C to 20.0 mg/ml for urea. Less toxic
concentrations were selected to include survivals over the
descending portion of the survival curve for each compound and
were normally 1:1.5 or 1:2 to 1:5 dilutions of the LC_{80-90}
concentration. Thus, the transforming activities of all compounds
were tested over coincident ranges of 3T3 cell survival; i.e.,
between 10 to 20% and 100% survival. The activities of these
model compounds in the 3T3 cell transformation assay are shown in
Table 1. Of the 27 compounds tested, 12 were found to induce
statistically significant ($p < 0.05$) increases in the frequency of
transformed foci compared to their concomitant negative control
(spontaneous) frequencies, and each of these 12 compounds was a
known or suspect carcinogen. The maximum responses for these
12 compounds varied from 0.49 focus/dish for auramine to
5.9 foci/dish for 7,12-dimethylbenz(a)anthracene (DMBA) and
occurred at cell survivals ranging between approximately
5 and 24%. In contrast to these observations, 15 of the compounds
tested did not induce statistically significant increases in
transformation frequency relative to their concurrent negative
controls. Among these compounds were several known or suspect
carcinogens: e.g., amitrole, benzene, dimethylnitrosamine,
diphenylnitrosamine, actinomycin D, progesterone, and
testosterone.

The relationship between cell survival and transformation
activity patterns observed among the tested model compounds was
analyzed, and typical results are shown in Figure 1. These data
show that dose-dependent and statistically significant responses
were obtained for N-methyl-N'-nitro-N-nitrosoguanidine (MNNG) and
3-MCA (Figures 1A and 1B) but not for acetone or benzene
(Figures 1C and 1D). In the case of MNNG treatments, a
dose-related increase was observed between 0.5 and 1.0 µg/ml, but
treatments with 2.0 and 4.0 µg/ml resulted in an inverse response,
apparently due to the increasing toxicities of these treatments
(Figure 1A). The response for 3-MCA described in Figure 1B was
typical of those observed for other polycyclic hydrocarbons; the
frequency of morphological transformants tended to saturate rather
than decrease as the treatment toxicity increased. In contrast to
the MNNG and 3-MCA observations, no evidence of dose-dependency or
relation to the toxicity of treatment was observed for acetone or
benzene (Figures 1C and 1D). When the cell survivals were
evaluated in relation to the treatment conditions resulting in
maximum observed absolute transformation frequencies for the
15 chemicals evaluated as inactive, no evidence for
survival-dependent response patterns was obtained even though the
tested survival ranges were similar to those employed for the
12 active compounds.

Table 1. Activities of Model Compounds in the BALB/c3T3 Transformation Assay

Chemical	Reported Carcinogenicity[a]	Dose Range Tested[b]	Transformation Frequency Maximum at/Dose[c]	Survival at the Maximum Response Dose (%)	Significance[d] t-Statistic	~ p value
Aflatoxin B1	R	0.1-1.0	4.51/0.1	18	+ 24.15	< 0.01
Acetoxy-2-acetylaminofluorene	U	0.25-2.0	0.62/2.0	5	+ 2.82	< 0.01
2-Acetylaminofluorene	O	12-120	0.57/40	24	+ 2.35	< 0.05
Auramine	R,M	0.31-1.25	0.49/1.25	20	+ 2.88	< 0.01
Benz(a)anthracene	R,M	30.0-300.0	1.48/300.0	19	+ 4.31	< 0.01
Benzo(a)pyrene (B[a]P)	R,M	3.0-12.0	1.12/12.0	18	+ 5.06	< 0.01
Cytosine arabinoside	U	0.008-0.08	2.72/0.08	20	+ 9.20	< 0.01
7,12-Dimethylbenz(a)anthracene	O	0.016-1.0	5.90/0.25	10	+ 17.92	< 0.01
Formaldehyde	U	0.5-2.5	1.27/2.5	22	+ 3.89	< 0.01
3-Methylcholanthrene	U	0.2-10.0	1.9/5.0	12	+ 7.80	< 0.01
N-Methyl-N'-nitro-N-nitrosoguanidine	R,M	0.5-3.0	1.50/1.0	20	+ 8.40	< 0.01
Mitomycin C	R,M	0.0025-0.025	0.51/0.025	20	+ 2.75	< 0.02
Acetone	U	0.5-4.0%	0.12/4.0%	20	+ 0.81	> 0.05
Actinomycin D	R	0.003-0.03	0.18/0.003	80	- 0.55	> 0.05
Amitrole	R,M	1.0-10.0[e]	0.21/1.0,3.0	60	+ 1.52	> 0.05
Benzene	H	500.0-4000.0	0.09/2000.0	75	- 1.09	> 0.05
Caffeine	U	120.0-1200.0	0.13/120.0	75	- 0.76	> 0.05
Dimethyl sulfoxide	U	0.5-4.0%	0.35/4.0%	20	- 0.05	> 0.05
Dimethylnitrosamine	R,M	0.25-20.0	0.46/10.0	30	+ 1.51	> 0.05
Diphenylnitrosamine	R,M	10.0-100.0	0.19/50.0	80	+ 0.09	> 0.05
Ethanol	U	0.5-6.0%	0.50/6.0%	5	+ 0.67	> 0.05
Naphthalene	U	15.0-150.0	0.29/50.0	50	+ 0.62	> 0.05
Phenanthrene	U	10.0-80.0	0.36/80.0	15	+ 0.92	> 0.05
Progesterone	M	15.0-120.0	0.14/60.0	55	- 1.25	> 0.05
Testosterone	M	12.5-125.0	0.26/42.0	50	+ 1.05	> 0.05
12-O-Tetradecanoyl-phorbol-13-acetate	U	0.1-4.0	0.34/0.50	100	- 0.001	> 0.05
Urea	U	2.5-20.0[e]	0.16/2.5	100	- 0.31	> 0.05

[a]Carcinogenicity data taken from IARC, NCI, and OSHA evaluation; R, rat; M, mouse; H, human; U, unevaluated; O, OSHA candidate carcinogen.
[b]Chemical concentrations given in µg/ml or µl/ml except as noted.
[c]Values are the maximum observed frequency of transformed foci per dish/the test material treatment that resulted in the maximum frequency.
[d]Significance calculated in relation to the concomitant negative control using Bailey's modification of Student's t-test (Bailey, 1959).
[e]Chemical concentration given in mg/ml.

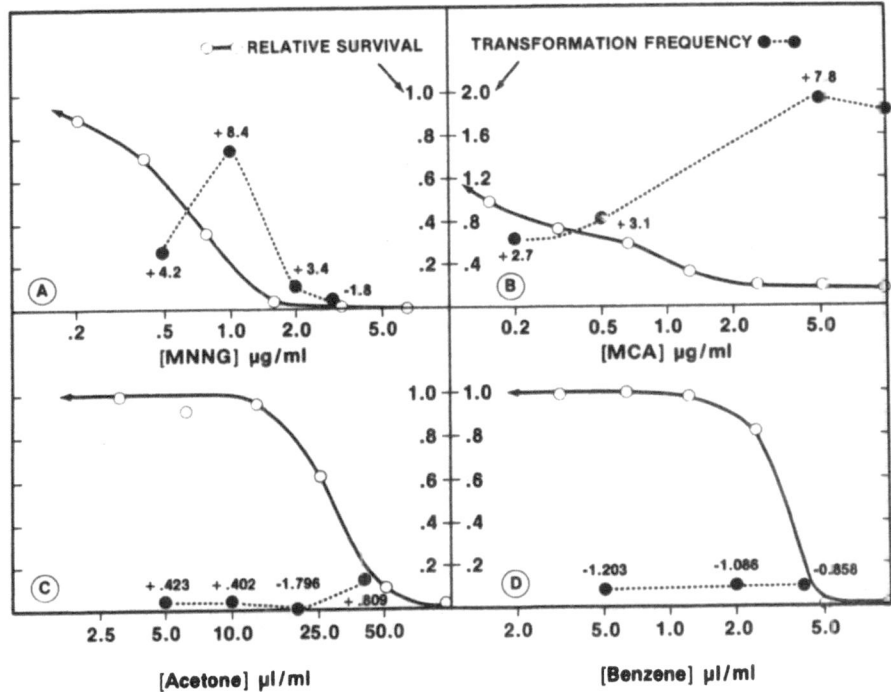

Figure 1.   Dose-response characteristics of BALB/c-3T3 cells.
Transformation frequencies for representative active
(MNNG, 3-MCA) and inactive (acetone, benzene) chemicals
in relation to the toxicity of the treatments. Signed
numbers are t-statistic values.

     The relation of the toxicity of treatment and the resultant
transforming activities of the 27 compounds tested is further
illustrated in Figure 2. These values were derived by
normalization of the data described in Table 1. To do this, the
observed maximum transformation frequency for each of the listed
compounds was assigned the value of 1.0 and the smaller
frequencies for the other treatments were thus expressed as
fractions of this value. As shown in Figure 2A, a nearly linear
relation (linear regression coefficient = 0.93) was observed when
the normalized transformation frequencies for active compounds
were plotted against survival and the maximum responses were
clustered in the 10 to 30% survival range. In contrast, when

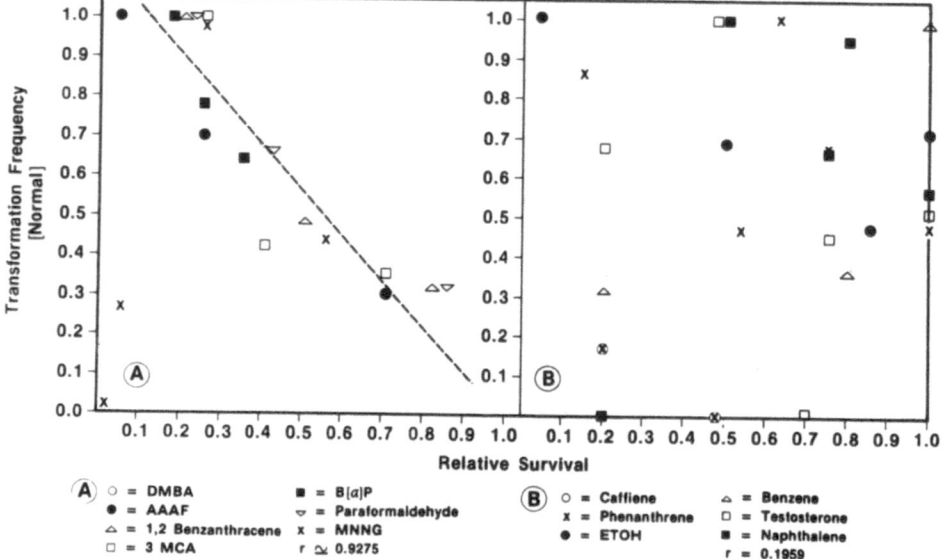

Figure 2.   The relation of the toxicity of treatment and
             transformation frequencies for active and inactive
             chemicals using normalized transformation frequencies.
             r is the linear regression coefficient for active
             chemicals (A) or inactive chemicals (B), respectively.

normalized data derived from the results of the inactive compounds
were similarly analyzed, no evidence for a linear relation between
transformation frequency and survival was observed (linear
regression coefficient = 0.19). These data showed that the
detection of chemical induction of 3T3 cell transformation
activity was critically dependent upon dose selection; in each
case where significant activity was observed, the maximum response
occurred within the dose range that resulted in survivals between
about 10 to 30%. Conversely, transformation activity was not
simply dependent upon chemical toxicity, since all of the tested
model compounds were not active in the same range of relative
survivals.

Phenotypic Coincidence Studies

To evaluate the coincidence between the morphologically transformed and malignant phenotypes, 12 foci of transformed cells were isolated from carcinogen-treated and untreated 3T3 cell cultures and tested for expression of the ar^+ phenotype and for transplantability in x-irradiated BALB/c mice. The ar^+ phenotype was selected because of the apparent relation between malignant potential and the ability of cells to grow in soft agar (Shin et al., 1975). The conditions of agar cloning were determined on the basis of the results of pilot studies (not shown) using transformed 3T3 cells of known malignant potential. Analyses of agar colony size distribution and frequencies among transformed and control (normal) 3T3 cell populations were conducted to develop evaluation criteria for this end point (see Materials and Methods). Female, weanling BALB/c mice were chosen for cell transplantation experiments because they are isogenic, readily available, and their husbandry is straightforward relative to athymic mice. Transplant recipients were x-irradiated 24 h prior to injection to achieve partial immunosuppression. The results of these phenotype coincidence studies are shown in Table 2. The frequency of the expression of the ar^+ phenotype in the control (untreated) 3T3 cell population was ~ 5×10^{-6} agar colonies per plated cell. In contrast, the ar^+ frequencies for the morphologically transformed isolates ranged between 1.0×10^{-1} for one 3-MCA-induced focus (MC3.4) to 1.5×10^{-3} for another 3-MCA-induced focus (MC12.4). Of the 12 tested isolates, 11 expressed the ar^+ phenotype with frequencies greater than 1×10^{-4}. One isolate, MC16.4, when tested at the fourth passage after isolation, exhibited an ar^+ frequency of approximately 2.0×10^{-5}. This isolate was, however, tumorigenic because each of the three mice injected with 1×10^7 MC16 cells developed progressively growing tumors at the injection site (Table 2).

These data were suggestive of a possible dissociation between the expression of the ar^+ phenotype and transplantability. However, the results of experiments (data not shown) evaluating the effect of serial passages on the quantitative expression of the ar^+ phenotype showed that ar^+ variants are enriched as a consequence of population expansion and, therefore, it is likely that retesting of MC16 after additional subcultivation would have resulted in an increase in the recovery of ar^+ variants. As shown in Table 2, expression of the ar^+ phenotype was not limited to morphological transformants induced by carcinogens. Transformed foci isolated from untreated cultures (SP1, SP2) expressed the ar^+ phenotype with frequencies similar to those observed for focal isolates obtained from carcinogen-treated cultures.

Table 2. Coincidence between the Morphologically Transformed, Anchorage-Independent (ar^+) and Tumorigenic Phenotypes among Isolated Transformed 3T3 Cell Foci

Cell Line[a]	Treatment	ar^+ Frequency[b]	Tumor Incidence[c]
3T3.6	Untreated control	5.0×10^{-6}	0/5
A1.14	B(a)P	6.3×10^{-2}	4/4
SP1.4	None[d]	1.5×10^{-3}	3/3
SP2.4	None[d]	2.0×10^{-3}	1/3
MC1.4	3-MCA	1.5×10^{-2}	4/4
MC3.4	3-MCA	1.0×10^{-1}	2/2
MC4.4	3-MCA	6.5×10^{-3}	1/3
MC12.4	3-MCA	1.5×10^{-3}	3/3
MC13.4	3-MCA	1.2×10^{-2}	3/3
MC16.4	3-MCA	2.0×10^{-5}	3/3
MC18.4	3-MCA	1.9×10^{-2}	2/3
MC20.4	3-MCA	6.5×10^{-3}	2/2
MC21.4	3-MCA	5.4×10^{-2}	2/2

[a]Decimal value is culture passage number of cell line at time of agar cloning and transplantation.
[b]Number of agar colonies > 0.1 mm diameter per plated cell.
[c]Number of animals bearing progressively growing tumors/number of animals in the transplant group.
[d]Spontaneous transformants.

The expression of the ar^+ phenotype correlated well with focus transplantability in irradiated BALB/c mice. Progressively growing tumors were observed in one or more transplant hosts for all 12 of the tested focal isolates (Table 2). No tumors were observed 120 days after injection of $\sim 2 \times 10^7$ control 3T3 cells per mouse. Tumor masses were obtained at necropsy from selected animals and were diagnosed as poorly differentiated sarcomas. The coincidence between the scored focus morphology, expression of the ar^+ phenotype, and transplantability was approximately 0.92 and 1.0, respectively, among the tested isolates, showing the 3T3 morphological transformation reliably predicted the acquisition of cellular malignant properties.

DISCUSSION

The studies described in the present report were undertaker.
to determine the response characteristics of the subclone of
3T3 cells used in our laboratory and to estimate the coincidence
between the cellular expression of tumorigenic potential and the
scored phenotype. The results of these studies describe and
demonstrate the significance of a uniform dose-selection
criterion, provide quantitative and qualitative data on the
sensitivity of these cells towards treatments with a variety of
chemicals, and show that the scoring criteria reliably predicted
the acquisition of cellular malignancy.

The effects of a variety of experimental variables on the
response of BALB/3T3 cells to chemical carcinogen treatments have
been previously described (Kakunaga, 1973, 1974). Among these
variables are the effect of initial cell plating density
(Kakunaga, 1973, 1974); the effect of serum concentration (in
C3H/10T1/2 cells; Bertram, 1977); the duration of chemical
treatment (Kakunaga, 1973); and the selection of suitably
responsive subclones (Kakunaga, 1973). The basic observations of
these studies have been reproduced in our laboratory (data not
shown) and elsewhere (Sivak et al., 1981), and have formed the
basis for the development and application of similar 3T3 cell
transformation assay study designs among the several laboratories
employing these cells. However, certain aspects of experimental
design have not previously been systematically studied, including
the relation of the toxicity of chemical treatment to the
induction of the transformed phenotype. To address this question,
we first examined the response characteristics of the 1-13, C-14
subclone of 3T3 cells in relation to the toxicity of treatments
with several model carcinogens. Our preliminary findings (not
shown) for direct-acting carcinogens (e.g., MNNG) and polycyclic
hydrocarbons (e.g., 3-MCA) showed that, in general, the 3T3 cell
monolayer could recover from treatments that resulted in nearly a
2-log reduction in survival as measured in a clonal toxicity test.
Since the target cell population could recover (i.e., form the
monolayer required for recognition of the transformed phenotype)
after treatments that resulted in as much as a 95% reduction in
relative survivals, we chose to evaluate the sensitivity of the
1-13, C-14 cells to carcinogen and noncarcinogen treatments using
doses that resulted in relative survivals ranging from between
~ 10 to 20% to nearly 100%. The results of these experiments
showed that the resolution of chemical transforming potential was
critically dependent upon the toxicity of treatment. Thus, the
maximal transformation frequencies observed in these experiments
were found to occur for treatments that resulted in between
5 and 24% survival and there was a nearly linear relation betweer.

the absolute frequency of transformation and the toxicity of
treatment. These data also showed that transformation was not
simply the result of cytotoxicity in that presumed noncarcinogens
(e.g., naphthalene) and some known rodent and human carcinogens
(e.g., amitrole, benzene) were not transforming even though they
were tested under similar conditions of relative survival to
treatment.

These observations provide the basis for a uniform
dose-selection criterion for the BALB/c-3T3 transformation assay
which does not depend on arbitrary exclusions as have been
employed by other investigators (Dunkel et al., 1981).
Accordingly, the single and limiting variable in dose selection,
within the constraints imposed by the general assay design and the
response characteristics of the target cells, is the cytotoxicity
of the test chemical.

The transforming activities of a total of 27 chemicals were
evaluated during the course of the present studies. Of these,
13 were identified as rodent or human carcinogens in IARC (IARC
Monographs, 1971-1981; IARC Working Group, 1980) or NCI and NTP
reports (NTP and NCI Technical Report Series, 1976-1981) and two
were listed as OSHA candidate carcinogens (Federal Register,
1980). The remaining 14 chemicals, although not yet evaluated by
these agencies, included materials generally accepted as
carcinogenic (e.g., AAAF and 3-MCA), as well as a total of eight
putative noncarcinogens: acetone, ethanol, dimethyl sulfoxide
(DMSO), caffeine, naphthalene, phenanthrene, 12-O-tetradecanoyl-
phorbol-13-acetate (TPA), and urea. Of the 27 chemicals
evaluated, a total of 12 were found to be active on the basis of
statistically significant ($p < 0.05$) increases in transformation
frequencies and dose-related responses (see Results). Of these
12 active chemicals, eight were listed as rodent carcinogens or
suspect human carcinogens by either the IARC, NCI, or OSHA
analyses.

Of the remaining four active chemicals, two, AAAF and 3-MCA,
are potent mutagens and are often employed as positive control
chemicals (3-MCA is listed as a candidate carcinogen by OSHA; see
Federal Register, 1980). Data on the rodent carcinogenicity of
formaldehyde was regarded as inconclusive by NIOSH, but this
important industrial chemical is listed as a candidate carcinogen
by the EPA (OSHR, 1980). Data on the rodent carcinogenicity of
cytosine arabinoside were regarded as inconclusive in an NCI
evaluation (NCI, 1978), but evidence that this chemical induces in
vitro cell transformation has been published (Kouri et al., 1975).

Not all of the known carcinogens assayed were found to be active when tested under the conditions of our protocol and evaluated by our criteria. These carcinogens included actinomycin D, amitrole, benzene, dimethylnitrosamine (DMN), diphenylnitrosamine (DPN), progesterone, and testosterone. At the same time, the known or putative noncarcinogens acetone, caffeine, DMSO, ethanol, naphthalene, phenanthrene, TPA, and urea were also negative, showing that the assay did not generate "false positive" data. The failure of the assay to detect the seven rodent and human carcinogens may be a reflection of limitations in the 3T3 target cells' intrinsic metabolic capacity. That is, these "false negative" results may have occurred, in part, because no exogenous metabolic system was employed in these assays. This hypothesis is supported by our finding that DMN induces the appearance of ouabain-resistant variants of 1-13, C-14 cells in the presence, but not in the absence, of S9 or rat liver cell activation (unpublished observations). Evaluated collectively, these data show that the BALB/c-3T3 target cells employed in these studies were sensitive to carcinogens from a variety of chemical classes such as mycotoxins, aromatic amines, and polycyclic hydrocarbons, but were insensitive to compounds representative of the nitrosamines and heterocyclics. In addition, no "false positives" were observed, showing that the responses of these target cells towards carcinogen treatments were specific within the constraints of their response characteristics.

The significance of cell transformation assays, in the context of their use in routine screening, lies in the relation between the in vitro observation (morphological transformation) and the expression of cellular malignancy. The data described in this report show a nearly unit coincidence between the scored and malignant phenotypes under the conditions of our protocol. In fact, the coincidence between the expression of cellular malignancy and the ar$^+$ phenotype (related to malignancy), in relation to the scored phenotype, was 1.0 and 0.92, respectively. These data are consistent with the previously published observations of Kakunaga (1973) who also found a good correlation between focus morphology and transplantability. In contrast, Sivak et al. (1981) reported a coincidence of 0.29 among seven isolated foci after as many as 25 serial passages. These differences between our observations and those of Sivak and his co-workers are likely due to our use of larger cell inocula and partially immunosuppressed animals. Our data show that the morphological response characteristics of the subclone of 3T3 cells used in our laboratory was highly predictive of cellular progression to malignancy.

ACKNOWLEDGMENTS

The authors gratefully acknowledge the expert technical
assistance of T. DelBalzo and thank M. Gonzalez for her help in
the preparation of this manuscript. We also thank Dr. Brian Myhr
and Dr. David Brusick for their helpful discussions and editorial
assistance.

REFERENCES

Bailey, N.T.J. 1959. Statistical Methods in Biology. Wiley &
    Sons, Inc.: New York. p. 50.

Bertram, J.S. 1977. Effects of serum concentration on the
    expression of carcinogen-induced transformation in the
    C3H/10T1/2, C18 cell line. Cancer Res. 37:514-523.

Berwald, Y., and L. Sachs. 1965. In vitro transformation of
    normal cells to tumor cells by carcinogenic hydrocarbons.
    J. Natl. Cancer Inst. 35:641-661.

Cahn, R.D., H.G. Coon, and M.B. Cahn. 1967. Cell culture and
    cloning techniques. In: Methods in Developmental Biology.
    F.H. Wilt and N.K. Wessels, eds. Thomas Y. Crowell Co.: New
    York.

Casto, B.C., W.J. Pieczynski, J.A. DiPaolo. 1974. Enhancement of
    adenovirus transformation by treatment of hamster embryo
    cells with diverse chemical carcinogens. Cancer Res.
    34:72-78.

DiPaolo, J.A., R.L. Nelson, and P.J. Donovan. 1971.
    Morphological, oncogenic, and karyological characteristics of
    Syrian hamster embryo cells transformed in vitro by
    carcinogenic polycyclic hydrocarbons. Cancer Res.
    31:1118-1127.

Dunkel, V.C., R.J. Pienta, A. Sivak, and K.A. Traul. 1981.
    Comparative neoplastic transformation responses of BALB/3T3
    cells, Syrian hamster embryo cells, and Rauscher murine
    leukemia virus-infected Fischer 344 rat embryo cells to
    chemical carcinogens. J. Natl. Cancer Inst. 67:1303-1315.

Evans, C.H., and J.A. DiPaolo. 1975. Neoplastic transformation
    of guinea pig cells in culture induced by chemical
    carcinogens. Cancer Res. 35:1035-1044.

Federal Register. 1980. List of substances which may be candidates for further scientific review and possible identification, classification, and regulation as potential occupational carcinogens. Fed. Reg. 45(157):53672-53679, August 12.

Freeman, A.E., P.J. Price, H.J. Igel, T.C. Young, J.M. Maryak, and R.J. Huebner. 1970. Morphological transformation of rat embryo cells induced by DEN and murine leukemia viruses. J. Natl. Cancer Inst. 44:65-78.

Heidelberger, C. 1981. Cellular transformation as a basic tool for chemical carcinogenesis. In: Advances in Modern Environmental Toxicology, Volume 1: Mammalian Cell Transformation by Chemical Carcinogens. N. Mishra, V. Dunkel, M. Mehlman, eds. Senate Press: Princeton Junction, NJ.

IARC. 1971-1981. Monographs on the evaluation of the carcinogenic risk of chemicals to humans, Volumes 1-20. International Agency for Research on Cancer; Lyons, France.

IARC Working Group. 1980. An evaluation of chemicals and industrial processes associated with cancer in humans based on human and animal data. Cancer Res. 40:1-12.

Kakunaga, T. 1973. A quantitative system for assay of malignant transformation by chemical carcinogens using a clone derived from BALB/3T3. Int. J. Cancer 12:463-473.

Kakunaga, T. 1974. Requirement for cell replication in the fixation and expression of the transformed state in mouse cells treated with 4-nitroquinoline-1-oxide. Int. J. Cancer 14:736-742.

Kakunaga, T. 1978. Neoplastic transformation of human diploid fibroblast cells by chemical carcinogens. Proc. Natl. Acad. Sci. USA 75:1334-1338.

Kouri, R.E., S.A. Kurtz, P.J. Price, and W.F. Benedict. 1975. 1-Beta-D-arabinofuranosyl cystosine-induced malignant transformation of hamster and rat cells in culture. Cancer Res. 35:2413-2419.

MacPherson, I. 1969. Agar suspension culture for quantitation of transformed cells. In: Fundamental Techniques in Virology. K. Habelman and N.R. Salzman, eds. Academic Press: New York.

Milo, G.E., and J.A. DiPaolo. 1978. Neoplastic transformation of human diploid cells in vitro after chemical carcinogen treatment. Nature 275:130-132.

NCI. 1978. Bioassays of Selected Cancer Chemotherapeutic Agents for Possible Carcinogenicity. DHEW publication no. (NIH) 78-1329. National Cancer Institute. Department of Health, Education and Welfare: Washington, DC.

NTP and NCI Technical Report Series. 1976-1981. Carcinogenesis Bioassay Program. National Toxicology Program, NIEHS and National Cancer Institute. TR 1-230. U.S. Government Printing Office: Washington, DC.

OSHR. 1980. EPA carcinogen assessment group carcinogens list referenced in cancer policy candidates notice. Chemicals having substantial evidence of carcinogenicity. Occupat. Safety and Health Rep. 10:305-306.

Reznikoff, C.A., J.S. Bertram, D.W. Brankow, and C. Heidelberger. 1973. Quantitative and qualitative studies of chemical transformation of cloned C3H mouse embryo cells sensitive to post-confluence inhibition of cell division. Cancer Res. 33:3239-3249.

Shin, S., V.H. Freedman, R. Risser, and R. Pollack. 1975. Tumorigenicity of virus-transformed cells in nude mice is correlated specifically with anchorage-independent growth in vitro. Proc. Natl. Acad. Sci. USA 72:4435-4439.

Sivak, A., M. Charest, L. Rudenko, D.M. Silviera, and A.M. Wood. 1981. BALB/c-3T3 as target cells for chemically induced neoplastic transformation. In: Mammalian Cell Transformation by Chemical Carcinogens. N. Mishra, V.C. Dunkel, and M. Mehlman, eds. Senate Press: Princeton Junction, NJ.

Traul, K.A., V. Kachevsky, and J.S. Wolff. 1979. A rapid in vitro assay for carcinogenicity of chemical substances in mammalian cells utilizing an attachment-independence end point. Int. J. Cancer 23:193-196.

A QUANTITATIVE, CLONAL ASSAY FOR CARCINOGEN-INDUCED ALTERATIONS OF

RESPIRATORY EPITHELIAL CELLS IN CULTURE

J. Carl Barrett, Thomas E. Gray, Marc J. Mass, and
David G. Thomassen

Environmental Carcinogenesis Group, Laboratory of Pulmonary
Function and Toxicology, National Institute of Environmental
Health Sciences, Research Triangle Park, NC 27709

INTRODUCTION

The identification of environmental carcinogens and the
assessment of the potential risks of these substances to humans
require experimental systems to measure quantitatively the
activity of carcinogens and promoters. Cell culture systems are
potentially very useful experimental models for such studies.
Short-term, inexpensive cell culture assays for carcinogens are
available. These assays have a high predictive ability for the
detection of known carcinogens with few false positive results
(Barrett et al., 1980). The end point measured, preneoplastic or
neoplastic transformation of the test cells, is relevant to the
carcinogenic process and is not predicated on a theoretical
correlation. These systems can also be used for mechanistic
studies on the cellular and molecular basis of neoplastic
development. The results with cells and tissues of different
species (including human) can be compared and contrasted under
similar experimental conditions.

The most relevant cell culture systems employ cells that are
from the target tissues for environmental carcinogens, i.e.,
epithelial cells. Unfortunately, most quantitative cell
transformation assays use fibroblasts, because methods for growing
these cells in vitro have been available for a number of years
(Barrett et al., 1980); however, malignant tumors of nonepithelial

cells account for only 10 to 20% of human malignant neoplasms.
Recently, methods for growing epithelial cells in culture have
been developed and a number of laboratories have demonstrated
induction of neoplastic transformation of these cells in culture
by various chemical carcinogens (Franks and Wigley, 1979; Harris,
1982). However, quantitation of the frequency of early,
carcinogen-induced changes on a per-cell basis has not been
achieved with epithelial cells. In this report we describe our
recent results on the growth of rat tracheal epithelial (RTE)
cells in culture and the use of these cells for quantitative
assays of carcinogen-induced cytotoxicity and preneoplastic
transformation.

Advantages of Carcinogenesis Studies with Rat Tracheal Epithelial Cells

Respiratory epithelium is an important target for
environmental carcinogens. Lung cancer in men accounts for 34% of
the cancer deaths in the U.S., and the incidence of this disease
is clearly related to exposure to a number of environmental
factors (American Cancer Society, 1980; Doll, 1978). Because of
the need to understand factors that influence the etiology and
pathogenesis of this prevalent human disease, model systems have
been developed using tracheal epithelium as the target tissue.
The development of both in vivo and in vitro models for
carcinogenesis studies with tracheal epithelium (see below),
combined with the environmental importance of this tissue, gives
this system a unique advantage over other epithelial systems.

A number of experimental models in laboratory animals have
been developed for studying carcinogenesis of respiratory
epithelium (for review see Nettesheim and Griesemer, 1978). The
best characterized models are the intratracheal instillation
technique of Saffiotti et al. (1968) using the Syrian hamster, and
the heterotopic tracheal graft model using Fisher strain 344 rats,
originally described by Kendrick et al. (1974). These models have
been characterized with respect to their target tissue's capacity
to metabolize and activate polynuclear aromatic hydrocarbons to
ultimate carcinogenic forms (Mass and Kaufman, 1982).

Nettesheim and co-workers, using the tracheal transplant
model, have studied the influence of host and environmental
factors on experimental respiratory carcinogenesis and the
induction, characteristics, and sequential histological changes of
carcinogen-induced preneoplastic lesions following controlled
exposure of the tracheas to carcinogens and cocarcinogens

(Nettesheim et al., 1977; Griesemer et al., 1977; Pal et al., 1977; Nettesheim and Griesemer, 1978; Nettesheim et al., 1981a,b).

An important observation that led to quantitative studies on the cellular alterations that occur during neoplastic development in rat tracheal epithelium was made by Marchok et al. (1977, 1978). These authors noted that tracheal epithelium exposed to carcinogens contained cells with an increased growth capacity in vitro. Both neoplastic and preneoplastic cells could be selected by their in vitro growth properties and the preneoplastic cells progressed to malignant cells following further growth in vitro. These cells formed differentiated carcinomas when injected into syngeneic animals (Marchok et al., 1978). Terzaghi and Nettesheim (1979) employed this in vitro selection technique for carcinogen-altered cells to quantitate the cellular changes occurring in tracheal epithelium in vivo during the process of carcinogenesis. They were able to identify cells in three different stages of preneoplastic progression and to quantitate the population dynamics of cells in these different stages.

Terzaghi et al. (1978) also developed an important technique for the characterization of cells isolated in vitro. The in vivo growth characteristics of normal, preneoplastic, and neoplastic epithelial cells cultured in vitro can be studied by repopulating "denuded" tracheal grafts. These grafts consist of rat tracheas from which the normal epithelium has been removed. These "denuded" tracheas can serve as growth chambers in vivo for inoculated epithelial cells when the grafts are transplanted subcutaneously into a syngeneic rat or nude mouse. When cell cultures of normal RTE cells are used to repopulate tracheas, mucociliary epithelium identical to that observed in situ is obtained; however, when carcinogen-altered cells are inoculated into denuded tracheas, epithelia with altered tissue architecture, including metaplastic, dysplastic, and invasive lesions, are formed. Thus, the histopathology of particular cell populations selected in vitro can be analyzed by this technique. By altering the substrate and/or mesenchymal components of the denuded trachea, this technique can also identify normal and preneoplastic cells on the basis of their abilities to repopulate modified tracheas (Terzaghi and Klein-Szanto, 1980).

Steele et al. (1979) demonstrated that rat tracheal organ cultures treated with carcinogens in vitro and then grown as explant cultures yielded preneoplastic and neoplastic cell lines in an analogous manner to in vivo exposed tracheas. These cell lines also formed differentiated carcinomas when injected into syngeneic animals. The effect of chemical carcinogens in this system was enhanced by treatment with tumor promoters (Steele et al., 1980). Recent improvements in the methods for the growth and differentiation of epithelial cells in vitro (Wu et al., 1981) allowed Pai et al. (1982) to transform normal rat tracheal cell cultures following carcinogen treatment. These authors were able to quantitate population changes in the carcinogen-treated cultures 18 days after treatment and estimated that the frequency of carcinogen-induced preneoplastic cells was ~ 10^{-2}/surviving cell.

The advantages of carcinogenesis studies with RTE cells are unique:

1) Respiratory epithelium is an important target for environmental carcinogens in humans.

2) Experimental in vivo models allow characterization of biochemical and biological responses of tracheal epithelium following controlled exposure of the cells to carcinogens and cocarcinogens.

3) Carcinogen-altered preneoplastic cells can be selected by enhanced growth in vitro.

4) Quantitation of the frequency of subpopulations of altered cells during neoplastic development in vivo can be made by selection techniques in vitro.

5) Isolated cell populations in vitro can be used to repopulate a "denuded" tracheal transplant, and thereby, their ability to grow and differentiate in vivo can be characterized.

6) Methods exist to examine growth and differentiation of normal tracheal epithelial cells in vitro.

7) Quantitative, clonal studies of carcinogen-induced preneoplastic cells in vitro are possible.

Excellent models of carcinogenesis of this environmentally important tissue are established and permit quantitative, controlled studies in vivo. The ability to perform in vivo to in vitro as well as in vitro to in vivo studies enables the quantitation of cellular changes in vivo and the histopathological characterization of cultured cells. The ability to propagate and to induce neoplastic transformation of normal RTE cells in culture makes this an attractive cell culture system for quantitative studies of carcinogenesis.

Requirements for a Quantitative Cell Culture Model for Chemical Carcinogenesis

The first requirement for an in vitro system for studies of chemical carcinogenesis is that the target cells must be capable of sufficient proliferation after carcinogen treatment to allow for fixation and expression of the carcinogen-induced events (Barrett et al., 1980). A second important requirement for a quantitative assay of cell transformation is that the number of cells exposed to the carcinogen and the number of cells surviving the treatment be known. The most reliable index of cell survival is colony formation since only cells that survive and proliferate can be considered in the quantitation of carcinogenic events. The most challenging task in the quantitation of the changes in the neoplastic progression of cells is the identification, selection, and quantitation of cells in the early, preneoplastic stages of this process. Cell transformation studies on fibroblasts in culture are, for the most part, predicated on morphological changes that occur following carcinogen treatment (Barrett and Elmore, 1982). Unfortunately, epithelial cells often do not display a morphological change. Thus, this marker is not as useful for studies on these cells. However, as mentioned above, carcinogen-altered, preneoplastic RTE cells can be distinguished from normal cells by the culture conditions required for their proliferation in vitro (Marchok et al., 1977; Terzaghi and Nettesheim, 1979). Thus, the potential exists to use differences in the ability to proliferate in vitro as a selective pressure to isolate and quantitate preneoplastic cells after transformation in culture.

A desirable feature of a model cell culture system for chemical carcinogenesis is that the target cells are normal and in an early passage. After the prolonged growth of cells in culture, aneuploid cell lines will often arise spontaneously. The use of aneuploid cell lines is important for studying some aspects of carcinogenesis; however, these cell lines are altered relative to normal, diploid cells and in some systems these cells have been shown to be preneoplastic (Barrett and Elmore, 1982). It is advantageous to employ normal, diploid cells since the early, preneoplastic changes in the cells, as well as their subsequent neoplastic progression, can be monitored.

Growth of Rat Tracheal Epithelial Cells on 3T3 Feeder Layers

With the requirements discussed above in mind, we have attempted to develop a quantitative cell culture model for chemical carcinogenesis studies using RTE cells. Proliferating

cultures of normal RTE cells were obtained by growing the cells on
lethally irradiated mouse 3T3 fibroblasts which serve as a feeder
layer for the epithelial cells. Our experiments were based on the
work of Howard Green and colleagues who demonstrated that the
feeder layers support the growth of normal human keratinocytes
(Rheinwald and Green, 1975) and tracheal cells (Green, personal
communication).

Rat tracheal epithelial cells were obtained from male
Fisher 344 rats (pathogen-free). After euthanasia of the rats,
their tracheas were surgically exposed and a small polyethylene
tube was inserted into each. The lumens of the tracheas were
washed with medium and then with a 1% solution of pronase. Each
trachea was filled to extension with pronase solution, tied off,
and then incubated to dissociate the tracheal epithelium
(Sonstegard et al., 1976; Terzaghi and Nettesheim, 1978; Wu et
al., 1981). The epithelial cells were collected, centrifuged,
washed, and then plated on feeder layers of 3T3 cells that had
been lethally irradiated with 5000 rads of γ-irradiation
(Rheinwald and Green, 1975). The cultures were maintained in
Ham's F-12 medium supplemented with 5% fetal bovine serum,
insulin, and hydrocortisone and incubated in 95% air and 5% CO_2 at
37°C. Rat tracheal epithelial cells grown on 3T3 feeder layers
have the following characteristics: The cells are epithelial in
nature as demonstrated by their morphology under phase-contrast
microscopy. Ultrastructural studies demonstrated the presence of
desmosomes and tight junctions, bundles of cytoplasmic filaments,
prominent microvilli, and cilia in these cells. Primary RTE cells
have a colony-forming efficiency of 5 to 10% and the cells grow
rapidly in culture with a population doubling time of < 20 h. The
cultures can be subcultured for up to 5 passages during which time
the cells undergo up to 22 cumulative population doublings. Thus,
the degree of proliferation of normal RTE cells on 3T3 feeder
layers is sufficient to allow for the fixation and expression time
required for carcinogen-induced cell transformation based on
previous studies with other cell types (Barrett et al., 1980).

Cytotoxicity Studies on Carcinogen-Treated Rat Tracheal Epithelial Cells in Culture

The high colony-forming efficiency (CFE) of RTE cells on
3T3 feeder cells permits one to determine the cytotoxic effect of
environmental toxins on respiratory cells in culture. An
illustration of this is presented in Figure 1, which is a dose
response of cell killing by a direct-acting mutagen and
carcinogen, N-methyl-N'-nitro-N-nitrosoguanidine (MNNG). The
response of RTE cells to a 4-h treatment with MNNG under three

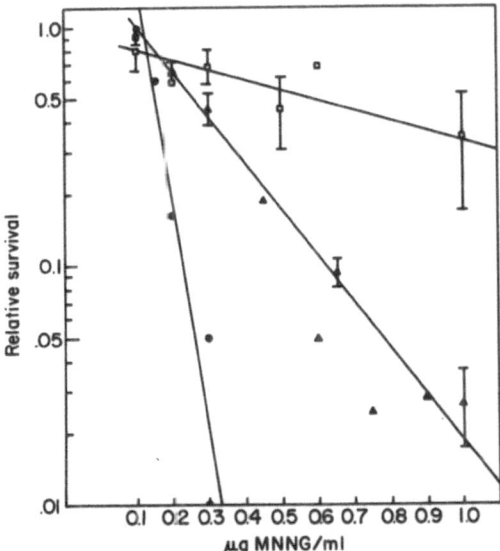

Figure 1.   Dose response of RTE cell killing by MNNG. Cell
            killing is measured as CFE in treated cultures relative
            to control cultures. RTE cells were treated for 4 h
            with MNNG (□) on 3T3 feeder layers in complete
            medium, with serum (▲) on 3T3 feeder layers in HEPES-
            buffered medium (pH 6.8), or (●) in HEPES-buffered
            medium (pH 6.8) in the absence of 3T3 cells. In the
            absence of feeder cells, RTE cell proliferation was
            supported by medium conditioned by 3T3 cells. Dishes
            were fixed, stained, and scored 7 to 10 days following
            treatment.

conditions is presented; RTE cells were treated with MNNG as
follows: on 3T3 feeder layers in complete medium supplemented
with serum, on 3T3 feeder layers in a HEPES-buffered medium

(pH 6.8), or in HEPES-buffered medium (pH 6.8) in the absence of
3T3 cells. In the latter condition the growth of the RTE cells
was supported by medium conditioned by 3T3 cells. A linear,
logarithmic decrease in cell survival was observed with increasing
dose of MNNG regardless of the treatment protocol (Figure 1).
Treatment in medium with serum (pH 7.4) was less toxic than
treatment in HEPES-buffered medium (pH 6.8). This is due to the
increased stability of MNNG at lower pH (Jacobs and DeMars, 1978).
The cell killing was also a function of cell density. Increased
cytotoxicity was observed when the RTE cells (1000 cells) were
treated in the absence of the large number of feeder cells
(2×10^5 cells). The cytotoxic response of RTE cells to MNNG
treatment was similar to the response of human skin fibroblasts
treated under similar conditions (Jacobs and DeMars, 1978).

Dose-response curves, which were linear when the logarithm of
the relative cell survival was plotted against dose, were obtained
following treatment of RTE cells with γ-irradiation, polycyclic
aromatic hydrocarbons (e.g., (±)-7β,8α-dihydroxy-9α,10α-epoxy-
7,8,9,10-tetrahydrobenzo(a)pyrene, and 7,12-dimethylbenz-
anthracene), direct-acting alkylating agents, and asbestos. These
results demonstrate the sensitivity of this assay for measuring
the cytotoxicity of environmental carcinogens on respiratory
epithelial cells and for comparing the response of these cells to
other cell types. The number of cells surviving the carcinogen
treatment can also be determined with accuracy which is essential
for calculating the frequency of carcinogen-induced cellular
alterations.

Selection and Quantitation of Carcinogen-Induced Alterations of Rat Tracheal Cells in Culture

Having established conditions for the growth of normal RTE
cells in culture and the quantitation of the cytotoxic effects of
carcinogens, the next step in the development of a quantitative
assay for cell transformation was the determination of conditions
for the selection of preneoplastic cells. The earlier work of
Terzaghi and Nettesheim (1979) demonstrated that preneoplastic
epithelial cells from rat tracheas exposed to carcinogens in vivo
could be selected by growth in standard tissue culture medium
(Ham's F-12) supplemented with fetal bovine serum, insulin, and
hydrocortisone. When normal RTE cells are grown under these
conditions, the cells attach, divide once or twice, and then
enlarge, appear squamous-like, and cease proliferation. The cells
either detach from the dish after 1 to 2 weeks or persist in the
culture for longer periods of time without cell division. This
process is analogous to the process of terminal differentiation

observed in cultures of rodent epidermal cells (Yuspa and Harris, 1974; Yuspa and Morgan, 1981; Sisskin and Barrett, 1981).

In contrast to the results observed with normal RTE cells, a small fraction of cells dissociated from carcinogen-exposed tracheal transplants proliferate and form large colonies of actively growing epithelial cells when placed in this culture medium (Terzaghi and Nettesheim, 1979). These colonies of altered cells were termed "epithelial foci" or "EF" by Terzaghi and Nettesheim (1979). Since colonies of normal epithelial cells can form under different culture conditions (i.e., with 3T3 feeder cells or medium conditioned by these cells), we have chosen not to use this terminology. Instead, we have termed cells that proliferate under conditions that are not permissive for the growth of normal cells as enhanced-growth (EG) variants. We believe that EG variants selected following carcinogen treatment of normal RTE cells in culture represent essentially the same preneoplastic cells that form the EF in the studies of Terzaghi and Nettesheim (1979).

The protocol we have developed for a quantitative assay for carcinogen-induced alterations is outlined in Figure 2. Normal RTE cells are plated onto 3T3 feeder layers and allowed 24 h for attachment. The cultures are treated and the cells are allowed to form colonies for 7 to 10 days. The cultures are then fixed, stained, and scored for the number of surviving colonies. For transformation experiments, the cells are allowed to proliferate for a fixation and expression time which is generally 7 days. At the end of the fixation and expression time, the culture conditions are changed to select for the preneoplastic cells with an enhanced-growth potential (EG variants). This selection is accomplished by either subculturing the cells and growing them in medium in the absence of 3T3 cells or by removing the 3T3 feeder cells from the cultures without subculturing the epithelial cells. The latter process involves treating the cultures with EDTA and physically removing the 3T3 cells by vigorous pipetting. The epithelial cells attach to the dishes more firmly than the 3T3 cells and very few are removed by this treatment.

A selection period of 4 to 5 weeks is allowed which permits the growth of EG variants into proliferating epithelial colonies. The cells in these colonies are heterogeneous, but all the colonies contain actively dividing cells tht are small and hyperchromatic with increased nuclear:cytoplasmic ratios (Figure 3).

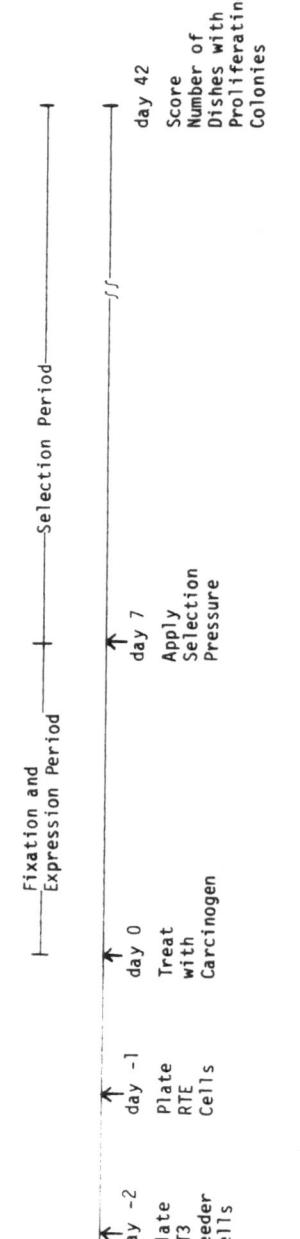

Figure 2.   Transformation and cytotoxicity protocol.

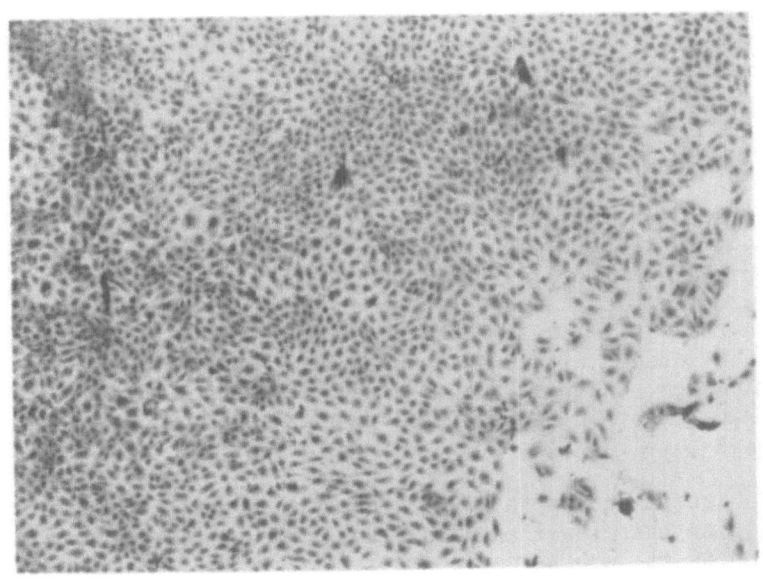

Figure 3. EG variant.

The number of dishes yielding colonies of EG variants is determined and the frequency of altered cells is calculated based on the Poisson distribution using the formula:

$$P_0 = e^{-n} \text{ and } n = (TF)(CFC)$$

where P_0 = fraction of the dishes without transformed colonies
    n = average number of transformation events per dish
CFC = colony-forming cells per dish, and
  TF = transformation frequency per surviving cell.

Using MNNG as a paradigmatic carcinogen, we examined the influence of this chemical on the transformation of rat tracheal cells using the protocol outlined in Figure 2. We found that the frequency of EG variants increased linearly with increasing dose of carcinogen from 0.3% with 0.1 µg/ml MNNG to 3.7% with 1.0 µg/ml MNNG. The frequency of cell transformation on a per-cell basis

was independent of the number of cells treated. The
transformation frequency was constant over a range of expression
times from 2 to 14 days. Longer expression times in some
experiments resulted in a decreased transformation frequency. The
transformation frequency was the same with the two selection
methods studied (subculturing and feeder removal).

CONCLUSIONS

     The studies summarized in this paper indicate that model
systems using tracheal epithelium have a unique advantage over
other epithelial systems for carcinogenesis studies because both
in vivo and in vitro analyses can be performed. Our studies
indicate that it is possible to do quantitative studies on
carcinogen-induced cytotoxicity and the early preneoplastic
alterations of respiratory epithelial cells in culture. We have
demonstrated that it is possible to quantitate these changes at
the clonal level.

     The determination that the frequency of early changes is high
and the fact that these changes occur within a few days after
treatment is of importance in understanding the mechanism of
neoplastic development. This frequency, which is higher than
values reported for mutation at single-gene loci, is similar to
that for the early morphological changes observed following MNNG
treatment of normal, diploid fibroblasts (Barrett and Ts'o, 1978).
As a result, it will be interesting to study early, preneoplastic
changes and their mechanism of induction in RTE cells and to
compare these alterations with changes known to occur in
fibroblasts (Barrett et al., 1980; Barrett, 1982). Further
comparisons of different cellular systems will be interesting.

     While these studies are still preliminary, we feel that this
system has already demonstrated its utility for studies on the
mechanisms of neoplastic transformation and the assessment of the
activity of chemical carcinogens on cells from an important target
tissue. We intend to continue studying the basis for these early,
preneoplastic changes to determine the effect of different
carcinogens and carcinogen doses on this process, and to study the
progression of these cells to the neoplastic state and the
conditions that can modify this process.

REFERENCES

American Cancer Society Facts and Figures. 1980. American Cancer
     Society: New York.

Barrett, J.C. 1982. Cell transformation, mutation, and cancer.
     In: The Use of Mammalian Cells for Detection of
     Environmental Carcinogens, Mechanisms and Application.
     C. Heidelberger, N. Invi, T. Kuroki, and M. Yamada, eds.
     Gann Monograph on Cancer Research. University of Tokyo
     Press: Tokyo, Japan.

Barrett, J.C., B.D. Crawford, and Paul O.P. Ts'o. 1980. The role
     of somatic mutation in a multistage model of carcinogenesis.
     In: Mammalian Cell Transformation by Chemical Carcinogens.
     N. Mishra, V.C. Dunkel, and M. Mehlman, eds. Senate Press,
     Inc.: New Jersey. pp. 467-500.

Barrett, J.C., and E. Elmore. 1982. Comparison of carcinogenesis
     and mutagenesis of mammalian cells in culture. In:
     Mutagenesis and Carcinogenesis. Handbook of Environmental
     Pharmacology. L.S. Andrews, R.J. Lorentzen, and W.G. Flamm,
     eds. Springer-Verlag: Berlin.

Doll, R. 1978. An epidemiologic perspective of the biology of
     cancer. Cancer Res. 38:3573-3583.

Franks, L.M., and C.B. Wigley, eds. 1979. Neoplastic
     Transformation in Differentiated Epithelial Cell Systems In
     Vitro. Academic Press: New York.

Harris, C.C., ed. (in press). Mechanisms of Chemical
     Carcinogenesis. A.R. Liss: New York.

Griesemer, R.A., P. Nettesheim, D.H. Martin, and J.E. Caton, Jr.
     1977. Quantitative exposure of grafted rat tracheas to
     7,12-dimethylbenz(a)anthracene. Cancer Res. 37:1266-1271.

Jacobs, L., and R. DeMars. 1978. Quantification of chemical
     mutagenesis in diploid human fibroblasts: induction of
     azaguanine-resistant mutants by N-methyl-N'-nitro-N-nitroso-
     guanidine. Mutation Res. 53:29-53.

Kendrick, J., P. Nettesheim, and A.S. Hammons. 1974. Tumor
     induction in tracheal grafts: a new experimental model for
     respiratory carcinogenesis studies. J. Natl. Cancer Inst.
     52:1317-1325.

Marchok, A.C., J.C. Rhoton, R.A. Griesemer, and P. Nettesheim.
     1977. Increased in vitro growth capacity of tracheal
     epithelium exposed in vivo to 7,12-dimethylbenz(a)anthracene.
     Cancer Res. 37:1811-1821.

Marchok, A.C., J.C. Rhoton, and P. Nettesheim. 1978. In vitro
    development of oncogenicity in cell lines established from
    tracheal epithelium preexposed in vivo to 7,12-dimethyl-
    benz(a)anthracene. Cancer Res. 38:2030-2037.

Mass, M.J., and D.G. Kaufman. 1982. Species differences in the
    activation of benzo(a)pyrene in the tracheal epithelium of
    rats and hamsters. In: Organ and Species Specificity in
    Chemical Carcinogenesis. R. Langenbach, S. Nesnow, and
    J. Rice, eds. Plenum Press: New York.

Nettesheim, P. and R.A. Griesemer. 1978. Experimental models for
    studies of respiratory tract carcinogenesis. In:
    Pathogenesis and Therapy of Lung Cancer: Lung Biology in
    Health and Disease, Volume 10. O. Lenfant and C.C. Harris,
    eds. Marcel Dekker, Inc.: New York and Basel. pp. 75-188.

Nettesheim, P., R.A. Griesemer, D.H. Martin, and J.E. Caton, Jr.
    1977. Induction of preneoplastic and neoplastic lesions in
    grafted rat tracheas continuously exposed to benzo(a)pyrene.
    Cancer Res. 37:1272-1278.

Nettesheim, P., A.J.P. Klein-Szanto, A.C. Marchok, V.E. Steele,
    M. Terzaghi, and D.C. Topping. 1981a. Studies of neoplastic
    development in respiratory tract epithelium. Act. Pathol.
    Lab. Med. 105:1-10.

Nettesheim, P., D.C. Topping, and R. Jamasbi. 1981b. Host and
    environmental factors enhancing carcinogenesis in the
    respiratory tract. Ann. Rev. Pharmacol. Toxicol. 21:133-163.

Pai, S.B., V.E. Steele, and P. Nettesheim. (MS). Quantitation of
    early cellular events during neoplastic transformation of
    tracheal epithelial cell cultures.

Pai, B.C., D.C. Topping, R.A. Griesemer, F.R. Nelson, and
    P. Nettesheim. 1978. Development of a system for controlled
    release of benzo(a)pyrene, 7,12-dimethylbenz(a)anthracene,
    and phorbol ester for tumor induction in heterotopic tracheal
    grafts. Cancer Res. 38:1376-1383.

Rheinwald, J.G., and H. Green. 1975. Serial cultivation of
    strains of human epidermal keratinocytes: the formation of
    keratinizing colonies from single cells. Cell 6:331-344.

Saffiotti, U., F. Cefis, and L.H. Kolb. 1968. A method for the
    experimental induction of bronchogenic carcinoma. Cancer
    Res. 28:104-124.

Sisskin, E.E., and J.C. Barrett. 1981. Inhibition of terminal differentiation of hamster epidermal cells in culture by the phorbol ester 12-0-tetradecanoylphorbol-13-acetate. Cancer Res. 41:593-603.

Sonstegard, K.S., E. Cutz, and V. Wong. 1976. Dissociation of epithelial cells from rabbit trachea and small intestine with demonstration of APUD endocrine cells. Am. J. Anat. 147:357-374.

Steele, V.E., A.C. Marchok, and P. Nettesheim. 1980. Enhancement of carcinogenesis in cultured respiratory tract epithelium by 12-0-tetradecanoylphorbol-13-acetate. Int. J. Cancer 26:343-348.

Steele, V.E., A.C. Marchok, and P. Nettesheim. 1979. Oncogenic transformation in epithelial cell lines derived from tracheal explants exposed in vitro to N-methyl-N'-nitro-N-nitroso-guanidine. Cancer Res. 39:3805-3811.

Terzaghi, M., and A.J.P. Klein-Szanto. 1980. Differentiation of normal and cultured preneoplastic tracheal epithelial cells in rats: importance of epithelial mesenchymal interactions. J. Natl. Cancer Inst. 65:1039-1048.

Terzaghi, M., and P. Nettesheim. 1979. Dynamics of neoplastic development in carcinogen-exposed tracheal mucosa. Cancer Res. 39:4003-4010.

Terzaghi, M., P. Nettesheim, and M.L. Williams. 1978. Repopulation of denuded tracheal grafts with normal, preneoplastic, and neoplastic epithelial cell populations. Cancer Res. 38:4546-4553.

Wu, R., J.W. Groelke, L.Y. Chang, M.E. Porter, D. Smith, and P. Nettesheim. 1982. Effects of hormones on the multiplication and differentiation of tracheal epithelial cells in culture. In: Cold Spring Harbor Conferences on Cell Proliferation, Volume 9. Growth of Cells in Hormonally Defined Media. Cold Spring Harbor Laboratory: Cold Spring Harbor, NY.

Yuspa, S.H., and C.C. Harris. 1974. Altered differentiation of mouse epidermal cells treated with retinyl acetate in vitro. Exp. Cell Res. 86:95-105.

Yuspa, S.H., and D.L. Morgan. 1981. Mouse skin cells resistant
    to terminal differentiation associated with initiation of
    carcinogenesis. Nature 293:72-74.

RECENT EXPERIENCE WITH THE STRAIN A MOUSE PULMONARY TUMOR BIOASSAY

MODEL

R.R. Maronpot,[1] H.P. Witschi,[2] L.H. Smith,[2] and J.L. McCoy[1]

[1]National Toxicology Program, Research Triangle Park, NC 27709, and [2]Oak Ridge National Laboratory, Oak Ridge, TN 37830

INTRODUCTION

Since its inception in November 1978, a major mandate of the National Toxicology Program (NTP) has been to identify toxic chemicals that must be controlled to prevent human disease. A consequent goal of this mandate is to support the development of efficient and economical methods that will identify biological effects and carcinogenic potential of chemicals. The strain A mouse pulmonary tumor test system has long been considered a useful in vivo animal model to predict potential carcinogenicity of chemicals (Shimkin and Stoner, 1975). This report communicates some recent experiences with the strain A mouse test system derived from separate contributions from two laboratories supported by the NTP.

MATERIALS AND METHODS

The mouse pulmonary tumor-induction bioassay was carried out essentially as described in the literature (Shimkin and Stoner, 1975) using groups of 20 or 30 inbred strain A mice. Both sexes of strain A/ST were used in one laboratory where 60 chemicals were submitted for testing. On the basis of unpublished observations of no sex difference in responsiveness, only male strain A/J mice

were used in the second laboratory where 16 chemicals and
5 complex mixtures have been tested to date.

Depending upon solubility, chemicals submitted for testing
were dissolved in 0.9% saline, corn oil, or tricaprylin. For each
series of tests there was a vehicle control group, an untreated
control group, and a positive (urethan) control group. The
maximum tolerated dose (MTD) was determined in preliminary
toxicity studies following which each chemical was given at the
MTD, MTD/2, and MTD/4 or MTD/5 in the bioassay. In the bioassay,
chemicals were given by intraperitoneal injection three times a
week for eight weeks. Mice were killed 16 weeks after the last
injection and grossly visible pulmonary tumors were counted. All
chemicals were tested as unknowns (coded samples) in the strain A
mouse bioassay. Carcinogenicity data from two-year rodent
bioassays are available for most of these chemicals tested in the
mouse pulmonary tumor test system.

RESULTS

Validation Studies

Of 60 chemicals submitted to Laboratory 1 for "blind" testing
in the strain A pulmonary tumor model, 54 have been tested
adequately in a two-year rodent bioassay and Technical Reports
have been issued. Consequently, six chemicals are excluded from
this present comparison because of insufficient two-year bioassay
data. The strain A mouse test system correctly predicted the
carcinogenicity or lack thereof for 20 (37%) of the 54 chemicals.
There were 7/16 (44%) "false positives" and 27/38 (71%) "false
negatives." These comparative results are presented in Table 1.
Of the 27 chemicals with "false negative" strain A test results,
14 were carcinogenic in rat and mouse in two-year bioassays,
8 were carcinogenic in rats only, and 5 were carcinogenic in mice
only.

Of the 54 chemicals examined in this comparison, 28 were
aromatic amines, 6 were aliphatic chlorides, and the remaining
20 were from a variety of chemical classes. The frequencies of
"false negatives" were 14/18 (78%) for the aromatic amines and 4/6
(67%) for the aliphatic chlorides. The frequency of "false
negatives" for the remaining chemicals was 9/14 (64%). Although
some chemical classes are either not represented or not adequately
represented among the 54 chemicals under investigation, these
findings are interpreted as an indication that the strain A mouse
test system "false negative" rate was not restricted to a single
chemical class.

Table 1. Comparative Results of Chemicals Tested in Strain A
Mouse Pulmonary Tumor Bioassay and Two-Year Rodent
Bioassay

Chemicals Positive[a] in Two-Year Bioassay and Pulmonary Tumor Bioassay	Chemicals Positive[a] in Two-Year Bioassay and Negative in Pulmonary Tumor Bioassay

Aromatic amines

3-Amino-9-ethylcarbazole·HCl	2-Aminoanthraquinone
4-Amino-2-nitrophenol	3-Amino-4-ethoxy acetanilide
2-Nitro-p-phenylenediamine	p-Chloroaniline
5-Nitro-o-toluidine	4-Chloro-m-phenylenediamine
	m-Cresidine
	p-Cresidine
	Direct black 38[b]
	Direct blue 6[b]
	Direct brown 95[b]
	4,4-'Methylenebis-(N,N-dimethyl-aniline)
	1,5-Naphthalenediamine
	5-Nitro-o-anisidine
	p-Nitrosodiphenylamine
	Phenazopyridine·HCl

Aliphatic chlorides

3-Chloromethyl pyridine·HCl	Aldrin
Sulfallate	1,1,2,2-Tetrachloroethane
	Tetrachloroethylene
	Toxaphene

Miscellaneous

Captan	Cupferron
N,N'-Dicyclohexyl thiourea	3,3'-Dimethoxybenzidine-4,4'-diisocyanate
2,4-Dinitrotoluene	2,5-Dithiobiurea
Hydrazobenzene	6-Nitrobenzimidazole
5-Nitroacenaphthalene	Nitrofen
	Pivalolactone
	p-Quinone dioxime
	Selenium disulfide
	1,1,3-Trimethyl-2-thiourea

(continued)

Table 1.   (continued)

Chemicals Negative in Two-Year Bioassay and Positive[a] in Pulmonary Tumor Bioassay	Chemicals Negative in Two-Year Bioassay and Pulmonary Tumor Bioassay

Aromatic amines

Anthranilic acid	p-Anisidine
4'-(Chloroacetyl)-acetanilide	Chloropropamide
p-Phenylenediamine·HCl	2,4-Dimethoxyaniline
	N-(1-Naphthyl)ethylenediamine
	4-Nitro-o-phenylenediamine
	Sulfisoxazole
	2,5-Toluenediamine sulfate

Miscellaneous

Diazinon	Tolbutamide
Parathion	3-Sulfolene
Triphenyltin hydroxide	
L-Tryptophan	

[a]Chemicals positive in the two-year bioassay had a statistically significant increase in tumors in treated animals versus concurrent controls. Equivocal results were considered negative. Chemicals positive in the pulmonary tumor bioassay had a statistically significant increase in tumor incidence and multiplicity in treated versus vehicle controls. Fiducial limit: $p = 0.05$.
[b]Chemicals positive in 13-week feeding study. Two-year bioassay not conducted.

Of 18 positive pulmonary tumor bioassays listed in Table 1, 5 were positive in both male and female strain A mice, 6 were positive only in males, and 8 were positive only in females.

To date, 16 of the 54 chemicals that were tested in Laboratory 1 have completed "blind" testing in the strain A mouse pulmonary tumor model in Laboratory 2. Comparative results from the two laboratories for these 16 chemicals are presented in Table 2. Eleven of these 16 chemicals were found carcinogenic in rats or mice in conventional two-year rodent bioassays. Of these

Table 2. Chemicals Tested in Strain A Mouse Pulmonary Tumor Bioassays in Two Laboratories

| Chemical | Strain A Test Results | | | Two-Year Bioassay Results | | | |
| | Laboratory 1[a] | | Laboratory 2[b] | Rat | | Mouse | |
	M	F	M	M	F	M	F
4-Amino-2-nitrophenol	+	+	–	+	+	–	–
Captan	+	–	–	–	–	+	+
4'-(Chloroacetyl)-acetanilide	+	–	–	–	–	+	–
m-Cresidine	–	–	–	+	+	–	–
2,4-Dimethoxyaniline	–	+	–	–	–	–	–
2,4-Dinitrotoluene	–	–	–	+[c]	+	–	–
Direct black 38	–	+	–	–	+[c]	–[c]	–[c]
5-Nitro-o-toluidine	–	–	–	+	–	+	+
p-Nitrosodiphenylamine	–	–	–	+	–	+	–
Phenazopyridine hydrochloride	–	+	–	–	+	–	–
p-Phenylenediamine dihydrochloride	–	–	–	+	–	–	–
Pivalolactone	–	–	–	–	+	–	–
p-Quinone dioxime	+	+	–	+	+	–	–
Sulfallate	+	–	+	–	+	+	+
2,5-Toluenediamine sulfate	–	–	–	–	–	–	–
Triphenyltin hydroxide	–	+	–	–	–	–	–

[a] Strain A/ST mice.
[b] Strain A/J mice.
[c] Tested in 13-week bioassay.

11 positive chemicals, Laboraory 2 produced 10 (91%) "false negatives" with the strain A pulmonary tumor model.

Complex Mixture Studies

Three shale oils and two coal gasification fractions were assayed in the strain A mouse lung tumor model and all were found positive. Four of these substances have been found carcinogenic in conventional mouse skin-painting studies and the fifth has not yet been tested in a long-term bioassay. The results of testing these complex mixtures are presented in Table 3. The strain A mouse pulmonary tumor test system was predictive of the carcinogenicity of these mixtures.

DISCUSSION

Two primary facts derive from data presented in this report. The first is that there was lack of congruity of results between the strain A pulmonary tumor bioassay and the two-year rodent carcinogenesis bioassay. The second fact is that there was a lack of consistency in strain A bioassay results from two separate laboratories. There are several possible explanations for this lack of congruity and consistency.

Failure of the strain A pulmonary tumor bioassay to give similar results to those obtained in the two-year rat and mouse carcinogenesis bioassay may be a reflection of different pharmacokinetic and metabolic considerations. Species differences in metabolism could account for results from eight chemicals which were negative in strain A mice and positive only in rats in the two-year bioassay. Strain differences in pharmacokinetics and metabolism may account for other situations in which there was lack of congruity. The B6C3F1 mouse was used in the two-year bioassay and the strain A mouse was used in the pulmonary tumor bioassay. Routes of chemical administration differed between the two bioassays. In the strain A pulmonary tumor bioassay, chemicals were given by intraperitoneal injection. In the two-year bioassay, chemicals were given by gavage or in the feed. Although not determined at present, it is quite probable that the total dose of chemical given differed between the two bioassays. In addition, the duration of exposure to chemicals differed between the two bioassays. All of these variables could impact on chemical disposition, detoxification, and activation and account for the lack of congruity between the two bioassays.

Table 3. Mouse Lung Tumor Bioassay of Complex Mixtures

Substance	Dosage Producing Carcinogenic Response (g/kg)	Number of Mice		% Mice At End with Lung Tumors	Number of Lung Tumors/Mouse (± SE)
		At Start	At End		
Shale oil	3.2	30	6	50	1.00 ± 0.60
Hydrotreated shale oil	1.25	30	12	50	1.25 ± 0.46
Hydrotreated shale oil residue	8.0	30	6	100	6.17 ± 1.20
	4.0	30	8	75	1.75 ± 0.63
UMD 19[a]	0.2	30	13	38	0.92 ± 0.44
UMD 83[a]	0.25	30	11	73	1.92 ± 0.75
	0.125	30	24	50	0.96 ± 0.31
Corn oil control	10.0	30	28	7	0.07 ± 0.05
Urethan control	1.0[b]	50	50	100	21.40 ± 1.40

[a]University of Minnesota at Duluth coal gasifier fraction.
[b]One injection.

Another factor which might explain the apparent failure of the pulmonary tumor bioassay relates to the relative sensitivity of the two bioassays. It is entirely conceivable that the so-called "false positives" in the pulmonary tumor bioassay represent situations wherein the pulmonary tumor bioassay is correct in predicting carcinogenicity while the two-year bioassay is relatively insensitive in its ability to detect carcinogenicity. In situations where an organ- or tissue-specific response is obtained in the two-year bioassay, the pulmonary tumor bioassay may be relatively insensitive because the target organ in the latter bioassay is the lung.

Consideration of what constitutes a positive response in each bioassay has a direct impact on any attempt to compare results. The pulmonary tumor bioassay is traditionally considered positive when there is a statistically significant increase in the incidence of tumor-bearing mice as well as a statistically significant increase in tumor multiplicity, with the latter carrying more weight. In the case of the results from Laboratory 1, the incidence and multiplicity of pulmonary tumors in negative and positive controls were less than what is typically reported in strain A bioassays. In situations such as this, perhaps there should be a modification of the traditional definition of what constitutes a positive or negative response. Criteria regarding interpretation of positivity in one sex versus both sexes are also needed. In the case of two-year bioassays, decisions as to what is a positive or negative response often represent difficult judgments requiring consideration of concurrent as well as historical control tumor incidences, information regarding chemical disposition and metabolism, carcinogenic activity of structurally related chemical species, pertinent genotoxicity data, etc.

Since there was considerable overlapping of the doses of each chemical administered in the two laboratories, lack of consistent agreement in the outcome of the pulmonary tumor bioassay between Laboratory 1 and Laboratory 2 cannot be explained by differences in the amount of chemical given. However, different substrains of the strain A mouse were used in the two laboratories. Laboratory 1 used strain A/ST while Laboratory 2 used strain A/J. Another possible explanation for the lack of concordance in results between the two laboratories is, in part, a consequence of Laboratory 2 having used only male mice while Laboratory 1 used both sexes in the pulmonary tumor bioassay. It is obvious from the data obtained from Laboratory 1 that there is a difference in pulmonary tumor responsiveness between male and female strain A mice.

In a certain sense it is "wrong" to attempt to validate any short-term bioassay by simply comparing its results to those of the two-year rodent carcinogenesis bioassay. The two-year bioassay itself has not been definitively validated and, in some instances, the short-term bioassay under investigation may be more predictive of carcinogenicity than the two-year rodent carcinogenesis bioassay. More appropriately, a validation exercise shold take into account all that is known about how the chemicals under test are metabolized in each bioassay model, organ specificity of the carcinogenic response, the degree of positivity of the response, available genotoxicity data, whether the bioassay under validation has been optimized for maximum sensitivity, etc.

In conclusion, while the data presented in this report are interpreted as indicative of a lack of concordance between the strain A mouse pulmonary tumor bioassay and the two-year rodent carcinogenesis bioassay, the issues mentioned above must be rigorously addressed before valid claims regarding the utility of the pulmonary tumor bioassay can be made.

REFERENCES

Shimkin, M.B. and G.D. Stoner. 1975. Lung tumors in mice: application to carcinogenesis bioassay. Adv. Cancer Res. 21:1-58.

INTEGRATED CHEMICAL AND BIOLOGICAL ANALYSIS OF ASPHALT AND PITCH FUMES

Philip S. Thayer,[1] Judith C. Harris,[1] Kenneth T. Menzies,[1] and Richard W. Niemeier,[2]

[1]Arthur D. Little, Inc., Cambridge, MA 02140, and [2]Division of Biomedical and Behavioral Science, National Institute for Occupational Safety and Health, Cincinnati, OH 45226

INTRODUCTION

Significant increased risks in developing cancer of the lung, upper respiratory tract, and upper gastrointestinal tract, including stomach cancer, have been demonstrated for individuals working 20 or more years in roofing operations (Hammond et al., 1976). These investigators have also documented trends of increased risk of prostate, bladder, or skin cancer, and leukemia in these workers. The occupational exposures incurred by roofers may be associated with high levels of polynuclear aromatic hydrocarbons (PAHs) arising from the heating and application of petroleum asphalts and coal tar pitch. These findings are consistent with results of other studies that have indicated that excessive occupational exposure to PAHs may be associated with increased mortality from various types of cancer. Examples of these occupations include chimney sweeps (Pott, 1775) and coke oven workers (Lloyd, 1971; Mazumdar et al., 1975; Redmond et al., 1972). Skin cancer mortality may be of borderline significance in many of these occupations only because deaths due to skin cancer are rare. However, skin cancer incidence may be excessive.

The roofing materials to which workers may be exposed include both pitches and asphalts. The principal source of pitch materials is the coke oven plant, where hot gases and vapors produced during the conversion of coal to coke are condensed to

yield coal tar among other byproducts; distillation of the coal
tar produces a variety of compounds, including coal tar pitch.
The carcinogenicity of neat coal tar pitch and aerosolized coal
tar, particularly PAHs, has been widely investigated and reviewed
(NIOSH, 1977a); however, the carcinogenicity of coal tar pitch
fumes as generated in roofing operations has not been
investigated.

Asphalts are usually obtained as the residue of the
distillation of petroleum crudes. Their chemical composition
varies depending on the crude oil used to manufacture them, the
refining process, and the physical specifications of the finished
product. Asphalts may also be derived from crude oil by solvent
precipitation and air-blowing. The carcinogenicity of asphalts is
not well documented primarily because of the confusion of many
authors concerning the terms asphalt, pitch, and tar (coal tar)
which are frequently used interchangeably and incorrectly.
Limited chemical analyses show the substances to be quite
different, especially in their proportions of PAHs and other known
carcinogenic chemicals (Wallcave et al., 1971; Puzinauskas and
Corbett, 1978; Bingham et al., 1979). Bingham and co-workers
(1979) and NIOSH (1977b) reviewed the literature concerning the
carcinogenic potential of asphalt.

In addition to the risks associated with exposure to the coal
tar pitch and/or petroleum asphalt fumes in the roofing
occupational environment, the added risk of sunlight exposure in
the production of cancer in this out-of-doors environment must be
considered. It was suspected that exposure to sunlight might
affect the skin cancer incidence in this population. Various
investigators working with experimental animal models have
observed that ultraviolet and/or visible light augments the
carcinogenic end point of PAH exposure (Santamaria et al., 1966;
Urbach, 1959).

NIOSH estimates that there are 12,000 roofing contractors
employing over 116,000 workers in the U.S. (NIOSH, 1974). Since
these workers have combined exposures to asphalt and coal tar
pitch fumes and sunlight, an experimental study was designed to
determine the relative importance of exposures to each and
combinations of these agents. The purpose of this investigation
was fourfold. The first objective was to assess by skin
application to mice the carcinogenic potential of condensed
volatiles from roofing asphalts and coal tar pitch materials that
were collected from fumes generated at the recommended application
temperatures of the materials. The studies and experimental
results that relate most directly to this objective are the focus
of this paper. Since the heating process in the kettle operations

is usually not controlled, it is common for the materials to be heated above their recommmended application temperatures (Thomas et al., 1975). Therefore, the design of this study incorporated a second objective: to assess the carcinogenic potential of the condensed volatiles collected from fumes generated at temperatures in excess of their recommended application temperatures. The third objective was to assess the effects of concomitant exposure to simulated sunlight on the carcinogenic outcome of the above materials. Because of the expected differing results from exposure to simulated sunlight a fourth objective was included: to assess the responses in a pigmented strain of mice and compare this to the responses in a nonpigmented strain. The experiments and results that relate to the effects of fume generation temperature, exposure to simulated sunlight, and pigmentation of the test strain, which are described in detail in the full report of this study (Thayer et al., 1981), are mentioned only briefly in this paper.

MATERIALS AND METHODS

There are four types of asphalt, Types I through IV, and three types of coal tar pitch, Types I through III, used for roof dampproofing and waterproofing. Only two types of each were used in this study and they were chosen on the basis of common use and extremes of the classifications. These are Type I and Type III asphalt, referred to in the industry as "dead level" and "steep," respectively, and Type I and Type III coal tar pitch, often referred to as "regular roofing" and "low fuming" pitch, respectively. These four materials are produced by several manufacturers according to physical specifications recommended by the American Society of Testing and Materials (ASTM) (ASTM, 1978).

The collection of fumes from a roofing material kettle in the field would be awkward at best due to the difficulty in controlling sample mixing, temperature, exposure to sunlight, and in collecting a condensed sample at a subambient temperature. Therefore, laboratory generation of fumes from an easily controlled glass generation system and collection of condensed material in a glass cryogenic system were used for the production of necessary amounts of fumes. A 12-1 round-bottom reaction flask containing 10 1 of pitch or asphalt was warmed in a forced air oven at 150°C. Above the softening point, a stainless steel multifinned stirring rod was inserted to permit uniform mixing (250 to 300 rpm). An electric heating mantle capable of reaching 450°C was utilized to heat the roofing material to the desired generation temperatures ± 5°C.

The fume collection system consisted of glass transfer tubes (20-mm OD) and 500-ml glass impingers (Ace Glass, Inc.) placed in three individual cryotraps containing ice (0°C) and dry ice/isopropanol (-77°C). An additional impinger containing a 50/50 mixture of cyclohexane/acetone was used after the cryotraps to provide additional collection through dissolution. Clean, dry air was pulled through the system at a rate of 10 lpm with a vacuum pump regulated by a needle valve restrictor. The air was cleaned using a high efficiency filter (Filterite 0.45-μm Microflow cartridge), silica gel, and granular activated charcoal for removal of particulates, water, and organic vapors. The air was then preheated to 100°C and pulled through the reaction flask where it entrained volatiles. The generation/collection system was contained in a laboratory that received only yellow light filtered through cellulose acetate-butyrate filters to reduce exposure to ultraviolet light.

After the fumes were condensed and collected in the sampling train and individual impingers and transfer lines were weighed, all material was quantitatively transferred to a large flask with an excess of cyclohexane/acetone solvent mixture used to assist complete transfer. The solvent was removed at reduced pressure with a water aspirator and at a temperature of 50°C. The materials (fumes) remaining after removal of solvents were weighed and dissolved in a 50/50 (v/v) cyclohexane/acetone mixture and combined. The choice of this solvent for collection, rinsing, and preparation of skin-painting solutions was the result of the following rationale: observations of the poorer solubility of roofing material fumes in either solvent alone; the low boiling point of each solvent; the compatibility of each solvent with respect to PAH stability; the lack of absorption of light in the simulated sunlight range; and the inactivity of each with respect to mouse skin bioassay.

A total of 144 successful daily generations were conducted to provide sufficient condensed material from the heating of 4 different roofing materials at 2 different temperatures to permit the preparation of about 4 l of 8 skin-painting solutions. The mass emission rates of volatiles from the roofing materials increased in the order: asphalt at 232°C; asphalt at 316°C; pitch at 232°C; pitch at 316°C. Each batch was analyzed for selected PAH and heteroaromatic compounds content to obtain a fingerprint of the relatively important chemicals.

To determine the concentrations of selected PAHs in skin-painting solutions, combined gas chromatographic/mass spectrometric (GC/MS) (Finnigan Model 4023) analysis was used. A glass capillary column (25-m coated with SP 2250) capable of high

resolution was generally employed to permit separation and
quantitation of closely related isomers such as benzo(e)pyrene and
benzo(a)pyrene (B[a]P). Compounds were considered to be
identified when the retention time for the chemical of interest
relative to the retention time of an internal standard matched the
relative retention time observed in a calibration standard, and
the mass spectrum of the PAHs of interest matched that obtained
from a calibration standard. There were considerable variations
in both the absolute and relative concentrations of specific PAHs
between one generation and another. Table 1 shows the mean
concentrations of 18 selected chemicals in the skin-painting
solutions prepared from the fumes of each roofing material. These
data represent the 232°C vapor generation experiments. The 316°C
experiments yielded a 2 to 10 times higher mass emission rate of
volatiles but the concentration of the skin-painting solution was
adjusted to be comparable to those for the 232°C materials. The
distribution among the 18 PAH compounds was similar for both
generation temperatures.

The solar simulator arrangement developed for this study used
a 15-cm Atlas 6.5-kW xenon arc, water-cooled, quartz-enveloped
burner that was located about 53 cm above midway between two
turntables. Short-wavelength ultraviolet radiation below 290 nm
was blocked using filters. Light measurements were made to
validate the geometric concept of the turntable arrangement.
Included in the design was an automatic exposure control detector
which maintained a constant exposure by adjusting the exposure
time to correct for instantaneous lamp intensity. This
integrating system eliminated the effects of fluctuations and low
changes in intensity of the arc due to aging. Two enclosures, one
containing the light source and another surrounding the light
source and the turntables, were constructed of aluminum (former)
or aluminum framing covered with sheet masonite (latter).

The male mice used in these bioassays were of two strains:
nonpigmented Swiss CD-1 (Charles River) and pigmented C3H/HeJ
(Jackson Laboratories). Upon arrival, at 6 weeks of age, all
animals were quarantined for a 6- to 9-week period. Animals were
housed individually in stainless steel, suspended metal cages and
provided food and water ad libitum except during the exposure
period. Each test group consisted of 50 formally randomized mice
of a given strain.

Each mouse in groups 1 through 32 received 50 µl of the
appropriate test material twice weekly. Each animal in the
negative (1:1 cyclohexane/acetone) control group received 50 µl of
the vehicle twice weekly. The positive control group animals each
received 50 µl of 0.01% B(a)P (5 µg of B[a]P in

Table 1. Mean Concentration of PAHs in Skin-Painting Solutions (μg/ml)

Chemical	Analytical Ion	Asphalt Type I 232°C	Type I 316°C	Type III 232°C	Type III 316°C	Pitch Type I 232°C	Type I 316°C	Type III 232°C	Type III 316°C
Naphthalene	128	22	4.4	17	49	> 1800	1770	288	> 620
Fluorene	166	36	22	39	28	x[a]	740	x[a]	x[a]
Carbazole	167	20	1.4	6.3	--[b]	1980	1450	540	1400
Anthracene/Phenanthrene	178	180	53	300	69	> 960	2960	> 2580	> 5200
Fluoranthene	202	86	10	97	7.3	> 2940	2350	> 960	> 2800
Pyrene	202	70	9.0	63	7.7	> 2070	1790	> 720	> 2300
Benz(a)anthracene	228	11	10	7.6	5.7	570	330	330	800
Chrysene/Triphenylene	228	25	19	13	14	460	300	290	710
Benzofluoranthenes	252	1.8	4.0	5.2	--[b]	230	230	250	250
Benzo(e)pyrene	252	5.5	8.3	3.6	1.4	42	51	45	46
Benzo(a)pyrene	252	2.2	1.9	2.9	--[b]	96	85	102	90
Indenopyrene	276	2.7	3.1	2.2	--[b]	33	1.7	11	6.8
Benzoperylene	276	0.8	1.5	0.8	--[b]	28	2.0	7.2	0.7
Dibenzanthracenes	278	1.6	--[b]	1.8	--[b]	12	--[b]	4.1	--[b]
Coronene	300	--[b]	--[b]	--[b]	--[b]	--[b]	--[b]	--[b]	--[b]
Dibenzopyrenes	302	--[b]	--[b]	--[b]	--[b]	--[b]	--[b]	--[b]	--[b]

[a] Indicates species not determined.
[b] Indicates species not detected (< 0.5 μg/ml).

1:1 cyclohexane/acetone twice weekly. In accordance with the specifications of the NIOSH test protocol, the dose of test material was limited to a level of 50% total solids content (25 mg solids/50 µl solution applied) or to a level of 0.01% B(a)P content (5 µg B[a]P/50 µl solution applied). The former constraint governed in the case of the asphalt fume, which had relatively low PAH content. In the case of the high-PAH-content pitch fume samples, the B(a)P concentration constraint determined the quantity of fume material applied at each dose.

Mice were observed daily for mortality, evidence of systemic toxicity, and gross appearance of tumors. The mice in a specific group were treated and observed until 85% of a group died, or until 18 months had elapsed. Mice found dead were necropsied. Those that were moribund were killed and necropsied and when groups were killed, those remaining animals were necropsied. All tissues mentioned in the National Cancer Institute (NCI) guidelines were examined, collected, and preserved in formalin. Gross diagnosis of a skin carcinoma was based on a lesion that upon palpation was attached to underlying tissues, which generally indicated invasion of connective tissue or muscle layers. All tumors were excised and fixed in buffered 10% formalin for microscopic examination. The distribution of tumors between dermal sarcomas, papillomas, and epidermal carcinomas was noted particularly.

More detailed descriptions of the experimental procedures and observations are provided in a report by Thayer et al. (1981).

RESULTS AND DISCUSSION

A complete presentation and discussion of the results of this study is beyond the scope of this paper. The interested reader is referred to the full report (Thayer et al., 1981), which includes a thorough statistical analysis of the data comparing tumorigenicity of the asphalt vs. pitch fumes generated at the two temperatures. This paper presents an abbreviated discussion of the specific activity of the various fume preparations in relation to available information concerning their chemical composition.

In general, survival of all groups of mice was satisfactory and comparable through 11 months for CD-1 mice and 10 months for C3H/HeJ mice. Survival was greater than 80% for all groups at the appearance of the first papilloma (effective number of animals at risk) except in two groups when the ventilation system in the solar simulator failed temporarily and the temperature within the chamber rose to over 40°C. After 10 to 11 months, the

experimental groups showed sharper declines in survival because of
termination of moribund animals to ensure preservation of tumors
for histopathological examination. The control groups of both
strains showed good survival throughout the study, approximately
as expected. Animal weights reflected very little effect of the
treatments on weight gain.

Tumor incidence, both as observed grossly during the life of
the animals and histopathologically, is graphically presented in
Figure 1. Mean time to tumor data is graphically presented in
Figure 2.

The pooled incidences of tumor-bearing animals expressed as a
percentage of the effective total for malignant tumors (carcinomas
or fibrosarcomas) given in Figure 1 show that the CD-1 mice have a
moderately high total tumor incidence, but quite a low incidence
of malignant tumors. The asphalts produced a much lower total
tumor incidence in the CD-1 as compared to the C3H/HeJ strain.
The fraction of tumors that were malignant was comparable for both
asphalts and pitches. The data show high activity of pitches on
both strains of mice and of asphalts on the C3H only.

The time-incidence data were used for calculation of latent
period (mean time to tumor). These data show the lower
sensitivity of the CD-1 mouse strain, with longer latent periods
than the C3H mice for all test materials except B(a)P. An
inhibitory effect of the simulated sunlight is manifested in the
data, since the latent period for the solar groups is longer than
for the corresponding nonsolar group.

Table 2 summarizes the relationship between the total dose of
test material applied and the mean time to tumor for the
experimental animals. Table 3 presents a comparable summary of
the data relating to incidence of malignant tumors in test
populations.

The pooled data in Table 2 show no straightforward
relationship between the mean time to tumor and the cumulative
dose of either B(a)P or total PAHs. In the case of the pitch
materials, for example, the total dose of B(a)P applied was about
37% lower than in the positive controls but the mean time to tumor
was about 33% shorter. This could be an indication that other
components of the pitch fume samples accelerate the expression of
the carcinogenic effect. Chemical characterization (Table 4) of a
direct extract of the pitch used in these studies and of a pitch
fume sample generated from a full-scale roofing kettle (not the
identical fume material used in these carcinogenicity studies)
according to EPA-RTP Level 1 procedures (Lentzen et al., 1978)

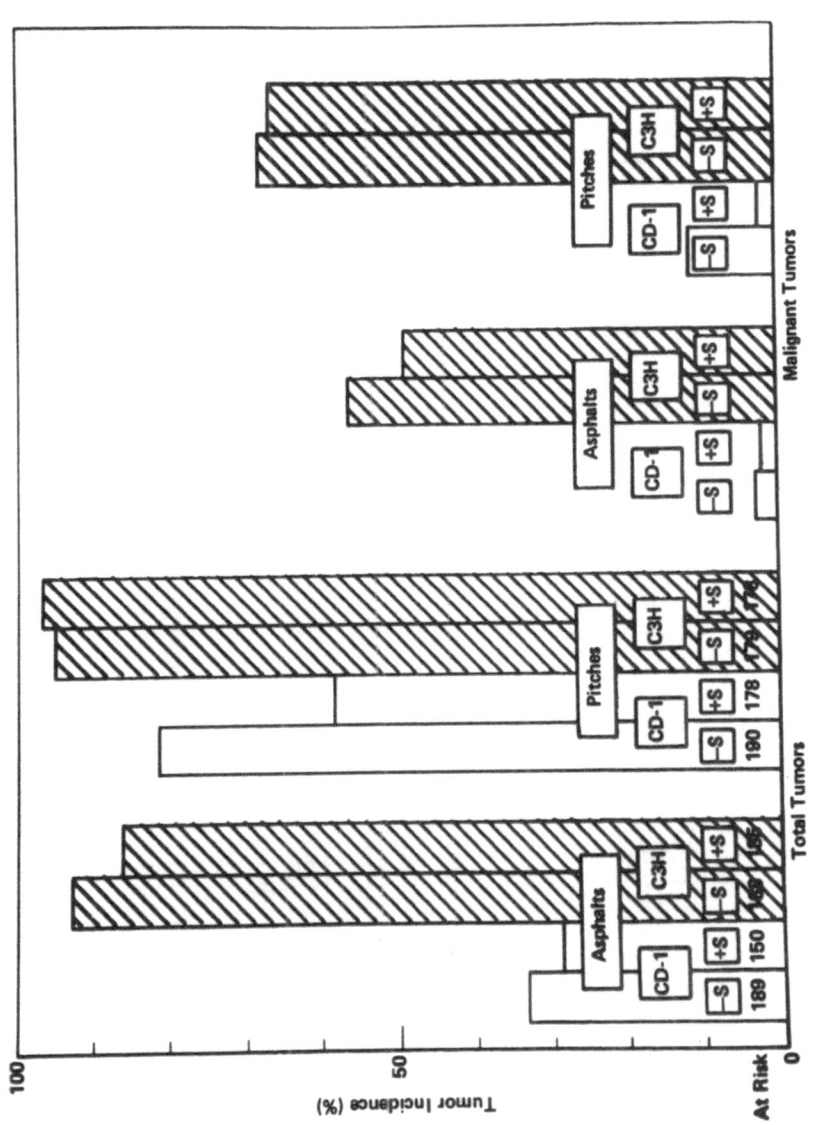

Figure 1. Tumorigenic effects of asphalt and pitches in CD-1 and CBH mice based on tumor incidence (pooled data).

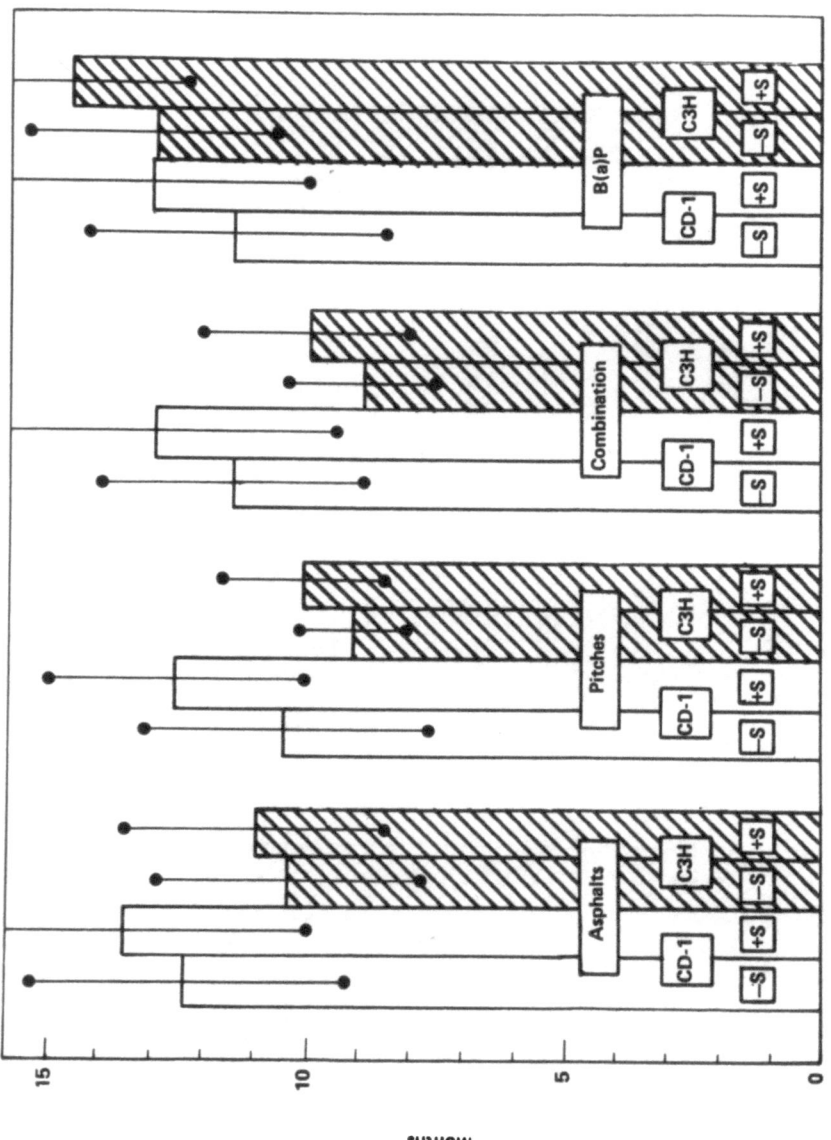

Figure 2. Tumorigenic effects of asphalt and pitches in CD-1 and CBH mice based on mean time to tumor (pooled data).

Table 2. Time of 50% Tumor Incidence (TI) as a Function of Total Dose of Test Material Applied[a]

Preparation	Mean Mass Emission Rate (g/h)	Amount of Material per Application Total Solids (mg)	PAH (μg)	B(a)P (μg)	Time to 50% TI (months)	Number of Applications	Total Dose Applied Total Solids (mg)	PAH (μg)	B(a)P (μg)
Asphalts									
Type I – 232°C	2.1 ± 1.2	25	23.2	0.11	12	104	2600	2.4	11.4
Type I – 316°C	33 ± 16	25	7.4	0.095	9.3	81	2025	0.60	7.7
Type III – 232°C	3.1 ± 2.6	25	28.0	0.145	11	96	2400	2.7	13.9
Type III – 316°C	27 ± 13	25	9.1	< 0.025	9.5	83	2075	0.76 <	2.1
Coal tar pitches									
Type I – 232°C	38 ± 10	3.9	> 560	4.8	9.7	84	328	> 47.0	403
Type I – 316°C	79[b]	2.75	603	4.25	9	78	214	47.0	332
Type III – 232°C	17 ± 12	4.2	> 306	5.1	9	78	328	> 23.9	398
Type III – 316°C	115[b]	1.5	> 711	4.5	8.8	77	116	54.7	346
B(a)P	--	--	5.0	5.0	13.7	119	--	0.6	595

[a]Data for C3H mice; no simulated solar exposure.
[b]Calculated on the basis of one generation experiment.

Table 3. Incidence of Malignant Tumors as a Function of Total Dose of Test Material Applied[a]

Preparation	Total Applications	Total Applied Solids (mg)	Total Applied PAH (µg)	Total Applied B(a)P (µg)	% Malignant Tumor-Bearing	Ratio of % Malignant Tumor-Bearing to Total µg Applied Solids	Ratio PAH	Ratio B(a)P
Asphalts								
Type I - 232°C	153	3825	3.55	16.8	44.9	1.2×10^{-5}	12.6	2.67
Type I - 316°C	138	3450	1.02	13.1	66.0	1.9×10^{-5}	64.7	5.04
Type III - 232°C	155	3875	4.34	22.5	53.2	1.4×10^{-5}	12.3	2.36
Type III - 316°C	137	3425	1.25	< 3.4	62.2	1.8×10^{-5}	50.0	> 18.3
Coal tar pitches								
Type I - 232°C	118	460	> 66.1	566	67.4	1.5×10^{-4}	< 1.02	0.12
Type I - 316°C	126	347	76.0	536	73.9	2.1×10^{-4}	0.97	0.14
Type III - 232°C	118	496	> 36.1	602	64.3	1.3×10^{-4}	< 1.78	0.11
Type III - 316°C	118	177	> 83.9	531	67.4	3.8×10^{-4}	< 0.80	0.13
B(a)P	149			595	64.3			0.11

[a]Data for C3H mice; no solar exposure. See Table 4 for dose per application.

Table 4. Results of Level 1 Chemical Characterization
of a Roofing Pitch and a Pitch Fume Sample[a]

Organic Compound Category	Approximate Concentration (mg/g)	
	Pitch Extract	Pitch Fume
Aliphatic hydrocarbons	--	8
Fused aromatic hydrocarbons	510	840
Heterocyclic sulfur compounds	31	83
Heterocyclic nitrogen compounds	15	180
Nitriles	16	10
Amines	8	--

[a]Source: Arthur D. Little, Inc., unpublished data.

suggests that heterocyclic sulfur and nitrogen compounds and
organic nitriles are present in the pitch samples at levels of
1 to 10% of the PAH content. Species of these types may
contribute to the carcinogenic effect of the roofing pitch fume
materials.

In the case of the asphalt fume samples, Table 2 indicates
that the mean time to tumor is appreciably shorter than in the
B(a)P controls, although the total B(a)P dose applied is more than
an order of magnitude lower than in the controls. This
observation, which suggests that other components in the asphalt
fume samples are significantly affecting carcinogenic activity, is
also evident in the data of Table 3. The incidence of malignant
tumors per microgram of B(a)P applied is much higher for asphalt
fume samples than for the B(a)P controls or for the pitch fume
samples (which are comparable to the B[a]P controls). This is a
clear indication that tumor incidence in these studies is not a
simple linear function of the B(a)P concentration of the
preparations applied. Although complete chemical characterization
data of the asphalt and pitch fume samples are not available, the
PAH analysis results have been supplemented by nuclear magnetic
resonance (NMR) analysis of these materials. The latter results
imply that the asphalt fume material is < 1% aromatic, > 99%
aliphatic, with straight-chain, unbranched materials
predominating. In contrast, NMR analysis of the pitch fume
indicated > 90% aromatic content. The high aliphatic content of

the asphalt fume sample is interesting in light of earlier observations (Horton et al., 1975, 1966; Bingham and Falk, 1969) that some alkyl hydrocarbons may exhibit cocarcinogenic activity.

CONCLUSIONS

The data generated in this study suggest that the carcinogenic activity of roofing pitch fume materials in skin-painting experiments may be understandable, to a first order approximation, in terms of the B(a)P content of those materials, although there is some indication of enhancing effects by other, unidentified components of the sample. The carcinogenic activity of the asphalt fume materials in these experiments, on the other hand, cannot even approximately be explained on the basis of their B(a)P content. It can be hypothesized, although it has not been demonstrated, that cocarcinogenic effects of aliphatic hydrocarbons, which are major components of the asphalt fumes generated in this study, may be partly responsible for the enhanced activity.

ACKNOWLEDGMENTS

The fume generation, PAH analyses, and skin-painting carcinogenicity studies reported here were conducted under NIOSH contract 210-78-0035. The supplementary Level 1 analyses of pitch samples and NMR analyses of pitch and asphalt fumes were supported by EPA contract 68-01-3111.

REFERENCES

ASTM. 1978. 1978 Annual Book of Standards, Part 15, Roads and Paving Materials; Bituminous Materials for Highway Construction, Waterproofing and Roofing. American Society for Testing and Materials: Philadelphia.

Bingham, E., H.L. Falk. 1969. Environmental carcinogens: the modifying effect of cocarcinogens on threshold response. Arch. Environ. Health 19:779-783.

Bingham, E., R.P. Trosset, and D. Warshawsky. 1979. Carcinogenic potential of petroleum hydrocarbons: a critical review of the literature. J. Environ. Pathol. Toxicol. 3:483-563.

Hammond, E.C., I.J. Selikoff, P.L. Lawther, and H. Seidman. 1976.
Inhalation of benzpyrene and cancer in man. N.Y. Acad. Sci.
271:116-124.

Horton, A.W., D.T. Denman, and R.P. Trosset. 1957.
Carcinogenesis of the skin: the accelerating properties of
aliphatic and related hydrocarbons. Cancer Res. 17:758-66.

Horton, A.W., M.J. Burton, R. Tye, and E. Bingham. 1963.
Composition vs. Carcinogenicity of Distillate Oil. ACS
Division of Petroleum Chemical Preprints 8. No. 4C:59-55.
American Chemical Society: Washington, DC.

Lentzen, D.E., D.E. Wagoner, E.D. Estes, and W.F. Gutknecht.
1978. IERL-RTP Procedures Manual: Level 1 Environmental
Assessment, 2nd Edition. EPA-600/7-78-201.
U.S. Environmental Protection Agency: Research Triangle
Park, NC.

Lloyd, J.W. 1971. Long-term mortality study of steelworkers,
Volume V: respiratory cancer in coke plant workers.
J. Occup. Med. 13:53-67.

Mazumdar, S., C.K. Redmond, W. Sollecito, and N. Sussman. 1975.
The epidemiological study of exposure to coal tar pitch
volatiles of coke oven workers. J. Air Pollut. Cont. Assoc.
25:382.

NIOSH. 1974. National Occupational Hazards Survey, Volume I:
Survey Manual, DHEW Pub. No. 74-127. Volume II: Data
Editing and Data Base Development, DHEW Pub. No. 77-213.
Volume III: Survey Analysis and Supplemental Tables, DHEW
Pub. No. 78-114. National Institute for Occupational Safety
and Health: Washington, DC.

NIOSH. 1977a. Criteria for a Recommended Standard--Occupational
Exposure to Coal Tar Products, DHEW Pub. No. 78-107.
National Institute for Occupational Safety and Health:
Washington, DC.

NIOSH. 1977b. Criteria for a Recommended Standard--Occupational
Exposure to Asphalt Fumes, DHEW Pub. No. 78-106. National
Institute for Occupational Safety and Health: Washington,
DC.

Pott, P. 1775. Cancer Scrote. In: Chirurgical Observations
    Relative to the Cataract, the Polypus of the Nose, the Cancer
    of the Scrotum, the Different Kinds of Ruptures and the
    Mortification of the Toes and Feet. L. Hawes, W. Clark, and
    R. Collins, eds. London. pp. 63-68.

Puzinauskas, V.P., and L.W. Corbett. 1978. Differences between
    Petroleum Asphalt, Coal-Tar Pitch and Road Tar. Research
    Report 78-1. Asphalt Institute: College Park, MD. 31 pp.

Redmond, C.K., A. Ciocco, J.W. Lloyd, and H.W. Rush. 1972.
    Long-term mortality study of steelworkers, Volume VI:
    mortality from malignant neoplasms among coke oven workers.
    J. Occup. Med. 14:621-629.

Santamaria, L., G.G. Giordano, M. Alfici, and F. Cascione. 1966.
    Effects of light on 3,4-benzpyrene carcinogenesis. Nature
    210:824-825.

Thayer, P.S., K.T. Menzies, and P.C. von Thuna. 1981. Roofing
    Asphalts, Pitch and UVL Carcinogenesis. Final report on
    National Institute for Occupational Safety and Health
    contract 210-78-0035.

Thomas, J.F., and M. Mukai. 1975. Evaluation of Emissions from
    Asphalt Roofing Kettles with Respect to Air Pollution.
    Research Report 75-2 (RR-75-2). Asphalt Institute: College
    Park, MD.

Urbach, F. 1959. Modification of ultraviolet carcinogenesis by
    photoactive agents. J. Invest. Dermat. 32:373-378.

Wallcave, L., H. Garcia, R. Feldman, W. Lyinksky, and P. Shubik.
    1971. Skin tumorigenesis in mice by petroleum asphalts and
    coal tar pitches of known polynuclear aromatic hydrocarbon
    content. Toxicol. Appl. Pharmacol. 18:41-52.

MOUSE SKIN CARCINOGENESIS: APPLICATION TO THE ANALYSIS OF COMPLEX MIXTURES

S. Nesnow,[1] L.L. Triplett,[2] and T.J. Slaga[2]

[1]Carcinogenesis and Metabolism Branch (MD 68), U.S. Environmental Protection Agency, Research Triangle Park, NC 27711, and [2]Biology Division, Oak Ridge National Laboratory Oak Ridge, TN 37830

INTRODUCTION

Current information suggests that chemical carcinogenesis is a multistage process; in this regard, one of the best studied models is the two-stage carcinogenesis system in mouse skin. Skin tumors can be induced by the sequential application of a subthreshold dose of a carcinogen (initiation phase) followed by repetitive treatments with a noncarcinogenic tumor promoter (promotion phase). The initiation phase requires only a single application of either a direct or indirect carcinogen at a subthreshold dose and is essentially irreversible. The promotion phase is brought about by repetitive treatments after initiation and is initially reversible but later irreversible. The mouse skin system can be used not only to determine the tumor-initiating and tumor-promoting activities of a compound: if the agent is given repeatedly by itself, one can also determine if it is a complete carcinogen (i.e., if it has both tumor-initiating and tumor-promoting activities).

Recently the generality of two-stage tumor induction has been demonstrated in a number of experimental carcinogenesis systems (e.g., liver, bladder, lung, colon, esophagus, stomach, mammary gland, pancreas, lung cells in culture) (Slaga et al., 1978). A wide variety of agents (e.g., diet, bile acids, hormones, saccharin, L-tryptophan, phenobarbital, polychlorinated biphenyls, polybrominated biphenyls, butylated hydroxytoluene) has been used successfully as promoters (Slaga et al., 1978).

Among rodent species, mice are generally more sensitive than rats and hamsters to skin carcinogenesis by either the complete carcinogenesis protocol or the initiation/promotion protocol (Slaga and Fischer, in press; Stenback, 1980). In mice, the complete carcinogenesis protocol gives rise to a low number of papillomas followed by a high incidence of squamous cell carcinomas, whereas the initiation/promotion protocol gives rise to a large number of papillomas followed by a moderate incidence of squamous cell carcinomas. In rats, both the complete carcinogenesis and initiation/promotion protocols give rise to basal cell carcinomas and very few papillomas and squamous cell carcinomas. In hamsters, the complete carcinogenesis protocol produces mainly squamous cell carcinomas and some melanomas, whereas the initiation/promotion protocol produces mainly melanomas.

Polycyclic aromatic hydrocarbons (PAHs) are one of the major classes of chemical carcinogens that have skin tumor-initiating or complete carcinogenic activity on mouse skin and have been studied extensively in this system. Over 100 PAHs, PAH derivatives, and PAH metabolites are known to be mouse skin tumor initiators or complete carcinogens (Nesnow et al., 1981; Pereira, 1982). Among these, moderate to strong initiators or complete carcinogens include 7,12-dimethylbenz(a)anthracene (DMBA), 3-methylcholanthrene, benzo(a)pyrene (B[a]P), 7-methylbenz(a)anthracene, 5-methylchrysene, dibenz(a,h)anthracene, dibenzo(a,h)pyrene, dibenzo(a,i,)pyrene, dibenzo(a,e)pyrene benzo(a)phenanthrene, dibenzo(a,j)anthracene, benzo(c)chrysene, benzo(g,h,i)perylene, dibenzo(a,c)naphthacene, and 11-methylcyclopenta(a)phenanthren-17-one. Besides PAHs, many other chemicals and chemical classes are known to be tumor initiators or complete carcinogens on mouse skin (Table 1) (Nesnow et al., 1981; Pereira, 1982).

Tumor Initiation

When appropriately tested, known skin complete carcinogens show skin tumor-initiating activity (Slaga et al., 1982). In the two-stage mouse skin system, initiation is the only stage that requires the presence of the complete carcinogen, and the measured complete carcinogenic potency of a chemical reflects its capacity for tumor initiation. There is both a good qualitative and quantitative correlation between the complete carcinogenic and tumor-initiating activities of several chemical carcinogens in mouse skin (Slaga et al., 1982). This is true when one considers the number of papillomas per mouse at early times (10 to 20 weeks) or the final carcinoma incidence after tumor initiation (Slaga et al., 1982).

Table 1. Chemicals Other Than Polycyclic Aromatic Hydrocarbons That Are Positive as Skin Tumor Initiators or Complete Carcinogens[a]

Class	Chemical(s)
Aldehyde	Malonaldehyde
Aziridine	β-Hydroxy-1-ethylaziridine
Carbamate	Urethane Vinyl carbamate N-Butyl-N-nitrosourethane
Epoxide, diepoxide	Glycidaldehyde 1,2,3,4-Diepoxybutane 1,2,4,5-Diepoxypentane 1,2,6,7-Diepoxyhexane Chloroethylene oxide 1,2-Epoxybutyronitrile
Haloalkylether	Bis(chloromethyl) ether α,α-Dichloromethyl methyl ether Chloromethyl methyl ether
Haloaromatic	2,3,4,5-Tetrachloronitrobenzene 2,3,4,6-Tetrachloronitrobenzene 2,3,5,6-Tetrachloronitrobenzene Pentachloronitrobenzene
Haloalkyl ketone, acid	Chloroacetone 3-Bromopropionic acid
Hydroxylamine	N-Acetoxy-4-acetamidobiphenyl N-Acetoxy-2-acetamidofluorene N-Hydroxy-2-aminonaphthalene N-Acetoxy-N-acetamidophenanthrene N-(4-Methoxy)benzoyloxypiperadine N-(4-Nitro)benzoyloxypiperadine N-Acetoxy-2-acetamidostilbene
Lactone	Propiolactone

See footnote at end of table.

(continued)

Table 1. (continued)

Class	Chemical(s)
Multifunctional	Triethylenemelamine 4-Nitroquinoline-N-oxide
Natural product	Aflatoxin B_1 Sterigmatocystin
Nitrosamide	N-Methyl-N'-nitro-N-nitrosoguanidine
Sulfonate	Allyl methylsulfonate
Sultone	1,3-Propanesultone
Urea	N-Nitrosomethylurea N-Nitrosoethylurea

[a]Sources of data: Nesnow et al., 1981; Pereira, 1982.

It is possible that a carcinogen lacking promoting ability would not be detected when tested as a complete carcinogen. It has been reported that a number of chemical compounds -- for example, benz(a)anthracene, dibenz(a,c)anthracene, chrysene, triethylenemelamine, urethane, B(a)P 7,8-dihydrodiol-9,10-epoxide, benz(a)anthracene 3,4-dihydrodiol-1,2-epoxide -- have mouse skin tumor-initiating activity but either lack or have very weak complete carcinogenic activity (Nesnow et al., 1981; Slaga et al., 1982).

Experiments with B(a)P and DMBA using different stocks and strains of mice (Reiners et al., in press) suggest the following ranking for sensitivity to two-stage (initiation/promotion) carcinogenesis: SENCAR >> CD-1 > ICR/Ha Swiss > BALB/c > C57BL/6 > C3H > DBA/2. The SENCAR mouse was derived by crossing Charles River CD-1 mice with skin tumor-sensitive mice (originally derived from Rockland mice) and selecting for sensitivity to DMBA-TPA two-stage carcinogenesis for 8 generations starting with the F_1 cross as described by Boutwell et al. (Slaga et., 1982; Boutwell, 1964). The mice developing the earliest and most papillomas after initiation/promotion treatment were selected for each breeding. It is important to emphasize the limitations of this subjective ranking. Firstly, only responses to B(a)P and DMBA were

considered. Secondly, dose-response data for the carcinogen and/
or promoter were not available for many of the mouse strains and
stocks. Despite these limitations, however, the differences
between mice at the extremes of the ranking are significant.

SENCAR mice are between 10 and 20 times more sensitive than
CD-1 mice to DMBA tumor initiation (DiGiovanni et al., 1980).
However, SENCAR mice are only between three and five times more
sensitive than CD-1 mice to B(a)P tumor initiation (DiGiovanni et
al., 1980). SENCAR mice are two to three times more sensitive
than CD-1 mice to TPA promotion (Reiners et al., in press).

Between SENCAR and C57BL/6 mice there is an even greater
difference in sensitivity to two-stage skin carcinogenesis.
C57BL/6 mice are very refractory to two-stage skin carcinogenesis
by B(a)P-TPA. Even high initiating doses of B(a)P (1600 nmol) and
high promoting doses of TPA (10 µg) are very ineffective in
causing skin tumors (Slaga and Nesnow, unpublished data).
However, C57BL/6 mice do respond to complete carcinogenesis by
B(a)P (Reiners et al., in press). Such unequal susceptibility to
complete and two-stage carcinogenesis within a stock or strain of
mice strongly suggests that the promotional phases of complete and
two-stage carcinogenesis are dissimilar. In addition, differences
in sensitivity to initiation and promotion between mice may be due
to alterations in the promotional phase of two-stage
carcinogenesis. In this regard, we recently found that benzoyl
peroxide is an effective promoter in C57BL/6 and SENCAR mice
(Slaga et al., unpublished data). For some reason, TPA is not an
effective promoter in C57BL/6 mice.

Complete Carcinogenesis

Complete carcinogenesis in mouse skin refers to the
production of tumors (mainly carcinomas) after repeated
application of a carcinogen for terms of up to 1 yr. Compounds
possessing both tumor-initiating activity and tumor-promoting
activity will produce tumors in this regimen. An examination of
six mouse stocks and strains using B(a)P and DMBA as the
carcinogens (Reiners et al., in press) suggests the following
ranking for sensitivity to complete carcinogenesis:
SENCAR > CD-1 > C57BL/6 > BALB/c > ICR/Ha Swiss > C3H.

Mouse skin complete carcinogenesis was used in the early
1900s to identify carcinogens in organic extracts of organic
particulate samples and to try to link these materials to the
etiology of human cancer (Table 2). Passey (1922) found that
ether extracts of coal chimney soot produced both "warts"

Table 2. Skin Carcinogenesis by Organic Extracts of Particulates: Historical Summary

Particulate Source	Mouse Strain	Tumor Type	References
Ambient	Swiss ICR C57 Black	Carcinoma; papilloma Carcinoma; papilloma	Wynder and Hoffman, 1965 Kotin et al., 1954
Coal chimney soot	"White"	Carcinoma; papilloma	Passey, 1922 Campbell, 1939
Diesel engine	C57 Black A	--- Carcinoma; papilloma	Kotin et al., 1955 Kotin et al., 1955
Gasoline engine	C57 Black Swiss	Carcinoma; papilloma Carcinoma; papilloma	Kotin et al., 1954 Wynder and Hoffmann, 1962
Industrial carbon black	Swiss	Carcinoma; papilloma	Von Haam and Mallette, 1952
Oil shale soot	"White"	Carcinoma; papilloma	Vosamae, 1979
Road dust	---	Carcinoma; papilloma	Campbell, 1939

(papillomas) and "cancers" (carcinomas) when applied repetitively
to the depilitated backs of mice. Campbell (1934) confirmed these
results with coal soot and also reported the mouse skin
carcinogenic activity of road dust extracts. Remarkably, Campbell
(1939) also observed both dermal and lung tumors in mice exposed
only to road dust particulates.

Kotin and coworkers examined the carcinogenic effects of
organic extracts of air particulate samples from the Los Angeles
area. These extracts produced both malignant and benign tumors
when applied repeatedly to the backs of C57BL/6 mice (Kotin et
al., 1954). Wynder and Hoffmann (1965) expanded these studies by
administering air particulate extracts from the Detroit area to
Swiss ICR mice. Both groups concluded that B(a)P alone could not
account for all the carcinogenic activity observed. Kotin et al.
(1954) and Wynder and Hoffman (1962, 1963, 1965) also studied the
carcinogenic effects of organic extracts of particulate emissions
from gasoline engines; both concluded that these extracts produce
tumors on mice. Kotin et al. (1955) studied extracts from
particulates isolated from a diesel engine. In contrast to the
positive results reported for air particulate and gasoline engine
emissions, these investigators found C57BL/6 mice to be refractory
to the diesel extracts. However, strain A mice did produce tumors
when treated repetitively with the diesel mixtures. Von Haam and
Mallette (1952) and Vosamae (1979) were able to induce tumors in
mice by applying extracts of industrial carbon black and oil shale
soot, respectively.

Most of the studies described above used limited numbers of
mice of various strains, did not explore potential sex
differences, and used limited numbers of doses. These studies
were not designed to produce comparative dose-response data.

A summary of our previous and current work is found in
Table 3. These studies were designed to produce extensive
dose-response data, using the same mouse strain (SENCAR), on the
tumorigenic and carcinogenic activities of organic extracts from a
variety of particulate emission sources: diesel and gasoline
vehicles, coke oven, and roofing tar. Earlier publications of
results from these experiments reported papilloma formation after
6 months in a tumor-initiation protocol (Nesnow et al., 1981;
Slaga et al., 1980; Slaga et al., 1981; Nesnow et al., 1982;
Nesnow et al., in press).

Table 3.   Skin Carcinogenesis by Organic Extracts of Particulates:
           Previous and Current Work

Particulate Source	Mouse Strain	Tumor Type
Coke oven	SENCAR	Carcinoma; papilloma
	C57 Black	--
Diesel engine	SENCAR	--
	C57 Black	--
Diesel vehicles	SENCAR	Carcinoma; papilloma
	C57 Black	--
Diesel oil furnace	SENCAR	--
	C57 Black	ND[a]
Gasoline vehicles	SENCAR	Carcinoma; papilloma
	C57 Black	--
Roofing tar	SENCAR	Carcinoma; papilloma
	C57 Black	--
B(a)P	SENCAR	Carcinoma; papilloma
	C57 Black	Carcinoma

[a]Not determined.

MATERIALS AND METHODS

Sample Generation and Isolation

     The details of sample generation and isolation were reported
previously (Huisingh et al., 1980). Vehicle emission samples were
obtained from a 1973 preproduction Nissan-Datsun 220C (Nissan), a
1976 prototype Volkswagen TurboRabbit (VW Rabbit), and a 1977 Ford
Mustang II-302 V-8 with a catalyst and EGR (Mustang). Each
vehicle was mounted on a chassis dynamometer and driven in a
repeated highway fuel economy test (HWFET) cycle of 10.24 mi, at
an average speed of 48 mph, and for a running time of 12.75 min.
The Nissan and VW Rabbit were fueled with the same batch of No. 2
diesel fuel. The Mustang was fueled with unleaded gasoline.
Particulate samples were collected with a dilution tunnel in which

the hot exhaust was diluted, cooled, and filtered through Pallflex Teflon-coated fiberglass filters.

Topside code oven samples were collected from the top of a coke oven battery at Republic Steel, Gadsden, AL, using a massive air volume sampler. Because of the topside ambient location and local wind conditions, an unknown portion of this emission sample contained particles from the local urban environment. The coke oven main sample was collected from a separator located between the gas collector and the primary coolers within the coke oven battery.

The roofing tar emission sample was collected from a conventional tar pot with an external propane burner. Pitch-based tar was heated to from 182° to 193°C, and emissions were collected with a 1.8-m stack extension and Teflon socks in a baghouse.

Only one vehicle or source was used for each sample; therefore, each sample may be unrepresentative of the particular technology.

All samples were Soxhlet extracted with dichloromethane, which was then removed by evaporation under dry nitrogen gas.

Tumor Experiments

Seven- to 9-week-old SENCAR mice (Boutwell, 1964) bred at Oak Ridge National Laboratory were used. There were 80 animals (40 of each sex) per treatment group. Animals were housed in plastic cages (10 per cage) under yellow light with hardwood-chip bedding, fed Purina chow and water ad libitum, and maintained at 22° to 23°C with 10 changes of air per hour. All mice were shaved with surgical clippers 2 days before the initial treatment, and only those mice in the resting phase of the hair cycle were used.

Under the tumor initiation protocol, all samples at all doses were applied as a single topical treatment in 0.2 ml spectral-quality acetone, except for the 10-mg dose, which was administered in 5 daily doses of 2 mg. Beginning 1 week after treatment, 2.0 µg TPA in 0.2 ml acetone was administered topically twice weekly. Under the complete carcinogenesis protocol, samples were administered in 0.2 ml acetone weekly (or twice weekly for the highest dose level) for 50 to 52 weeks. Under the tumor promotion protocol, mice were first initiated with 50.5 µg B(a)P in 0.2 ml acetone and then treated weekly (or twice weekly for the highest dose level) for 34 weeks with the sample.

Skin tumor formation was recorded weekly, and papillomas of > 2 mm in diameter and squamous cell carcinomas were included in the cumulative total if they persisted for 1 week or longer. The number of mice with tumors, the number of mice surviving, and the total number of tumors were determined and recorded weekly. At 6 months the number of papillomas per surviving animal was recorded for statistical purposes. The tumors were histologically verified; also, nondermal tumors were histopathologically identified.

Statistical Analysis

Tumor incidence analyses were carried out on the papilloma data obtained at 6 months and on the cumulative number of animals with carcinomas at 1 yr. The data were fitted to a probit model, taking into account the numbers of spontaneous tumors occurring in the control groups (Finney, 1971; Hasselblad et al., 1980). From the model, the dose that would produce a 50% tumor incidence in excess of the control rate and the 95% confidence limits were estimated.

The papilloma scores at 6 months were also subjected to tumor multiplicity analysis by a nonlinear Poisson model (Stead et al., 1981). From the model, the number of papillomas per mouse for a dose of 1 mg and the 95% confidence intervals for these activities were estimated.

RESULTS

Tumor Initiation

The tumor initiation experiments were designed to compare the relative tumorigenic activities of the diverse complex mixtures (Nissan, VW Rabbit, Mustang, topside coke oven, coke oven main, and roofing tar extracts) and B(a)P. Animals were scored at 6 months for papillomas and at 1 yr for carcinomas. The carcinoma data represent the cumulative number of animals with carcinomas at 1 yr in each treatment group regardless of survival. The papilloma data have been previously presented and this report discusses the carcinoma data.

The B(a)P, topside coke oven, coke oven main, Nissan, roofing tar, and Mustang samples produced a 20% or greater carcinoma incidence at the highest dose level applied (Table 4). In general, samples that produced a papilloma response greater than 5 papillomas per mouse at 6 months (Nesnow et al., in press)

Table 4. Carcinoma Incidence in SENCAR Mice after Single Administration of Complex Mixture Extracts or B(a)P followed by TPA[a]

Dose (μg/mouse)[c]	Sex	B(a)P	Mice with Carcinomas (%)[b]					
			Topside Coke Oven	Coke Oven Main	Roofing Tar	Nissan	VW Rabbit	Mustang
2.52	M	5						
2.52	F	5						
12.6	M	20						
12.6	F	23						
50.5	M	25						
50.5	F	20						
100	M	30	0	10	5	0	0	5
100	F	25	8	25	10	5	0	13
500	M		5	54	10	13	0	0
500	F		15	54	18	10	0	10
1,000	M		15	53	5	20	3	5
1,000	F		3	48	15	13	3	10
2,000	M		13	48	13	13	5	15
2,000	F		10	45	15	15	6	13
3,000	M							5
3,000	F							20
10,000	M		13	55	23	36	5	
10,000	F		20	65	48	31	10	

[a] Mice were initially treated with single doses of extract or with B(a)P, followed by twice-weekly applications of TPA (2 μg/treatment).
[b] Cumulative number of tumors on living and moribund animals after 52 weeks.
[c] Mice treated once with 0.2 ml acetone followed by twice-weekly applications of TPA (2 μg/treatment) gave a 5% and 0% tumor incidence for male and female animals, respectively.

produced a carcinoma response of 0.15 to 0.65 carcinomas per mouse, with 13% to 65% of the animals bearing at least 1 tumor at 1 yr. Coke oven main produced the largest incidence: 65% of the animals bore carcinomas at the 10-mg dose. The B(a)P-treated animals responded at much lower doses (12.6 µg); however, the maximal incidence was 30% at 100 µg.

Complete Carcinogenesis

Four agents were examined for their ability to act as complete carcinogens in the SENCAR mouse skin system: B(a)P, coke oven main extract, roofing tar extract, and Nissan extract. Weekly applications of 50.5 µg B(a)P produced a carcinoma incidence of greater than 93%, with almost one carcinoma per mouse. Higher doses did not increase the tumor multiplicity. No carcinomas were observed in the control animals (Table 5).

The coke oven main sample also produced a strong complete carcinogen response in both male and female mice. Male mice seemed to be more sensitive: 98% of the males bore approximately one carcinoma, while only 75% of the females responded. The roofing tar sample produced a significant response only at the highest dose applied (4 mg/mouse/week), with 25% to 28% of the mice bearing tumors (Table 5). The Nissan sample was essentially inactive as a complete carcinogen at the doses applied.

Tumor Promotion

The coke oven main and roofing tar samples were applied weekly to mice previously initiated with a single dose of 50.5 µg B(a)P. The coke oven main sample was 1/1000 as active as TPA (Figure 1). The roofing tar was also active as a tumor promoter and produced a dose-related effect up to the highest dose applied. Mice treated with only a single dose of B(a)P produced no tumors.

DISCUSSION

The SENCAR mouse, specifically bred for increased sensitivity to two-stage (initiation/promotion) carcinogenesis, has demonstrated its responsiveness to carcinogen (DiGiovanni et al., 1980; Slaga et al., 1980; Hennings et al., 1981). Of the mouse strains and stocks examined, the SENCAR mouse was most sensitive to the tumor-initiating and complete carcinogenic effects of B(a)P and DMBA (Reiners et al., in press; DiGiovanni et al.,1980).

Table 5.   Carcinoma Incidence in SENCAR Mice after Weekly
           Administration of Complex Mixture Extracts or B(a)P

			Mice with Carcinomas (%)[a]		
Dose (μg/mouse/week)[b]	Sex	B(a)P	Coke Oven Main	Roofing Tar	Nissan
12.6	M	10			
12.6	F	8			
25.2	M	63			
25.2	F	43			
50.5	M	93			
50.5	F	98			
100	M	80	5	0	0
100	F	90	5	0	0
200	M	80			
200	F	93			
500	M		36	0	0
500	F		30	0	0
1,000	M		48	3	0
1,000	F		60	0	0
2,000	M		82	3	0
2,000	F		78	8	0
4,000[c]	M		98	25	3
4,000[c]	F		75	28	5

[a]Cumulative number of tumors on living and moribund animals after
52 weeks.
[b]Mice treated weekly with 0.2 ml of acetone did not develop any
tumors.
[c]Twice-weekly treatments of 2000 μg/mouse.

     These studies of the effects of complex mixtures and B(a)P on
SENCAR mouse skin are the most extensive to date, and the results
confirm the applicability of this mouse strain to the analysis of
complex mixtures.

     The B(a)P, coke oven main, and roofing tar samples were
positive in both sexes as tumor initiators, producing both
papillomas and carcinomas (Nesnow et al., in press), and were also
positive as tumor promoters and complete carcinogens. In general,
those agents that produced a strong tumor initiation papilloma

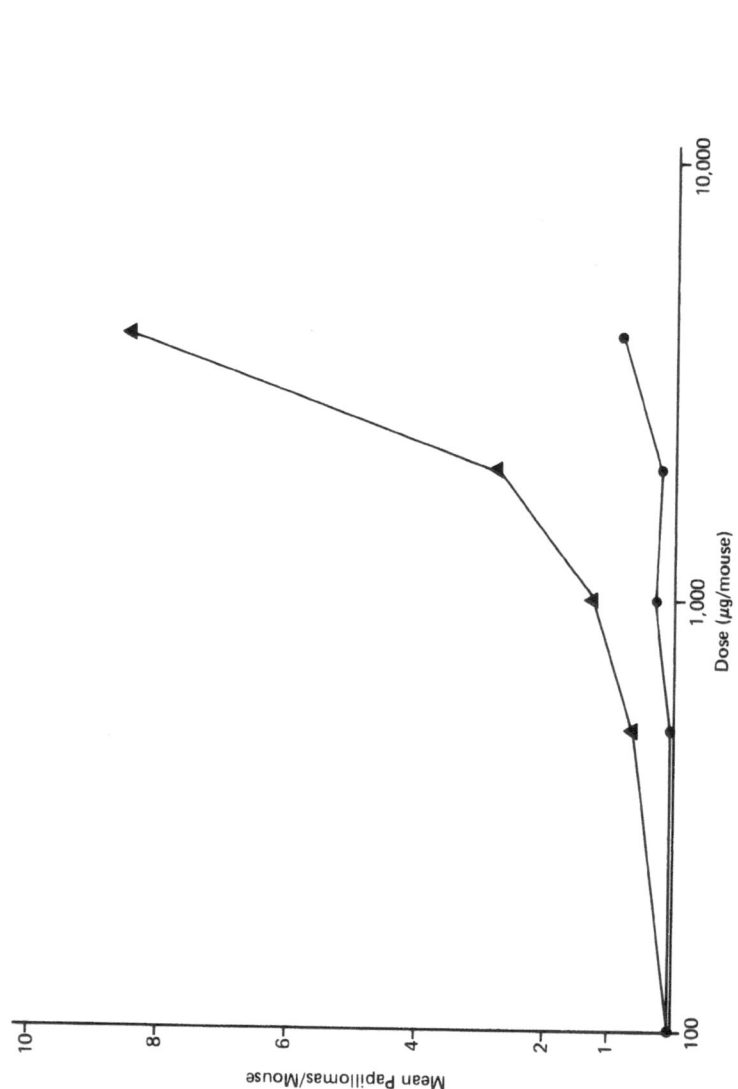

Figure 1. Tumors observed following repetitive administration of coke oven main (△) or roofing tar extract (●) to SENCAR mice initiated with B(a)P. Mice (80 M/F) were initiated with 50.5 µg B(a)P and subsequently treated weekly (or twice weekly at the highest dose) with extract. Mice initiated with 50 µg B(a)P and promoted with acetone did not produce tumors. Mice initiated with 50 µg B(a)P and promoted with TPA (2 µg, twice weekly) produced an average of 5.4 papillomas/mouse.

response also produced carcinomas in the same animals when scored at 1 yr. The two diesel engine samples were positive as tumor initiators, as was the unleaded gasoline engine sample. Additional work with emission extracts from other diesel vehicles and engines has demonstrated their activity as mouse skin tumor initiators (Nesnow et al., 1981; Nesnow et al., 1982; Nesnow et al., in press).

Of the strong tumor initiators -- i.e., B(a)P, coke oven main, roofing tar, Nissan -- only Nissan was not also a complete carcinogen at the doses tested. Kotin et al. (1955) obtained tumors from diesel particulate extract on strain A mice; this indicates that higher doses might induce tumors in SENCAR mice. However, on a weight basis, coke oven main and roofing tar are much more active than Nissan as complete carcinogens.

Of the strong tumor initiators, only Nissan seemed not to possess tumor-promoting activity. The presumed lack of tumor-promoting activity in the Nissan sample was probably a function of the composition of the Nissan mixture. The skin tumorigenesis results indicate that the coke oven main sample was a stronger tumor promoter than the roofing tar sample. Chemical fractionation and mutagenesis studies show that both the chemical composition and genetically active components of diesel, roofing tar, and coke oven main samples are significantly different (J. Lewtas, personal communication).

Chemicals that seem to be only tumor initiators on mouse skin may also possess complete carcinogenic activity when administered by other routes to mice and rats. Urethane (IARC, 1954) and triethylenemelamine (IARC, 1975) are both probably pure mouse skin tumor initiators: repeated applications of these agents on mouse skin do not yield tumors. However, urethane administerered intraperitoneally, subcutaneously, or orally to mice produces a variety of lesions, including lung, liver, and lymphoid tumors. Urethane administered orally to rats also produces multiple tumors (IARC, 1954). Triethylenemelamine produces lung tumors in mice after intraperitoneal injection and muscle tumors in rats after subcutaneous injection (IARC, 1975).

It is compelling to postulate that the B(a)P in these complex mixtures accounts for their tumorigenic activity, since mouse skin is exquisitely sensitive to this agent. The results presented here reveal that a single application of less than 5 µg B(a)P as a tumor initiator yields a 50% tumor incidence. However, the relationship between the B(a)P content of each mixture and the papilloma response for each mixture is not linear (Figure 2). Probably none of the activity of the coke oven sample as a tumor

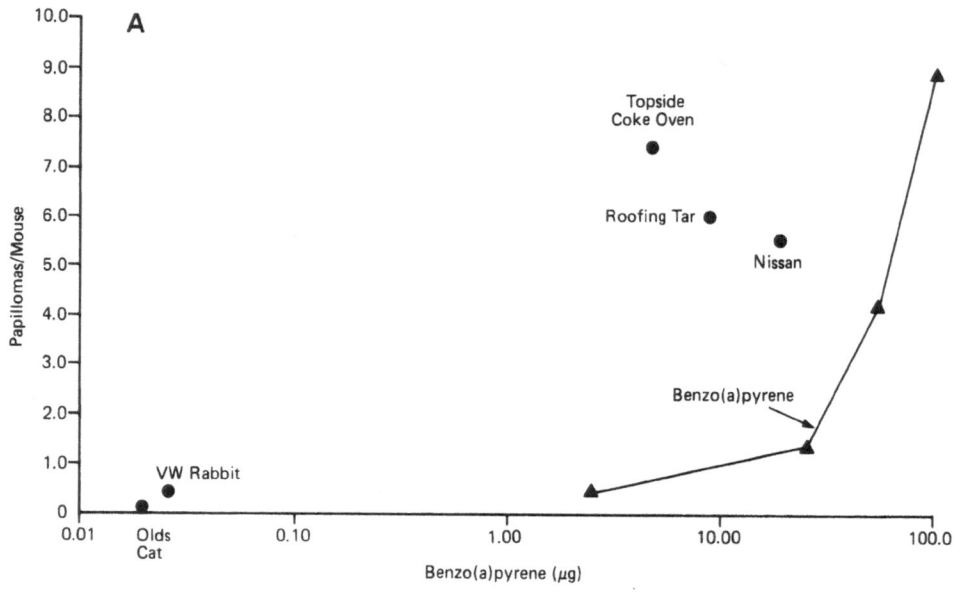

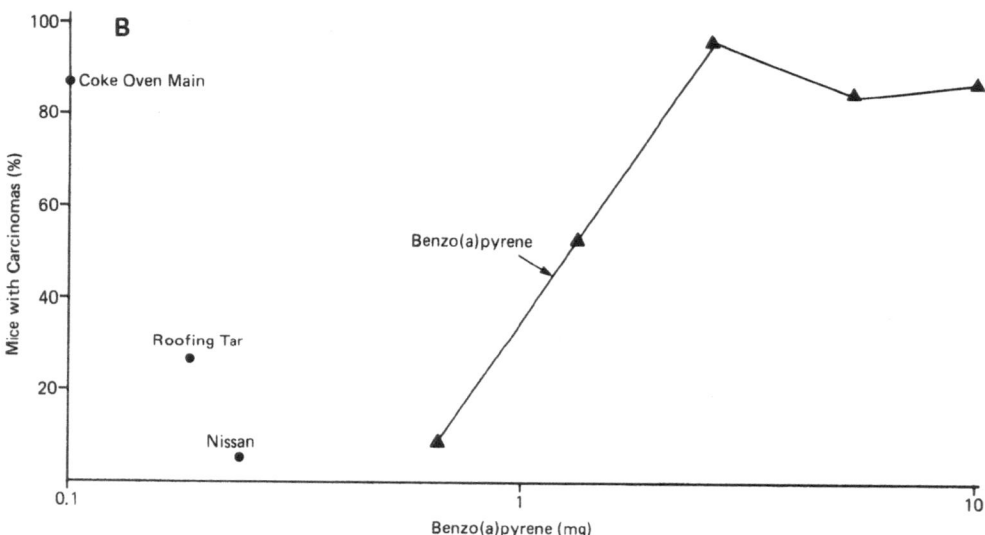

Figure 2.   B(a)P dose response in SENCAR mice and tumorigenic
            effect of extracts at levels of B(a)P determined in the
            extracts.   (a) Papilloma formation at 6 months after
            single administration of extract followed by twice-
            weekly application of TPA.   (b) Carcinoma incidence
            at 1 year after administration of extract.   B(a)P
            values are total level administered after 1 year.

initiator (Figure 2a) or a complete carcinogen (Figure 2b) can be explained by B(a)P content. Even the B(a)P level in the Nissan sample (11 µg/10 mg extract) can account for only 20% to 30% of the papilloma response elicited by the Nissan sample (Figure 2a). This conclusion has also been stated by Kotin et al. (1954) and Wynder and Hoffmann (1965), based on the analysis of other complex mixtures. Other components of these mixtures may plan an important role in their tumorigenic activities. For example, β-propiolactone, a mouse skin tumor initiator, has been identified in diesel exhaust particulate extracts (Menster and Sharkey, 1977).

Quantitative methods for the analysis of tumor data are many and employ tumor incidence, tumor multiplicity, and tumor latency data. Statistical methods have been employed using Poisson and other distribution assumptions, as well as both univariate and multivariate analytical approaches (Gart et al., 1979; Totter and Finamore, 1978; Drinkwater and Klotz, 1981; Holland et al., 1981). We chose to apply a nonlinear Poisson model to the papilloma incidence data (Table 6). This model assumes a Poisson distribution of tumors, that tumor multiplicity is related to dose, that the response may be nonlinear, and that there is a background response. Results based on papilloma data (Nesnow et al., in press) from the nonlinear Poisson model suggest the following ranking: topside coke oven > Nissan > roofing tar > VW Rabbit = Mustang. The values calculated are only estimates and in some cases the Poisson assumption made to derive the estimates is only partially fulfilled.

A probit model (Table 6) was chosen to evaluate the tumor incidence data. The probit model examines animals with tumors (regardless of multiplicity) and animals without tumors. Results based on papilloma data (Nesnow et al., in press) from the probit analysis suggest the following ranking: B(a)P > coke oven main > topside coke oven > Nissan = roofing tar.

The carcinoma incidence data (tumor initiation and complete carcinogenesis protocols) were applied to the probit model (Table 7). With the coke oven main sample, the estimated values of TD_{50} are similar for both the tumor initiation (single application) and complete carcinogenesis (weekly application) protocols. However, B(a)P as a tumor initiator was much less effective in producing carcinomas than was B(a)P as a complete carcinogen. When the comparison was made on the total amount applied to the mice, the tumor initiation protocol was more sensitive than the complete carcinogenesis protocol for both agents.

Table 6. Nonlinear Poisson and Probit Model Estimates Based on Papilloma Incidence at 6 Months in SENCAR Mice in the Tumor Initiation Protocol.

Sample	Sex	Nonlinear Poisson		Probit	
		Papillomas/ Mouse at 1 mg	95% Confidence Intervals	Dose for 50% Papilloma Incidence (TD_{50}) (mg)	95% Confidence Intervals
B(a)P	M	NC[a]		0.0036	0.0021-0.0062
	F	NC[a]		0.0091	0.0057-0.015
Coke oven main	M	ND[b]		0.079	0.027-0.23
	F	ND[b]		0.19	0.14-0.28
Topside coke oven	M	2.2[c]	2.00-2.40	0.30	0.22-0.40
	F	2.0[c]	1.90-2.20	0.42	0.31-0.58
Nissan	M	0.49[c]	0.38-0.63	1.60	1.2-2.2
	F	0.68[c]	0.57-0.79	1.50	1.1-1.9
Roofing tar	M	0.38[c]	0.30-0.49	1.8	1.2-2.7
	F	0.44[c]	0.35-0.55	2.1	1.5-2.8
VW Rabbit	M	0.21	0.14-0.30	NC[d]	
	F	0.17	0.11-0.25	NC[d]	
Mustang	M	0.17	0.12-0.24	NC[d]	

[a] Not calculated, since data were obtained at a lower dose range.
[b] Not determined.
[c] The distribution of tumors at some dose levels was not a Poisson distribution, as the variances exceeded the means.
[d] Not calculated, since tumor incidence did not exceed 50%.

Table 7. Comparison of Probit Model Estimates Based on Cumulative Carcinoma Incidence at 1 Year in SENCAR Mice in the Tumor Initiation and Complete Carcinogenesis Protocols

Sample	Sex	Tumor Initiation[a]		Complete Carcinogenesis[b]	
		Dose for 50% Carcinoma Incidence (TD50) (mg)	95% Confidence Intervals	Dose for 50% Carcinoma Incidence (TD50) (mg)	95% Confidence Intervals
B(a)P	M	> 0.10	—	0.025	0.017–0.037
	F	> 0.10	—	0.029	0.023–0.036
Coke oven main	M	1.9	0.83–4.4	0.78	0.62–0.99
	F	1.5	0.57–3.7	0.93	0.69–1.2

[a] Estimates based on single dose administration.
[b] Estimates based on weekly dose administration.

CONCLUSIONS

   Mouse skin tumor initiation/promotion and complete
carcinogenesis have been used for many years to study the
mechanisms of chemical carcinogenesis, especially with regard to
multistage carcinogens. The early history of mouse skin
carcinogenesis includes its use in elucidation of the pathological
effects of complex mixtures. Our studies have concentrated on
providing a quantitative data base of cancer data in mice that
could be applicable to predicting adverse human health effects.

ACKNOWLEDGMENTS

   The authors thank C. Evans, A. Stead, and J. Creason for
statistical analyses and J. Croteau for preparation of this
manuscript. The research was sponsored by the U.S. Environmental
Protection Agency, under contract no. 79D-X0526, and by the Office
of Health and Environmental Research, U.S. Department of Energy,
under contract no. 7405 eng-26 with the Union Carbide
Corporation.

REFERENCES

Boutwell, R.K. 1964. Some biological aspects of skin
   carcinogenesis. Progr. Exp. Tumor Res. 4:207-250.

Campbell, J.A. 1939. Carcinogenic agents present in the
   atmosphere and incidence of primary lung tumours in mice.
   Brit. J. Exp. Path. 20:122-132.

Campbell, J.A. 1934. Cancer of skin and increase in incidence of
   primary tumours of lung in mice exposed to dust obtained from
   tarred roads. Brit. J. Exp. Path. 15:287-294.

DiGiovanni, J., T.J. Slaga, and R.K. Boutwell. 1980. Comparison
   of the tumor-initiating activity of 7,12-dimethylbenz(a)-
   anthracene and benzo(a)pyrene in female SENCAR and CD-1 mice.
   Carcinogenesis 1:381-389.

Drinkwater, N.R., and J.H. Klotz. 1981. Statistical methods for
   the analysis of tumor multiplicity data. Cancer Res.
   41:113-119.

Finney, P.J., 1971. Probit Analysis. Cambridge University Press:
   Cambridge, MA.

Gart, J.J., K.C. Chu, and R.E. Tarone. 1979. Statistical issues in interpretation of chronic bioassay tests for carcinogenicity. J. Natl. Cancer Inst. 62:957-974.

Hasselblad, V., A.G. Stead, and J.P. Creason. 1980. Multiple probit analysis with a nonzero background. Biometrics 36:659-663.

Hennings, H., D. Devor, M.L. Wenk, T.J. Slaga, B. Former, N.H. Colburn, G.T. Bowen, E. Kjell, and S.H. Yuspa. 1981. Comparison of two stage epidermal carcinogenesis initiated by 7,12-dimethylbenz(a)anthracene or N-methyl-N'-nitro-N-nitrosoguanidine in newborn and adult SENCAR and BALB/c mice. Cancer Res. 41:773-779.

Hoffmann, D., and E.L. Wynder. 1963. Studies on gasoline engine exhaust. J. Air Pollut. Contr. Assoc. 13:322-327.

Hoffmann, D., E. Theisz, and E.L. Wynder. 1965. Studies on the carcinogenicity of gasoline exhaust. J. Air Pollut. Contr. Assoc. 15:162-165.

Holland, J.M., D.A. Wolf, and B.A. Clark. 1981. Relative potency estimation for synthetic petroleum skin carcinogens. Environ. Health Perspect. 38:149-155.

Huisingh, J.L., R.L.Bradow, R.H. Jungers, B.D. Harris, R.B. Zweidinger, K.M. Cushing, B.E. Gill, and R.E. Albert. 1980. Mutagenic and carcinogenic potency of extracts of diesel and related environmental emissions: Study design, sample generation, collection, and preparation. (abstr.). In: Health Effects of Diesel Engine Emissions. Proceedings of an International Symposium, Volume 2. W.E. Pepelko, R.M. Danner, and N.A. Clarke, eds. EPA-600/9-80-057b. U.S. Environmental Protection Agency: Washington, DC. pp. 788-800.

IARC Monographs on the Evaluation of the Carcinogenic Risk of Chemicals to Man, Volume 9. 1975. International Agency for Research on Cancer: Lyons, France. pp. 95-105.

IARC Monographs on the Evaluation of the Carcinogenic Risk of Chemicals to Man, Volume 7. 1974. International Agency for Research on Cancer: Lyons, France. pp. 111-140.

Kotin, P., H.L. Falk, and M. Thomas. 1955. Aromatic hydrocarbons, III: presence in the particulate phase of diesel-engine exhausts and the carcinogenicity of exhaust extracts. AMA Arch. Ind. Health 11:113-120.

Kotin, P., H.L. Falk, and M. Thomas. 1954a. Aromatic hydrocarbons, II: presence in the particulate phase of gasoline-engine exhausts and the carcinogenicity of exhaust extracts. AMA Arch. Ind. Hyg. Occup. Med. 9:164-177.

Kotin, P., H.L. Falk, M.P. Mader, and M. Thomas. 1954b. Aromatic hydrocarbons, I: presence in the Los Angeles atmosphere and the carcinogenicity of atmospheric extracts. AMA Arch. Ind. Hyg. Occup. Med. 9:153-163.

Menster, M., and A.G. Sharkey, Jr. 1977. Chemical Characterization of Diesel Exhaust Particulates. PERC/RI-77/5. ERDA Technical Information Center: Oak Ridge, TN.

Nesnow, S., L. Triplett, and T.J. Slaga. (in press). Comparative tumor-initiating activity of complex mixtures from environmental particulate emissions on SENCAR mouse skin. J. Natl. Cancer Inst.

Nesnow, S., C. Evans, A. Stead, J. Creason, T.J. Slaga, and L.L. Triplett. 1982. Skin carcinogenesis studies of eemission extracts. In: Toxicological Effects of Emissions from Diesel Engines. J. Lewtas, ed. Elsevier-North Holland: New York. pp. 295-320.

Nesnow, S., L.L. Triplett, and T.J. Slaga. 1981. Tumorigenesis of diesel exhaust, gasoline exhaust, and related emission extracts on SENCAR mouse skin. In: Short-Term Bioassays in the Analysis of Complex Environmental Mixtures II. M.D. Waters, S.S. Sandhu, J.L. Huisingh, L. Claxton, and S. Nesnow, eds. Plenum Press: New York. pp. 277-297.

Passey, R.D. 1922. Experimental soot cancer. Brit. Med. J. 2:1112-1113.

Pereira, M.A. 1982. Skin tumorigenesis research data base. J. Amer. Coll. Toxicol. 1:47-82.

Reiners, J., K. Davidson, K. Nelson, M. Mamrack, and T.J. Slaga. (in press). Skin tumor promotion: a comparative study of several stocks and strains of mice. In: Organ and Species Specificity in Chemical Carcinogenesis. R. Langenbach, S. Nesnow, and J.M. Rice, eds. Plenum Press: New York.

Slaga, T.J., and S.M. Fischer. (in press). Strain differences and solvent effects in mouse skin carcinogenesis experiments using carcinogens, tumor initiators and promoters. Progr. Exp. Tumor Res.

Slaga, T.J., A. Sivak, and R.K. Boutwell, eds. 1978. Carcinogenesis: A Comprehensive Survey, Volume 2: Mechanisms of Tumor Promotion and Cocarcinogenesis. Raven Press: New York.

Slaga, T.J., L.L. Triplett, and S. Nesnow. 1981. Mutagenic and carcinogenic potency of extracts of diesel and related environmental emissions: two stage carcinogenesis in skin tumor sensitive mice (SENCAR). Environ. Int. 5:417-424.

Slaga, T.J., S.M. Fischer, L.L. Triplett, and S. Nesnow. 1982. Comparison of complete carcinogenesis and tumor initiation and promotion in mouse skin: the induction of papillomas by tumor initiation-promotion, a reliable short term assay. J. Amer. Coll. Toxicol. 1:83-100.

Slaga, T.J., L.L. Triplett, and S. Nesnow. 1980a. Mutagenic and carcinogenic potency of extracts of diesel and related environmental emissions: two-stage carcinogenesis in skin tumor sensitive mice (SENCAR). (abstr.). In: Health Effects of Diesel Engine Emissions. Proceedings of an International Symposium, Volume 2. W.E. Pepelko, R.M. Danner, and N.A. Clarke, eds. EPA-600/9-80-057b. U.S. Environmental Protection Agency: Washington, DC. pp. 874-897.

Slaga, T.J., G.L. Gleason, G. Mills, L. Ewald, P.P. Fu, H.M. Lee, and R.G. Harvey. 1980b. Comparison of the skin tumor-initiating activities of dihydrodiols and diol-epoxides of various polycyclic aromatic hydrocarbons. Cancer Res. 40:1981-1984.

Stead, A.G., V. Hasselblad, J.P. Creason, and L. Claxton. 1981. Modeling the Ames test. Mutation Res. 85:13-27.

Stenback, F. 1980. Skin carcinogenesis as a model system; observations on species, strain and tissue sensitivity to 7,12-dimethylbenz(a)anthracene with and without promotion from croton oil. Acta Pharmacol. Toxicol. 46:89-97.

Totter, J.R., and F.J. Finamore. 1978. Dose response to cancerogenic and mutagenic treatment. Environ. Int. 1:233-244.

Von Haám, E., and F.S. Mallette. 1952. Studies on the toxicity and skin effects of compounds used in the rubber and plastics industries, III: carcinogenicity of carbon black extracts. Ind. Hyg. Occup. Med. 6:237-242.

Vosamae, A.I. 1979. Carcinogenicity studies of Estonian oil shale soots. Environ. Health Perspect. 30:173-176.

Wynder, E.L., and D. Hoffmann. 1965. Some laboratory and epidemiological aspects of air pollution carcinogenesis. J. Air Pollut. Control Assoc. 15:155-159.

Wynder, E.L., and D. Hoffman. 1962. A study of air pollution carcinogenesis, III: carcinogenic activity of gasoline engine exhaust condensate. Cancer 15:103-108.

SESSION 5

DEVELOPMENT OF SHORT-TERM BIOASSAYS: TERATOLOGY

FROG EMBRYO TERATOGENESIS ASSAY: <u>XENOPUS</u> (FETAX) -- A SHORT-TERM ASSAY APPLICABLE TO COMPLEX ENVIRONMENTAL MIXTURES

James N. Dumont,[1] T. Wayne Schultz,[1] Michelle V. Buchanan,[2] and Glen L. Kao[2]

[1]Biology Division and [2]Analytical Chemistry Division, Oak Ridge National Laboratory, Oak Ridge, TN 37830

INTRODUCTION

There is no question that a rapid and inexpensive screening tool is needed to assess potential teratogenicity. The now classical Ames test (Ames, 1975) and other tests screen for potential mutagenicity and carcinogenicity. Although a number of relatively rapid bioassays, ranging from the invertebrates through lower vertebrates to mammals, as well as cell, organ, and embryo culture systems, have been used to examine teratogenicity, none has been adequately validated or widely accepted for routine use. The tier-testing approach, which uses rapid screens to detect potential hazard and indicate need for further testing, has been useful and expedient. Comparable tests, however, are not yet applicable for teratogenesis testing despite the fact that there are a number of methods available. Among these is one which we submit may be a valid model for preliminary assessment of potential teratogens. This model, referred to as FETAX (Frog Embryo Teratogenesis Assay: <u>Xenopus</u>), has been applied to examine the relative teratogenic risk of a variety of chemicals and complex mixtures. The mixtures we have tested come from coal-conversion and shale-oil technologies and their effects have been compared to those of similar materials derived from natural petroleums.

THE FETAX MODEL

Background

One reason the amphibian embryo is attractive as an assay system is that it has been and remains a classical model for experimental embryological studies. Unlike some of the other teratogenesis screens that use cell or organ cultures, the frog embryo is an intact developing system which undergoes events (e.g., cleavage, gastrulation, organogenesis) comparable to those of other vertebrates, including mammals.

The following are other attributes of the model which make it attractive: 1) The assay is rapid; end points are routinely reached at 96 h. 2) It is relatively inexpensive. 3) Methodologically, the protocol is simple. 4) Large numbers of embryos (3,000 to 5,000) are routinely available for testing from a single spawning. 5) The developmental stages are easily observed and scored (Nieuwkoop and Faber, 1975). 6) Dose levels of teratogens are easily controlled. 7) Pulsed exposures to specific developmental stages are easy, as are prolonged exposures beginning at specific stages. 8) FETAX allows direct exposure of the embryo to either the teratogen or its metabolites and has the potential for in vitro metabolic activation by microsomal (S9) preparations. Despite concern that the jelly coat surrounding the embryo may act as a permeability barrier (Manson, 1980), our experience has been that it does not (Davis et al., 1981). In any case, the jelly can be easily removed by manual (Rugh, 1952) or chemical means (Gusseck and Hedrick, 1971). 9) The embryos are compatible with low concentrations of solubility vehicles including DMSO, propylene glycol, acetone, and alcohol. 10) The embryos are functional organisms. They are motile at 96 h so assessment of the functional state of the nervous and muscular systems can be made by simple observations of swimming ability. Prenatal development includes all stages prior to metamorphosis--stages when organ development and growth occur. Postnatal development can be considered to begin with metamorphosis and conclude with young adults in 2 to 3 months. 11) Finally, end points are easy to recognize and score. We routinely use the following: mortality, anatomical deformities, growth, developmental stage attained, motility (behavior), and pigmentation (see End Points below).

Methods

The routine method employed is a static, 96-h exposure of embryos obtained from a pair of Xenopus injected 8 to 10 h earlier with gonadotropin to induce ovulation and amplexus (Browne et al., 1979). Healthy appearing, normally developing embryos, in the mid- to late blastula stage of development are selected manually with the aid of a dissecting microscope. Groups of 25 (or 50) embryos are placed in 100 ml (or 200 ml) of artificial pond water (5% DeBoers salt solution) in 90- x 50-mm (or 125- x 65-mm) covered Pyrex dishes. Exposure is at room temperature for a period of 96 h. At 96 h the surviving embryos are scored for the degree of pigmentation and motility and fixed in 6% glutaraldehyde in 0.1 M phosphate buffer (pH 7.3). They are then counted and the number of survivors and the number of abnormals recorded. Some embryos may be fixed and prepared for paraffin sectioning. These data form the basis for quantitative evaluations of the end points.

End Points

A variety of end-point data can be collected from the FETAX system. Those routinely obtained are outlined below.

1) The LC_{50} and LC_{100}, parameters which indicate embryo lethality, and the EC_{50} and the EC_{100}, i.e., concentrations that induce terata in 50% (or 100%) of the surviving embryos. Since the EC end points reflect abnormals among survivors, they represent a more realistic evaluation of teratogenesis. We do not rule out the possibility that mortality prior to 96 h may be caused by developmental rather than toxic effects. The LC and EC values are calculated by log/PROBIT transformation and linear regression methods. 2) The NOEC, i.e., no observable effects concentration, is also determined. 3) The developmental stage attained is determined against a standard table of development (Nieuwkoop and Faber, 1975). In cases of grossly malformed embryos, precise determinations of stage are impossible since morphological characteristics of several stages may occur in the same embryo. 4) Growth (length) attained is determined by measuring, with an occular micrometer, the length of only normal-appearing embryos. This method, which selects against observable terata, gives a more realistic estimate of normal growth. 5) Evaluations of motility (behavior) and pigmentation are ranked from 0 to 4, where 4 represents normal (control) swimming activity or degree of pigmentation. These facets of development are related to neuronal control involving both the central and peripheral nervous systems. Thus, they provide clues to adverse effects on nervous and muscular system development and function. 6) Finally, gross anatomical observations are made simply with the aid of a dissecting microscope. These observations allow detection of gross anatomical defects in major

organ systems, notably gut, eyes, and brain, but can be augmented by a more detailed study with light or electron microscopy.

The end points enumerated above provide a range of data which can be integrated into comparatively precise and detailed statements of relative potential teratogenic hazard. Other observations are possible including examinations for chromosomal damage as well as biochemical parameters such as RNA, DNA, and protein synthesis, and specific enzyme levels (Edmisten et al., 1981; J.A. Bantel, Oklahoma State University, personal communication).

The above points set forth some of the characteristics of FETAX which make it attractive as a potential screen for teratogens. The usefulness of the system lies in its potential to establish relative teratogenic risks and thus set priorities for further testing in higher-level assays.

Teratogenic Index

An embryonic Teratogenic Index (TI) has been developed to permit comparisons of diverse compounds or mixtures with regard to their inherent teratogenic risks. The TI simply defines the relationship between the 96-h LC_{50} and the 96-h EC_{50}, i.e., between lethality and the number of surviving abnormal embryos. Thus, $TI = LC_{50}/EC_{50}$.

The Teratogenic Index makes it possible to relate teratogenic risk of highly toxic (lethal) materials to the risk of those which are relatively non-embryolethal. It allows for comparison between known and suspected teratogens. In this regard, and for validation and standardization purposes, we have tested a series of about 20 known mammalian teratogens with FETAX. Some of these materials, such as trypan blue, hydroxyurea, and retinoic acid are well established as "blatant" mammalian teratogens. Others such as alcohol, saccharin, and aspirin are more subtle teratogens (or considered nonteratogenic). On the basis of TIs derived from these standardization compounds, we have set a TI value of 2.0 or greater as one which would indicate the need for further testing. TI values between 1.5 and 2.0 suggest to us materials that should be treated with suspicion and caution as potential teratogens and tested further in other screening systems. TI values below 1.5 reflect compounds which are more embryolethal, that is, are coeffective teratogens (Johnson, 1981) whose toxicity and teratogenicity are difficult to delineate and whose lethality may be more pertinent to risk assessment than teratogenicity.

MATERIALS AND METHODS

Synfuel Materials

The fossil-fuel samples were obtained from the USEPA/USDOE Fossil Fuel Research Materials Facility, Oak Ridge National Laboratory (Griest et al., 1980). Those whose effects are reported herein include five standard reference materials (Comparative Research Materials, or CRMs). They are CRM-1, a coal-derived fuel-oil blend; CRM-2, a shale-derived crude; CRM-4, a coal-gasifier electrostatic-precipitator tar; CRM-3 and CRM-5, an aromatic and an aliphatic natural petroleum crude.

The approach taken to testing complex mixtures from fossil-fuel materials involves the preparation of aqueous extracts of the material. This is done by slow-stirring an 8:1 mixture of water:material for 16 h at room temperature. After 16 h, the mixture is poured into a separatory funnel where the aqueous phase separates and is removed and filtered through glass wool. This material is referred to as the aqueous extract (AE) and is the 100% stock from which dilutions are made. Aqueous materials derived directly from synfuel plants are themselves teratogenic (Dumont and Schultz, 1980).

Chemical Analysis

Two sets of aqueous extracts (AEs) were prepared for chemical analyses: one immediately following the extraction procedure and the second, identical to the first, except that it was allowed to stand for 96 h under conditions of the embryo assay procedure. This was done to determine what, if any, materials might be lost or altered over the 96-h embryo testing period. For gas chromatography/mass spectrometry (GC-MS), the AEs of CRMs 1, 2, 3, and 5 were adjusted to pH 11 with 50% NaOH and extracted three times with 60-ml aliquots of methylene chloride to isolate the basic and neutral components. They were then adjusted to pH 2 with concentrated HCl and similarly extracted with methylene chloride. The AE of CRM 4 (tar) was adjusted to pH 12 with 50% NaOH and passed through 2 g of XAD-2 (Isolab, Akron, OH). The effluent was adjusted to pH 2 with concentrated HCl and passed through a second XAD-2 column. The organics trapped in these colums were stripped with 15-ml acetone, which was then evaporated to 1 ml.

Mass spectra were generated using a Hewlett-Packard 5985A GC-MS with electron-impact ionization at 70 eV. A 25-m fused silica capillary column (J & W, Inc., Orangeval, CA), coated with

SE-52 was used in the GC-MS interface. The column temperature was maintained at 50°C for 5 min and then programmed to increase to 250°C at 2°C/min. Helium was used as the carrier gas at 30 lbs of head pressure. A splitless injector was used to introduce the samples.

RESULTS AND DISCUSSION

Biological Effects

     Table 1 summarizes the effects of the CRM aqueous extracts. CRM-1 and CRM-4 appear to be the most severe embryotoxic hazard with LC_{50}s of 1.48% and 0.83%, respectively. These are followed by CRM-2 with LC_{50} of 6.97%. The CRM-3 and CRM-5 are essentially nontoxic; 90% of CRM-5 extract does not cause any observable effects.

Table 1.   Summary of Effects of Aqueous Extracts of Comparative Research Materials

Sample	96 h			Teratogenic Index
	LC_{50} (%)	EC_{50} (%)	EC_0 (%)	
CRM 1	1.48	0.96	0.10	1.54
CRM 2	6.97	3.36	0.50	2.07
CRM 3	33.38	31.10	10.00	1.07
CRM 4	0.83	0.48	0.01	1.73
CRM 5	--	--	90.00	< 1.00

     Cursory examination of EC_{50} values suggests that the coal-derived materials are the most teratogenic, followed by the shale, aromatic, and aliphatic petroleums. However, since the LC and EC values of the coal material are relatively close, they may be considered as coeffective teratogens (Johnson, 1981), that is, their major effects as teratogens may be masked by the embryo death they cause. The TIs of these materials show that the extract of the shale-derived crude has the highest index number,

2.07, while those of the coal materials are somewhat lower, i.e., 1.54 and 1.73. On the basis of our standardization studies with known teratogens, we suggest that the teratogenic potential of the shale crude is greater than that of either the coal crude or the gasifier tar. So, in any ranked scheme for testing, shale crude would warrant first priority, followed closely by gasifier tar and coal crude. The concentrations at which no observable effects were noted are also included in Table 1.

The effects of the extracts of the CRMs on growth and development are summarized in Table 2. The data are based on the 96-h EC_{50} concentrations. In general, the parameters indicated are more severely affected at higher concentrations. The coal materials most severely reduce the final size attained and CRM-1 most severely inhibits developmental stage attained. Interestingly, the shale crude (CRM-2), while the most teratogenic, has little effect on growth, even though development is markedly retarded. The petroleum products have little effect on development, although the aromatic CRM-3 does reduce growth.

Table 2. Developmental Effects of Aqueous Extracts on Comparative Research Materials

Sample	Growth (%)	Stage	Pigmentation	Motility
Control	100	46/47	++++	++++
CRM-1	81	44	+++	+++
CRM-2	96	45/46	+++	+++
CRM-3	90	46	++++	++++
CRM-4	87	45/47	+++	+++
CRM-5[a]	100	46/47	++++	++++

[a]90% aqueous extract.

Pigmentation and motility are also reduced by some of these materials. These characteristics may not represent classical teratological syndromes, but they do reflect developmental events which involve nervous or muscular activity. As such, they represent, in a subtle way, developmental damage. Data from other studies (Browne and Dumont, 1980; Davis et al., 1981; Schultz et al., 1982) document cellular damage correlated with impaired motility.

The gross effects caused by 96-h exposure to the AEs are shown in Figure 1. Control embryos (Figure 1a) (Stage 46/47) are streamlined with a broad dorsal and ventral tail fin. Their eyes are large and well differentiated. The intestine has undergone torsion into 2 to 3 coils; hind-limb buds are present; the central nervous system is well differentiated; otic anlage has developed; chondrification of the visceral skeleton has been completed; nephric structures have appeared; and spleen, gonads, liver, and other major organ differentiation has been completed.

The embryos pictured in Figure 1b were exposed to 2% AE of CRM-1. They present grossly distorted profiles; the tail is short and often kinked, the head is misshapen, the eyes of some are small, and coiling of the gut is abnormal. Histologically, the small optic cup contains masses of undifferentiated cells and the lens frequently develops external to the cup. The four embryos shown in Figure 1c were exposed to 0.5% CRM-4 AE. These are also short and have kinked tails. Development of the head, eyes, and gut is abnormal. Many have epidermal blisters, especially along the dorsal aspect of the tail, and appear edemic. We believe this is an acute toxic effect, but the possibility exists that it may reflect developmental problems with the renal system. Details of the effects from CRM-4 are reported elsewhere (Schultz et al., 1982).

The embryos in Figures 1d and 1e were exposed to 10% and 7.5%, respectively, of CRM-2 extract. A common characteristic is the extreme dorsal flexion of the tail. This phenomenon is dose dependent and, in high concentrations, the tail becomes tightly curled over the back of the embryo (Figure 1d). Major organ systems (gut, eye) and cephalic/facial features are also adversely affected.

Finally, Figure 1f is an embryo exposed to 50% extract of CRM-3, the natural petroleum crude. The effects of this material are less severe than those of the others tested. There are distortions, both gross and histological, of the normal appearance.

We wish to emphasize the following point: all materials studied, both complex mixtures and the known teratogens, produce specific and characteristic terata. In other words, it is possible to identify, on the basis of the appearance of the embryo, the teratogen to which it was exposed. We believe that this reflects the embryo's response to specific teratogenic

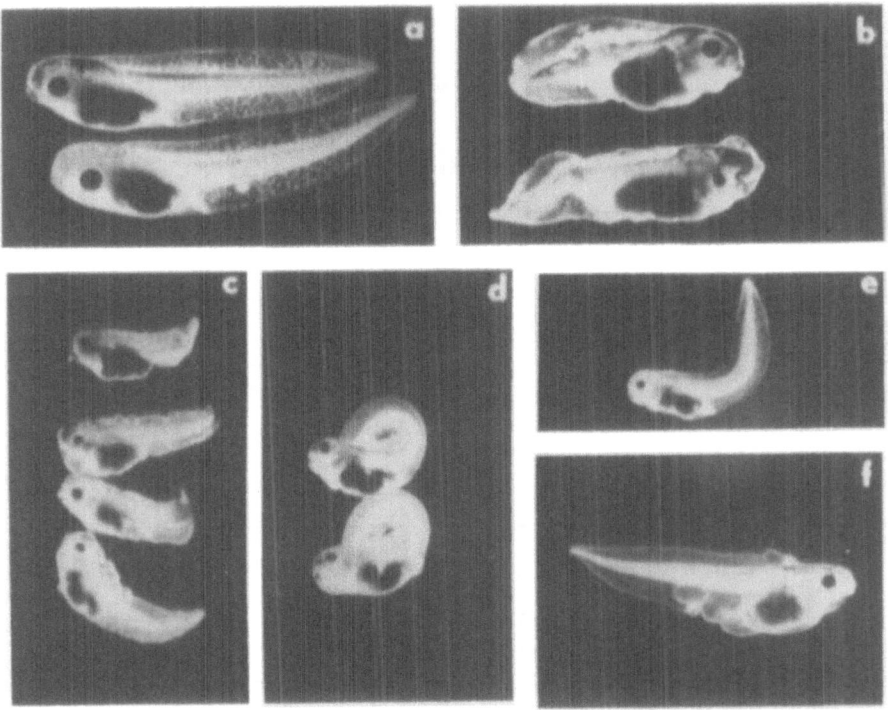

Figure 1.   96-h Xenopus embryos: a) control, 8x; exposed to b) 2%
            extract of CRM-1, 8x; c) 0.05% extract of CRM-4, 5x;
            d) 10% extract of CRM-2, 5x; e) 7.5% extract of CRM-2,
            5x; f) 50% extract of CRM-3, 8x.

insults and supports our view that the FETAX system is a valid one
for teratogenic screening and teratological studies.

Chemical Analyses

     Since the materials tested are truly complex mixtures, it is
difficult to assess which are responsible for the observed
biological effects. However, it is useful to identify the major
chemical species and begin to draw some correlations between

chemical composition and biological effects. Table 3 presents the
weight percents of the major organic elements of the raw,
unextracted CRMs and Table 4 the GC-MS data for the aqueous
extracts. The data from the 0-h and 96-h extracts showed similar
qualitative composition but only small changes in the quantities
of the more volatile components. Only the 0-time samples were
analyzed.

Table 3. Weight Percent of Major Organic Elements in CRMs

Element	CRM-1	CRM-2	CRM-3	CRM-4	CRM-5
C	87.33	85.97	85.65	81.67	85.69
C_A[a]	62.50	32.90	25.90	51.50	16.50
H	8.77	11.44	11.60	8.24	13.44
N	0.85	1.93	0.60	1.22	0.03
S	0.33	0.74	1.87	0.43	0.13
O	2.73	2.01	0.29	8.26	0.72

[a]Carbon as aromatic carbon.

Concerning the major organic elements, the total carbon
content is essentially the same for all samples, but the content
of carbon as aromatic carbon (C_A) is highest in CRM-1 and CRM-4.
On the other hand, the hydrogen content, which also reflects
aromaticity, is lowest in these samples. The GC-MS profiles show
that CRM-1 and CRM-4 contain similar components, although the
quantities and degree of alkyl substitution varies. Both of these
CRMs have high oxygen content, an indication of acidic components.
GC-MS profiles confirm the presence of acids, including phenol and
traces of carboxylic acids. These compounds are highly water
soluble and would be expected in high concentrations in aqueous
extracts. Phenols are very embryolethal in the FETAX system
(Dumont, unpublished observations), and their presence, along with
mono- and diaromatics, in the coal-derived materials probably
accounts for the higher embryotoxicity of these materials. Bantle
(personal communication) and Edmisten et al. (1981) have also
shown that naphthalene is teratogenic to Xenopus embryos. Our
extracts appear to contain both teratogenic and toxic (lethal)
agents that act coeffectively, causing lethality to the extent
that full expression of teratogenic potential is masked.

Table 4. Comparative Composition of Aqueous Extracts as Determined by GC-MS

Compound	CRM-1	CRM-2	CRM-3	CRM-4	CRM-5
Bases					
Pyridines	C_1, C_2, C_5[a]	C_2-C_7	--[b]	C_3, C_4	--
Tetrahydroquinolines	--	--	--	C_0, C_1, C_3	--
Quinolines	C_1	--	--	C_0, C_1	--
Neutrals					
Benzenes	C_4, C_5	--	C_2	C_4-C_6	C_2, C_3
Indanes	C_2, C_3	--	--	C_1	--
Indenes	--	--	--	C_1	--
Tetrahydronaphthalenes	--	--	C_0	--	CO
Naphthalenes	C_0-C_3	--	C_0	C_0	CO
Indoles	C_0-C_1	--	--	C_0	--
Furans	--	C_3	--	C_3, C_5	--
Thiophenes	--	--	C_2-C_5	--	C_3-C_5
Acids					
Phenols	C_0-C_4	C_0, C_3	--	C_0-C_6	--
Hydroxyindans	C_0	--	--	--	--
Hydroxynaphthalenes	--	--	--	C_0, C_1	--
Hydroxyfluorenes	--	--	--	C_0	--
Hydroxybiphenyls	C_0-C_2	--	--	--	--
Cyclic alkanes	--	--	t[c]	--	t
Cyclic ketones	--	t	--	--	--
Carboxylic acid	t	t	--	--	--

[a] C_n, n = number of alkyl carbons.
[b] Not detected.
[c] Trace, molecular weight not determined.

The shale (CRM-2) and tar (CRM-4) materials contain the highest content of nitrogen -- an indication of aromatic amines and azaarines, both of which, but especially the azaarines, have been shown to be teratogenic in FETAX (Dumont et al., 1979; Davis et al., 1981). Analyses of the CRM-2 extracts show substantial amounts of alkyl-substituted pyridines, furans, phenols, and trace amounts of other acid components.

Aqueous extracts of CRM-3 and CRM-5, the natural petroleum crudes, are similar in composition. One, the aromatic CRM-3, has the highest elemental sulfur content; AEs of both are dominated by neutral components. Virtually no acidic (< 1 ppb) or basic components are present. Neither of these crudes is very toxic or teratogenic.

Although only inferences can be drawn about which components might be teratogens, the presence of acidic species is almost certainly responsible for the embryolethal effects. The thiophenes and furans have not been examined for teratogenicity. It will be interesting to examine some of these classes of compounds, specifically those which have heteroatom substitution, in an effort to identify active teratogenic compounds.

ACKNOWLEDGMENTS

We thank Rhonda Epler for excellent technical assistance. Research was supported jointly by the USEPA under interagency agreements 79-D-X0533 and 79-D-X0521 and by the Office of Health and Environmental Research, USDOE, under contract W-7405-eng-26 with the Union Carbide Corporation.

REFERENCES

Ames, B.N., J. McCann, and E. Yamasaki. 1975. Methods for detecting carcinogens and mutagens with the Salmonella/-mammalian-microsome mutagenicity test. Mutation Res. 31:347-364.

Browne, C.L. and J.N. Dumont. 1979. Toxicity of selenium to developing Xenopus laevis embryos. J. Toxicol. Environ. Health 5:699-709.

Browne, C.L. and J.N. Dumont. 1980. Cytotoxic effects of sodium selenite on Xenopus laevis tadpoles. Arch. Environ. Contamin. Toxicol. 9:181-191.

Davis, K.R., T.W. Schultz, and J.N. Dumont. 1981. Toxic and teratogenic effects of selected aromatic amines on embryos of the amphibian Xenopus laevis. Arch. Environ. Contamin. Toxicol. 10:371-391.

Dumont, J.N., T.W. Schultz, and R.D. Jones. 1979. Toxicity and teratogenicity of aromatic amines to Xenopus laevis. Bull. Environ. Contamin. Toxicol. 22:159-166.

Dumont, N.J. and T.W. Schultz. 1980. Effects of coal-gasification sour water on Xenopus laevis embryos. J. Environ. Sci. Health A15:127-138.

Edmisten, G.E., M.B. Couch, C.L. Courchesne, and J.A. Bantle. 1981. The use of Xenopus laevis as a test organism to measure the toxicity and teratogenicity of aquatic borne compounds. Presented at the second annual meeting of the Society of Environmental Toxicology and Chemistry, Arlington, VA. p. 131.

Griest, W.H., D.L. Coffin, and M.R. Guerin. 1980. Fossil Fuels Research Matrix Program. ORNL/TM-7346. Oak Ridge National Laboratory: Oak Ridge, TN.

Gusseck, D.J. and J.L. Hedrick. 1971. A molecular approach to fertilization, I: disulfide bonds in Xenopus laevis jelly coat and a molecular hypothesis for fertilization. Develop. Biol. 25:337-347.

Johnson, E.M. 1981. Screening for teratogenic hazards: nature of the problems. Ann. Rev. Pharmacol. Toxicol. 21:417-429.

Manson, J.M. 1980. In vitro teratogenicity tests. In: Proceedings of a Workshop on Methodology for Assessing Reproductive Hazards in the Workplace. Department of Health and Human Services publication no. 81-100. National Institute of Occupational Safety and Health: Washington, DC. 423 pp.

Neiuwkoop, P.D. and J. Faber. 1975. Normal table of Xenopus laevis (Daudin), 2nd ed. North Holland: Amsterdam.

Rugh, R. 1952. Removal of jelly capsules. In: Experimental Embryology. Burgess Publishing Co.: Minneapolis, MN. p. 15.

Schultz, T.W., J.N. Dumont, B.R. Clark, and M.V. Buchanan. 1982. Embryotoxic and teratogenic effects of aqueous extracts of tar from a coal gasification electrostatic precipitator. Teratogen. Carcinogen. Mutagen. 2:1-11.

HUMAN SERUM TERATOGENICITY STUDIES USING <u>IN VITRO</u> CULTURES OF RAT

EMBRYOS

Norman W. Klein, Clare L. Chatot, John D. Plenefisch, and
Sean W. Carey

Department of Animal Genetics and Genetics and Cell Biology
Section, University of Connecticut, Storrs, CT 06268

INTRODUCTION

Our objective is to identify those conditions that constitute
reproductive risks to man. We are particularly concerned with
those conditions that cannot be or have not been identified by
present methodologies. These conditions constitute the majority
of factors causing fetal wastages and birth defects. In our test
system, we use intact rat embryos that are cultured <u>in vitro</u> for
2 days. During culture, embryos of the particular stage selected
(9.5 days) undergo rapid growth and extensive organogenesis. In
comparison to test systems involving cell cultures or bacteria,
embryo cultures at present require considerable expenditures of
time, skill, and supplies, but our findings to date strongly
suggest that this system may have a number of distinct advantages.
First, whole-embryo culture provides the test with the entire
repertoire of processes involved in embryonic development.
Second, whole-rat embryos can be cultured on high levels of blood
serum, an advantage, since the maternal circulatory system has
been shown to closely approximate the fetal milieu. Finally, they
can be cultured on serum from human subjects, which provides a
direct and unique evaluation of the principal organism of our
concern. In regard to this last point, it is important to
recognize that there is a large range of teratogenic responses and
sensitivities to teratogens dependent upon both individual and
species differences. By testing serum from individual human

subjects, these important considerations can be addressed in the
appropriate species.

METHODS

The basic procedures used for culturing rat embryos of
9.5 days of pregnancy -- removing Reichert's membrane and
preparing immediately centrifuged, heat-inactivated blood
serum -- were developed by New and his associates (New, 1978). In
addition, we used their recommended 30-rpm culture rotation rate.
Our modifications were introduced to permit standardized testing
of large numbers of human serum samples using minimal serum
volume. For example, all collaborators who contributed serum
samples received our selected brand of syringes, centrifuge tubes,
and prewashed bottles. All serum samples were filtered in our
laboratory for sterility. The embryo culture bottles were
designed to handle either 3 embryos in 2 ml of media or 2 embryos
in 1.3 ml of media (Klein et al., 1978). Media consisted of 90%
serum and 10% H_2O. The 10% water was found to facilitate yolk-sac
expansion, which in turn allowed for a higher frequency of normal
embryo curvature in controls. The water also provided the volume
needed to adjust serum glucose levels to 3 mg/ml and allowed us to
add antibiotics and teratogens without changing the serum
concentration. It should be noted that blood glucose levels were
determined and were adjusted to 3 mg/ml for all studies with human
or monkey serum.

Although some laboratories distributed all of the embryos
from a single rat between control and experimental groups, we
selected only a very narrow developmental stage (mid head fold)
from many rats for experimentation and distributed embryos at
random. This is a costly practice, as we averaged only three
embryos of the appropriate stage per pregnant rat. We feel that
this was justified because in an early study (Klein et al., 1980),
we observed a range of cultured embryo responses to the teratogen
cadmium that was highly stage specific. Nevertheless, the extent
to which other teratogens show such stage specificity and the
importance of maternal influences on the embryo responses has not
been fully evaluated.

One final point on methodology concerns the gassing of
cultures. We pregas prior to embryo addition, immediately after
addition, and then every 4 p.m. and 9 a.m. for the subsequent four
gassings. Cultures were gassed with 5% CO_2 throughout but with
differing percentages of O_2 and N_2 as follows: day 1: pregas,

zero time, and 4 p.m., O_2 (5%), N_2 (90%); day 2: 9 a.m., O_2 (15%), N_2 (80%), 4 p.m., O_2 (30%), N_2 (65%); day 3: 9 a.m., O_2 (40%), N_2 (55%).

Early Studies

     The feasibility of using rat embryo cultures to monitor serum for teratogenic agents was first evaluated using serum drawn from rats at different time intervals following injections with the known teratogen cadmium chloride, a direct-acting teratogen, or cyclophosphamide, a teratogen requiring metabolic activation (Klein et al., 1980). With both substances, cultured embryo responses (lethality, abnormalities, and protein accumulations) could be related in a reproducible manner to the time interval between teratogen administration and sacrifice as well as to the dosage administered. The value of passing teratogens through animals in this manner was clearly demonstrated with cyclophosphamide, as this drug was completely inactive when added directly to serum from an uninjected animal. In addition, using serum taken from rats previously injected with cadmium also produced effects that were distinct from directly added cadmium. Although the basis for this difference in embryo responses to cadmium was not studied, this observation itself suggested that the approach could be useful in studying indirect actions of teratogens, such as interference with maternal liver function.

     Following these rat injection studies, we found that rat embryos could be cultured on human serum if the serum was supplemented with glucose (Chatot et al., 1980). Furthermore, sera from patients receiving frequent medication for cancer or epilepsy caused abnormalities and reduced embryo growth when used as culture media. These observations provided the basis for our present and rather extensive program of human testing. Currently, this includes a study of cigarette smokers used as a model approach to identifying occupational reproductive risks, a study concerning the ability of rat embryos to predict a serum donor's reproductive history of fetal wastage, and studies involving comparisons of drug teratogenicity and identifications of potential immunological reproductive problems. Of these four studies, only the drug teratogenicity and immunological studies involve "analysis of complex mixtures" and will be considered in separate sections. The other two studies will be presented in brief to provide an overall picture of the variety of studies under consideration.

Occupational risk substances were identified by taking serum samples from a single individual both before and after the potential exposure. With the many possible variables that could influence human serum composition, we felt that this approach alone incorporated the necessary "controls." Thus, the cigarette smokers were required to quit smoking, and blood samples were drawn at frequent intervals both during smoking and after quitting. A total of 11 subjects participated in the study. For five subjects, the frequencies of abnormal rat embryos dropped sharply in serum samples drawn between the 3rd and 5th day after quitting, and maximum embryo protein contents were achieved on serum samples of days 8 to 10. This latter observation was of particular interest, as reduced fetal size has been the problem most frequently associated with mothers who smoke. Of these 5 subjects, 2 returned to smoking 14 days and 21 days, respectively, after quitting, and samples drawn subsequent to this time produced increased abnormality frequencies and decreased embryo size. Six subjects did not show this pattern of improved embryo growth and development after quitting; factors such as diabetes or occupations involving chemical exposure could account for the lack of embryo growth and development in four of these cases. For the final two subjects, serum from one supported excellent embryo growth and development before and after quitting, while the last subject showed increased embryo abnormalities and decreased embryo size after quitting.

No attempt has been made to analyze serum samples from this study. Clearly such analysis would be essential to establish possible causes for the observed embryo responses and for the observed differences between individuals. We have chosen instead to direct our analytical efforts to situations where causes for reproductive problems are completely unknown (such as fetal wastage) rather than to the identification of particular teratogenic substances in an already defined teratogenic situation, as with maternal cigarette smoking. Nevertheless, for the five subjects who followed an improving pattern after quitting, evidence from the literature suggests that the rapid clearance of nicotine and carboxyhemoglobin could be related to the initial decrease in abnormalities, and the much slower clearance of serum thiocyanate to the longer period before increased protein contents were seen. Nicotine and carboxyhemoglobin have biological half-lives of 1 and 4 h, respectively, while serum thiocyanate has a half-life of about 14 days (Armitage et al., 1975; Landaw, 1973; Butts et al., 1974).

The finding that rat embryo cultures might be useful in the study of spontaneous abortions and fetal wastages in general was fortuitous. Our original intent was to incorporate primate metabolism into a teratogenicity test by using serum from monkeys that previously had been given a substance under study. Before starting such a program, we tested the ability of monkey serum to support the growth and development of rat embryos. Of the 18 monkeys screened in the first study, serum from 2 caused embryo abnormalities. We then learned that these latter 2, unlike the other 16, had failed to have a successful pregnancy in over 2 years of breeding. This study with Macaca mulatta was then extended to a group of Macaca nemestrina that had precisely recorded reproductive histories. Testing serum from 12 monkeys having excellent reproductive records and 14 with histories of fetal wastage, we found that approximately 80% of the cultured rat embryo responses coincided with reproductive histories (Klein et al., 1982). That is, high risk monkey serum produced abnormalities in cultured rat embryos, and low risk serum supported normal embryo development.

At present, the studies concerning fetal wastage are progressing in two directions. First, serum samples from these high risk monkeys are being analyzed to determine the basis for their interference with cultured rat embryo development. Based on these findings, monkeys will receive the appropriate treatment to determine whether the cultured embryo analysis coincides with the reproductive problem of the serum donor. Second, serum samples from human subjects with histories of fetal wastage are being tested to determine if rat embryo cultures can also be used in the analysis of their problems.

Drug Teratogenicity: Identification of Nutritional Deficiencies

For several reasons, one of the first large-scale studies attempted with human subjects involved anticonvulsant drugs. First, such drugs are being used extensively during pregnancy; second, epileptic mothers have approximately twice the incidence of babies with birth defects as compared to nonepileptics; third, comparative anticonvulsant teratogenicity studies have not provided a clear consensus for the preferred drug; and finally, numerous studies have suggested species specificity in drug metabolism, making direct testing of human subjects particularly important. Serum samples were tested with rat embryo cultures from a group of 120 subjects receiving drug treatments with one of four drugs and also from a group of eight epileptics not receiving medication. Whether evaluating frequencies of abnormal embryos or frequencies of abnormal samples, phenobarbital was found to be less teratogenic than either phenytoin, valproic acid, or carbamazepine. Indeed, phenobarbital gave frequencies comparable to nondrug-treated epileptics, suggesting that phenobarbital itself was not teratogenic in this test. That these findings were in agreement with human epidemiological findings (Fedrick, 1973; Shapiro et al., 1976) and clinical suggestions (Golbus, 1980) provided an important validation for the test. However, our subsequent observations questioned the value of applying generalities when dealing with individual human subjects.

Our suspicions were first aroused by the inability to relate the teratogenicity of a serum sample to the concentration of the parent drug in the sample. Considering the possibility that teratogenicity was related to a metabolite that could be protein bound, serum samples were dialyzed against a balanced salt solution to completely remove the parent drug. Because dialysis also removed such components as amino acids and vitamins, a supplement was developed for dialyzed human serum. The supplement was also added to undialyzed serum as a control for direct effects of the supplement on dialyzed samples. Surprisingly, this simple addition of the supplement to some undialyzed teratogenic samples overcame the teratogenicity. This was the situation for 32 out of 53 teratogenic samples in this study. The entire 90 teratogenic samples could not be tested because sufficient serum volumes were not always available. At this time, it has not been possible to determine if a relationship exists between drug treatment and observed nutritional responses.

It must now be determined if the nutritional status of the donor is actually reflected by the embryo responses to the serum. If this can be answered positively, cultured embryos may provide a unique means of identifying specific nutritional problems. The broad distribution of multivitamins for use during pregnancy may not be the simple answer, as nutrients such as vitamin A may be teratogenic when provided in excess. Preliminary findings suggested that rat embryos do identify donor problems, since not all samples were responsive to supplements and those that were did not all respond to the same supplement components. Finally, for two individuals, we not only defined their specific serum nutrient deficiency by embryo culture but we also confirmed our findings with serum chemical analysis. The remaining definitive experiment will require the reversal of a donor reproductive problem based on cultured rat embryo responses to the particular supplement.

Embryo-Lethal Serum -- A Toxic Substance

The detection of nutritional deficiencies as previously discussed may provide a very useful application of rat embryo cultures to human reproductive problems. In addition, these cultures may provide the basis of an assay used to identify specific teratogenic substances in serum. This possibility was first examined when serum samples obtained from a 20-year-old woman proved to be embryo lethal. Based on this case, we have developed a protocol that we hope may be applied generally to the analysis of human serum for embryo-toxic substances (Carey et al., MS).

The serum obtained from this subject was found to kill rat embryos by the end of the 48-h culture period, and additional samples drawn over the next year maintained a comparable lethal activity. Small amounts of the embryo-lethal serum, 5 to 10% by volume, when added to human control serum (serum drawn from a healthy male) retained the ability to cause severe abnormalities. This observation, together with a lack of any response to vitamin supplementation, suggested that the embryo-lethal serum contained an embryo-toxic factor rather than a deficiency. The embryo-toxic property of the serum was not dialyzable but was sensitive to digestion by trypsin. That a protein may be involved led us to ammonium sulfate fractionation and subsequent localization of embryo-toxic activity in the 0 to 30% saturation range. SDS polyacrylamide gel analysis revealed predominantly heavy- and light-chain polypeptides of IgG and IgM in this fraction. Purified IgG isolated by DEAE-Affi Gel Blue chromatography showed an embryo toxicity similar to unfractionated embryo-lethal serum. At present, we are attempting to determine the reactive antigens through immobilization of various tissue-IgG complexes on nitrocellulose. Eventually, we hope to study the mechanism by which this antibody interferes with embryo survival and ultimately determine the relevance of antibody reactivities to human reproduction. However, the primary intent in presenting this particular study was to stress the potential value of using rat embryo cultures to identify unknown causes of reproductive failure that are expressed in serum.

ACKNOWLEDGMENTS

This chapter is dedicated to the memory of Alan Kolber.

Scientific contribution number 957, Storrs Agricultural Experiment Station, University of Connecticut, Storrs, CT 06268. This research was supported by U.S. Department of Energy contract EV03139 (Office of Health and Environmental Research).

REFERENCES

Armitage, A.K., C.T. Dollery, C.F. George, T.H. Houseman, P.J. Lewis, and D.M. Turner. 1975. Absorption and metabolism of nicotine from cigarettes. Br. Med. J. 4:313-316.

Butts, W.C., M. Kuehneman, and G.M. Widdowson. 1974. Automated method for determining serum thiocyanate, to distinguish smokers from nonsmokers. Clin. Chem. 20/10:1344-1348.

Carey, S.W., C.L. Chatot, and N.W. Klein. (MS). The use of rat embryo cultures in teratological testing: human serum analysis and identification of embryo toxic IgG.

Chatot, C.L., N.W. Klein, J. Piatek, and L.J. Pierro. 1980. Successful culture of rat embryos on human serum: use in the detection of teratogens. Science 207:1471-1473.

Fedrick, J. 1973. Epilepsy and pregnancy: a report from the Oxford Record Linkage Study. Br. Med. J. 2:442-448.

Golbus, M.S. 1980. Teratology for the obstetrician: current status. Obstet. Gynecol. 55:269-277.

Klein, N.W., P.P. Minghetti, S.K. Jackson, and M.A. Vogler. 1978. Serum protein depletion by cultured rat embryos. J. Exp. Zool. 203:313-318.

Klein, N.W., M.A. Vogler, C.L. Chatot, and L.J. Pierro. 1980. The use of cultured rat embryos to evaluate the teratogenic activity of serum: cadmium and cyclophosphamide. Teratology 21:199-208.

Klein, N.W., J.D. Plenefisch, S.W. Carey, W.T. Fredrickson, G.P. Sackett, T.M. Burbacher, and R.M. Parker. 1982. Serum from monkeys with histories of fetal wastage causes abnormalities in cultured rat embryos. Science 215:66-69.

Landaw, S.A. 1973. Effects of cigarette smoking on total body burden and excretion rates of carbon monoxide. J. Occup. Med. 15:231-235.

New, D.A.T. 1978. Whole-embryo culture and the study of mammalian embryos during organogenesis. Biol. Rev. 53:81-122.

Shapiro, S., S. Hartz, V. Siskind, A. Mitchell, D. Slone,
    L. Rosenberg, R. Monson, O. Heinonen, J. Idanpaan-Heikkila,
    S. Haro, and L. Saxen. 1976. Anticonvulsants and parental
    epilepsy in the development of birth defects. Lancet
    1:272-275.

A TERATOLOGY TEST SYSTEM WHICH UTILIZES POSTNATAL GROWTH AND VIABILITY IN THE MOUSE

Neil Chernoff and Robert J. Kavlock

Developmental Biology Division, Health Effects Research Laboratory, U.S. Environmental Protection Agency, Research Triangle Park, NC 27711

INTRODUCTION

The necessity of testing chemical agents for their potential to induce teratogenic effects has led to the establishment of standard teratology protocols (WHO, 1967; Health Protection Branch, 1977). These involve the exposure of pregnant animals to exogenous agents during the period of major embryonic organogenesis and the subsequent examination of the fetuses for soft visceral tissue and skeletal anomalies. These protocols are labor intensive and require highly trained personnel, factors which contribute to the considerable expense of such testing. A variety of potential screens, including developmental anomalies in invertebrates and lower vertebrates (Johnson, 1980; Davis et al., 1981), growth inhibition of mammalian cell cultures (Freese et al., 1979; Wilk et al., 1980), effects on organ culture (Kochar and Aydelotte, 1974), effects on cultured mammalian embryos (Kochar, 1975; Chatot et al., 1980), and in vivo embryotoxicity (Marhan and Jelinek, 1979), have been discussed as possible screens that would allow the prioritization of environmental agents for testing by standard methodologies. There have also been several excellent reviews of this general area in recent years (Wilson, 1978; Barrach and Neubert, 1980). There are severe disadvantages of previously proposed screens, however, including the lack of maternal/fetal interactions throughout development, the complexities of maintaining culture systems, and difficulties in the delivery of water-insoluble compounds.

The underlying hypothesis for the current study was that most prenatal insults would manifest themselves postnatally as reduced viability or impaired growth. The experiments reported were initiated to determine the utility of these simple measures of postnatal toxicity as a potential screening system. A wide variety of agents was tested, and results were compared with those noted in standard test procedures when possible.

Compounds were administered at or near maternal minimally toxic dose (MTD) since we wanted to maximize our chances of identifying those chemicals which exhibited greater toxicity to the conceptus than to the adult.

METHODS

Animal Husbandry

Female mice of the CD-1 strain (Charles River Laboratories, MA) were used throughout. Animals were individually housed in solid-bottom cages with wood shavings for bedding in temperature-controlled bio-clean rooms (20 to 24°C) with a 12-h light and dark cycle, and fed commercial lab chow and water ad libitum.

Establishment of Dose Levels

In studies to determine the MTD, nonpregnant female mice were housed five per cage. Animals were dosed (0.5 ml/day) for five consecutive days at one of five dose levels. Each dose level consisted of 10 animals. The MTD was considered that dose which resulted in either significant weight reduction during the treatment period, mortality, or other signs of toxicity.

Teratology Screening Procedures

In the definitive studies, pregnant mice were singly housed and received one of a variety of chemicals (Table 1) by either gavage (0.5 ml/day) or intraperitoneal injection (0.2 ml/day). The route of administration was generally oral, although other routes were used based upon studies which previously demonstrated the teratogenic potential of the specific compound or related compounds. Animals were dosed on days 8 through 12 (or single days within that period), which is within the period of major embryonic organogenesis. Compounds were administered at or near the MTD level as determined in preliminary experiments. The dose

Table 1. Maternal and Perinatal Effects of Prenatal Administration of Test Compounds.

Block	Compound	Dose (mg/kg)	Route	Vehicle	Day(s)	No. Treated	No. Died	No. Preg.	Maternal Weight Change	No. Live Day 1	Avg. Weight Day 1	No. Live Day 3	Avg. Weight Day 3
1	Control		oral	H_2O	8-12	24	0	21	7.3 ± 0.5	10.7 ± 0.5	1.60 ± 0.02	10.5 ± 0.5	2.05 ± 0.05
	Sodium salicylate	800	oral	H_2O	8-12	24	0	22	8.1 ± 0.4	11.0 ± 0.3	1.51 ± 0.02[a]	10.4 ± 0.5	1.90 ± 0.05[a]
2	Control		oral	c.o.	8-12	24	0	19	6.5 ± 0.4	10.2 ± 0.5	1.53 ± 0.03	9.1 ± 0.8	1.99 ± 0.05
	Carbaryl	100	oral	c.o.	8-12	23	0	18	6.1 ± 0.5	9.8 ± 0.8	1.58 ± 0.04	9.5 ± 0.8	2.15 ± 0.07
	Dinoseb	15	oral	c.o.	8-12	23	1	12	6.8 ± 0.5	10.3 ± 0.8	1.54 ± 0.04	9.9 ± 0.9	2.00 ± 0.10
3	Control		oral	c.o.	8-12	25	0	12	7.4 ± 0.5	9.8 ± 0.8	1.68 ± 0.03	9.3 ± 0.7	2.33 ± 0.09
	Toxaphene	75	oral	c.o.	8-12	25	2	11	4.2 ± 0.8[a]	10.3 ± 0.6	1.54 ± 0.03[a]	9.3 ± 1.1	2.11 ± 0.09
	Vitamin A	160,000 U	oral	c.o.	8-12	25	0	15	5.0 ± 0.4[a]	7.5 ± 0.7[a]	1.60 ± 0.05[a]	3.8 ± 0.7[a]	2.26 ± 0.11
4	Control		oral	H_2O	8-12	24	0	18	6.6 ± 0.3	10.1 ± 0.7	1.64 ± 0.04	9.3 ± 0.5	2.26 ± 0.08
	Cacodylic acid	600	oral	H_2O	8-12	24	7	9	3.9 ± 0.9[a]	8.0 ± 1.9	1.42 ± 0.04[a]	7.1 ± 1.8	4.89 ± 0.05[a]
	Caffeine	200	oral	H_2O	8-12	24	1	15	2.9 ± 0.5[a]	9.1 ± 0.8	1.59 ± 0.06	7.6 ± 1.1	2.05 ± 0.11[a]
5	Control		oral	c.o.	8-12	25	0	15	6.9 ± 0.3	10.3 ± 0.6	1.72 ± 0.04	10.2 ± 0.6	2.35 ± 0.08
	Endrin	2	oral	c.o.	8-12	25	1	17	3.5 ± 0.9[a]	9.7 ± 0.5	1.61 ± 0.03[a]	9.3 ± 0.5	2.26 ± 0.07
	Kepone	20	oral	c.o.	8-12	25	4	15	2.7 ± 0.7[a]	8.8 ± 0.4[a]	1.56 ± 0.03[a]	7.3 ± 0.9[a]	2.00 ± 0.09[a]
6	Control		i.p.	H_2O	8	24	0	12	7.7 ± 0.4	9.7 ± 0.4	1.72 ± 0.03	9.7 ± 0.5	2.55 ± 0.07
	5-Bromodeoxyuridine	400	i.p.	H_2O	8	24	1	15	–	5.0 ± 0.8[a]	1.67 ± 0.04	4.9 ± 0.9[a]	2.43 ± 0.11
7	Control		oral	c.o.	8-12	24	1	16	6.5 ± 0.3	11.6 ± 0.5	1.53 ± 0.03	10.6 ± 0.6	1.89 ± 0.07
	Benomyl	200	oral	c.o.	8-12	24	1	18	5.7 ± 0.3	5.8 ± 1.2[a]	1.37 ± 0.04[a]	3.9 ± 1.1[a]	1.81 ± 0.13
8	Control		oral	100 mM K_3PO_4	8-12	25	0	21	7.0 ± 0.3	10.5 ± 0.5	1.68 ± 0.02	10.4 ± 0.5	2.17 ± 0.05
	Ethylenediaminetetra acetic acid (EDTA)	1,000	oral	100 mM K_3PO_4	8-12	25	0	17	6.9 ± 0.5	9.8 ± 0.7	1.72 ± 0.03	9.0 ± 0.9	2.28 ± 0.07
	Ethylenethiourea (ETU)	300	oral	H_2O	8-12	24	3	17	7.4 ± 0.5	9.9 ± 0.5	1.73 ± 0.04	10.0 ± 0.6	2.31 ± 0.07
10	Control		i.p.	H_2O	11	25	0	21	8.9 ± 0.9	9.3 ± 0.7	1.75 ± 0.03	9.3 ± 0.7	2.38 ± 0.06
	6-Aminonicotinamide	10	i.p.	H_2O	11	25	0	21	–	6.9 ± 0.8[a]	1.37 ± 0.03[a]	4.4 ± 0.8[a]	1.87 ± 0.10[a]
11	Control		oral	c.o	8-12	24	0	16	6.7 ± 0.6	9.6 ± 0.9	1.74 ± 0.04	9.5 ± 0.9	2.44 ± 0.08
	Decamethrin	10	oral	c.o	8-12	25	0	14	8.4 ± 0.7	11.4 ± 1.0	1.73 ± 0.05	11.4 ± 1.0	2.27 ± 0.07
	Ethylenebisisothiocyanate (EBIS)	200	oral	c.o	8-12	23	4	15	8.9 ± 0.4	11.4 ± 0.7	1.70 ± 0.02	11.3 ± 0.7	2.22 ± 0.05
12	Control		oral	c.o	8-12	25	1	14	6.9 ± 0.4	8.9 ± 0.6	1.71 ± 0.03	8.9 ± 0.6	2.43 ± 0.08
	Cyclohexanone	800	oral	c.o	8-12	24	2	18	7.2 ± 0.5	10.9 ± 0.6	1.63 ± 0.03	10.7 ± 0.5	2.34 ± 0.06
	Hexachlorocyclopentadiene	45	oral	c.o	8-12	25	3	16	7.3 ± 0.4	9.9 ± 0.7	1.68 ± 0.03	9.9 ± 0.7	2.36 ± 0.04
13	Control		oral	c.o.	8-12	25	0	17	5.9 ± 0.4	9.6 ± 0.4	1.68 ± 0.02	9.6 ± 0.4	2.27 ± 0.04
	1,2,4-Trichlorobenzene	130	oral	c.o.	8-12	25	0	18	6.7 ± 0.3	9.6 ± 0.5	1.72 ± 0.07	9.6 ± 0.5	2.35 ± 0.11
14	Control		oral	c.o.	8-12	25	0	15	6.4 ± 0.5	11.1 ± 0.8	1.64 ± 0.04	10.7 ± 0.7	2.21 ± 0.06
	Chlordane	50	oral	c.o.	8-12	25	3	17	5.5 ± 0.4	9.9 ± 0.9	1.63 ± 0.03	9.9 ± 0.9	2.19 ± 0.07
	Lindane	25	oral	c.o.	8-12	23	0	13	5.6 ± 0.6	9.7 ± 0.9	1.69 ± 0.03	9.7 ± 0.9	2.30 ± 0.09

(continued)

Table 1. (continued)

Block	Compound	Dose (mg/kg)	Route	Vehicle	Day(s)	No. Treated	No. Died	No. Preg.	Weight Change	No. Live Day 1	Avg. Weight Day 1	No. Live Day 3	Avg. Weight Day 3
15	Control		oral	c-o	8-12	25	3	16	7.7 ± 0.6	10.8 ± 0.7	1.61 ± 0.03	10.4 ± 0.7	2.08 ± 0.06
	Mirex	7.5	oral	c-o	8-12	24	3	14	8.4 ± 0.8	7.2 ± 1.4a	1.49 ± 0.04a	4.1 ± 1.3a	1.73 ± 0.12a
16	Control		oral	c-o	8-12	25	5	13	7.4 ± 0.7	10.6 ± 0.4	1.59 ± 0.02	10.8 ± 0.4	2.11 ± 0.03
	Nitrofen	200	oral	c-o	8-12	25	1	17	7.9 ± 0.4	8.3 ± 0.7	1.42 ± 0.04a	1.9 ± 0.7a	1.55 ± 0.14a
	2,4,5-Trichlorophenoxy-acetic acid	130	oral	c-o	8-12	25	1	18	7.1 ± 0.9	8.6 ± 1.1	1.51 ± 0.06	7.1 ± 1.2a	2.14 ± 0.09
17	Control		oral	H2O	8-12	40	1	21	7.3 ± 0.3	9.9 ± 0.5	1.67 ± 0.03	9.8 ± 0.5	2.22 ± 0.06
	Congo red (oral)	1,000	oral	H2O	8-12	30	3	22	6.7 ± 0.3	10.1 ± 0.5	1.63 ± 0.03	9.3 ± 0.6	2.01 ± 0.07a
18	Control		oral	c-o.	8-12	40	1	29	6.3 ± 0.3	9.7 ± 0.4	1.65 ± 0.03	9.6 ± 0.4	2.07 ± 0.06
	Kelevan	125	oral	c-o.	8-12	30	4	19	7.3 ± 0.4	9.9 ± 0.7	1.61 ± 0.02	9.5 ± 0.7	1.97 ± 0.05
	Nickel chloride (oral)	100	oral	H2O	8-12	30	3	20	7.2 ± 0.3	11.1 ± 0.4	1.66 ± 0.03	10.7 ± 0.4	2.07 ± 0.05
19	Control		i.p.	H2O	8	40	0	24	0.3 ± 0.1	10.9 ± 0.5	1.59 ± 0.03	10.6 ± 0.6	2.13 ± 0.04
	Congo red (i.p.)	300	i.p.	H2O	8	30	6	15	-1.8 ± 0.2a	7.6 ± 1.5a	1.61 ± 0.05	7.4 ± 1.5a	2.00 ± 0.10
	Chicago sky blue	250	i.p.	H2O	8	30	5	17	-2.1 ± 0.2a	0a	–	0a	–
	Trypan blue	250	i.p.	H2O	8	30	0	12	-3.2 ± 0.2a	0a	–	0a	–
20	Control		i.p.	H2O	8	40	0	21	0.6 ± 0.1	9.3 ± 0.8	1.65 ± 0.04	9.1 ± 0.8	2.13 ± 0.06
	Alcian blue	250	i.p.	H2O	8	30	3	17	-2.6 ± 0.2a	3.3 ± 1.3a	1.66 ± 0.04	3.5 ± 1.4a	1.95 ± 0.03
	Diazo red	150	i.p.	H2O	8	30	2	16	0.2 ± 0.2	10.2 ± 0.9	1.50 ± 0.03a	9.3 ± 0.9	1.80 ± 0.07a
	Nickel chloride (i.p.)	30	i.p.	H2O	8	30	0	12	-1.4 ± 0.5a	6.2 ± 2.4a	1.63 ± 0.07	6.2 ± 2.4	2.11 ± 0.18
21	Control			b		40	0	33	2.0 ± 0.1	10.8 ± 0.3	1.58 ± 0.02	9.5 ± 0.6	1.98 ± 0.05
	Aminopterin	20	i.p.	NaHCO3 buffer	11	30	1	23	0.4 ± 0.1	0.8 ± 0.6a	1.71 ± 0.01	0.8 ± 0.6a	2.15 ± 0.13
	Cortisone	300	i.m.	saline	12	30	1	25	1.2 ± 0.2a	4.9 ± 0.9a	1.25 ± 0.04a	1.3 ± 0.5a	1.63 ± 0.14a
22	Control			b		40	0	32	6.7 ± 0.5	11.3 ± 0.3	1.62 ± 0.02	11.1 ± 0.3	2.15 ± 0.04
	Sodium selenite	10	oral	H2O	8-12	30	2	23	4.7 ± 0.5a	9.4 ± 0.6a	1.70 ± 0.02	9.4 ± 0.6a	2.17 ± 0.04
	Thalidomide	2,000	oral	c-o.	8-12	30	0	27	8.1 ± 0.4	10.5 ± 0.7	1.70 ± 0.02	10.3 ± 0.7	2.21 ± 0.05
	2,4,5-Trichlorophenol	800	oral	c-o.	8-12	30	4	18	6.5 ± 0.3	9.8 ± 0.6a	1.65 ± 0.04	9.8 ± 0.6	2.12 ± 0.06
23	Control		oral	Me cellulose	8-12	30	2	12	14.1 ± 0.6	10.0 ± 0.5	1.69 ± 0.04	10.1 ± 0.4	2.44 ± 0.07
	Lithium carbonatec	400	oral	Me cellulose	8-12	25	0	17	5.9 ± 0.4	9.1 ± 0.5	1.83 ± 0.03	9.1 ± 0.5	2.64 ± 0.06
	Lithium carbonate (poly)c	400	oral	Me cellulose	8-12	25	0	16	5.6 ± 0.5	7.8 ± 0.8a	1.81 ± 0.03	7.8 ± 0.8a	2.71 ± 0.07

Values are means ± standard error.

a Significantly different from concurrent control value, p < 0.05.

b Control group was divided equally among the different treatments. Statistical analysis did not reveal any significant differences related to the treatment vehicles, so the control data presented are the total of the treatments.

c An earlier test with LiCO3 did not detect any treatment-induced difference. Since it was administered as a thick suspension and since it was shown to be a teratogen, we decided to retest it after subjecting the solution to polytronning to reduce particle size. This later replicate (presented here) also contained a group treated as before. The polytronned group showed significant reductions in no. live on PD 1 and PD 3, when compared to both the control and non-polytronned groups. We feel that our initial negative results were, therefore, vehicle rather than compound related.

levels for seven known teratogens (aminopterin,
6-aminonicotinamide, benomyl, 5-bromodeoxyuridine, cortisone,
2,4,5-trichlorophenoxyacetic acid [2,4,5-T], and sodium
salicylate) were taken directly from previous studies in our
laboratory or the literature. Two or three compounds and a
control group were assayed concurrently. A treated group
generally consisted of 24 to 30 dams and the control group
contained 24 to 40. The change in maternal weight during the
treatment period was calculated. Dams were allowed to give birth,
and the litters were counted and weighed on postnatal days 1 and 3
(PD1 and PD3). Mice in our facility usually deliver on the
evening of day 19 of gestation, and day 20 of gestation was
therefore defined as PD1. Dead pups recovered from the nest were
necropsied and abnormalities noted. Dams which had not given
birth by PD3 were killed and their uteri examined for the presence
of implantation sites.

All data analyses compared treatment groups and their
concurrent controls and were performed using analysis of variance
procedures available in the General Linear Model Procedure on
version 79.2B of the Statistical Analysis System (SAS User's
Guide, SAS Institute, Cary, NC). When a significant treatment
effect was detected by ANOVA, individual group means were compared
using the Student's t-test on least-square means. Since our
a priori hypothesis was that treatments could only reduce litter
size, one-tailed tests were used for the analysis of the number of
live pups on days 1 and 3. To correct for differences in pup
weights due to differing litter sizes, the number of live pups on
day 1 was used as a covariate in the analyses of postnatal body
weights. Dams which had nidation sites but no record of having
given birth were considered to have no live pups on PD1 and PD3.

RESULTS AND DISCUSSION

In the studies to determine MTD levels, toxic effects on the
dam were noted for 19 of 33 compounds tested. MTDs could not be
determined for thalidomide, $LiCO_3$, sodium salicylate, Congo red
(oral), and EDTA. In the case of thalidomide, the dose used in
the final test was determined to be an effect dose as measured by
significantly reduced activity in a residential maze (Gray et al.,
1981).

The results of the definite screen studies are shown on
Table 1. The average litter in the control group contained
10.2 pups weighing an average of 1.65 g each on PD1. Control pup
survivorship to PD3 was 98.9%, and the average pup weighed 2.20 g
at that time. Litter size was more variable than pup weight among

the control groups. The coefficients of variation for the mean of
the control group data were 7.0% and 12.1% for litter size on PD1
and PD3, respectively, while the coefficients of variations for
pup weights were 3.8% and 7.8%, respectively.

In general, litter size reductions of 12% or more relative to
the concurrent control group were necessary to obtain statistical
significance for pup weights. Due to the smaller variability in
pup weights as compared to litter sizes, smaller reductions were
necessary to obtain statistical significance for this parameter.
There were statistically significant reductions in pup weights in
all cases where decreases were greater than 7% on PD1 or by 10% on
PD3 relative to the concurrent control value. For the twelve
compounds which produced statistically significant reductions in
pup weight on PD1, eight continued to show significant effects on
PD3. In only two instances (caffeine and oral Congo red) did an
effect on pup weight appear on PD3 when no effect was noted on
PD1. It should be noted that in a few instances, decreased
fertility rates or increased maternal death decreased the sample
size of some groups (e.g., cacodylic acid) so that relatively
large effects on litter size (greater than a 20% reduction) were
not significantly different from control values.

The correlation between the effects reported in standard
tests and those found in the screen is presented in Table 2. When
compounds of known teratogenic potential in the mouse are divided
into three groups: those producing major malformations or lethal
effects; those resulting in fetal toxicity in the form of either
reduced weight or increased incidence of minor anomalies; and
those that have no discernible effects on fetal mouse development,
a strong correlation between historical and the present test
results is evident. Of the 15 compounds which were known to
produce fetal lethality or teratogenic effects by a similar route
of administration, 13 resulted in significant reductions in litter
size or postnatal viability and the remaining two (cacodylic acid
and sodium salicylate) reduced birth weight. There were,
therefore, no false negatives in the screen. Reduced fetal weight
was the only effect reported in the literature for both endrin and
sodium selenite. This effect was approximated for endrin (reduced
weight on PD1), while sodium selenite reduced litter size. Two
compounds (ETU and decamethrin), whose sole reported effect in
standard tests was an increased incidence of supernumerary ribs,
produced no effect in the screen. Of the nine compounds for which
no effects were detected by standard test procedures, six produced
no effects in the screen. The remaining three compounds which
were reported to be negative in the literature but which produced
positive effects in the screen were toxaphene and caffeine
(reduction in litter weight), and 2,4,5-trichlorophenol (reduced

Table 2.   Correlation of Screen Results with Those Obtained in
           Standard Tests Utilizing the Mouse[a]

Standard Results	Screen Results		
	Reduced Litter Size or Viability	Reduced Weight	No Effects
Teratogens	Aminopterin 6-AN Benomyl 5-BrdU Cortisone Kepone LiCO$_3$ Mirex NiCl$_2$ Nitrofen 2,4,5-T Trypan blue Vitamin A	Cacodylic acid Sodium salicylate	
Fetotoxins	Sodium selenite	Endrin	ETU Decamethrin
No effect	2,4,5-Tri-chlorophenol	Caffeine Toxaphene	Carbaryl Chlordane Dinoseb EBIS HCCPD Thalidomide

[a]References to published data for most compounds tested may be
found in Shepard, 1980. Other references are: benomyl,
R.J. Kavlock et al., 1982; cacodylic acid, N. Chernoff and
E.H. Rogers, 1975; carbaryl, F.J. Murray et al., 1979;
decamethrin, R.J. Kavlock et al., 1979; endrin, R.J. Kavlock et
al. 1981; ethylenebisisothiocyanate/ethylenethiourea, N. Chernoff
et al., 1979; sodium selenite, T. Nobunaga et al., 1977;
thalidomide, W.J. Scott et al., 1977; toxaphene, N. Chernoff and
B.D. Carver, 1976; 2,4,5-trichlorophenol, R.D. Hood et al., 1979.
Data on mirex and nitrofen are either in press or being prepared
for publication.

litter size). It is important to note that "false negatives" may indicate the presence of functional deficits not readily identifiable in standard teratology test procedures and therefore, may represent true toxic responses.

Examination of the dead pups recovered from the nest indicated that a number of the compounds produced gross terata and other morphological alterations. Vitamin A and benomyl both produced encephalocele and hemimelia, while 5-BrdU produced a high incidence of encephaloceles. Cleft palate was found in litters treated with 2,4,5-T and nitrofen. Pups exposed to mirex and nitrofen exhibited a high incidence of grossly distended abdomens, a condition that was usually fatal by PD3. While in this study the determination of positive effects was not enhanced by necropsy of the non-surviving pups, we feel that this procedure may aid in the identification of teratogenic potential of those compounds, which may induce low rates of defects in the absence of statistically significant effects on pup growth or viability.

The primary function of a teratology screen should be the prioritization of compounds of unknown developmental toxicity for subsequent in-depth testing by standard procedures. The present screen allows such a prioritization based upon an in vivo mammalian system. This prioritization may be accomplished by dividing compounds into basic categories: those that induce perinatal death should be tested as rapidly as possible; those which induce only perinatal weight changes would be given a lower testing priority; and those producing no effect, given the lowest priority. We feel that the test has proven to be an accurate reflection of results obtained by standard mouse teratology tests.

Users may, of course, wish to alter the specific design of such a screen according to their specific requirements. While we chose to test at an MTD level, thereby maximizing the chance of identifying developmentally toxic compounds, other dosing regimens, perhaps based upon some multiple of theoretical human exposure levels or some fraction of the LD_{50}, are equally plausible. It is worth noting that of the 17 compounds which produced maternal toxicity in the screen, 3 did not result in any perinatal effects, demonstrating that compounds may be tested at MTD levels without necessarily resulting in such effects. Similarly, we chose to terminate dosing at least seven days prior to parturition, giving the dams a longer recovery period than in a standard teratology test where dosing continues until day 17. It was felt that this longer recovery period would lessen the chances that maternal toxicity would confound the neonatal results (McClain and Hoar, 1980). As is the case in selection of the test dose, the duration of exposure or the route of administration can be altered to better reflect specific compound usage.

The results obtained in these studies indicate that regardless of the precise experimental conditions employed, the proposed test provides a simple but accurate picture of the _in vivo_ response of the mammalian embryo/fetus. It must be remembered that for purposes of extrapolating experimental results to humans, this or any other screen is only as relevant as the experimental species employed. Compounds which are not teratogenic in the mouse but teratogenic in other species by standard tests (thalidomide, ethylenethiourea, carbaryl) are not positive in the proposed test. The proposed screen might be less sensitive for the compounds that result solely in anomalies not incompatible with life (i.e., supernumerary ribs), but more sensitive in those cases where a compound might result in a functional organ impairment not readily apparent by fetal necropsy.

REFERENCES

Barrach, H., and D. Neubert. 1980. Significance of organ culture techniques for evaluation of prenatal toxicity. Arch. Toxicol. 45:161-187.

Chatot, C.L., N.W. Klein, J. Piatek, and L.J. Pierro. 1980. Successful culture of rat embryos on human serum: use in the detection of teratogens. Science 207:1471-1473.

Chernoff, N., and B.D. Carver. 1976. Fetal toxicity of toxaphene in rats and mice. Bull. Environ. Cont. Toxicol. 15:660-664.

Chernoff, N., and E.H. Rogers. 1975. Proceedings of the Substitute Chemical Program: First Year of Progress, Volume II. Toxicological Methods and Genetic Effects Workshop. pp. 197-204.

Chernoff, N., R.J. Kavlock, E.H. Rogers, B.D. Carver, and S. Murray. 1979. Perinatal toxicity of Maneb, ethylene-thiourea, and ethylenebisisothiocyanate sulfide in rodents. J. Toxicol. Environ. Health 5:821-834.

Davis, K.R., T.W. Schultz, and J.N. Dumont. 1981. Toxic and teratogenic effects of selected aromatic amines on embryos of the amphibian Xenopus laevis. Arch. Environ. Cont. Toxicol. 10:371-391.

Freese, E., B.C. Levin, R. Pearch, T. Sreevalsan, J.J. Kaufman, W.S. Koski, and N.M. Semo. 1979. Correlation between the growth inhibitory effects, partition coefficient and teratogenic effects of lipophilic acids. Teratology 20:413-440.

Gray, L.E., Jr., R.J. Kavlock, N. Chernoff, J.A. Gray, and J. McLamb. 1981. Perinatal toxicity of endrin in rodents, III: alterations of behavioral ontogeny. Toxicology 21:187-202.

Health Protection Branch. 1977. The testing of chemicals for carcinogenicity, mutagenicity, and teratogenicity. Ministry of Health and Welfare, Canada.

Hood, R.D., B.L. Patterson, G.T. Thacker, G.L. Sloan, and G.M. Szczech. 1979. Prenatal effects of 2,4,5-T, 2,4,5-trichlorophenol, and phenoxyacetic acid in mice. J. Environ. Sci. Health C13:189-204.

Johnson, E.M. 1980. A subvertebrate system for rapid determination of potential teratogenic hazards. J. Environ. Pathol. Toxicol. 4:153-156.

Kavlock, R.J., N. Chernoff, R. Baron, R. Linder, E. Rogers, and B. Carver. 1979. Toxicity studies with decamethrim, a synthetic pyrethroid insecticide. J. Environ. Pathol. Toxicol. 2:751-765.

Kavlock, R.J., N. Chernoff, L.E. Gray, Jr., J.A. Gray, and D. Whitehouse. 1982. Teratogenic effects of benomyl in the Wistar rat and CD-1 mouse, with emphasis on the route of administration. Toxicol. Appl. Pharmacol. 62:44-54.

Kavlock, R.J., N. Chernoff, R.C. Hanisch, J. Gray, E. Rogers, and L.E. Gray. 1981. Perinatal toxicity of endrin in rodents, II: fetotoxic effects of prenatal exposure in rats and mice. Toxicology 21:141-150.

Kochar, D.M., and M.B. Aydelotte. 1974. Susceptible stages and abnormal morphogenesis in the developing mouse limb, analyzed in organ culture after transplacental exposure to vitamin A (retinoic acid). J. Embryol. Exp. Morphol. 31:721-734.

Kochar, D.M. 1975. Assessment of teratogenic response in
    cultured postimplantation mouse embryos: effects of
    hydroxyurea. In: New Approaches to the Evaluation of
    Abnormal Embryonic Development. D. Neubert and H.J. Merker,
    eds. George Thieme: Stuttgart. pp. 250-277.

Marhan, O., and R. Jelinek. 1979. Efficiency of embryotoxicity
    testing procedures, I: a compromise approach. Toxicol.
    Lett. 4:385-388.

McLain, R.M., and R.M. Hoar. 1980. The effect of flunitrazepam
    on reproduction in the rat: the use of cross-fostering in
    the evaluation of postnatal parameters in rat reproduction
    studies. Toxicol. Appl. Pharmacol. 53:92-100.

Murray, F.J., R.E. Staples, and B.A. Schwetz. 1979. Teratogenic
    potential of hexachlorocyclopentadiene in mice and rabbits.
    Appl. Pharmacol. 51:81-89.

Nobunaga, T., H. Satoh, and T. Suzuki. 1979. Effects of sodium
    selenite on methylmercury embryotoxicity and teratogenicity
    in mice. Toxicol. Appl. Pharmacol. 47:79-88.

Scott, W.J., R. Fradkin, and J.G. Wilson. 1977. Non-confirmation
    of thalidomide-induced teratogenesis in rats and mice.
    Teratology 16:333-336.

Shepard, T.H. 1980. Catalogue of Teratogenic Agents. Johns
    Hopkins University Press: Baltimore.

Wilk, A.L., J.H. Greenberg., E.A. Horigan, R.M. Pratt, and
    G.R. Martin. 1980. Detection of teratogenic compounds using
    differentiating embryonic cells in culture. In Vitro
    16(4):269-276.

Wilson, J.G. 1978. Review of in vitro systems with potential for
    use in teratogenicity screening. J. Environ. Pathol.
    Toxicol. 2:149-167.

World Health Organization. 1967. Principles For the Testing of
    Drugs For Teratogenicity. WHO Tech. Rep. Series No. 364:
    Geneva.

SESSION 6

INTEGRATED ASSESSMENT

MUTAGENICITY OF PULP AND PAPER MILL EFFLUENT: A COMPREHENSIVE

STUDY OF COMPLEX MIXTURES

George R. Douglas,[1] Earle R. Nestmann,[1] A.B. McKague,[2]
O.P. Kamra,[3] E.G.-H. Lee,[2] J.A. Ellenton,[4] R. Bell,[1]
D. Kowbel,[1] V. Liu,[2] and J. Pooley[1]

[1]Mutagenesis Section, Department of National Health and
Welfare, Ottawa, Canada, [2]British Columbia Research
Vancouver, British Columbia, Canada, [3]Biology Department
Dalhousie University, Halifax, Nova Scotia, Canada, and
[4]Canadian Wildlife Service, Department of the Environment
Ottawa, Canada

INTRODUCTION

Pulp and paper mill effluents are complex mixtures of
dissolved lignin and cellulose degradation products and other
substances extracted during the pulping process. The toxicity of
these effluents to fish has been documented (Howard and Walden,
1971; Wande, 1976; Walden and Howard, 1977). A number of
investigators have shown that effluents and process streams from
pulp and paper mills are mutagenic in Salmonella (Ander et al.,
1977; Eriksson et al., 1979; Douglas et al., 1980) and cause
chromosome damage in mammalian cells (Douglas et al., 1980). Over
300 compounds have been identified in studies on various pulp and
paper mill effluents (reviewed in CPAR, 1978). Because of the
extensive use of chlorine in the bleaching process, many of these
compounds contain chlorine substitutions. Since first
chlorination-stage liquors are consistently among the most
mutagenic by-products of the pulping process, it has been
suggested that chlorinated substances are responsible for a major
portion of the mutagenicity found (Bjørseth et al., 1979; Nestmann
et al., 1979; Nestmann et al., 1980; McKague et al., 1981; Rapson
et al., 1980).

In our laboratories, we have begun a coordinated program of investigation, consisting of three basic components: 1) From the list of known chemical constituents, selected compounds were tested using the Salmonella/mammalian-microsome assay and in Saccharomyces cerevisiae. Compounds identified as mutagens in this initial screen are being tested further in a battery of in vitro mammalian cell assays. 2) In addition to the study of known constituents, studies are in progress on first chlorination-stage effluent as a model complex mixture using a battery of microbial and mammalian in vitro and in vivo mutagenicity tests. 3) In order to provide new data on the components responsible for the mutagenicity, fractions of the above effluent were tested for mutagenicity in Salmonella and the most mutagenic fractions subjected to chemical analysis. Compounds so identified were then tested for determination of their mutagenicity in Salmonella. These approaches, taken together, are designed to better characterize the course of the mutagenicity in pulp and paper effluents and to gain insight into the mechanisms involved in the mutagenicity of mixtures.

MATERIALS AND METHODS

Mutagenicity Assays

The Salmonella/mammalian-microsome assay was carried out according to the method recommended by Ames et al. (1975), and assays for reverse mutation and gene conversion in strains of Saccharomyces cerevisiae followed the method of von Borstel et al. (1981) for treating exponential-phase cultures. In vitro mammalian cell assays in wild-type Chinese hamster ovary (CHO) cells were performed as described by Douglas et al. (1981) to detect the following end points: cytotoxicity, detected as survival of colony-forming ability; DNA damage, detected by alkaline sucrose gradient (ASG) sedimentation; chromosome aberrations; and sister-chromatid exchanges (SCE). In addition, the CHO/HGPRT⁻ forward mutation assay was performed essentially as described by O'Neill and Hsie (1979). For all in vitro assays, Aroclor 1254-induced rat-liver homogenate (S9) mix was prepared according to the method published by Ames et al. (1975). For the detection of in vivo genetic effects, the mouse bone marrow micronucleus assay was carried out according to the revised procedure described by Salamone et al. (1980).

Studies on Chlorination-Stage Effluent

Concentration of chlorination-stage effluent. Work was performed on samples of first chlorination-stage effluent from a bleached kraft pulp mill in interior British Columbia, using a C_DEHDED bleaching sequence. The sample of effluent used for large-scale fractionation was obtained when bleaching of softwood pulp at Kappa number 24 was performed with a percent chlorine to chlorine dioxide ratio of 7.2 to 0.3 and a temperature of 46°C. Residual chlorine was 0.12 g/l. Effluent volume was approximately 36 m^3 per ton of pulp. Before fractionation, the effluent was tested for mutagenicity in Salmonella strain TA100, immediately after adjustment to pH 7 and filtration through a 0.45-μm filter.

Columns of XAD-2 resin (3 1) were prepared by washing with methanol (2 1), ether (6 1), methanol (4 1), and water (30 1). Using three such columns, a total of 105 1 of the chlorination-stage effluent was processed, 35 1/column at a rate of 3 1/h/column. The effluent eluate from each column was combined, giving successive 25-1 portions of total eluate which were bioassayed for mutagenicity. A sample (25 1) of the fresh water supplied to the mill acidified to pH 1.6 was passed through a 3-1 column of the resin. Each column was eluted with ether (10 1), followed by methanol (3 1) to recover adsorbed material. Solvents were removed on a rotary evaporator using a bath temperature below 50°C. The ether eluate was redissolved in ether (500 ml), dried (MgSO4), and evaporated. Bioassays were performed on aliquots of dimethyl sulfoxide (DMSO) solutions of the ether eluate.

Fractionation of the ether eluate on silica gel. A portion of the ether eluate (4.5 g) from the chlorination-stage sample equivalent to 25 1 of effluent was separated into five fractions by column chromatography on silica gel (300 g, 60/200 mesh). The column was eluted successively with hexane (1.5 1), hexane:ether 9:1 (1.5 1), hexane:ether 1:1 (1.5 1), ether (2 1), and methanol (1.5 1) to remove compounds of increasing polarity. Solvents were evaporated under reduced pressure and the mutagenicity of each fraction was determined as previously described.

Instrumental analyses. Gas chromatography (GC) was performed on a Hewlett-Packard gas chromatograph model 5830A equipped with a 0.25-mm ID x 14.5-m WCOT OV-101 glass capillary column and flame ionization detector (FID). Linear gas flows of 30 to 35 cm/s helium and various split ratios were used at the injector inlet. Gas chromatography/mass spectrometry (GC/MS) was performed on a Finnigan 4000 Series instrument which had an open split interface, including a 0.1-mm ID quartz transfer line to the MS. The GC column was a fused silica capillary SE 54, 0.32-mm ID x 30 mm.

Mass spectra were compared with those stored in the EPA/NIH
spectral library.

GC and GC/MS analyses of silica gel fractions. Five-
milligram samples were dissolved in methanol (0.1 to 0.2 ml) and
2 μl analyzed before and after methylation with diazomethane by GC
on OV using FID. Temperature was programmed from 90 or 100°C to
250°C at 4°/min. Fractions 3 and 4 were analyzed by GC/MS using a
similar temperature program.

Analysis of ether extracts of chlorination-stage samples.
Aliquots (250 ml) of the chlorination-stage sample under detailed
study above and a second fresh sample were treated with NaCl
(25 g) and extracted with ether (1 x 50 and 2 x 25 ml). The
extracts were combined, washed with water (2 x 10 ml), dried
(MgSO$_4$), and evaporated. Residues were dissolved in ether or
methanol (0.2 to 0.3 ml) and analyzed by GC and GC/MS as described
above. The concentrations of chloroacetones were determined using
methyl heptadecanoate internal standard.

Effect of pH on chloroacetones in the chlorination-stage
sample and tetrachloroacetone in water. Samples of effluent
(250 ml) were stirred at pH 2, 7, and 10, respectively, for 1 h at
room temperature. NaCl (25 g) was added and the samples were
extracted with ether as described above. Recovery of
tetrachloroacetone was determined by GC using methyl
heptadecanoate internal standard.

RESULTS AND DISCUSSION

Screening Compounds Identified in Effluent

Microbial assays. Fifty-eight compounds were selected for
mutagenicity screening using microbial systems. Table 1 shows
eight compounds that were tested in all three bacterial and yeast
assays and detected as mutagens in at least one. With the
standard Salmonella assay, using strains TA1535, TA100, TA1537,
TA1538, and TA98, with and without metabolic activation (S9),
three compounds (neoabietic acid, tetrachloropropene, and
pentachloropropene) were found to induce histidine reversion
mutations. These three and five others (7-oxodehydroabietic acid,
chloromuconic, acid, acetovanillone, dichlorocatechol, and
dichloroguaiacol) induced tryptophan revertants in exponential
cultures of yeast strain XV185-14C, a strain described in detail
elsewhere (von Borstel, 1981). None of these eight compounds
induced gene conversion in growing cells of strain D7, derived by
Zimmerman et al. (1975). Table 1 also shows that three additional

Table 1.   Summary of Positive Responses in Bacteria and Yeast

Chemical Category	Compound	Reversion in Salmonella	Conversion in Yeast D7	Reversion in Yeast XV185-14C
Carboxylic acid	Chloromuconic acid	−	−	+
Phenolics	Acetovanillone	−	−	+
	Dichlorocatechol	−	−	+
	Dichloroguaiacol	−	−	+
Resin acids	7-Oxodehydroabietic acid	−	−	+
	Neoabietic acid	+	−	+
Chlorinated aliphatic hydrocarbons	1,1,2,3-Tetrachloro-2-propene	+	−	+
	1,1,2,3,3-Pentachloropropene	+	−	+
	Dichloromethane[a]	+	NT[b]	NT
	1,2-Dichloroethane[a]	+	NT	NT
	1,1,1-Trichloroethane[a]	+	NT	NT

[a]Tested only in Salmonella using a method modified for assaying the mutagenicity of volatile compounds.
[b]Not tested.

compounds (dichloromethane, 1,2-dichloroethane, and
1,1,1-trichloroethane) were mutagenic in a modified Salmonella
assay using desiccators as exposure chambers for these volatile
solvents. These compounds were not tested in the yeast assays,
but dichloromethane has been shown previously to induce gene
convertants in strain D7 (Callen et al., 1980). These data
illustrate the importance of using more than one assay in the
screening of chemicals for mutagenicity and the utility of yeast
as an assay system.

Table 2 lists the remaining 34 compounds that were found to
be nonmutagenic in the bacterial and yeast test systems. Of these
compounds, trans-anethole was reported previously to be mutagenic
in Salmonella with the use of a different activation system, that
is, liver microsomes and cytosol S13 (Swanson et al., 1979).
Previous work has shown that tetrachloroethane induced gene
convertants in strain D7 (Callen et al., 1980), whereas it did not
in this study (Table 2). Callen et al. (1980) showed that the
genetic effects at short exposure times disappeared as growth
periods were extended and, therefore, limited their treatment time
with certain chlorinated aliphatics to 1 h at 37°C. The exposure
time in the present study was 24 h at 30°C, which probably
accounts for the discrepancy in findings for tetrachloroethane.
Table 2 also shows an additional 13 nonmutagenic compounds that
were tested in Salmonella, but not with the yeast assays. Of
these, chloroform was found previously to induce gene conversion
in yeast strain D7 (Callen et al., 1980).

More detailed descriptions of these results in Salmonella
have been published previously (Nestmann et al., 1979; Nestmann et
al., 1980). A preliminary report of the yeast data has been
published (Nestmann and Lee, 1981), and a more detailed manuscript
for publication is in preparation.

Mammalian in vitro assays. 1,1,2,3-Tetrachloro-2-propene
(TCP) and 1,1,2,3,3-pentachloropropene (PCP) were tested in CHO
cells using assays to detect the following: cytotoxicity,
chromosome aberrations, SCE, and DNA damage. The results of the
first three assays are shown in Figure 1. For both compounds, the
addition of S9 mix reduced cytotoxicity, and the dose-related
increases in SCE without S9 were concomitantly reduced with the
addition of S9 mix. However, while a similar S9-related reduction
in the effective concentration range for the induction of
chromosome aberrations was observed with PCP, such was not the
case with TCP. The addition of S9 mix enhanced, rather than
reduced, the potency of TCP with respect to chromosome
aberrations, an effect opposite to that observed for cytotoxicity.
In addition to these effects, neither TCP nor PCP caused any DNA

Table 2. Compounds with Negative Responses in Bacteria and Yeast

---

Aldehydes
  O-Chlorobenzaldehyde
  Veratraldehyde

Aromatics
  Acenaphthene
  trans-Anethole
  3,4-Dimethoxyacetophenone
  Ethylbenzene
  4-Hydroxy-3-methoxypropiophenone
  Toluene[a]
  Trimethoxychlorobenzene

Carboxylic Acids
  Dichlorostearic acid
  cis-9,10-Epoxystearic acid
  trans-9,10-Epoxystearic acid[a]
  Glucoisosaccharinic acid
  Homovanillic acid
  Vanillic acid

Resin Acids
  Abietic acid[a]
  Chlorodehydroabietic acid
  Dehydroabietic acid
  Dichlorodehydroabietic acid
  Isopimaric acid
  Levopimaric acid
  Pimaric acid
  Sandaracopimaric acid

Chlorinated Aliphatics
  Chloroform[a]
  1,1,2,2-Tetrachloroethane

Esters of Fatty Acids
  Trilaurin[a]
  Triolein[a]
  Tristearin[a]

Ketones
  2-Butanone[a]

Phenolics
  Acetosyringone
  m-Cresol[a]
  o-Cresol[a]
  p-Cresol[a]
  2,6-Dichlorophenol
  Eugenol
  Guaiacol
  Isoeugenol
  Syringol
  2,3,4,5-Tetrachlorocatechol
  Tetrachloroguaiacol[a]
  3,4,5-Trichlorocatechol
  Trichloroguaiacol[a]
  2,4,5-Trichlorophenol

Quinones
  Choranillic acid
  2,5-Dichloro-3,6-disulfo-
    hydroquinone
  Tetrachloro-o-benzoquinone

---

[a]Tested only in Salmonella.

lesions detectable as single-strand discontinuities in alkaline sucrose gradients (data not shown).

These data, as well as indicating the mutagenic activity of TCP and PCP, reveal other important effects: Because of the differential action of S9, chromosome aberrations induced by TCP or its metabolites are caused by different DNA lesions than those

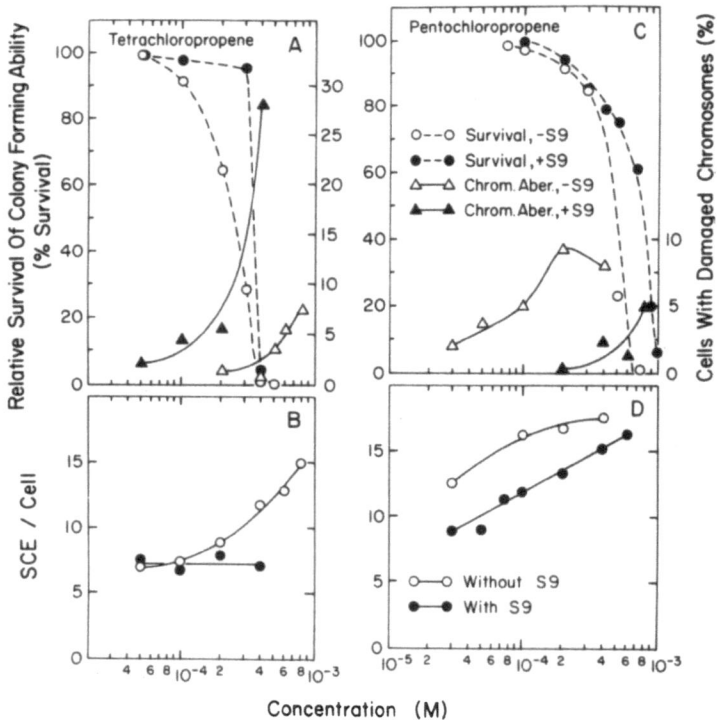

Figure 1.   Effect of 1,1,2,3-tetrachloro-2-propene (A,B) and
             1,1,2,3,3-pentachloropropene (C,D) on survival,
             chromosome aberrations, and SCE in CHO cells.

inducing SCE; the lesions causing chromosome aberrations may be
different for TCP and PCP; and the DNA lesions causing chromosome
aberrations and SCE with both compounds are different from those
capable of inducing lesions detectable by the ASG technique.   The
results of these CHO assays are in general accord with the

corresponding microbial data (Figure 1; Nestmann et al., 1980; Rannug et al., 1980; Douglas et al., 1980).

The importance of TCP and PCP, or any other chemical, in terms of human health, is largely determined by their persistence in the environment and the potential for human exposure. After biotreatment, the mutagenicity of pulp and paper mill effluents is generally much attenuated; nevertheless, such biotreated effluents can still induce chromosomal damage in CHO cells (Douglas et al., 1980). Furthermore, certain chlorinated compounds have a tendency to persist in the environment, increasing their chances of exposure to man. While there are little or no data on environmental levels of PCP, TCP has been identified in finished drinking water (International Joint Commission, 1981). It is not known if the origin of this contamination is pulp and paper mills, or other industrial or natural sources. Nevertheless, the data presented here suggest TCP and PCP may pose a genetic hazard. Estimation of any hazard will require further studies _in vivo_ and should include the determination of exposure levels. The CHO data on 1,1,2,3-tetrachloro-2-propene have been published in more detail elsewhere (Ellenton et al., 1981).

Mutagenicity-Directed Fractionation of Chlorination-Stage Effluent

Mutagenicity of chlorination-stage effluent and concentrate. The chlorination-stage effluent gave a dose-related mutagenic response in _Salmonella_ strain TA100 without metabolic activation (Table 3). Such a response is typical, and is reduced by the addition of S9 mix (Ander et al., 1977; Eriksson et al., 1979; Douglas et al., 1980). After passage through XAD-2 resin, mutagenicity bioassays showed that most of the mutagenicity was removed (Table 3).

Elution of the resin with ether followed by methanol and evaporation of the solvents gave a brown ether eluate (18.9 g) and a dark brown methanol eluate (4.3 g). A pale yellow oil (150 mg) was obtained after evaporation of the ether eluate from the mill supply water sample. A DMSO solution of the ether eluate was mutagenic up to a concentration of 10 µl/plate (equivalent to 2 ml of original effluent) above which toxicity to the bacteria occurred (Table 4). Mutagenicity of the methanol eluate was near background and the ether eluate from mill supply water was equivalent to background (Table 4). Virtually all the mutagenicity removed from the sample by the resin was, therefore, recovered in the ether eluate and was not derived from acidified mill water or extracted from the resin.

Table 3. Response[a] of S. typhimurium Strain TA100 to
          Chlorination-Stage Effluent in the Absence
          of S9 Mix

Concentration (ml/plate)	Effluent	First 25-liter XAD-2 Eluate	Last 25-liter XAD-2 Eluate
0	121	139	139
0.25	287	166	157
0.5	666	174	218
1.0	1089	307	182
Sodium azide (10 µg)	1067	1012	1012

[a]Means of three replicates per concentration.

Table 4. Response[a] of S. typhimurium to Strain TA100 to Solvent
          Eluates of Chlorination-Stage Effluent Adsorbed to
          XAD-2 Resin

Concentration[b] (µl/plate)	Ether Eluate	Methanol Eluate	Acidified Mill Supply Water
0	136	136	136
5	941	149	145
10	1097	174	152
25	326	210	144
Sodium azide (10 µg)	1083	1083	1083

[a]Mean of three replicates per concentration.
[b]1 µl = 200 µl of original effluent.

Fractionation of the ether eluate on silica gel. Ninety-six
percent of the material was recovered after fractionation of the
ether eluate on silica gel. The weights of the recovered
fractions and the solvents used are shown in Table 5. Although
dose-related mutagenicity was observed using strain TA100 in

Table 5.   Fractions of XAD-2 Extract Recovered from Silica Gel

Fraction	Solvent	Weight of Fraction (g)
1	Hexane	0.03
2	Hexane:Ether 9:1	0.01
3	Hexane:Ether 1:1	0.16
4	Ether	1.55
5	Methanol	2.55

Fractions 3, 4, and 5 (Figure 2), Fraction 3, containing compounds of moderate polarity, was by far the most mutagenic on a unit-weight basis.

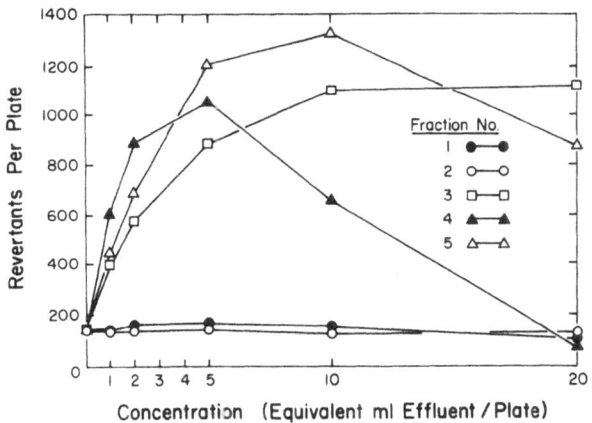

Figure 2.   Response of S. typhimurium strain TA100 to silica gel fractions of XAD-2 extract without S9 mix.

GC and GC/MS analyses. Preliminary analysis by GC revealed a large number of components in Fraction 3 (not shown). After methylation, a major peak at 2.16 min disappeared and a new peak appeared at 3.73 min. Other changes in the chromatogram after methylation were relatively minor. No evidence was obtained for the presence of significant amounts of neoabietic acid, a known mutagenic resin acid (Nestmann et al., 1980). Fractions 4 and 5 were less amenable to analysis by GC. Methylation of Fraction 4 revealed the presence of trichloro- and tetrachlorocatechol, known constituents of chlorination-stage effluent (Lindström and Nordin, 1976); however, peak heights in chromatograms of these two fractions were very low. Trichloro- and tetrachlorocatechol are not mutagenic (Rapson et al., 1980; Räsänen et al., 1977). Since Fraction 3 contained the most mutagenicity on a per-gram basis, this material was the first to be analyzed by GC/MS.

Mass spectra were obtained for 14 components in Fraction 3. Tetrachloroacetone (1,1,3,3,-isomer), diethyl succinate, diethyl methylsuccinate, 3- or 4-chlorobenzyl alcohol, and threo-9,10(10,9)-chlorohydroxystearate were identified by matching spectra from the EPA/NIH reference library, manual interpretation of mass spectra, and comparison of GC retention times with standards. Tetrachloroacetone was identified as the component that gave a major peak at 2.16 min and disappeared on methylation, because of the observation that the tetrachloroacetone standard also reacted with diazomethane to produce a new component whose retention time on GC was identical to that observed for the new peak, which appeared at 3.73 min in the chromatogram of Fraction 3 after methylation. It is known that haloaldehydes and ketones with electron withdrawing groups in the α-position react with diazomethane to form epoxides (Adams, 1954) which would result in similar shifts in GC retention times. Additional spectral data will be required to interpret the remaining mass spectra from Fraction 3. Manual interpretation of GC/MS data from Fraction 4 resulted in the identification of 7-oxodehydroabietic acid. Subfractionation of Fraction 3 on silica gel revealed a compound that was identified by X-ray crystallography as a new compound, 1-oxa-6,10-trichlorospiro[4,5]dec-6-en-8,9-dione (McKague et al., 1981).

Analysis of ether extracts of original chlorination-stage effluent. Ether extracts of the original effluent sample were analyzed for tetrachloro- and other chloroacetones. 1,1,3-Trichloro-, 1,1,3,3-tetrachloro-, and pentachloroacetone (Figure 3) were identified by GC/MS and comparison of retention times with standards. Subsequently, hexachloroacetone was identified using an electron capture detector on the GC instead of a flame ionization detector.

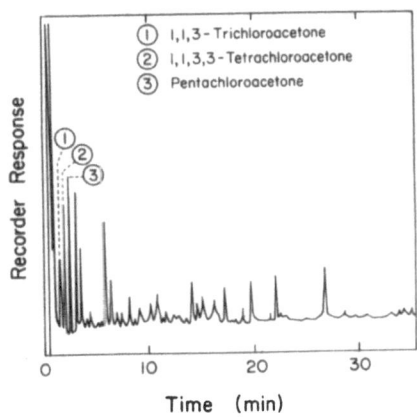

Figure 3. Gas chromatogram of ether extract from chlorination-
stage effluent.

Effect of pH on chloroacetones. Comparison of gas
chromatograms obtained after pH adjustment of the
chlorination-stage sample showed significant changes had occurred
in the composition of the constituents. Tetrachloro- and
pentachloroacetone were clearly present in the extract at pH 2.
After 1 h at pH 7, the peak due to pentachloroacetone had nearly
disappeared, while treatment for 1 h at pH 10 resulted in the
disappearance of both tetrachloro- and pentachloroacetone.
Recoveries of pure tetrachloroacetone from distilled water were
68% at pH 2, 40% at pH 7, and 0% at pH 10. The observation that
tetrachloroacetone disappeared more slowly than pentachloroacetone
at pH 7 agrees with the results of the analysis of Fraction 3, in
which only tetrachloroacetone was recovered, presumably because of
the greater pH lability of pentachloroacetone. The pH lability of
chloroacetones is in accord with the same phenomenon observed
generally for the mutagenicity of pulp and paper mill effluents
(Eriksson et al., 1979; Douglas et al., 1980), but provides only

circumstantial evidence for the contribution of chloroacetones to
the mutagenicity of the effluent. 1,1,3-Trichloroacetone,
1,1,3,3-tetrachloroacetone, pentachloroacetone, and
hexachloroacetone all show dose-related increases in mutagenicity
of Salmonella strain TA100 (Figure 4), trichloroacetone being the
most potent. The reduction in potency observed with increasing
chlorine substitution correlates with the lability of these
compounds at neutral and higher pH. Therefore, this trend in the
loss of mutagen potency may in part be due to degradation of the
test chemicals at the more or less neutral pH prevailing during
the conduct of the mutagenicity tests. Reduced mutagenicity was
also observed using strain TA98 for all four chloroacetones (not
shown). The presence of S9 mix reduced the mutagenicity in all
cases (not shown).

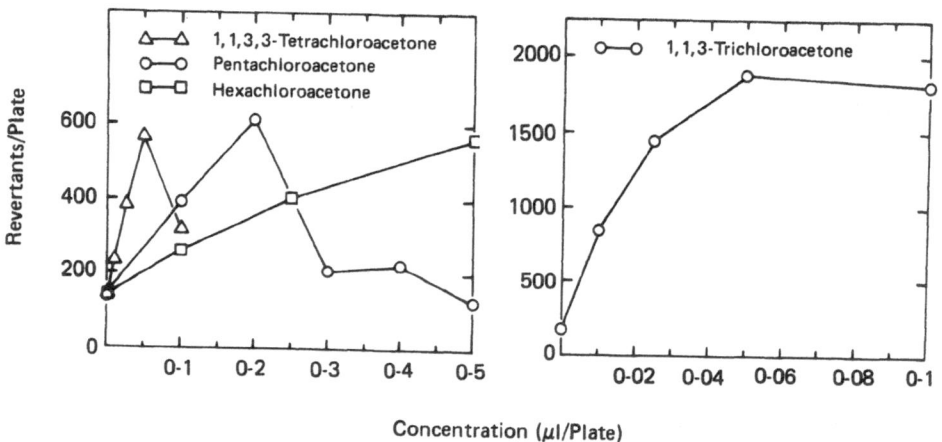

Figure 4.   Response of S. typhimurium strain TA100 to
             chloroacetones (without S9 mix).

The occurrence of alkaline labile chloroacetones in pulp and
paper mill effluents has been detected (Wong and Mortimer, 1979),
and their presence as mutagenic components has been reported
(Stockman et al., 1980; Kringstad et al., 1981; McKague et al.,
1981). Chloroacetones have also been found in finished drinking
water and can be derived from the chlorination of humic acid
(Christman et al., 1980) and resorcinol (Rook, 1980).

Furthermore, chloroacetones have been implicated in the formation
of chloroform (Rook, 1980; Gurol et al., 1982), a carcinogen
(National Cancer Institute, 1976) also found in drinking water
(Bellar et al., 1974; Rook, 1974). Some of the data on the
identification and mutagenicity of chloroacetones have been
published elsewhere (McKague et al., 1981).

Mutagenicity of chloroacetone mixtures. To gain insight into
the contribution of the individual chloroacetones to the overall
mutagenic activity of the chlorination-stage effluent used in the
present study, a mixture of the mutagenic chloroacetones in the
concentrations present (Table 6) was tested in Salmonella.

Table 6. Concentrations of Chloroacetones Found in First
Chlorination-Stage Effluent

Chloroacetones	mg/l
1,1,3-Trichloroacetone	2.1
1,1,3,3-Tetrachloroacetone	2.4
Pentachloroacetone	5.2
Hexachloroacetone	1.1

Table 7 shows that the mutagenicity of the mixture of
chloroacetones was enhanced over the expected response based on
the additive effect of the individual components. The response
(with the background subtracted) for 1 ml of the effluent per se
in strain TA100 is 960 rev/plate. Thus, at face value, even
without correcting for suboptimal recoveries, it would seem that
chloroacetones are a major contributor to the mutagenicity of the
chlorination-stage effluent examined in the present study.
1,1,3-Trichloroacetone, being the most mutagenic in Salmonella
strain TA100, appears to contribute the most of any of the
chloroacetones tested to the effect observed with the
chloroacetone mixture. However, despite the very appealing nature
of such a comparison, extreme caution must be exercised in drawing
general conclusions. For example, the addition of another
component to the mixture of chloroacetones might reduce the
mutagenicity to less than an additive response. Furthermore, the
activity of the effluent sample might be an underestimate because
of differences in handling the effluent and the mixture prior to

Table 7. Response[a] of $\underline{S.\ typhimurium}$ Strain TA100 to a Mixture[b]
of Chloroacetones for 1-ml Equivalent of Chlorination-
Stage Effluent without S9 Mix

Mixture	Expected[c]
312	161

[a]Revertants per plate with background subtracted.
[b]Based on concentrations in Table 6.
[c]Expected additive response estimated from initial slopes of
individual dose-response curves.

testing. Recently, Zochlinski and Mower (1981) reported an
interaction between hexachloroacetone and the solvent DMSO that
resulted in an enhancement of mutagenicity and toxicity when
tested in Salmonella. We have confirmed the existence of such an
interaction, which appears to be similar to the reaction of
trichloroacetic acid and DMSO, reported previously (Nestmann et
al., 1980). The extent to which these findings bear upon the
above interpretation has yet to be determined. However, based on
the relative reactivity and lability of chloroacetones, which
increase with chlorine substitution, we would predict that any
possible interactions with DMSO would diminish correspondingly
with increasing chlorine substitution in penta-, tetra-, and
trichloroacetone.

While chloroacetones seem to be quite ubiquitous in pulp and
paper mill effluents, their levels and, thus, their importance as
mutagenic constituents may vary widely from source to source. For
example, Kringstad et al. (1982) found levels of chloroacetones
6 to 120 times lower than in the present study. Such differences
illustrate the diversity and variation in the operation of
different pulp and paper mills, and they serve to emphasize the
need for caution in generalizing about the relative importance of
individual components. Nevertheless, as mentioned above,
chloroacetones are found in the environment outside pulp and paper
mills, and have been implicated in the formation of chloroform, a
carcinogen (National Cancer Institute, 1976). Therefore, further
studies are warranted to determine their importance as
environmental contaminants. Two other compounds,
3-chloro-4-dichloromethyl-5-hydroxy-2(5H)-furanone (Holmbom et
al., 1981) and 2-chloropropenal (Kringstad et al., 1982), have
been identified as major components of the mutagenicity of

chlorination-stage effluents; however, environmental levels of these substances are as yet unknown.

Mutagenicity of additional compounds identified. The following compounds, which were identified as being present in chlorination-stage effluent, gave negative responses when tested in S. typhimurium strains TA98 and TA100 with and without S9 mix: 3- or 4-chlorobenzyl alcohol; succinic acid; threo-9,10(10,9)-chlorohydroxystearic acid; and 7-oxodehydroabietic acid. Tri- and tetrachlorocatechol have been reported previously to be negative in the Salmonella/mammalian-microsome assay (Rapson et al., 1980; Räsänen et al., 1977). 7-Oxodehydroabietic acid is mutagenic in yeast (Table 1).

Genetic Activity of Effluent Extract in Microbial and Mammalian Assays

Microbial assays. For this portion of the study, bleached kraft chlorination-stage effluent from the same mill as the fractionation study was adsorbed to XAD-2 resin and eluted with ether, as described above. A preliminary experiment using Salmonella strains TA100 and TA98 and an amount of extract ranging from 5 to 5000 µg/plate showed no mutation induction at 5 µg, complete lethality at 5 mg, and dose-related mutagenicity at 50 and 500 µg (data not shown). The dose range was reduced and results from one of the two confirming experiments using all five standard strains (±S9) are shown in Table 8. Mutagenic responses are found for strains TA1535 and TA100, which revert by base substitution mutation, and for strain TA98, which reverts mainly by frameshift mutation. The extract apparently is not mutagenic in strains TA1537 and TA1538, which revert through frameshift mutations. In the absence of metabolic activation mix (S9), bacterial lethality occurs in strains TA1535 and TA98 at the dose 500 µg/plate, as shown by reductions in the numbers of revertants at this dose. Survival in strain TA100 is reduced by the highest dose (1 mg/plate). The presence of S9 appears to reduce the mutagenic and bactericidal effects of the extract, as shown by lower numbers of revertants for each nontoxic dose. More mutants are found in the presence of S9 at higher doses due to deactivation of the lethal effects observed in the absence of S9. To determine whether this is an enzymatic process or merely a random deactivation of active molecules by their binding to S9 proteins, another experiment was performed using normal S9 mix, S9 mix without the co-factors glucose-6-phosphate and NADP, and heat-inactivated S9 mix (58°C for 10 min). The same amounts of extract per plate were used as in Table 8 with strain TA100 as the test organism. All three S9 preparations showed the same results

Table 8. Mutagenicity of Effluent Extract in the Salmonella/Mammalian-Microsome Assay

Histidine Reversions/Plate[a]

Dose[b] (mg/plate)	TA1535 -S9	TA1535 +S9	TA100 -S9	TA100 +S9	TA1537 -S9	TA1537 +S9	TA1538 -S9	TA1538 +S9	TA98 -S9	TA98 +S9
0	9	10	62	82	5	7	9	19	11	25
0.05	33	14	336	138	8	11	8	18	48	30
0.20	61	24	880	374	9	12	14	15	142	52
0.50	10	71	1255	834	12	11	NG	20	7	120
1.00	NG	179	NG	1374	NG	11	NG	19	NG	185
Positive controls										
MNNG	1345		1400							
2NF							1951		1582	
2AA	13	455	140	1827	13	236	138	1160	161	2011
9AA					696					

aAverages of duplicate plates. NG means that no growth of revertants or background was observed on the plates.
bExtract and positive controls dissolved in dimethyl sulfoxide (spectrophotometric grade; Aldrich) which served as the negative (solvent) control.

as found with normal S9 mix in Table 8 (data not shown),
indicating that the S9-mediated deactivation of effluent extract
mutagenicity is not enzymatic but random binding to S9 proteins.
The above data confirm previous observations on the mutagenic
properties of chlorination-stage effluent (Douglas et al., 1980;
Rannug et al., 1981).

Yeast assays for gene conversion and mitotic recombination
with effluent extract were performed with growing cells, as
described in a previous section, according to the method of
von Borstel et al. (1981). The only difference was the addition
of sodium thiosulfate (3% final concentration) for 30 min to the
control and treated tubes, after the 24-h treatment period (at
30°C) and before washing the cells (2 times) with water. The
results from an experiment showing induction of gene convertants
in strain D7 are presented in Table 9. Compared to the background
level of convertants with the solvent control, the maximal
response (expressed as convertants/10^6 survivors) is a factor of
3.6 higher (1 mg extract/ml). The next dose (2 mg/ml) resulted in
a marked reduction of convertants, and survival was reduced below
detectable limits. These decreases were probably due to the
lethal effects of toxic chemicals in the effluent concentrate.
These results were confirmed in a subsequent experiment using the
same dose levels (data not shown). Table 10 shows results of a
separate experiment using strain D7 to assay for mitotic
recombinants. Dose-response increases were found for both
recombinants (twin-spot colonies) and other aberrant colonies (due
to nonspecific genetic events including DNA damage, aneuploidy,
and gene conversion), with maximum responses of 7.8-fold higher
twin-spots and 60-fold more aberrant colonies (per 10^3 survivors)
than for the DMSO control.

Mammalian assays. The chlorination-stage extract was tested
in CHO cells to detect the induction of chromosome aberrations,
SCE, and ASG-detectable DNA damage. Assays were conducted in the
absence of S9 mix only because of the well-demonstrated,
direct-acting mutagenicity of such material. Figure 5 shows the
effect of this mixture on SCE and chromosome aberrations. Both
SCE and chromosome aberrations showed dose-related increases;
however, the increase in SCE was proportionally much less than the
increase in the frequency of cells with chromosome aberrations.
The observation that concentrations above 25 mg/l were
sufficiently toxic to prevent recovery of second-division cells
(showing sister-chromatid differentiation) perhaps accounts for
the fact that a greater effect enhancement of SCE was not
obtained.

Table 9. Induction of Gene Conversion in Strain D7 with Effluent Extract

Dose (mg/ml)	Convertants[a] (per ml plated)	Survivors[b] (x 10^{-8} per ml)	Survival (%)	Convertants (per 10^6 survivors)
(Water control)	1379	1.88	100.0	7.34
(DMSO control)	1377	1.28	68.1	10.8
0.50	1919	2.05	109.0	9.36
0.70	3386	0.96	50.8	35.5
1.0	2055	0.53	28.0	39.0
2.0	24	0.47	0[c]	--
(EMS 1.18)	17880	1.60	85.1	112.0

[a]Counted on tryptophan omission medium in quintuplicate (0.2 ml/plate).
[b]Assayed on plates containing tryptophan in quintuplicate (0.2 ml/plate).
[c]Survival was reduced to a level below the detectable limits of the dilution plated.

Table 10. Induction of Mitotic Recombinants in Strain D7 with Effluent Extract

Dose (mg/ml)	Survival (%)	No. of Colonies Scored[a]	Total No. Aberrant Colonies	Aberrant Colonies per 10^3 Survivors	Total No. Twin-Spots	Twin-Spots per 10^3 Survivors
(Water control)	100	20768	3	0.14	1	0.05
(DMSO control)	90.3	18744	3	0.16	1	0.05
0.50	48.7	10120	10	0.99	2	0.20
0.70	37.3	7744	8	1.03	3	0.39
1.0	14.2	29488	33	1.12	9	0.31
2.0	2.0	210	2	9.52	0	0
(EMS 1.18)	78.0	4048	59	14.6	12	2.96

aAll colonies scored on YEPD medium (0.2 ml/plate), 40 plates/dose, except for the positive control (10 plates).

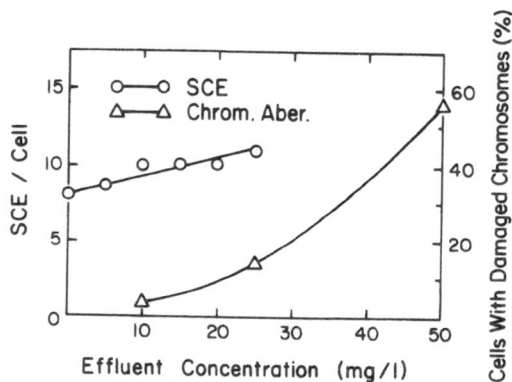

Figure 5.   Effect of chlorination-stage effluent on chromosome
             aberration and SCE in CHO cells without S9 mix.

        Table 11 shows the effect of chlorination-stage extract on
the induction of DNA lesions detectable by the ASG technique. As
with TCP and PCP, the extract caused no increase in this type of
damage. As well, the DNA lesions leading to the formation of
chromosome aberrations and SCE by this extract are not the type
detectable by ASG sedimentation. As with TCP and PCP, such a
relationship has been found in the case of triallate, a
thiocarbamate herbicide (Douglas et al., 1981). The results of
two preliminary experiments to determine if the chlorination-stage
extract induces HGPRT⁻ forward mutations are equivocal. While the
level of mutant phenotypes is increased up to five times the
background, a consistent dose-related effect has not been observed
(not shown). Rannug et al. (1981) have published data from
experiments using Chinese hamster V-79 cells that are in general
agreement with the preliminary findings which must be confirmed
through further careful experimentation.

        Table 12 shows the results obtained in a preliminary
experiment to determine the mutagenic activity of the
chlorination-stage extract <u>in vivo</u>. B6C3 mice given a single i.p.
injection at a dose equivalent to 80% of the LD_{50} were sacrificed

Table 11. Effect of Chlorination-Stage Extract on ASG-Detectable DNA Lesions

Concentration (mg/l)	ASG-Detectable Sites
1	0
10	0
25	0.056
50	0
100	0
2×10^{-3} M EMS	1.83

Table 12. Micronucleated Polychromatic Erythrocytes (per 1000) in Bone Marrow of B6C3 F_1 Mice after Treatment with Chlorination-Stage Effluent Extract[a]

Test Substance	Sex	Time after i.p. Injection (h) 24	48	72
Extract (200 mg/kg)	F	2.3	3.3	4.4
	M	3.3	5.3	6.0
Solvent control (DMSO)	F			1.5
	M			3.0
Untouched control	F			3.5
	M			3.5
Cyclophosphamide (75 mg/kg)	F	18.5		
	M	41.5		
DMBA (40 mg/kg)	F		12.5	
	M		9.0	

[a]2000 Polychromatic erythrocytes/group were scored for the presence of micronuclei (500/animal x 4).

at different times to determine if there was an optimal interval
for the expression of micronuclei in bone marrow polychromatic
erythrocytes. Taking male and female separately, there was no
statistically significant (based on Poisson statistics) effect at
any time point, although the reponses at 72 h appeared higher.
However, if the male and female data were pooled, the 72-h results
were significant at p < 0.02. These preliminary data must be
confirmed by subsequent experiments, but suggest that the
cytogenetic effects of the extract in vivo are much less than that
found in vitro. There are a number of possible reasons for this
difference in potency, including the action of metabolism on the
extract and/or pharmacokinetic factors that prevent high
concentrations of the extract or its metabolites from reaching the
bone marrow. Before generalizing about this reduction in potency
in vivo and its implications in assessing hazard compared to the
in vitro data, further experimentation will be necessary.

The chlorination-stage effluent under investigation in this
study exhibits a diversity of mutagenic or related activities.
Nevertheless, there are differences in its effect on seemingly
closely related end points which indicate the selectivity of its
action. Given that the effluent is a complex mixture of different
substances, such specificity would seem less likely unless there
are only a few components that contribute to its mutagenicity.

ACKNOWLEDGMENTS

The authors thank Ms. Kathy Nesbitt for her perseverance and
skill in typing the manuscript.

REFERENCES

Adams, R. 1954. Organic Reactions. John Wiley and Sons: New
    York. pp. 364-429.

Ames, B.N., J. McCann, and E. Yamasaki. 1975. Methods for
    detecting carcinogens and mutagens with the Salmonella/
    mammalian-microsome mutagenicity test. Mutation Res.
    31:347-365.

Ander, P., K.E. Eriksson, M.-C. Kolar, K. Kringstad, U. Rannug,
    and C. Ramel. 1977. Studies on mutagenic properties of
    bleaching effluents, Part 1. Svensk Papperstidn. 80:454-459.

Bellar, T.A., J.J. Lichlenberg, and R.C. Kroner. 1974. The occurrence of organohalides in chlorinated drinking waters. J. Am. Water Works Assoc. 66:703.

Bjørseth, A., G.E. Carlberg, and M. Møller. 1979. Determination of halogenated organic compounds and mutagenicity testing of spent bleach liquors. Sci. Total Environ. 11:197-211.

Callen, D.F., C.R. Wolf, and R.M. Philpot. 1980. Cytochrome P-450 mediated genetic activity and cytotoxicity of seven halogenated aliphatic hydrocarbons in Saccharomyces cerevisiae. Mutation Res. 77:55-63.

Christman, R.F., J.D. Johnson, F.K. Pfaender, D.L. Norwood, and M.R. Webb. 1980. Chemical identification of aquatic humic chlorinated products. In: Water Chlorination: Environmental Impact and Health Effects, Volume 3. R.L. Jolley, W.A. Brungs, and R.B. Cumming, eds. Ann Arbor Science: Ann Arbor. pp. 75-83.

CPAR Project Report No. 678-1. 1978. Biological characteristics of pulp mill effluents, Part 1. Environmental Protection Service, Department of the Environment: Ottawa, Ontario, Canada (available upon request).

Douglas, G.R., E.R. Nestmann, J.L. Betts, J.C. Mueller, E.G.-H. Lee, H.F. Stich, R.H.C. San, R.P. Brouzes, A.L. Chmelauskas, H.D. Paavila, and C.C. Walden. 1980. Mutagenic activity in pulp mill effluents. In: Water Chlorination: Environmental Impact and Health Effects, Volume 3. R.L. Jolley, W.A. Brungs, and R.B. Cumming, eds. Ann Arbor Science: Ann Arbor. pp. 865-880.

Douglas, G.R., E.R. Nestmann, C. Grant, R. Bell, J. Wytsma, and D.J. Kowbel. 1981. Mutagenic activity of diallate and triallate determined by a battery of in vitro mammalian and microbial tests. Mutation Res. 85:45-56.

Ellenton, J.A., G.R. Douglas, and E.R. Nestmann. 1981. Mutagenic evaluation of 1,1,2,3-tetrachloropropene, a contaminant in pulp mill effluents, using a battery of in vitro mammalian and microbial tests. Canad. J. Genet. Cytol. 23:17-25.

Eriksson, K.E., M.-C. Kolar, and K. Kringstad. 1979. Studies on mutagenic properties of bleaching effluents, Part 2. Svensk Paperstidn. 82:95-104.

Gurol, M.D., S. Myers, A. Wowk, and I.H. Suffet. (in press).
    Kinetics and mechanism of haloform formation: chloroform
    formation and trichloroacetone. In: Water Chlorination:
    Environmental Impact and Health Effects, Volume 4.
    R.L. Jolley, W.A. Brungs, and R.B. Cumming, eds. Ann Arbor
    Science: Ann Arbor.

Holmbom, B.R., R.H. Voss, R.D. Mortimer, and A. Wong. 1981.
    Isolation and identification of an Ames-mutagenic compound
    present in kraft chlorination effluents. Tappi 64:172-174.

Howard, T.E., and C.C. Walden. 1971. Toxicity and BOD
    characteristics of kraft pulp mill wastes. Pulp Paper Mag.
    Canada 62:T3-T14.

International Joint Commission. 1981. Annual Report, Committee
    on the Assessment of Human Health Effects of Great Lakes
    Water Quality.

Kringstad, K.P., P.O. Ljungquist, F. de Sousa, and L.M. Stromberg.
    (in press). Contributions of some chlorination aliphatic
    compounds to the mutagenicity of spent kraft pulp
    chlorination liquor. In: Water Chlorination: Environmental
    Impact and Health Effects, Volume 4. R.L. Jolley,
    W.A. Brungs, and R.B. Cumming, eds. Ann Arbor Science: Ann
    Arbor.

Kringstad, K.P., P.O. Ljungquist, F. de Sousa, and L.M. Stromberg.
    1981. Identification and mutagenic properties of some
    chlorinated aliphatic compounds in the spent liquor from
    kraft pulp chlorination. Environ. Sci. Technol. 15:562-566.

Lindström, K. and J. Nordin. 1976. Gas chromatography-mass
    spectrometry of chlorophenols in spent bleach liquors.
    J. Chromatog. 128:13-26.

McKague, A.B., E.G.-H. Lee, and G.R. Douglas. 1981.
    Chloroacetones: mutagenic constituents of bleached kraft
    chlorination effluent. Mutation Res. 91:301-306.

McKague, A.B., S.J. Rettig, J. Trotter, and G.R. Douglas. 1981.
    Structure of a trichlorocyclohexenone from bleached kraft
    chlorination-stage effluent. Canad. J. Chem. 59:3372-3375.

National Cancer Institute. 1976. Carcinogenesis Bioassay of
    Chloroform. Natl. Tech. Inf. Service No. PB264018/AS.

Nestmann, E.R., I. Chu, T.I. Matula, and D.J. Kowbel. 1980.
     Short-lived mutagen in Salmonella produced by reaction of
     trichloroacetic acid and dimethyl sulphoxide. Canad. J.
     Genet. Cytol. 22:35-40.

Nestmann, E.R., and E.G.-H. Lee. 1981. Mutagenicity of compounds
     identified in pulp mill effluents in Saccharomyces
     cerevisiae. (abstr.). Environ. Mutagen. 3:392.

Nestmann, E.R., E.G.-H. Lee, T.I. Matula, G.R. Douglas, and
     J.C. Mueller. 1980. Mutagenicity of constituents identified
     in pulp and paper mill effluents using the Salmonella/
     mammalian-microsome assay. Mutation Res. 79:203-212.

Nestmann, E.R., E.G.-H. Lee, J.C. Mueller, and G.R. Douglas.
     1979. Mutagenicity of resin acids identified in pulp and
     paper mill effluents using the Salmonella/mammalian-microsome
     assay. Environ. Mutagen. 1:361-369.

O'Neill, J.P., and A.W. Hsie. 1979. CHO/HGPRT mutation assay:
     Experimental procedure and adaptation for mutagen screening.
     In: Mammalian Cell Mutagenesis: The Maturation of Test
     Systems. A.W. Hsie, J.P. O'Neill, and V.K. McElheny, eds.
     Cold Spring Harbor Laboratory: Cold Spring Harbor, NY.
     pp. 311-318.

Rannug, U. 1980. Mutagenicity of effluents from chlorine
     bleaching in the pulp and paper industry. In: Water
     Chlorination: Environmental Impact and Health Effects,
     Volume 3. R.L. Jolley, W.A. Brungs, and R.B. Cumming, eds.
     Ann Arbor Science: Ann Arbor. pp. 851-863.

Rannug, U., D. Jenssen, C. Ramel, K.-E. Eriksson, and
     K. Kringstad. 1981. Mutagenic effects of effluents from
     chlorination bleaching pulp. J. Tox. Environ. Health
     7:33-47.

Rapson, W.H., M.A. Nazar, and V.V. Butsky. 1980. Mutagenicity
     produced by aqueous chlorination of organic compounds. Bull.
     Environ. Contam. Toxicol. 24:590-596.

Räsänen, L., M.L. Hattula, and A.U. Arstila. 1977. The
     mutagenicity of MCPA and its soil metabolites, chlorinated
     phenols, catechols and some widely used slimicides in
     Finland. Bull. Environ. Contam. Toxicol. 18:565-571.

Rook, J.J. 1974. Formation of haloforms during chlorination of
     natural waters. Water Treat. Exam. 23:234-243.

Rook, J.J. 1980. Possible pathways for the formation of
    chlorinated degradation products during chlorination of human
    acids and resorcinol. In: Water Chlorination:
    Environmental Impact and Health Effects, Volume 3.
    R.L. Jolley, W.A. Brungs, and R.B. Cumming, eds. Ann Arbor
    Science: Ann Arbor. pp. 85-98.

Salamone, M., J. Heddle, E. Stuart, and M. Katz. 1980. Towards
    an improved micronucleus test: studies on 3 model agents,
    mitomycin C, cyclophosphamide and dimethylbenzanthracene.
    Mutation Res. 74:347-356.

Stockman, L., L. Strömberg, and F. de Sousa. 1980. Mutagenic
    properties of bleach plant effluents: present state of
    knowledge. Cell. Chem. Technol. 14:517-526.

Swanson, A.B., D.D. Chambliss, J.C. Blomquist, E.C. Miller, and
    J.A. Miller. 1979. The mutagenicities of safrole,
    estragole, eugenol, trans-anethole, and some of their known
    or possible metabolites for Salmonella typhimurium mutants.
    Mutation Res. 60:143-153.

von Borstel, R.C. 1981. The yeast Saccharomyces cerevisiae: an
    assay organism for environmental mutagens. In: Short-Term
    Tests for Chemical Carcinogens. H.F. Stich and R.H.C. San,
    eds. Springer-Verlag: New York. pp. 161-170.

von Borstel, R.C., M.M. Shahin, and R.D. Mehta. 1981. Protocol
    for a haploid yeast reversion test for assaying mutagens.
    In: Short-Term Tests for Chemical Carcinogens. H.F. Stich
    and R.H.C. San, eds. Springer-Verlag: New York.
    pp. 171-174.

Walden, C.C., and T.E. Howard. 1977. The toxicity of pulp and
    paper mill effluents, a review of regulations and research.
    Tappi 50:122-125.

Wande, C.C. 1976. The toxicity of pulp and paper mill effluents
    and corresponding measurement procedures. Water Res.
    10:639-664.

Wong, A., and R. Mortimer. 1979. Isolation and identification of
    toxicants present in effluents derived from the pulping and
    bleaching of western red cedar. CPAR Project Report
    No. 711-2. Environmental Protection Service, Department of
    the Environment: Ottawa, Ontario, Canada (available upon
    request).

Zimmerman, F.K., R. Kern, and H. Rasenberger. 1975. A yeast strain for simultaneous detection of induced mitotic crossing over, mitotic gene conversion, and reverse mutation. Mutation Res. 28:381–388.

Zochlinski, H., and H. Mower. 1981. The mutagenic properties of hexachloroacetone in short-term bacterial mutagen assay systems. Mutation Res. 89:137–144.

EVALUATION IN MAN AND ANIMALS OF TESTS FOR THE DETECTION OF POPULATION EXPOSURES TO GENOTOXIC CHEMICALS

Jonathan B. Ward, Jr.,[1] Marvin S. Legator,[1]
Michael A. Pereira,[2] and Lina W. Chang[2]

[1]Division of Environmental Toxicology, Department of
Preventive Medicine and Community Health, University of Texas
Medical Branch, Galveston, TX 77550 and [2]Health Effects
Research Laboratory, U.S. Environmental Protection Agency
Cincinnati, OH 45219

INTRODUCTION

Need for Genetic Monitoring of Populations Exposed to Environmental Mutagens

Over the past several decades, our increasing use of and
reliance upon fossil fuels and synthetic chemicals have resulted
in rapidly increasing human contact with chemicals not previously
found in the environment. Acute toxic effects of chemical
exposure have usually been recognized quickly and steps taken to
minimize hazards. The effects of chronic exposures or delayed
effects of exposure have been more difficult to recognize.

The cellular genetic apparatus is susceptible to damage
produced by a wide variety of chemical and physical agents present
in the environment. Heritable mutations can produce various
adverse health effects. Cancer is the major outcome associated
with somatic cell mutation (Green, 1979), while adverse
reproductive outcomes are the major concern in germinal cell
mutation (Wilson, 1977). It should be noted that the outcomes of
heritable mutation are not usually observable immediately
following exposure to environmental chemicals which cause, or
contribute to, their appearance.

Toxic effects on the genetic apparatus differ from toxicity in other cellular systems in a significant way. The major manifestations of mutation in man and higher organisms include reproductive effects (infertility, spontaneous abortion, and birth defects) and cancer. These effects result from mutations produced in a single cell with which the mutagenic agent has interacted. Furthermore, mutations, once fixed, are irreversible. Until the appearance of these "outcomes," the specific damage causing them remains undetectable. The medical consequences of mutagen exposure may not be observed until the conception of the next generation or until a lengthy period of latency preceding malignancy has occurred.

Two basic approaches have been used in evaluating the mutagenic potential of chemicals or complex mixtures of environmental samples. The first uses laboratory tests with animals, or in vitro test systems in which the test organism is exposed to controlled levels of an agent, and cancer, mutation, or various manifestations of genetic damage are observed. This approach allows the toxic potential of an agent to be explored in detail but does not actually define what occurs in exposed humans. The second approach uses epidemiological methods to look for excessive incidence of deleterious outcomes such as cancers, birth defects, and spontaneous abortions in human populations. The latency inherent in the appearance of these outcomes and the complexity of typical human environments frequently make it difficult or impossible to document exposures that might have contributed to the incidence of disease. In addition, increases over high background incidence rates can be detected only by studying relatively large populations. Epidemiologic studies describe events in human populations only after the appearance of outcomes. As a result, such studies offer no protection to the population studied and permit long periods of exposure to other populations before identifying a hazard.

A significant problem not directly addressed by either of these approaches is the detection of human exposure to mutagenic substances at an early stage, well before outcomes are manifested and in time to reduce the probability of disease by reduction of exposure. Observations made in animals suggest a possible solution to the problem. Although fixed heritable mutations occur with a low frequency (10^{-5} to 10^{-6} per cell generation) and are technically difficult to demonstrate in intact animals, other manifestations of damage to the genetic apparatus occur at much higher frequencies following exposure to genetically toxic agents. The types of damage observed may include chromosome damage, physical damage to DNA such as strand breaks or cross-links, induced abnormalities in sperm morphology and number, and the excretion of mutagenic substances detectable by using in vitro mutagenicity tests to analyze urine samples. These end points can be evaluated in specimens of human blood, semen, or urine.

A growing number of studies have been conducted in human populations exposed to mutagenic agents in which parameters such as chromosome damage or sperm morphology were evaluated, with the result that elevated frequencies of abnormalities were observed. For example, the International Agency for Research on Cancer has categorized 18 chemicals as carcinogenic for humans, 18 as probably carcinogenic for humans, and an additional 18 as suspected but with insufficient data for classification (IARC, 1979). Among these compounds, 5 in the first group: arsenic (Beckman et al., 1979; Nordenson et al., 1978), benzene (Tough et al., 1970; Picciano et al., 1979), bis (chloromethyl) ether and chloromethyl methyl ether (Zundova and Landa, 1979), and vinyl chloride (Funes-Cravioto et al., 1975; Ducatman et al., 1975) have been evaluated for their ability to produce chromosome damage in man, and were found positive. Similarly, 3/3 compounds tested in the second category: cyclophosphamide (Etteldorf et al., 1976), epichlorohydrin (Kucerova et al., 1977; Picciano, 1979), and thiotepa (Silezneva and Korman, 1973) have been checked and shown to cause chromosome damage in man. In the third category, 4/4 compounds tested: chloroprene (Katosova, 1973; Sanotskii, 1976), ethylene oxide (Garry et al., 1979), lead (Forni and Secci, 1972; Garza-Chapa et al., 1977), and styrene (Meretoja et al., 1977) caused chromosome damage in man. The limited data available suggest that human cytogenetic analysis can accurately detect exposure to carcinogens.

It must be clearly understood that these changes are not adverse health outcomes in themselves. Rather, these are observable changes which we know from controlled animal studies, supported by the limited human data available, to be produced by agents capable of producing heritable mutations and genetic disease. The presence of increased levels of these changes is an indication of an individual's exposure to a mutagenic agent. The exposure may produce widespread genetic damage, the bulk of which is resolved through repair processes or cell death. Only a small residual of the initial damage may result in fixed mutations or other alterations which contribute to the process of neoplasia. Abnormalities such as chromosome aberrations are, in fact, by-products of the initial interaction of mutagenic agents with cellular DNA.

In developing a strategy for human genetic damage monitoring, several important conditions must be considered in study design. First, results must be interpreted on a group, rather than an individual, basis. Individual responses to the same exposure can differ substantially for intrinsic biological reasons, such as metabolic characteristics and variations in DNA repair. In addition, human exposures to specific agents are not isolated events, but are components of complex environments that can differ markedly among individuals. Factors such as nutritional status, life-style, health status, and use of medications can influence responses.

To make meaningful associations between observed genetic damage and specific chemicals or environmental complex mixtures, a population with a common exposure experience must be carefully identified. Whenever possible, environmental measurements should be made to determine both the identity and concentrations of agents in the study environment. Confounding environmental factors and interindividual variations must be controlled by the selection and simultaneous evaluation of a control population matched for as many variables as possible, other than the exposure of interest. Individual variations over time can be controlled by sampling the exposed and control populations more than once. As many tissues and end points as possible should be examined, but tests inappropriate for the type of exposure or which are redundant should not be included. For example, radiations do not produce metabolites, so tests for excreted mutagens or adduct formation would be inappropriate.

Human genetic monitoring could be a particularly appropriate technique in situations involving complex environmental mixtures. In these circumstances, the reconstruction of environmental conditions in laboratory tests may be difficult or impossible. Furthermore, the effects in man of interactions between agents may not be reproducible in animal or in vitro systems. Only by examining human populations exposed to the specific environment may the presence of a genetic risk be documented.

Available Tests Applicable to Man

Cytogenetic tests. Since the early 1960s, cytogenetic analysis in animals and humans has played a central role in most programs designed to detect mutagenic agents. The cytogenetic effects in man of known or suspected human carcinogens have been discussed above. The human lymphocyte is long lived in the peripheral blood (Normal et al., 1966; Buckton et al., 1977; Dolphin et al., 1973; Bloom et al., 1966), tolerant of some types of genetic damage when not proliferating (Nowell, 1965), and is easily obtained. Consequently, it is a very suitable target cell in which to look for cumulative damage resulting from long periods of exposure. The techniques for culturing lymphocytes and analyzing for chromosome aberrations are now well established (Evans and O'Riordan, 1977).

The development of the BudR/Hoechst 33258 and Giemsa staining technique for visualizing sister chromatids (Latt, 1974) has facilitated the development of assays for induced DNA damage resulting in sister chromatid exchanges (SCEs). A wide variety of chemical mutagens have been shown to induce SCEs in several types of mammalian cells, including human lymphocytes (Latt et al., 1977).

Hemoglobin alkylation. Most mutagenic and carcinogenic chemicals are electrophilic agents or are converted to electrophilic agents in vivo. Alkylating agents react with nucleophilic centers in DNA (guanine-N-7, guanine-O-6, adenine-N-3, etc.), but also react with nucleophilic centers in proteins, such as cysteine-S and -N-1, or -N-3 of the imidazole ring of histidine. Ehrenberg and his colleagues have developed techniques and principles to determine the degree of alkylation of specific nucleophilic sites in macromolecules and to calculate the tissue-specific dose of an agent in an exposed animal (Ehrenberg and Osterman-Golkar, 1980). Techniques using radiolabeled compounds have been used in animals to determine the degree of alkylation of amino acids of hemoglobin in erythrocytes. The dose (time integral of concentration of the chemical) can be calculated from the degree of alkylation (Osterman-Golkar et al., 1976; Osterman-Golkar et al., 1977). More recently, non-isotopic techniques have been developed for measuring the degree of hemoglobin alkylation in man (Calleman et al., 1979). Alkylated hemoglobin can accumulate over the life span of the erythrocyte, allowing exposures accumulated over time to be detected. The technique is relatively insensitive to confounding influences of incidental exposures or biological effects not related to the exposure of primary interest. This is because the end point measured is the formation of a specific adduct to the target amino acid. Consequently, the determination of hemoglobin alkylation can be a powerful technique for evaluating chronic occupational exposures to specific chemicals.

Sperm analysis. Sperm in a semen specimen is the one germinal cell type available in large number and not requiring any invasive procedures. Several different observations can be made on sperm that may reveal the impact of mutagenic activity on their development. Over the last few years, two assays have been developed which detect abnormalities in human sperm from donors who have been exposed to mutagenic agents.

One procedure detects Y-chromosome nondisjunction in sperm
(Kapp and Jacobson, 1980). The background and basis for the test
are described by Kapp and Jacobson (1980). The Y chromosome
appears as a fluorescent spot in quinacrine-stained sperm. In
good preparations, about 40 to 50% of sperm heads are observed to
contain a Y-fluorescent (YF) body as expected on genetic grounds.
Sperm containing two fluorescent bodies (YFF) have been observed
in about $0.7 \pm 0.7\%$ of sperm from donors with no known chemical
exposures. The presence of two YF bodies indicates the occurrence
of nondisjunction of the Y chromosome in meiotic anaphase II (Kapp
and Jacobson, 1980). Exposures to several known mutagenic agents,
including adiramycin, X irradiation, and the nematocide
dibromochloropropane (DBCP), have produced increases in YFF
frequency.

The second assay detects agents that increase the frequency
of sperm with abnormal morphologies (Wyrobeck and Bruce, 1978).
The normal human sperm-head shape is distinctive and changes are
easily recognized. In mice, several lines of evidence indicate
that sperm-head shaping is under rigorous genetic control.
Studies of strains with different head shapes and hybrids among
them indicate that about 10 genetic regions on the X and Y
chromosomes as well as autosomes control sperm morphology. In
addition, sperm-head abnormalities are induced in mice by X rays
and several clinical mutagens. Spermatocytes and late
spermatogonia are the most sensitive cell types. The
f_1 generation of treated males also has increased rates of
abnormality, suggesting that induced abnormalities are heritable.
Sperm abnormalities do not correlate with the presence of
chromosome abnormalities such as translocations, indicating that
point mutations may be responsible for the occurrences (Wyrobek,
1979).

Urine analysis. The analysis of urine in animals for
excreted mutagens using various microbial-indicator organisms is a
widely used technique (Gabridge et al., 1969; Durston and Ames,
1974; Commoner et al., 1974). A number of studies in humans has
also evaluated the urinary excretion of mutagens following
exposures to various drugs (Siebert and Simon, 1973; Minnich et
al., 1976; Legator et al., 1975). The appearance of mutagens
usually occurs rapidly following exposure and diminishes quickly
when exposure ends (Connor et al., 1977).

DNA filter elution. Alkaline filter elution of DNA detects
damage based on the fact that alkaline denaturing and unwinding of
double-stranded DNA and the release of single strands occurs at a
rate inversely proportional to single-strand size (Kohn et al.,
1976). In practice, DNA from mutagen-damaged cells is eluted

through a filter under alkaline conditions and collected in fractions. The rate of DNA elution, determined from the DNA content of each fraction, is more rapid if the DNA contains single-strand breaks than for undamaged DNA.

Alkaline elution has been used successfully to detect carcinogens in whole animal studies (Petzgold and Swenberg, 1978; Parodi et al., 1978). Parodi et al. (1978) recently adapted the alkaline elution technique for use with the microfluorometric method of Kissane and Robins (1958) for quantitation of extremely small amounts of DNA. It is also possible to determine the degree of DNA-interstrand linkage and DNA-protein linkage caused by a test agent (Ross and Shipley, 1980). Because the alkaline elution technique can now be employed with mammalian cells whose DNA has not been labeled by incorporation of isotopic precursors, the technique appears to be ideally suited to the monitoring of human populations for exposure to genetically active agents.

Rationale for Coordinated Testing in Man and Animals

By using several of the methods described above, a battery of tests can be developed for the investigation of human exposures to mutagenic environments. Careful test selection can allow the efficient observation of several types of genetic end points. Both somatic and germinal cells can be observed, and genetic damage at the chromosome level and at the molecular level can be detected. In well-defined circumstances, a single test might be suitable for monitoring human exposure to specific chemicals.

We are conducting a study in which a human population is being monitored for possible genetic effects resulting from exposure to formaldehyde. The study was designed to coordinate the human monitoring with parallel assays in animals and in vitro for several reasons. Unless the chemical nature of the human environment is well described and the principal agents have been previously well studied, it may not be possible to predict in advance which tests are most appropriate for detecting exposure. Preliminary studies of individual agents or environmental samples in animals may greatly improve the selection of tests for human monitoring. In the presence of exposure to a complex mixture even when its components have been identified, human monitoring alone may not identify the component most responsible for mutagenic activity. Animal studies can be used to evaluate the activities of individual agents. One of the primary reasons for conducting animal tests in coordination with human monitoring is to establish the dose response for specific end points in the animal. The magnitude of effects in human subjects can be related to responses

in animals in order to estimate the exposure level. When the
exposure level to humans is known, this comparison can be used to
evaluate and determine differences in animal and human
susceptibility. In addition, comparison of human and animal
effects may provide an indication of degree of risk posed by the
human exposure (Pereira et al., 1981).

For some tests that must be modified to detect the effects of
specific agents, animal or in vitro studies may be necessary to
establish methods and conditions for conducting human studies.
For example, urine testing for the presence of mutagens may
require the use of extraction or concentration procedures. Animal
or in vitro studies may be needed to optimize or validate these
procedures and may facilitate the identification of active
metabolites. Hemoglobin alkylation monitoring in humans requires
that a specific adduct be identified and that techniques for its
isolation and measurement be established. Preliminary studies in
vitro or in animals are necessary to accomplish these objectives.

The primary objective of this study was to develop and
improve techniques for monitoring human exposures to mutagenic
substances. A major means to this end was the development of a
study design that would coordinate human, animal, and in vitro
studies using a population with a definable exposure to a
mutagenic substance. A population exposed occupationally to
formaldehyde was chosen as the test case for several reasons.
Evidence accumulated over the last several years indicated that
formaldehyde is mutagenic in some test systems (Auerbach et al.,
1977) and produces DNA damage in many organisms, including
mammalian cells (Ross and Shipley, 1980). Furthermore, recent
studies suggest that formaldehyde is carcinogenic in rats
(Swenberg et al., 1980). Beyond that, several practical
considerations made the study of formaldehyde exposure appealing
as an initial case. An exposed population was available within
our institution and both the institution and individuals involved
were prepared to cooperate in a study. A large population of
individuals from whom to select a control population was readily
available. Methods for environmental monitoring for formaldehyde
were established and made available to us; radiolabeled
formaldehyde, which would be needed particularly for the
development of hemoglobin alkylation techniques, was commercially
available. The direct alkylating activity of formaldehyde has the
practical advantage of making the development of a hemoglobin
alkylation assay simpler than would be the case if metabolic
activation was required. This paper will concentrate particularly
on the study design and the way in which it meets the criteria
discussed earlier and on the protocol established for the
coordinated acquisition and processing of samples.

METHODS

Study Design

The study was designed as a joint project between the authors at the University of Texas Medical Branch (UTMB) and at the Health Effects Research Laboratory (HERL) of the U.S. Environmental Protection Agency (EPA) in Cincinnati. Human studies were to be performed at UTMB, while coordinated studies in animals would be conducted at HERL. The Autopsy Service in the Department of Pathology was identified as a test population and approval of the institution for the study was secured. Exposed and control populations were recruited and subjects matched in pairs. The populations were sampled in matched pairs three times at intervals of about two months for blood, urine, and semen. Blood provided lymphocytes for cytogenetic analysis and DNA damage assays, and erythrocytes for hemoglobin alkylation determination. Urine was assayed for the presence of substances mutagenic to Salmonella typhimurium. The sperm count and viability were determined in semen, as well as the frequency of morphologically abnormal sperm. The frequency of nondisjunction of the Y chromosome was also measured. Over the same time period, an environmental monitoring program was conducted obtaining both area and individual data on formaldehyde concentration in air. Studies of formaldehyde binding to hemoglobin, both in vitro and in animals, were undertaken in both laboratories to identify the alkylation adducts formed, to standardize a method for the quantitation of the predominant adduct, and to determine the rate constant of alkylation. At HERL, animal studies were also initiated to evaluate end points equivalent to those evaluated in man. The studies included analysis of chromosome aberrations, sister chromatid exchanges and the micronucleus test in bone-marrow cells of mice, alkylation of hemoglobin in erythrocytes, analysis of urine for excreted mutagens, and sperm morphology analysis. The tests included in the overall study and the information transferred between laboratories are diagrammed in Figure 1.

Population Selection

The formaldehyde-exposed human population was recruited from the Autopsy Service at UTMB by the use of posters and return response cards. Each respondent completed a questionnaire providing information on age, sex, work area, estimated exposure and job description, other sources of exposure to mutagens, smoking, drinking and recreational drug-use habits, health history, and medication use. A control population was recruited from the UTMB community using posters and return response cards

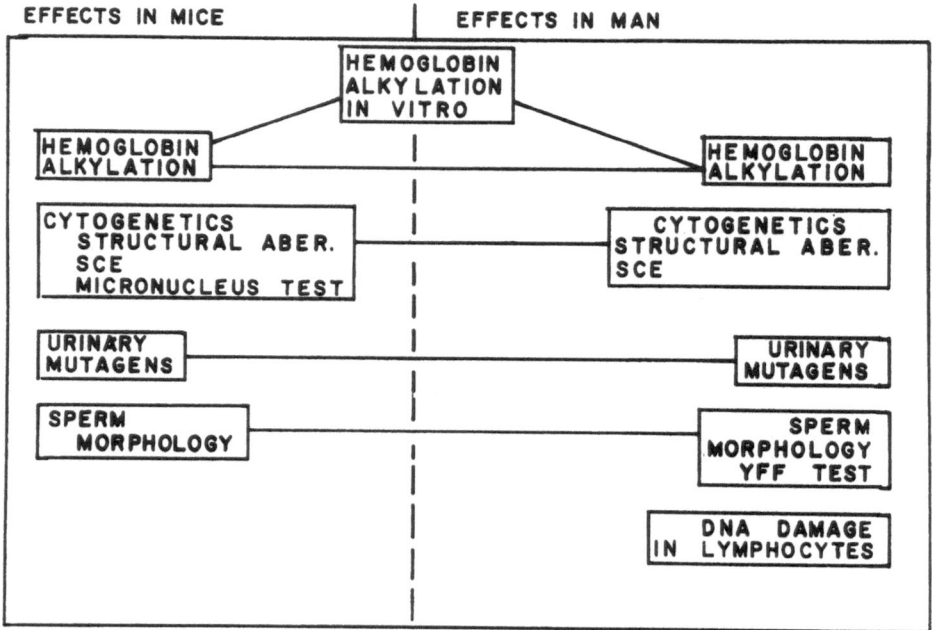

Figure 1.   Coordinated assays for genetic damage conducted in mice
            and in human subjects. Mouse studies are conducted at
            HERL and human studies are done at UTMB. Lines
            indicate information transfer for comparison of animal
            and human results.

supplemented by publicity in the University newspaper.
Respondents completed questionnaires similar to those used by the
exposed subjects. Each exposed subject was paired with a control
subject on the basis of information obtained from the
questionnaires. Subjects were matched primarily on age and sex,
then smoking and drinking habits. Control recruits were excluded
as much as possible if they had unusual environmental exposures, a
recent history of formaldehyde exposure, or recent or
long-standing use of medications likely to interfere with results.

Sample Acquisition

A sample-processing protocol was devised to allow samples to be collected in a uniform manner within the constraints of the manpower and equipment available in the laboratory. Two matched pairs of subjects were sampled in each session and two sampling sessions were conducted weekly. Subjects were provided containers on the first day of the session for urine and semen. The urine specimen was returned later in the day. The semen specimen was collected by the subject at home the following morning and returned within 2 h to the laboratory. The blood sample was split, half being allowed to settle for the collection of lymphocyte-rich plasma (buffy coat) for cytogenetic analysis. The other half was used for DNA damage and hemoglobin alkylation studies. Each pair of subjects was sampled in March or April, 1981, again in June or July, 1981, and again in August or September, 1981. A few new subjects were added to the study in August and December, 1981, and are still being sampled.

Environmental Monitoring

Room and personal air sampling was conducted with DuPont P200 pumps and two midget impingers containing distilled water connected in series with Tygon tubing. The pumps were calibrated before and after sampling with a Kurz model 541S-A mass flow meter. Water from the impingers was analyzed for formaldehyde using the chromotropic acid method recommended by NIOSH (1977). Area samples were taken by locating the sampling apparatus in various locations in the autopsy suite and collecting for an average of 4.5 h. Personal samples were obtained by attaching the impingers to the subject's shirt or coat collar so that breathing zone air could be obtained. The subject carried the pump on his (her) belt or in a lab coat pocket. Duration of sampling was 2-8 h, with an average of 4.3 h. One pathologist, one resident, and one technician each were sampled for five consecutive days. Other exposed subjects were sampled one or two times during the period of biological sampling.

Genetic Monitoring Tests

Cytogenetics. Lymphocytes were cultured from the buffy coat of gravity-sedimented whole heparinized blood. Cultures were stimulated with phytohemagglutinin and cultured for 72 h. Bromodeoxyuridine (10 μM) was added to cultures to be used for SCE analysis. Cultures were blocked with Colcemid (0.08 μg/ml) for 2 h prior to harvest. Cells were fixed, mounted, and analyzed for

structural aberrations by the method of Evans and O'Riordan (1977)
and for SCE analysis essentially by the Giemsa method of Latt et
al. (1977).

DNA damage assay. DNA damage in the form of single-strand
breaks or alkali-labile sites was assayed by the alkaline filter
elution technique of Parodi et al. (1978). Lymphocytes were
separated from whole blood on Ficoll-Hypaque gradients and washed
in minimum essential medium prior to elution. The DNA eluted was
collected in 1-ml fractions and assayed by the method of Kissane
and Robins (1958). Results were expressed as the percent of DNA
remaining on the filter at each fraction.

Hemoglobin alkylation. The identification and quantitation
of adducts formed by formaldehyde with hemoglobin were
accomplished by the techniques developed by Ehrenberg and
Osterman-Golkar (1980). Preliminary studies in vitro using
^{14}C-formaldehyde followed the techniques of Osterman-Golkar et al.
(1976) and Pereira and Chang (1981). After treatment of
erythrocytes with two 100-µl portions of 5 mg/ml sodium
borohydride to stabilize formaldehyde adducts, globin was prepared
and hydrolyzed as described previously (Truong et al., 1977). In
the first experiment, adducts were detected by amino acid
analysis. In subsequent experiments, the hydrolysate was
chromatographed on Dowex 50WX4 columns eluted with 1 and 2 M HCl
prior to amino acid analysis. Globin from test subjects was
prepared in the same way and stored frozen. The nonisotopic
techniques for quantitating adducts followed those used by
Calleman et al. (1979).

Urine analysis. Urine samples of approximately 200 ml were
analyzed for pH and specific gravity, then centrifuged to remove
sediment and filter-sterilized with a 0.45-µm Nalgene filter. A
100-ml portion was incubated overnight at 37°C with 1000 units of
β-glucuronidase. Samples were extracted on XAD-2 resin by the
method of Yamasaki and Ames (1977). The organic fraction was
eluted from the XAD-2 column with acetone, concentrated by
evaporation, and taken up in DMSO. The concentrate was tested for
mutagenicity by the method of Ames et al. (1975).

Sperm evaluation. Sperm count, viability, and motility were
determined on receipt of the semen samples. Sperm count was
determined by Coulter counter or microsopically. Viability was
determined by trypan blue exclusion at 4 and 24 h, and motility
was determined by microscopic observation at 2 h. Air-dried sperm
smears were prepared for morphology studies and the determination
of Y-body frequency. Slides for morphology studies were stained
with Papanicolaou stain and evaluated by the method of Wyrobek and

Bruce (1978). Slides for Y-body analysis were stained with quinacrine mustard immediately prior to examination and evaluated by fluorescence microscopy using the criteria of Kapp and Jacobson (1980).

RESULTS

A study of this type has many facets which progress at different rates. While the project is not completed at this point, some objectives have been substantially met. We will present results from several parts of the study which indicate the way in which the different parts of the study interact with each other.

Hemoglobin Alkylation In Vitro

In vitro studies of ^{14}C-formaldehyde binding to hemoglobin in intact erythrocytes were conducted to identify the primary alkylation products and to determine the best procedure for handling blood samples from the test populations. A primary concern was whether formaldehyde adducts would be stable enough to remain bound to hemoglobin through isolation and hydrolysis and whether the adducts could be chemically stabilized after formation. In an initial experiment, ^{14}C-formaldehyde was added to suspensions of washed erythrocytes at concentrations of 150 and 300 mM. After 30 min, 25-μl portions of 5 mg/ml sodium borohydride were added 3 times at 30-min intervals to each suspension. After a total of 2 h of incubation, the globin was prepared and the radioactivity in the washed membranes and globin determined. Reduction with sodium borohydride enhanced binding to both globin and membranes at both formaldehyde contrations. In globin, the binding was enhanced 32-fold at 150 mM formaldehyde and 22-fold at 300 mM. Binding to globin was twofold or more greater than binding to membranes at both concentrations.

Using borohydride reduction, the binding of formaldehyde to erythrocytes was measured over a range from 150 to 1000 mM. Dose-related binding to both globin and membranes was observed up to 1000 mM formaldehyde. Once again, binding to globin was greater than to membranes by about twofold.

To identify the alkylation products in globin, the sample treated at 750 mM formaldehyde was subjected to acid hydrolysis. The hydrolysate was analyzed on a Beckman 121 M amino acid analyzer using a long-column physiological fluid program. The radioactivity eluted in 3 peaks, two of which correspond closely

in retention time with 1- and 3-methyl histidine. An earlier
eluting peak had a retention time consistent with 5-methylcystine.
Preparative chromatography of the same hydrolysate on a
Dowex 50WX4 column with 1 M and 2 M HCl produced two minor and two
major radioactive fractions. Colorimetric amino acid analysis of
the peak fractions indicated that the largest peak contained
histidine which elutes just prior to 1- and 3-methyl histidine.
Under the conditions used here, the principal formaldehyde
alkylation products of hemoglobin appear to be 1- and 3-methyl
hystidine, although N-methyl lysine is a possible product.

Cytogenetic Effects of Formalin Exposure in Animals

An initial cytogenetic study was conducted to determine the
effect of formalin on bone-marrow cells in mice. In addition, the
effects of methanol, which is present in commercial formalin at a
concentration of about 10% to 15%, were investigated. Mice were
treated in groups of 15 by oral gavage with single doses of water,
100 mk/kg formalin, 1000 mg/kg methanol, or 100 mg/kg
cyclophosphamide as a positive control. Twenty-eight hours after
treatment, 1 mg/kg Colcimid was administered, and 2 h later the
animals were sacrificed and bone marrow was prepared from the
femurs. Analysis of metaphase chromosomes for structural
aberrations produced the results shown in Table 1. Treatment with
formalin or methanol produced a large increase in exchanges
(Robertsonian translocations) and in aneuploid cells as compared
to the negative control. While break damage predominated in the
cyclophosphamide-treated animals, neither formaldehyde nor
methanol produced a significant increase in breaks.

Human Population Monitoring

Description of the study population. The population
initially recruited for the study consisted of 16 exposed and
17 control subjects. Each subject in the exposed group was
matched with a control subject. As of this writing, a total of
107 blood samples, 104 urine specimens, and 64 semen specimens
have been obtained from this population. A small number of
subjects have not been sampled three times yet, and few new
subjects can be expected to enter the study as new residents
rotate onto the autopsy service. For this reason and because of
the time required to score some assays, the data are not
completely compiled for most of the biological tests at this
point.

Table 1. Structural Chromosome Aberrations in Male $B_6C_3F_1$ Mice Treated with Formaldehyde

Treatment	Dose (mg/kg)	n	Cells Scored	Aneuploid	Breaks	Exchanges	Gaps	% Aberrant w/o Gaps	% Aberrant w/Gaps
Water	-	15	655	6	1	10	3	1.7	2.1
Formalin	100	15	651	101	3	40	2	43	45
Methanol	1000	15	667	65	3[a]	24	5	27	32
Cyclophosphamide	100	15	513	3	32	17	8	55	63

[a] plus 8 severely damaged cells.

Environmental monitoring. The individuals in the exposed group fall into four job categories. Laboratory technicians were responsible for general duties including preparation of solutions and preservation of samples obtained from autopsy. Residents and faculty pathologists are responsible for conducting autopsies, including examination of preserved tissues. Residents serve on the Autopsy Service on a three-month rotation and work full-time while there. Faculty pathologists usually have other departmental duties and work part-time in the Autopsy Service. The fourth group are individuals who handle preserved specimens but are generally only in the autopsy suite for twice-weekly conferences. Area sampling indicated that the main autopsy room had an average formaldehyde level of 2.13 ppm over 13 sampling periods, while other rooms in the suite average 0.13 ppm. Personal monitoring data indicated that exposure was very strongly related to certain activities. Tissue sectioning and preservation produced an average exposure in 6 samples of 3.28 ppm, while general laboratory duties produced an exposure of only 1.03 ppm (7 samples) and autopsies 0.92 ppm (6 samples). Residents had the highest average exposure as a group (1.58 ppm in 10 samples), while pathologists had an average exposure of 1.24 (10 samples) and technicians only 0.57 ppm. The overall impression of exposure developed from monitoring data was that baseline levels were around 0.1 to 1.0 ppm but that certain activities could produce transient exposure to levels in the 1 to 5 ppm range. The magnitude of exposure would then depend on the type and frequency of activities of the individuals in the study.

Sperm analysis. A preliminary analysis of sperm tests for abnormal morphology and Y-body frequency is presented in Table 2. The data are compiled as the mean of all tests for each individual. Only those samples were analyzed for which both members of a matched pair exist. A total of 10 pairs was available for analysis. Values for the frequency of one and two Y bodies were very similar in the two groups. The percentage of abnormal sperm was somewhat higher in the exposed group than in the control; however, a single individual with an abnormal morphology value of 39.8% accounts for a large portion of the differences.

DISCUSSION

The object of this project has been to identify a battery of tests which can detect genetic damage in man resulting from exposure to environmental mutagens. Animal and in vitro studies have been conducted in coordination with the human tests for two purposes. First, the animal and in vitro tests are needed to

Table 2. Sperm Analysis in Formaldehyde-Exposed and Control
Individuals: Morphological Abnormalities and Y-Body
Frequency[a]

Exposed	No. Subjects	No. Samples	YF	YFF	% Abnormal Sperm
Exposed	10	22	46.8 ± 3.3	1.7 ± 0.9	14.1 ± 9.9
Control	10	22	44.4 ± 2.3	1.6 ± 0.8	10.5 ± 4.3

[a]Results are means ± standard activation.

identify genetic effects of particular agents and to assist in
establishing methods and conditions for the human assays. Second,
they provide information on genetic effects produced under
controlled laboratory conditions which can be used as a standard
in interpreting the effects observed in man.

The various tests used in this investigation respond to a
variety of different effects. Genetic damage at both the
molecular level and chromosome level are detected. Both germinal
and somatic cells are evaluated. In addition, while the
cytogenetic, alkylating, and sperm effects may persist for a
period of several weeks following exposure, DNA strand breakage
and urinary mutagenic excretion are effects that will quickly
disappear once exposure ends. Thus, the likelihood is maximized
that genetic damage, if it occurs, will be observed in at least
one test.

In interpreting results from these tests, several points
should be kept in mind. Interindividual and temporal variation as
well as confounding environmental factors make interpretation on
an individual basis unjustifiable. The study design incorporates
the use of matched exposed and control populations, repeat
testing, and multiple end points to control variation. By
evaluating results on a population basis, variations among
individuals can be averaged out, allowing the effect of the
environmental exposure to be analyzed for the exposed group as a
whole. In addition, it must be recognized that the end points
observed are not the health outcomes with which we are concerned.
While observed effects like chromosome damage are qualitatively
related to the appearance of outcomes such as cancer, no
quantitative relationship has been established between the two
events. For this reason, results of genetic damage tests should

not be viewed as usable in quantitative risk assessment. Rather, these techniques might be thought of as a group film badge capable of providing an early indication that a particular environment contains a genetically hazardous agent. Where a monitoring system of this type indicates a hazard, a more detailed evaluation of the environment should be made to identify mutagenic components within it and determine the best ways to reduce human exposure to them.

Another point of interpretation which should be recognized is that we cannot expect these tests to identify individuals who are, for genetic or constitutional reasons, at unusually high risk of injury from exposure to environmental mutagens. This view basically grows out of the same considerations that require analysis on a population basis and prevent quantitative risk assessment.

Several weaknesses in this approach to human genetic risk evaluation must be acknowledged. The first, already discussed, is that it does not measure heritable mutation in man; therefore, the effects observed cannot be quantitatively related to adverse health outcomes. Second, the sensitivities of the various tests of human exposures to mutagens are not well established. In fact, enlarging the baseline data collection for these assays and evaluating sensitivities are major objectives of this study. We know that some tests such as cytogenetics have been used successfully in several circumstances to detect human exposures to carcinogens. A statistical analysis by Whorton et al. (1979) indicates that a 1% increase in chromosome aberration rate over a background rate of 2% could be detected with 80% certainty at an error rate (p value) of 5% in a population of less than 50 exposed and 50 control individuals. The sperm morphology and YFF tests need additional work for validation in man and to define the relationship between observed effect and genetic damage more thoroughly. Since all of these tests require the systemic distribution of the active chemical for detection, this battery may not detect effects of chemicals that are very rapidly bound to tissues at point of exposure or rapidly metabolized.

The study of the formaldehyde-exposed population described here is not yet completed, so several results remain to be analyzed in detail. Environmental monitoring indicates that formaldehyde exposures in this population take the form of repeated acute episodes rather than a continuous exposure. The results of sperm tests described here do not indicate any significant effect of formaldehyde exposure. However, we cannot draw any conclusions about the human genetic hazards of formaldehyde until the study is complete. Even then, the small sample size and relatively low time-weighted average exposure of

the group under study may limit the sensitivity of the study.
Cytogenetic effects were clearly observed in mice. Methanol was
tested because formaldehyde is a metabolite of methanol and
methanol is present at 10 to 15% in the formalin preparation used
in this study. The clastogenic effect of formalin appeared to be
equal to or greater than that of methanol alone. Results of a
dose-response study currently in progress comparing formalin and
methanol should clarify the extent to which methanol contributes
to chromosome damage produced by formalin.

REFERENCES

Ames, B.N., J. McCann, and E. Yamasaki. 1975. Methods for
    detecting carcinogens and mutagens with the
    Salmonella/mammalian-microsome mutagenicity test. Mutation
    Res. 31:347-364.

Auerbach, C., M. Moutschen-Dahmen, and J. Moutschen. 1977.
    Genetic and cytogenetical effects of formaldehyde and related
    compounds. Mutation Res. 39:317-362.

Beckman, G., L. Beckman, I. Nordenson, and S. Nordstrom. 1979.
    Chromosomal aberrations in workers exposed to arsenic. In:
    Genetic Damage Caused by Environmental Agents. Academic
    Press: New York. pp. 205-211.

Bloom, A.D., S. Neriishi, N. Kamada, T. Iseki, and R.J. Kehha.
    1966. Cytogenetic investigation of survivors of the atomic
    bombings of Hiroshima and Nagasaki. Lancet 11:672-674.

Buckton, K.E., P.G. Smith, and W.M. Court-Brown. 1967.
    Estimation of life span from studies on males treated with
    X rays for ankelosing spondylitis. In: Human Cytogenetics.
    H.J. Evans, W. Court-Brown, and A.S. McLean, eds. North
    Holland: Amsterdam. pp. 106-114.

Calleman, C.J., L. Ehrenberg, B. Jansson, S. Osterman-Golkar,
    D. Segerbach, K. Svennson, and C.A. Wachtmeister. 1978.
    Monitoring and risk assessment by means of alkyl groups in
    hemoglobin in persons occupationally exposed to ethylene
    oxide. J. Environ. Pathol. Toxicol. 2:427-442.

Commoner, B., A.J. Vithayathil, and J.I. Henry. 1974. Detection
    of metabolic carcinogen intermediates in urine of
    carcinogen-fed rats by means of bacterial mutagenesis.
    Nature 249:850-852.

Connor, T.H., M. Stoeckel, J. Evrard, and M.S. Legator. 1977.
    The contribution of metronidazole and two metabolites to the
    mutagenic activity detected in urine of treated humans and
    mice. Cancer Res. 37:629-633.

Dolphin, G.W., D.C. Lloyd, and R.J. Purrot. 1973. Chromosome
    aberration analysis as a dosimetric technique in radiological
    protection. Health Physics 25:7-15.

Ducatman, A.K., K. Hirschhorn, and I.V. Selikoff. 1975. Vinyl
    chloride exposure and human chromosome aberrations. Mutation
    Res. 31:163-168.

Durston, W.E., and B.N. Ames. 1974. A simple method for the
    detection of mutagens in urine: studies with the carcinogen
    2-acetylaminofluorene. Proc. Natl. Acad. Sci. USA
    71:737-741.

Ehrenberg, L., and S. Osterman-Golkar. 1980. Alkylation of
    macromolecules for detecting mutagenic agents. Teratogen.
    Carcinogen. Mutagen. 1:105-127.

Etteldorf, J.N., C.D. West, J.A. Pitcock, and D.L. Williams.
    1976. Gonadal function testicular history and meiosis
    following cyclophosphamide therapy in patients with nephrotic
    syndrome. J. Pediat. 88:206-212.

Evans, H.J., and M.L. O'Riordan. 1977. In: Handbook of
    Mutagenicity Test Procedures. B.J. Kilby, M.S. Legator,
    W. Nichols, and C. Ramel, eds. Elsevier Scientific
    Publishing Co.: New York. pp. 261-274.

Forni, A., and G.C. Secci. 1972. Chromosome changes in
    preclinical and clinical lead poisoning: correlation with
    biochemical findings. Proceedings of the International
    Symposium Environmental Health Aspects of Lead, Amsterdam.

Funes-Cravioto, F., B. Lambert, J. Lindsten, L. Ehrenberg,
    A.T. Natarajan, and S. Osterman-Golkar. 1975. Chromosome
    aberrations in workers exposed to vinyl chloride. Lancet
    1:459.

Gabridge, M.G., A. Denuzio, and M.S. Legator. 1969. Microbial
    mutagenicity of streptozotocin in animal-mediated assays.
    Nature 221:68-70.

Garry, V.F., J. Hozier, D. Jacobs, R.L. Wade, and D.G. Gray. 1979. Ethylene oxide: evidence of human chromosomal effects. Environ. Mutagen. 1:375–382.

Garza-Chapa, R., C.H. Leal-Garza, and G. Molina-Ballestros. 1977. Archivos de Investigacion Mexicol (Mexico) 8:11–20.

Green, M.H.L. 1979. Mutagenic consequences of chemical reaction with DNA. In: Chemical Carcinogens and DNA, Volume II. P.L. Grover, ed. CRC Press, Inc.: Boca Raton, Florida. pp. 95–132.

IARC, 1979. IARC Monographs on the evaluation of the carcinogenic risk of chemicals to humans. International Agency for Research on Cancer, Lyons, France. IARC Monographs Supplement 1.

Kapp, R.W., Jr., and C.B. Jacobson. 1980. Analysis of spermatozoa for Y chromosome non-disjunction. Teratogen. Carcinogen. Mutagen. 12:193–212.

Katosova, L.D. 1973. Cytogenetic analysis of peripheral blood of workers engaged in the production of chloroprene. Gigiera. fuda i professional nye Zabolevaniia 10:30–32.

Kissane, J.M., and E. Robins. 1958. The fluorometric measurement of deoxyribonucleic acid in animal tissues with special reference to the central nervous system. J. Biol. Chem. 233:184–188.

Kohn, K.W., L.C. Erickson, R.A.G. Ewig, and C.A. Friedman. 1976. Fractionation of DNA from mammalian cells by alkaline elution. Biochemistry 15:4629–4637.

Kucerova, M., V.S. Zhurkor, L. Polivkova, and J.E. Ivanove. 1977. Mutagenic effect of epichlorohydrin, II; analysis of chromosomal aberrations in lymphocytes of persons occupationally exposed to epichlorohydrin. Mutation Res. 48:355–360.

Latt, S.A. 1974. Localization of sister chromatid exchanges in human chromosomes. Science 185:74–76.

Latt, S.A., J.W. Allen, W.E. Rogers, and L.A. Juergens. 1977. In vitro and in vivo analysis of sister chromatid exchange formation. In: Handbook of Mutagenicity Test Procedures. B.J. Kilbey, M.S. Legator, W. Nichols, and C. Ramel, eds. Elsevier Scientific Publishing Co.: New York. pp. 275–291.

Legator, M.S., T.H. Connor, and M. Stoeckel. 1975. Detection of mutagenic activity of metronidazole and niridazole in body fluids of humans and mice. Science 188:1188-1189.

Meretoja, T., H. Vainio, M. Sorse, and H. Harkonen. 1977. Occupational styrene exposure and chromosomal aberrations. Mutation Res. 56:193-197.

Minnich, V., M.E. Smith, D. Thompson, and S. Kornfeld. 1976. Detection of mutagenic activity in human urine using mutant strains of Salmonella typhimurium. Cancer 31:1253-1258.

NIOSH, 1977. National Institute of Occupational Safety and Health Manual of Analytical Methods. 2nd edition, Volume 1.

Nordenson, I., G. Beckman, L. Beckman, and S. Nordstrom. 1978. Occupational and environmental risks in and around smelter in northern Sweden, II; chromosomal aberrations in workers exposed to arsenic. Hereditas 88:47-50.

Normal, A., M.S. Sasaki, and R.E. Ottoman. 1966. Elimination of chromosome aberrations from human lymphocytes. Blood 27:706-714.

Nowell, P.C. 1965. Unstable chromosome changes in tuberculin-stimulated leukocyte cultures from irradiated patients; evidence for immunologically committed, long-lived lymphocytes in human blood. Blood 26:798-804.

Osterman-Golkar, S., L. Ehrenberg, D. Segerback, and I. Hallstrom. 1976. Evaluation of genetic risks of alkylating agents, II: haemoglobin as a dose monitor. Mutation Res. 34:1-10.

Osterman-Golkar, S., D. Hultmark, D. Segerback, C.J. Calleman, R. Gothe, L. Ehrenberg, and C.A. Wachtmeister. 1977. Alkylation of DNA and proteins in mice exposed to vinyl chloride. Biochem. Biophys. Res. Commun. 76:259-266.

Parodi, S., M. Taningher, L. Santi, M. Cavanna, L. Sciaba, A. Maura, and G. Brambilla. 1978. A practical procedure for testing DNA damage in vivo, proposed for a pre-screening of chemical carcinogens. Mutation Res. 54:39-46.

Pereira, M.A., L.-H.C. Lin, and L.W. Chang. 1981. Dose dependence of 2-aminofluorene binding to liver DNA and hemoglobin in mice and rats. Toxicol. Appl. Pharmacol. 60:472-478.

Pereira, M.A., and L.W. Chang. (in press). Binding of bromoform and chloroform to mouse and rat hemoglobin. Chem.-Biol. Int.

Petzgold, G.L., and J.A. Swenberg. 1978. Detection of DNA damage induced in vivo following exposure of rats to carcinogens. Cancer Res. 38:1589-1594.

Picciano, D.J. 1979. Cytogenetic investigation of occupational exposure to epichlorohydrin. Mutation Res. 66:169-173.

Picciano, D.J. 1979. Cytogenetic study of workers exposed to benzene. Environ. Res. 19:33-38.

Ross, W.E., and N. Shipley. 1980. Relationship between DNA damage and survival in formaldehyde-treated mouse cells. Mutation Res. 79:277-283.

Sanotskii, I.V. 1976. Aspects of the toxicity of chloroprene: immediate and long-term effects. Environ. Health Perspect. 17:85-93.

Siebert, D., and A. Simon. 1973. Genetic activity of metabolites in the ascitic fluid and in the urine of a human patient treated with cyclophosphamide: induction of mitotic gene conversion in Saccharomyces cerevisiae. Mutation Res. 21:257-262.

Silezneva, T.G., and N.P. Korman. 1973. Analysis of chromosomes of somatic cells in patients treated with anti-tumor drugs. Soviet Genet. 9:1575-1579.

Swenberg, J.A., W.D. Kerns, R.I. Mitchell, E.J. Gralla, and K.L. Pavkov. 1980. Induction of squamous cell carcinomas of the rat nasal cavity by inhalation exposure to formaldehyde vapor. Cancer Res. 40:3398-3402.

Tough, I.M., P.G. Smith, W.M. Court-Brown, and D.G. Hardner. 1970. Chromosome studies on workers exposed to atmospheric benzene: the possible influence of age. Eur. J. Cancer 6:49.

Truong, L., J.B. Ward, Jr., and M.S. Legator. 1978. Detection of alkylating agents by the analysis of amino acid residues in hemoglobin and urine, 1; the in vivo and in vitro effects of ethylmethanesulfonate, methyl methanesulfonate, hycanthone methanesulfonate, and naltrexone. Mutation Res. 54:271-281.

Whorton, E.B., Jr., D.B. Bee, and D.J. Kilian. 1979. Variations in the proportion of abnormal cells and required sample sizes for human cytogenetic studies. Mutation Res. 64:79–86.

Wilson, J.G. 1977. Current status of teratology; general principles and mechanisms derived from animal studies. In: Handbook of Teratology, Volume 1. J.G. Wilson and F.C. Fraser, eds. Plenum Press: New York. pp. 47–74.

Wyrobek, A.J., and W.R. Bruce. 1978. The induction of sperm shape abnormalities in mice and humans. In: Chemical Mutagens: Principles and Methods for Their Detection, Volume 5. A. Hollaender and F.J. deSerres, eds. Plenum Press: New York. pp. 257–285.

Wyrobek, A.J. 1979. Changes in mammalian sperm morphology after x-ray and chemical exposures. Genetics 92:s104–s119.

Yamasaki, E., and B.N. Ames. 1977. Concentration of mutagens from urine with the non-polar resin XAD-2: cigarette smokers have mutagenic urine. Proc. Natl. Acad. Sci. USA 74:3555–3559.

Zundova, Z., and K. Landa. 1977. Genetic risk of occupational exposures to haloethers. Mutation Res. 46:242–243.

BIOLOGICAL ACTIVITY OF AIR PARTICULATE EXTRACTS (APE) IN

SHORT-TERM TEST

Hélène M. Coulomb,[1] Yves A. Courtois,[2] Françoise Boizard Callais,[3] Bernard Festy,[3] and Ivan Chouroulinkov[1]

[1]Institut de Recherches Scientifiques sur le Cancer, B.P. No. 8, 94802-Villejuif, France, [2]Laboratoire d'Hygiène de la Ville de Paris, and [3]Université Paris V, France

INTRODUCTION

Due to the increase of lung cancer incidence in developed countries (Higgins, 1976; Higginson, 1976, 1979), systematic controls of air quality, physicochemical and biological characterizations, are recommended. With the recent progress in technology, the physicochemical characterization of air pollution is, in our opinion, relatively easy to realize. In contrast, the methodology to evaluate the carcinogenic and mutagenic potentials of complex mixtures still requires improvement. For these investigations, short-term tests have been useful. The Salmonella/microsome mutagenicity test (developed by Ames et al.) has been shown to be especially effective in evaluating the mutagenic activity of air particulate extracts (Tokiwa et al., 1977; Talcott and Wei, 1977; Commoner et al., 1978; Teranishi et al., 1978; Coviaux et al., 1980). Nevertheless, for a better evaluation of carcinogenic potential, one mutagenicity test is not sufficient; a battery of tests is better and more suitable.

For this purpose, we undertook a periodic biological characterization of air particulate extracts (APE) with the Salmonella/microsome mutagenicity test, the 6-thioguanine-resistant (TGr) mutants test, the sister chromatid exchange (SCE) test, the cell transformation (hamster embryo cells) test, and a short-term skin test. In the transformation assay, no transformed colonies were observed; for this reason, we

reported only the cytotoxic effect. One major difficulty in
studying the biological activity of APE in several short-term
tests is to get enough material, which implies having a high
quantity of particulate matter. At least two types of sampling
were performed to get the 100 to 200 mg of APE required for the
tests: a long period of sampling (month) with low delivery (1 to
1.5 m^3/h), and a short period of sampling (days) with high
delivery. (30 to 40 m^3/h). Long-time sampling was performed for
the cutaneous test on mouse skin, which needed 1.5 g of APE;
shorter-time sampling was performed for the other biological
tests, which needed 100 mg of APE.

MATERIALS AND METHODS

Sampling

The air particles were collected in an urban site in the
center of Paris in the Laboratoire d'Hygiène de la Ville de Paris
(LHVP) using a Teflon wool plug with 12.5 cm^2 of filtration area.
The plug was protected by a funnel, kept in darkness during the
filtration, and elevated at 12 m above street level to avoid the
influence of individual emissions. Air-suspended particles were
collected for 1 month (January, 1980), and during a week randomly
chosen in a month (February to June, 1980). A battery of
16 suction pumps (Austen F65DE) was used with a delivery of
1.5 m^3/h, controlled with gas counters (Gallus 6/20).

Extraction

Each plug was extracted twice in acetone (Normapur) at room
temperature for 20 min with gentle shaking (Hughes et al., 1980).
The acetone extract was filtered (No. 3-Prolabo), evaporated to
dryness, and weighed. The APE were dissolved in organic solvent
(acetone or DMSO) prior to biological assays. Control plugs
extracted as described above were not mutagenic in the
Salmonella/microsome test.

Bacterial Mutagenicity Assay

Bacterial strains used were Salmonella typhimurium
histidine-dependent (strain TA98) and nitroreductase-deficient
(TA98NR) (Ames et al., 1975; Rosenkranz and Speck, 1975). The
liver S9 fraction was prepared from male Sprague-Dawley rats of
about 200 g, pretreated with Aroclor 1254 (500 mg/kg) injected
intraperitoneally. Bacteria were treated (with and without S9

mix), with APE dissolved in DMSO. Linear dose-response curves
were established, and the genotoxicity was expressed as induced
rev/mg of APE. Bacterial strains were systematically tested with
mutagenic control chemicals 4-nitroquinoline-N-oxide (4-NQO),
2-nitrofluorene (2-NF), and benzo(a)pyrene (B[a]P) as shown in
Table 1.

Table 1.   Number of Spontaneous or Induced Revertants in Ames Test
           per µg of Chemical

Compound	S9 Mix	TA98	TA98NR
Spontaneous revertants	−	25	10
Spontaneous revertants	+	35	25
4-Nitroquinoline-n-oxide	−	1050	900
1-Nitrofluorene	−	350	35
Benzo(a)pyrene	+	560	

Induction of 6-Thioguanine-Resistant (TGr) Mutants in V79 Cells

This test was carried out according to the method of
Abbondandolo et al. (1977). Dose-response curves were established
for the time of expression, 6 days.

Induction of Sister Chromatid Exchanges (SCEs)

The 10^5 cells were seeded in 100-mm dishes with 10 ml Eagle's
MEM (Gibco) supplemented with 10% inactivated fetal calf serum
(FCS) and incubated for 24 h in a humidified incubator at 37°C in
the presence of 5% CO_2. The medium was then replaced with
serum-free MEM containing APE. After 1 h of treatment, the medium
was changed by complete medium containing 5-bromodeoxyuridine
(BrdU, 5 µg/ml) for 20 to 37 h in darkness. Colcemide (30µg/ml)
was added into the dishes 24 h before the cells were collected for
hypotonic treatment (KCl;, 0.04 M) and fixed. Differential
staining of SCE was performed according to the method previously
described (Cheng et al., 1980).

Cell Transformation on Hamster Embryo Cells

The 3×10^2 secondary (1st passage) hamster embryo cells in
5 ml Dulbecco modified Eagle's MEM (D-MEM, Gibco) supplemented
with 10% inactivated FCS were seeded in 60-mm dishes with a feeder
layer of 60,000 hamster X-irradiated (4000 R) cells, and left in a
humidified incubator at 37°C in the presence of 5% CO_2. The
following day, 3 ml of complete medium containing B(a)P (1 µg/ml)
as positive control or APE dissolved in acetone (0.25% in the
medium) were added and cells were incubated for 7 days. The cells
were then washed with PBS and fixed and stained with Giemsa
(DiPaolo et al., 1969). The cloning efficiency (CE) was
determined by dividing the average number of colonies per plate by
the number of cells seeded per plate multiplied by 100. The
cytotoxicity was determined by the relative CE of the treated to
the control cells (expressed in percent).

Short-term Test on Mouse Skin

This test was described by Lazar et al. (1963). Mice were
treated twice on closely clipped dorsal skin area with acetone or
with APE. Eight days after the first treatment, the mice were
killed and treated skin areas were fixed for histological
examination. The thickness of the epidermis and the number of
sebaceous glands were determined.

RESULTS

Extraction

Figure 1 shows that the concentration of particulates ($\mu g/m^3$)
collected weekly on Teflon wool in Paris decreases from February
to June. Results presented in Table 2 show that the extraction
efficiency of APE is about 25 to 35% except for the last week,
corresponding to June (87%). This increase in June is probably
due to an artifact during the extraction at the end of which we
observed an oily fraction.

Bacterial Mutagenicity Assay

In the Salmonella/microsome test, the genotoxicity of APE is
higher in winter (February, March) than in spring (April, May,
June) in both strains, and this activity is increased when S9 mix
is added. Moreover, results also show a decrease in the
genotoxicity of APE when TA98NR is used (Table 3).

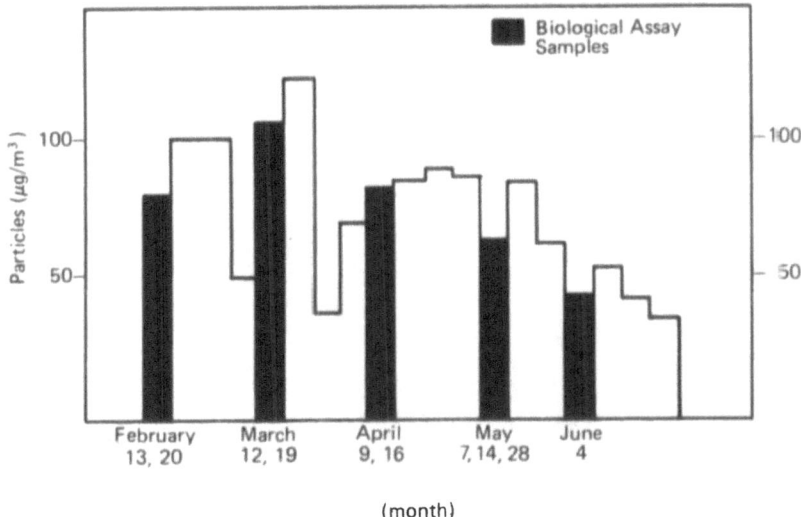

Figure 1. Weekly variations of air particles ($\mu g/m^3$) collected on
Teflon wool in Paris.

Table 2. Organic Extracts of Air Particulates Collected Weekly in
Paris for Biological Assays

Collection Period (1980)	Volume (m^3)	Particulates ($\mu g/m^3$)	Acetone Extract ($\mu g/m^3$)	Extraction Efficiency (%)
02/13–02/20	4551	80.5	27.0	33.6
03/12–03/19	4857	107.0	34.3	32.0
04/09–04/16	4853	82.1	18.6	22.6
05/07–05/14	4778	63.5	22.8	35.9
05/28–06/04	4705	43.5	38.1	87.5

Table 3. Mutagenic Activity of Air Particulate Extracts from
          7-Day Samples in Salmonella/Mutagenicity Test[a]

Week	S9 Mix	TA98	TA98NR
02/13–02/20	–	410	110
	+	1300	425
03/12–03/19	–	240	100
	+	2000	150
04/09–04/16	–	25	5
	+	45	5
05/07–05/14	–	35	15
	+	55	5
05/28–06/04	–	45	10
	+	125	25

[a]Number of induced rev/mg of APE.

Induction of 6-TGr Mutants in V79 Cells

Table 4 gives the frequency of 6-TGr mutants in the control
and in the groups treated with APE (expression time, 6 days). The
APE from February and March induce mutations, but not those
corresponding to April, May, and June.

Induction of SCE in V79 Cells

The results presented in Table 5 show that APE significantly
increase the induction of SCEs in V79 cells. This increase is
progressive, the extracts from May and June being the most active.

Cell Transformation

We have not observed transformed colonies, and we have
analyzed only the cytotoxic effect of the extracts. Table 6 shows
that the extract from May is not cytotoxic at the doses used, the
extracts from February and April are cytotoxic at the higher dose
(80 µg/ml), and the June extract is very toxic (100%), even at
20 µg/ml.

Table 4. 6-Thioguanine-Resistant (TGr) Mutant Induction in V79 Cells by Air Particulate Extracts from 7-Day Samples[a]

	APE (µg/ml)			
Week	0	10	20	50
02/13–02/20	9.76	10.28	10.77	15.89
03/12–03/19	10.7	3.53	7.44	20.22
04/09–04/16	7.48	12.5	14.97	6.80
05/07–05/14	39	20.2	11.3	25
05/28–06/04	14.7	3.1	7.1	13.7

[a]Expression time, 6 days; positive control, EMS 20 mM/ml = 470 mutants/10^6 cells.

Table 5. SCE Induction in V79 Cells by Air Particulate Extracts from 7-Day Samples[a]

	Mean ± S.E. of SCE/Cell[a]		
Week	Control	APE (6 µg/ml)	Significance Level
02/13–02/20	5.96 ± 0.2	6.70 ± 0.46	$p < 0.05$
03/12–03/19	5.96 ± 0.2	6.9 ± 0.35	$p < 0.01$
04/09–04/16	5.46 ± 0.29	6.90 ± 0.31	$p < 0.01$
05/07–05/14	5.46 ± 0.29	7.33 ± 0.37	$p < 0.0001$
05/28–06/04	5.46 ± 0.29	9.13 ± 0.44	$p < 0.0001$

[a]Number of metaphases scored: 30.

Short-term Tests on Mouse Skin

In the hyperplasia test, APE from January showed a positive effect: a significant increase in the epidermal thickness, similar to the effect of 12-0-tetradecanoyl-phorbol-13-acetate (TPA) (Table 7). In contrast, we did not observe any significant effect when mouse skin was treated with APE from June. In

Table 6.   Cloning Efficiency (CE) and Toxicity (T) of Hamster
           Embryo Cells Treated with Air Particulate Extracts
           (APE) from 7-Day Samples

Week	APE (µg/ml)	Total No. of Colonies	CE (%)	T (%)
02/13–02/20[a]	0	937	31.2	0
	20	1106	37	0
	40	1013	33.8	0
	80	651	21.7	30.5
03/12–03/19	nt	nt	nt	nt
04/09–04/16[b]	0	896	30	0
	20	930	31	0
	40	903	30.1	0
	80	726	24.2	19.1
05/07–05/14[a]	0	937	31.2	0
	20	1072	35.2	0
	40	998	33.3	0
	80	1022	34	0
05/28–06/04[a]	0	937	31.2	0
	20	0	0	100
	40	0	0	100

[a]CE of B(a)P 1 µg/ml = 20.7%.
[b]CE of B(a)P 1 µg/ml = 24%.

sebaceous gland tests, we did not observe any significant
destruction.

DISCUSSION

In relation to our objectives of trying to improvise a method
for a systematic measurement of air quality, we must first discuss
the representativeness of the collection and the extraction
procedures. As we mentioned in Materials and Methods, two
sampling procedures were used (medium period: 7 days; long
period: 28 days). Being limited by the low capacity of the
sampling equipment, we have chosen an intermediate sampling time

Table 7. Effect of Air Particulate Extracts (APE) in Short-Term Skin Test (28-Day Samples)

Treatments	No. of Animals	Sebaceous Glands		Hyperplasia	
		M̄ ± S.E.	Destruction (%)	M̄ ± S.E.	Increasing (%)
APE January	15	16.1 ± 2.5	5.3	32.1 ± 3.6	191.1
Control	20	17.0 ± 1.37	0.0	16.8 ± 1.37	100.0
APE June	11	19.8 ± 1.2	0.0	16.6 ± 1.65	111.4
Control	20	19.6 ± 1.47	0.0	14.9 ± 0.91	100.0
B(a)P 0.067%	15	9.0 ± 1.83	47	18.7 ± 1.84	111.3
TPA 0.001%	20	18.6 ± 1.64	0.0	38.7 ± 4.27	230.4

to have a large quantity of particulates and to reduce artifacts (chemical transformation, formation, or loss of compounds) that can occur over a long period. We used the 28-day collection procedure only for skin tests, fully aware of the risks we mentioned above. In relation to the sampling period, the Teflon wool plug filter appears to be more appropriate than glass- or Teflon-fiber filters, since clogging decreases the airflow through these filters.

The choice of the solvent was in agreement with the work of Löfroth (1980) and our preliminary assays (unpublished), which showed that an acetone extract is more mutagenic in the Salmonella mutagenicity test than other solvent extracts (cyclohexane, toluene, etc.). In fact, the extraction efficiency of different acetone extracts is relatively homogeneous (25 to 35%), except for the June sample (87%) (Table 2). We have really no explanation for these results, except that the oily part of the extract seems similar to the paraffinic fraction of dichloromethane particulate diesel extract.

The genotoxic effects observed in the Salmonella mutagenicity tests of February and March extracts were higher than those of April, May, and June. This seasonal variation was also observed in Paris with monthly collected samples (Coviaux et al., 1980) and in Norway (Alfheim and Möller, 1979) with daily samples and as a function of the wind. We also observed a higher mutagenic activity when a metabolic activation system was added, suggesting the presence of indirect-acting compounds such as PAH. The decrease of the mutagenic activity in the TA98NR strain indicates the presence of nitrated compounds. We do not actually know if these compounds are normally present in the air or formed on the filter during the collection period. The results of the 6-TGr induction test can only be considered as qualitative indicators and are incomplete. Higher doses should be tested with a metabolic activation system.

In the SCE test, the extracts from May and June are more active than the others. These results are not in agreement with the results obtained in the bacterial test. A similar phenomenon was observed with food acetone extract (unpublished) indicating that SCEs are not induced by the same compounds present in the mixture as in the Salmonella mutagenicity test. The cytotoxicity results show different activities between the different APE, which indicates a discriminating capacity of this test. Even if we do not know which compounds are involved in this activity, the information can be used to evaluate the cytotoxic potential of air particulates for the lung epithelial cells. However, in this case, we need a large scale of concentrations to establish a dose response. The results from sebaceous glands and hyperplasia tests (Table 7) indicate no detectable carcinogenic potential (no effect in sebaceous gland test) but a promoting potential (positive effects in hyperplasia test) for the January sample. The promoting potential was not detected in our previous long-term studies (Chouroulinkov et al., 1980). It can be explained by a possible difference in the extract composition, and above all, by the low promoting dose used, if compared to standard doses for promotion assay with cigarette smoke condensates (a complex mixture).

CONCLUSION

In conclusion, for systematic quality evaluation of organosoluble air particulate extracts, especially for mutagenic and carcinogenic potentials, the Ames test seems to be necessary as a first step of evaluation. Nevertheless, the information furnished is not sufficient to cover all active compounds; a second mutagenicity test should be conducted. The SCE test seems more adequate than gene mutation induction tests in mammalian cells. A cell transformation test is necessary because it is closely related to carcinogenesis. But the short-term skin tests can be useful in a battery of tests to improve and to standardize the methods in the field of the carcinogenic activity of complex mixtures. Also, the cytotoxicity test can provide very interesting information about toxic compounds present in the air, and these toxic compounds can affect the lung epithelial cells and favor the activity of mutagenic and carcinogenic compounds.

ACKNOWLEDGMENTS

F. Coviaux and Y. Le Moullec are acknowledged for the sampling, and B.N. Ames and H.S. Rosenkranz for providing the Salmonella strains. The authors wish to thank all volunteers in the study and gratefully acknowledge the excellent technical assistance of S. Audu, A. Duverger, L. Huynh, and G. Moens, as well as the statistical analysis of M.T. Maunoury, and the secretarial assistance of L. Guglielmi and F. Maugain. This study was supported in part by Commission des Communautés Européennes Contrat No. EN 375 F (S).

REFERENCES

Abbondandolo, A., S. Bonatti, G. Corti, R. Fioro, N. Loprieno, and
    A. Mazzacaro. 1977. Induction of 6-thioguanine-resistant
    mutants in V79 Chinese hamster cells by mouse-liver
    microsome-activated dimethylnitrosamine. Mutation Res.
    46:365-373.

Ames, B.N., J. McCann, and E. Yamasaki. 1975. Methods for
    detecting carcinogens and mutagens with the Salmonella/
    microsome mutagenicity test. Mutation Res. 31:347-364.

Alfheim, I., and M. Möller. 1979. Mutagenicity of long range
    transported atmospheric aerosols. Sci. Total Environ.
    13:275-278.

Cheng, S.J., M. Sala, M.H. Li, M.Y. Wang, J. Pot-Deprun, and
    I. Chouroulinkov. 1980. Mutagenic, transforming and
    promoting vegetables from Linxian County, China.
    Carcinogenesis 1:685-692.

Chouroulinkov, I., Y. Le Moullec, and B. Festy. 1980. Action
    biologique d'extraits organiques des particules
    atmosphériques urbaines: effets carcinogènes. Pollut.
    Atmosph. 85:58-61.

Commoner, B., P. Madyastha, A. Brondson, and A.J. Vithayathil.
    1978. Environmental mutagens in urban air particulates.
    J. Toxicol. Environ. Health 4:59-77.

Coviaux, F., A. Person, Y. Courtois, Y. Le Moullec, and B. Festy.
    1980. Mesure d'hydrocarbures polycycliques aromatiques et
    détection de potentialitiés génotoxiques dans les particules
    en suspension de la région parissienne. VDI Berichte
    358:155-170.

Di Paolo, J.A., P. Donovan, and R. Nelson. 1969. Quantitative
    studies of in vitro transformation by chemical carcinogens.
    J. Natl. Cancer Inst. 42:867-874.

Higgins, I.T.T. 1976. Epidemiological evidence on the
    carcinogenic risk of air pollution. In: Environmental
    Pollution and Carcinogenic Risks, Volume 52. C. Rosenfeld
    and W. Davis, eds. IARC Scientific Publications. INSERM
    Symposia Series: Lyons. pp. 41-52.

Higginson, J. 1976. Importance of environmental factors in cancer. In: Environmental Pollution and Carcinogenic Risks, IARC Scientific Publications. INSERM Symposia Series: Lyons. pp. 15-24.

Higginson, J. 1979. Perspectives and future development in research on environmental carcinogenesis. In: Carcinogens: Identification and Mechanisms of Action. A.C. Griffin and C.R. Shaw, eds. Raven Press: New York. pp. 187-208.

Hughes, T.J., E. Pellizari, L. Littel, E.S. Sparacino, and A. Kolber. 1980. Ambient air pollutants: collection, chemical characterization and mutagenicity testing. Mutation Res. 76:51-83.

Lazar, P., C. Libermann, I. Chouroulinkov, and M. Guerin. 1963. Test sur la peau de souris pour la détermination des activités carcinogènes: mise au point méthodologique. Bull. Cancer 50(4):567-577.

Löfroth, G. 1980. Comparison of the mutagenic activity from diesel and gasoline powered motor vehicles to carbon particulate matter. In: Short-Term Bioassays in the Analysis of Complex Environmental Mixtures II. M.D. Waters, S.S. Sandhu, J.L. Huisingh, L. Claxton, and S. Nesnow, eds. Plenum Press: New York. pp. 319-336.

Rosenkranz, H.S., and W.T. Speck. 1975. Mutagenicity of metronidazole: activation by mammalian liver microsomes. Biochem. Biophys. Res. Commun. 66:520-525.

Talcott, R., and E. Wei. 1977. Airborne mutagens bioassayed in Salmonella typhimurium. J. Natl. Cancer Inst. 58:445-451.

Teranishi, K., K. Hamada, and H. Watanabe. 1978. Mutagenicity in Salmonella typhimurium mutants of benzene soluble organic matter derived from airborne particulate matter and its five fractions. Mutation Res. 56:273-280.

Tokiwa, H., K. Morita, H. Takeyoshi, K. Takahashi, and Y. Ohnishi. 1977. Detection of mutagenic activity in particulate air pollutants. Mutation Res. 48:237-248.

ON THE USE OF RELATIVE TOXICITY FOR RISK ESTIMATION

C.S. Dudney, P.J. Walsh, T.D. Jones, E.E. Calle, and
G.D. Griffin

Health Effects and Epidemiology Group, Health Studies
Section, Health and Safety Research Division, Oak Ridge
National Laboratory, Oak Ridge, TN 37830

INTRODUCTION

Risk assessment may be viewed as the integration of
comprehensive health and environmental studies for the estimation
of human health risk. There is increasing impetus to use risk
assessment to set priorities for research needs in health and
environmental areas as well as to provide the best quantification
of risk based on available data. Any general risk assessment
methodology includes two components: measurement or estimation of
exposure, and estimation of exposure-response relationships.
These two components are usually called exposure assessment and
health effects assessment.

Health effects assessment uses epidemiological,
toxicological, and basic biological, biochemical, and biophysical
data to estimate effects in humans to exposures of interest, or to
rank a set of materials according to estimated human effects.
Such methodology involves a number of very complex issues, many of
which will be addressed later. In our view, the output of a
useful health effects assessment must include estimates of the
following: 1) a human dose-response function, 2) the uncertainty
in the estimated effects, 3) the "most probable" range of risk as
well as the "upper limit" risk, 4) the predictive power of
biological test systems used, 5) comparative risks from other
materials or sources, and 6) data gaps, priorities, and research

needs. We have developed a method based upon relative
toxicological potency which incorporates these features. A
discussion of the approach and an example of its application
follow.

THE RELATIVE POTENCY APPROACH

A major need in environmental risk assessment methodology is
the development of practical approaches so that the large number
of chemicals which may represent a risk to human health can be
compared on a common scale. One way to compare chemicals on a
common scale is through use of relative potency. The question in
health effects assessment is whether dose to cells, tissues, or
organisms at risk can be quantified in such a way that relative
potency, RP, is given by

$$RP = \frac{\text{dose of test chemical}}{\substack{\text{dose of reference chemical which} \\ \text{results in the same degree of toxicity}}}$$

The biologically effective dose, BED, is then given by

$$BED = \frac{(D)\ (MFs)}{RP}$$

where MF refers to modifying factors such as biological
sensitivity, spatial distribution of dose, metabolism, etc.

If BED is appropriately defined, the ratio of response to
dose is constant within useful limits. The concept is similar to
the use of rem dose for ionizing radiation which is used to
compare different types of radiation on a common scale. The RP is
equivalent to the inverse of relative biological effectiveness
used in ionizing radiation dosimetry.

Our approach generalizes the above concepts to address
extrapolation problems using the relative predictive power of
subhuman biological testing systems. In addition, the approach
allows for uncertainties as to the proper dose-response function
for human risk estimation. In contrast to the case for ionizing
radiation, no one reference chemical can be selected at present.
Different reference chemicals may be used for different specific
biological end points and test chemicals. Dosimetric approaches

for which response/BED ~ constant will facilitate risk assessments
for chemicals.

The approach is also fundamentally similar to those that have
been used for estimating the health effects of steroid-like toxins
(EPA, 1981) and diesel emissions (Harris, 1981). However, the
approach is more general and less dependent on specific initial
assumptions, especially as regards dose-response analysis. The
approach provides a more general framework for identifying and
analyzing the sources of uncertainties and the effects of
necessary assumptions. However, the relative potency aspects of
the methodology will be the main focus of this paper.

Description of the Approach

The relative potency approach uses human data on reference
chemicals and subhuman data on both test and reference chemicals
to estimate human effects attributable to a test chemical.
Potency of a test chemical is reviewed at all levels of biological
organization (subcellular, cellular, animal, human) where data
exist. Comparison to one or more reference chemicals provides a
set of relative potency factors. A measure of uncertainty in this
process is defined by the scatter and range of independent risk
estimates, and in retrospect, the factors having the greatest
bearing on the scatter can be identified.

Development of relative potency factors. The role of
relative potency in the methodology is illustrated in Table 1.
Potency is defined as the dose necessary to yield a given level of
response. The ratio of test chemical potency to reference
chemical potency in the subhuman system is assumed to equal that
in the human so that the ratio of the dose of the test chemical
(d_t) to the dose of the reference chemical (d_r) that produces the
same level of response (R) in a subhuman system is used to
estimate, in the human system, that dose of the test chemical (D_t)
that is equivalent to the dose of the reference chemical (D_r) in
terms of human response. The ratio d_t/d_r is a relative potency
factor and may be extrapolated to the human system under the
assumption that $d_t/d_r = D_t/D_r$. The methodology will allow for an
empirical test of consistency of relative potency factors across
biological systems for all chemicals in the data base. For any
particular test and reference chemical, the consistency of
relative potency across subhuman systems may be defined, and for
any two reference chemicals, relative potency across all levels of
biological organization may be defined. Consistency in relative
potencies across biological systems would have important
implications for future risk assessment and research; if the

Table 1. Comparison of Relative Potencies

	Subhuman System A	Subhuman System B	Subhuman System C	Human Data
Test chemical	d_t	d'_t	d''_t	
Reference chemical	d_r	d'_r	d''_r	D_r
Relative potency	$\dfrac{d_t}{d_r}$ $\overset{?}{=}$	$\dfrac{d'_t}{d'_r}$ $\overset{?}{=}$	$\dfrac{d''_t}{d''_r}$	$\dfrac{D_t}{D_r} \pm \Delta \dfrac{D_t}{D_r}$
Reference chemical 2	d_{r2}	d'_{r2}	d''_{r2}	D_{r2}
Relative potency	$\dfrac{d_r}{d_{r2}}$ $\overset{?}{=}$	$\dfrac{d'_r}{d'_{r2}}$ $\overset{?}{=}$	$\dfrac{d''_r}{d''_{r2}}$ $\overset{?}{=}$	$\dfrac{D_r}{D_{r2}}$

d_t and d_r = Doses of test and reference chemicals producing same reponse (R) in subhuman test system.

d_t/d_r = Relative potency factor.

D_t/D_r = Relative potency at human level assuming $d_t/d_r = D_t/D_r$ for one test system.

$D_t/D_r \pm \Delta\,D_t/D_r$ = Mean potency and range of potencies for all subhuman test systems = $P_o \pm \Delta P$.

factors are reasonably constant, in vitro and short-term test systems may be used to estimate relative potency at the human level and preclude reliance on much more costly whole-animal data. A general lack of consistency in the relative potency factors would imply that each test system should be used as an independent estimator of relative potency at the human level and the differences in the resulting estimates would contribute to the range of uncertainty (see following section on estimation of uncertainty). If certain individual test systems are found to differ repeatedly in their relative potency factors from other test systems and/or from the human level relative potency, this methodology will suggest that those systems are an inappropriate basis for estimating human risk.

Selection of reference chemicals. Selection of reference chemicals to use in the health effects methodology will be

determined by the availability of data relating some measure or
estimate of exposure to the chemical and a disease end point in
human populations. Ideally, the information needed to establish
such estimates would include a set of well defined dose-response
curves that describe the functional relationships between exposure
to toxic materials of concern and disease response. Human
dose-response curves of this form are rare and certainly are not
available for the large number of potential pollutants of
interest. Even for those substances that have been associated
with increased disease induction in humans, the total induced
incidence or mortality, and the intensity, duration, and
conditions of exposure are usually very poorly known.

The International Agency for Research on Cancer's (IARC) list
of 26 chemicals associated with cancer induction in humans
provides a point of departure in the search for candidate
reference chemicals when cancer is the end point of interest.
Many of the chemicals will not be directly relevant to the test
chemicals to be assessed. However, a chemical should not be
rejected automatically because it is not closely related to the
test chemical. Exposure to chemicals used for medicinal purposes
may have resulted in more complete and accurate dose-response data
than occupational chemical exposures, and thus will provide
relative potency data at various levels of biological organization
that will be useful for dose-response modeling.

A National Academy of Sciences (NAS) panel reviewed the data
for a small number of chemicals for which they felt human exposure
and induced cancer incidences could be roughly estimated (National
Research Council, 1975). These included benzidine,
chloronaphazine (N,N-bis(2-chloroethyl)-2-naphthylamine),
diethylstilbesterol (DES), aflatoxin B_1, vinyl chloride, and
cigarette smoke. These substances, with the exception of
cigarette smoke, are a subset of the IARC list of human
carcinogens.

In addition to the IARC chemicals and cigarette smoke,
benzo(a)pyrene (B[a]P), coke oven emissions, and ionizing
radiation may be suitable reference agents. The Carcinogen
Assessment Group (CAG, 1978a,b) of the U.S. Environmental
Protection Agency has estimated dose-response functions for human
lung cancer for both B(a)P and benzene-soluble organics. Human
dose-response data on carcinogenesis after exposure to ionizing
radiation are also available (BEIR, 1980).

Deriving dose-response functions for reference chemicals.
Quantitative estimates of exposure and induced disease incidence
or mortality will usually be in the form of single estimates such

as an estimate of lifetime exposure to a cohort yielding a
lifetime probability of a disease or, at best, a narrow range of
exposure and response estimates. Consequently, the shape of the
dose response for most reference chemicals cannot actually be
constructed; rather, in order to extrapolate to low levels of
exposure, models must be used.

There is no consensus in the scientific community on which
models and methods to use for extrapolation across levels of
biological organization or in dose response at the human level.
Several papers in the Proceedings of the Third ORNL Life Sciences
Symposium on Health Risk Analysis (Richmond et al., 1981) offer,
in combination, excellent discussions of models, modeling, and
associated problems (e.g., Altshuler, Crump, Hoerger, Munro,
Thorslund, Totter). Only a brief overview will be provided here
with emphasis on some unique features of our approach.

The basic dilemma in choice of dose-response model is that
many models "fit" available data at higher dose levels but can
yield predictions that differ by orders of magnitude at low dose
levels of concern for health protection. This dilemma exists
because detailed mechanisms of disease are unknown or uncertain.
Thus, the parameters of most models are not linked to
mechanisms -- they are simply mathematical/statistical curve
fitting expedients. There are also the familiar statistical and
economic limitations associated with the large experiments that
would be needed to experimentally define response at lower dose
levels. This dilemma does not appear to be amenable to complete
resolution in the near future. However, there are general
modeling approaches and mechanistic theories that do provide some
resolution in choice of dose-response models, as illustrated by
the following.

Data on induction of lung cancer in humans due to exposure to
coal tar pitch volatiles (Mazumdar et al., 1975) have been fit
with several models. The results are illustrated in Figure 1.
One of the models is based on the initiation-promotion concept of
carcinogenesis (Jones et al., 1981). It is seen in Figure 1 that
the models diverge widely when extrapolated to low levels of
exposure. This divergence is artificial since logically all the
models should predict zero excess incidence at zero excess
exposure. Thus in Figure 2, the models are forced to pass through
the expected level of lung cancer incidence at zero exposure and
"fit" the rest of the data. Constraining the models to pass
through the expected level provides some resolution in the choice
of models as can be seen by comparing Figures 1 and 2. Our
general approach, then, is to "fit" available human data for each
reference chemical with several models to convey the range of

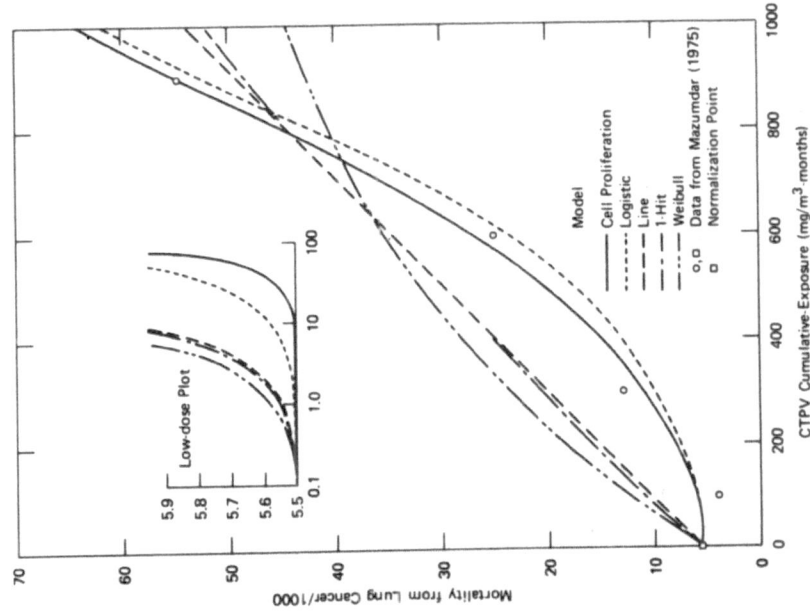

Figure 2. The models in Figure 1 were fit while constrained to predict the expected response at zero exposure.

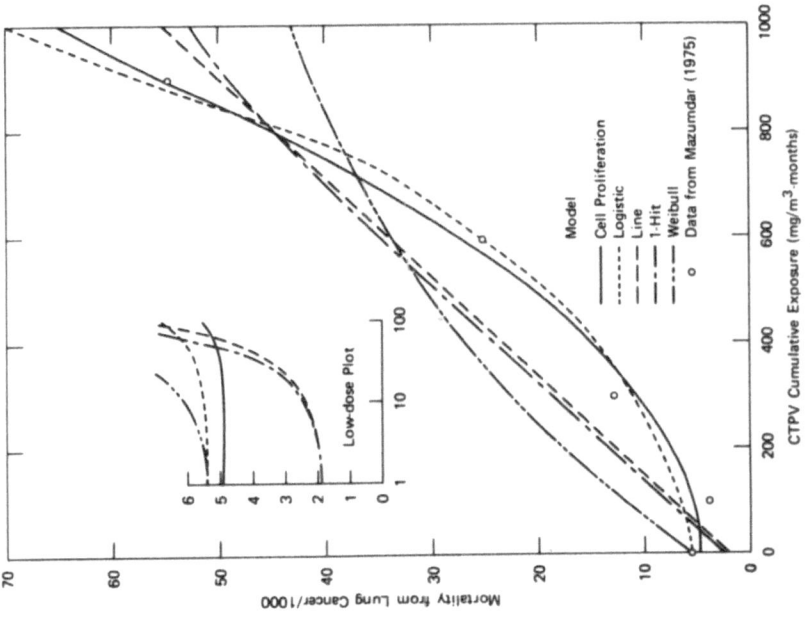

Figure 1. Lung cancer in coke oven workers was modeled without constraints.

uncertainty in dose-response extrapolation where all the models are constrained to predict the expected level at zero exposure. The approach is similar to that proposed by others (Guess et al., 1977; Peto, 1978) where added risk above that expected is assumed to be linearly related to dose except that we make no a priori assumption about the form of dose response. Munro and Krewski (1981) discuss these matters in more detail.

Estimation of Uncertainty

There are two general sources of uncertainty in the application of the relative potency approach. One is the range of values obtained for relative potencies for test chemicals in subhuman test systems and the application of this range at the human level. The other is the uncertainty in the human dose response. Recall that to obtain a single estimate of human response for a test chemical, the following information is needed: 1) a relative potency factor -- the ratio of the dose of the test chemical to the dose of a reference chemical that produces the same level of response in a given subhuman test system (d_t/d_r), 2) selection of a reference chemical for which estimates of human exposure and disease are available, and 3) dose/exposure response functions for the reference chemical. These items may be varied to provide a range of human response estimates for each test chemical.

The range in human response produced by the range in relative potencies and variation in dose response of the reference chemicals is illustrated in Figure 3. The range of doses of test chemicals that produce the equivalent human level response of the reference chemical is $D_t = D_r(P_0 \pm \Delta P)$ where P_0 is the mean potency and ΔP is the range of potencies. The two curves for the reference chemicals perhaps represent extreme ranges in curve shapes. The total range in estimated human health risk comes from the combination of the range in relative potency and the range of models used for dose response. In many cases, one can do relatively little to reduce the range in estimated human response by reducing the range in relative potencies unless the dose response questions can be resolved.

For a given reference chemical some resolution can be obtained as outlined in the previous section. The issue then becomes which reference chemical curve to use for a given test chemical. The issue can be resolved by looking for similarity in dose response between test and reference chemicals. Thus, it is important to look at several dose levels in subhuman systems. If there is similarity between dose-response functions for the test

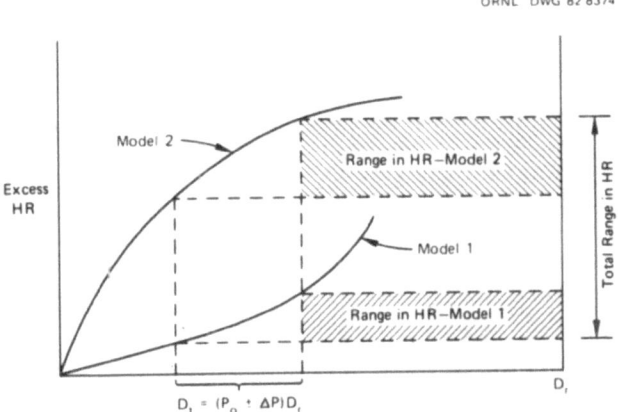

Figure 3.   Uncertainty in predicted risk derives from several
            factors.

and reference chemicals in subhuman systems, then the human level
dose response for that reference chemical should be used to model
the test chemical. If a choice of a particular reference chemical
is not possible, then an envelope of possible dose-response
relationships must be used as illustrated in Figures 2 and 3.
Since all possible dose-response relationships must converge at
the expected response at zero excess exposure, the range in
estimated human response may actually be smaller at exposure
levels near zero than at higher levels. Since human exposures
will probably be controlled at low levels, it is possible to
provide a reasonable range for the estimate of human response.

Examples of Relative Potency Ranges

     A key assumption made in this methodology is that observed
relative potency factors in subhuman tests are approximately equal
to human derived relative potency factors and, therefore, constant
across various levels of biological organization. It is possible
to begin comparing this assertion with observations in a limited
number of cases. One such case is the data base that Harris
(1981) has assembled in his study of diesel exhaust emissions.

This set of data is especially suitable for use in the relative potency approach for several reasons. For each test system, all eight chemical mixtures were tested in the same laboratory and comparison of potencies, in the form of relative potency factors, will not be confounded by differences in experimental protocol. For each mixture, aliquots of a common stock were tested in the various tests and the comparisons will not be confounded by differing chemical compositions of the test samples. Each experiment was considered to have been done reliably, reproducibly, and in a mammalian system. Harris (1981) has analyzed the data from each of the above experiments and presented a proportionality constant between experimental response and "dose" of chemical mixture. Using these constants, potency of each chemical mixture relative to either coke oven emissions or roofing tar emissions has been calculated. The results are presented in Figure 4.

It is clear that there is substantial variation in relative potency factors among biological systems. Screening-level test systems, including those used here, have differential sensitivities to various chemicals, and indeed, many were designed with the intent of detecting classes of carcinogens not detected in earlier systems. Ultimately, the variation in relative potency factors depends on such intrinsic sensitivity of test systems, and detailed knowledge of sensitivies will facilitate the risk assessment process. On the other hand, some patterns are evident in Figure 4. In general, emissions from the five vehicles can be ordered with A being least hazardous and B most hazardous, while all are less hazardous than coke oven or roofing tar emissions per unit mass of extract. The test systems can be compared with one another in those cases where engine emissions were tested in all four systems and potency factors relative to both coke oven and roofing tar emissions were calculated. Mouse lymphoma cell mutagenesis without metabolic activation yields the lowest relative potency factors. The highest relative potency factors come from mutagenesis with activation or from mouse skin painting experiments. These results probably reflect the chemical composition of engine exhaust emissions.

The consistency of the relative potency factors can be partially evaluated by comparison of human relative potency factors with subhuman factors. Harris (1981) has presented data indicating that the absolute potencies of roofing tar and coke oven emissions for inducing respiratory cancer in man are

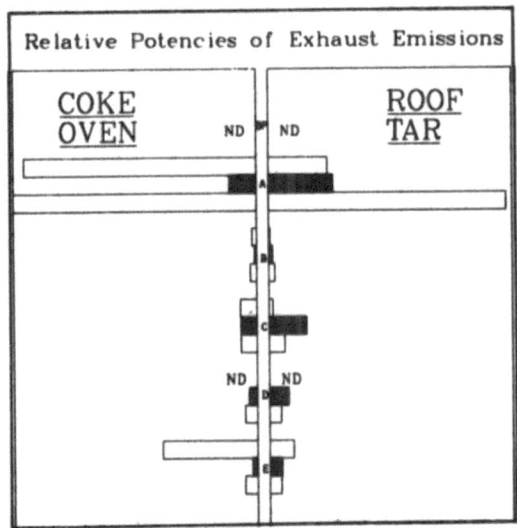

Figure 4.   Emissions from five vehicles, labeled A through E, were
             extracted and tested in three biological systems. In
             each set of three bars, the length of the topmost bar
             is proportional to the relative potency factor from
             mouse skin carcinogenesis experiments. The second bar
             is based on in vitro transformation data, and the third
             is based on mutation data in S9-supplemented L5178Y
             cells. The reference chemical is identified at the
             top. Benzo(a)pyrene was also tested but the potency
             factors were so low that they can't be seen on the
             scale. "ND" means no data are available.

$$1.30 \times 10^{-4}* \text{ and } 4.40 \times 10^{-4} \left(\frac{\mu g \text{ of extract}}{m^3} - yrs\right)^{-1},$$

---

*The published figure in Harris (1981) is 1.64×10^{-4}, but there
seems to have been an error in unit conversion. In any case, the
difference between these values is less than 25%.

respectively. From these numbers, we estimate that extract of
roofing tar emissions is 3.39-fold less potent than coke oven
emissions extract in inducing respiratory cancer. When these
extracts were tested in nonhuman systems, the potencies of roofing
tar extract relative to coke oven emissions extract were the
following: mouse skin painting, 3.9; in vitro transformation,
0.4; mutagenesis without activation, 2.3; mutagenesis with
activation, 1.0. While such comparison of relative potencies may
be confounded by the fact that the human subjects were truly
exposed to emissions which include particles and the nonhuman
tests involved only particle-free extracts, the comparison does
begin to identify test systems which more accurately reflect human
relative potency. In this case the most accurate system is the
mouse skin painting. It is also noteworthy that relative potency
is consistent within an order of magnitude.

Example Application: Relative Potency and Dose Response

When the appropriate data have been assembled, this
methodology provides a protocol for estimating human risk from
complex technological facilities. To illustrate the methodology,
a hypothetical example will be developed, namely the extent of
human respiratory cancer as a function of exposure to diesel
exhaust from vehicle C. The nonhuman testing data from Harris
(1981) and the data on lung cancer in coke oven workers (Mazumdar
et al., 1975) will be used in this example.

As can be seen in Figure 4, the potency of extract C relative
to coke oven emissions is reasonably constant. The range in
relative potency, considering only three tests, is 12.8 to 13.5
with the median being:

$$13.0 \ \frac{\mu g \ \text{vehicle C extract}}{\mu g \ \text{coke oven extract}}$$

We will use the median value in this example.

The data on lung cancer mortality versus cumulative exposure
to coke oven emissions (Mazumdar et al., 1975) were presented in
Figure 2. Five dose-response models were fit to the data with the
constraint that they must pass through the observed mortality for
zero exposure. As can be seen, two of the models clearly fit the
trend of the data while three do not. These two models have been
redrawn in Figure 5, where a relative potency of 13.0 has been
used to derive an equivalent exposure to vehicle C exhaust

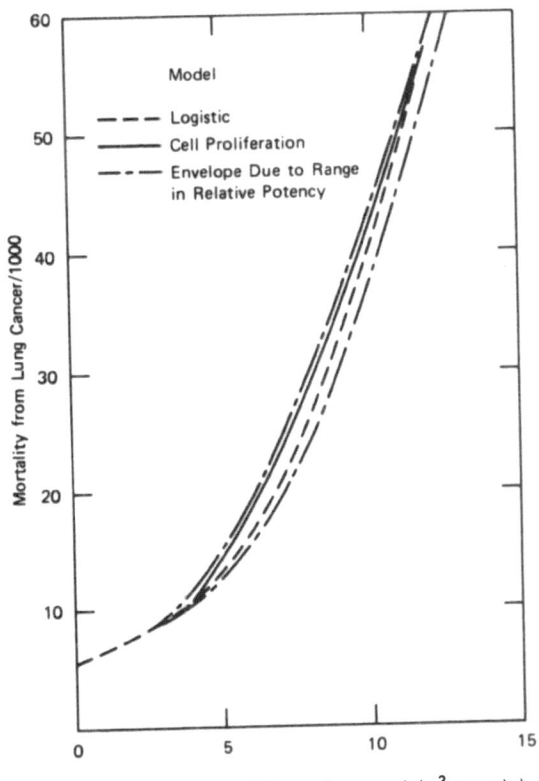

Figure 5. Risk due to exposure to exhaust from vehicle C was predicted by the relative potency approach.

emissions. While we acknowledge that such a hypothetical dose-response curve is not definitive, it is a reasonable approach to risk estimation for those chemicals which may be both toxic and beneficial but for which we do not yet have human data.

CONCLUSIONS

The relative potency approach for health effects analysis meets risk assessment requirements. Furthermore, it makes maximum use of available data on human incidence of chronic diseases. Data taken in any biological test system can be analyzed and provide input to the risk assessment process using the relative potency approach. This approach is similar to other health effects assessment protocols (Harris, 1981; EPA, 1981; BEIR, 1980) but is broader in its consideration of many possible reference chemicals. A major shortcoming of this general approach is the paucity of data which are particularly applicable in this approach. On the other hand, if further experimental and theoretical analysis confirms our preliminary observation that relative potency factors are reasonably consistent among biological systems, including man, then perhaps more attention should be given to including a set of reference chemicals in biological test protocols.

ACKNOWLEDGMENTS

Research was sponsored by the U.S. Environmental Protection Agency under Union Carbide Contract no. W-7405-eng-26 with the U.S. Department of Energy.

REFERENCES

BEIR (Committee on Biological Effects of Ionizing Radiation). 1980. The Effects on Populations of Exposure to Low Levels of Ionizing Radiation. National Academy Press: Washington, DC.

Guess, H., K. Crump, and R. Peto. 1977. Uncertainty estimates for low dose-rate extrapolations of animal carcinogenicity data. Cancer Res. 37:3475-3483.

Harris, J.E. 1981. Potential Risk of Lung Cancer from Diesel Engine Emissions. National Academy Press: Washington, DC.

Jones, T.D., G.D. Griffin, C.S. Dudney, and P.J. Walsh. 1981. Empirical observations in support of carcinogenic promotion as a tool for screening and regulation of toxic agents. In: Chemical Analysis and Biological Fate: Polynuclear Aromatic Hydrocarbons. M. Cooke and A.J. Dennis, eds. Battelle Press: Columbus, OH.

Mazumdar, S., C. Redmond, W. Sollecito, and N. Sussman. 1975. An epidemiological study of exposure to coal tar pitch volatiles among coke oven workers. J. Air Pollut. Control Assoc. 25:382-389.

Munro, I.C., and D.R. Krewsky. 1981. The role of risk assessment in regulatory decision making. In: Health Risk Analysis; Proceedings of the Third Life Sciences Symposium. The Franklin Institute Press: Philadelphia. pp. 443-460.

National Research Council, 1975. Pest Control: An Assessment of
    Present and Alternative Technologies, Volume I: Contemporary
    pest control practices and prospects, the report of the
    Executive Committee. National Academy of Sciences:
    Washington, DC.

Office of Technology Assessment. 1981. Assessment of
    Technologies for Determining Cancer Risks from the
    Environment. Report published and available from Congress of
    the United States.

Peto, R. 1978. Carcinogenic effects of chronic exposure to very
    low levels of toxic substances. Environ. Health Perspec.
    22:155-159.

Richmond, C.R., P.J. Walsh, and E.D. Copenhaver, eds. 1981.
    Health Risk Analysis; Proceedings of the Third Life Sciences
    Symposium. The Franklin Institute Press: Philadelphia.

U.S. Environmental Protection Agency. 1981. The Assessment of
    Risk to Human Reproduction and Development of the Human
    Conceptus from Exposure to Environmental Substances. Draft
    of a workshop report to be published in Advances in Toxicol.

PASSIVE SMOKING AND URBAN AIR POLLUTION: SALMONELLA/MICROSOME

MUTAGENICITY ASSAY OF SIMULTANEOUSLY COLLECTED INDOOR AND OUTDOOR

PARTICULATE MATTER

Göran Löfroth,[1] Lena Nilsson,[1] and Ingrid Alfheim[2]

[1]Department of Radiobiology, University of Stockholm
S-106 91 Stockholm, Sweden, and [2]Central Institute for
Industrial Research, Blindern, Oslo 3, Norway

INTRODUCTION

The mutagenic and carcinogenic potential of ambient airborne
particulate matter originating from mobile and stationary
combustion sources has attracted a renewed interest since it was
discovered that extracts of such particles contained substances,
detectable by the Salmonella mutagenicity assay, that were not
conventional polycyclic aromatic hydrocarbons (PAH) (Pitts et al.,
1977; Talcott and Wei, 1977). It is now indicated that the
responsible compounds, which can be present in both combustion
emissions and ambient air, might be nitrated or oxygenated PAH
derivatives (Alfheim et al., 1982), but the relative and absolute
health implications of these components and PAH are not known.

Tobacco smoking is known to cause several health problems
such as a higher risk of cancer. The major causative mechanism,
whether being initiation by mutagenic compounds, promotion, or
some other mechanism, is not known although it has been indicated
that promotion may be a major factor for the development of
bronchial cancer. Tobacco smoking is associated with the emission
of sidestream smoke, and this type of air pollution has been the
subject of several studies (see, e.g., Schmeltz et al., 1975;
Valentin et al., 1978). A continued interest can be expected
following the reports of Garfinkel (1981), Hirayama (1981), and

Trichopoulos et al. (1981) concerning the risk of cancer from exposure to sidestream tobacco smoke.

This study is an attempt to compare indoor air pollution from cigarette sidestream smoke to urban airborne particulate matter using the Salmonella/microsome mutagenicity assay as a tool. The result of the study shows that cigarette sidestream smoke may be a more signficant source of indoor air particulate mutagenicity than other combustion sources in an urban area.

MATERIALS AND METHODS

Sampling

    Sampling sites. Sampling was made at two large office buildings, I and II, in the Stockholm area and was performed simultaneously outdoors at the rooftop near the intake of fresh ventilation air and indoors in the ventilation chambers. The buildings are equipped with ventilation systems by which air is partly recirculated for energy conservation. The extent of recirculation is dependent on the outdoor temperature. Both recirculated and fresh air passes through a filter system for removal of some suspended particulate matter. The ventilation filters were either of the type Hi-Flo 45/Airosolve 45 (Office I in 1978) or Hi-Flo 85/Airosolve 85 (Office II in 1979; Office I in 1981). Indoor sampling was either performed at a site representing air leaving the offices or at sites before or after the ventilation filter representing air going into the offices. Building I engaged 300 to 400 persons during office hours, had a net volume of about 55,000 m^3, and a ventilation rate of 61 m^3/s, of which about 25% was exchanged for fresh air during the study periods. Building II had a similar net volume and ventilation rate per person, and the exchange for fresh air was 30 to 40%. The ventilation systems operated during the entire sampling periods.

    Particulate matter. Outdoor and indoor particulate matter was collected by high volume sampling on glass-fiber filter as has been described earlier (Löfroth, 1981). All samples have been collected for 24 h from early morning to the same time next day.

Particulate matter in cigarette sidestream smoke was sampled by successively smoldering 100 cigarettes of common brands in a funnel-shaped hood and collecting particulates on glass-fiber filter from the top of the hood with a high volume sampler run on reduced air flow. The time spent between cigarette and filter is estimated to about 1 min.

Sample Preparation

Glass-fiber filters were Soxhlet-extracted with acetone and the extract was converted into a dimethyl sulfoxide (DMSO) solution (Löfroth, 1980, 1981).

Fractionation with respect to polarity into basic, acidic, aliphatic, aromatic, and oxygenated fractions was made as described in an earlier study of motor vehicle exhaust (Löfroth, 1980).

Analyses

Mutagenicity was determined by the Salmonella plate incorporation method with bacterial cultures fully grown overnight as described by Ames et al. (1975) and applied to extracts of particulate matter (Löfroth, 1980, 1981). Assays were performed with the tester strains TA98 and TA100 obtained from Dr. B.N. Ames (Berkeley, CA). The microsome-containing liver supernatant (S9) was prepared from Aroclor 1254-induced male Sprague-Dawley rats and was used with necessary cofactors (S9-mix). The amount of S9 used was 20 or 50 μl/plate.

Analysis of PAH was performed by using a part of the glass-fiber filter with extraction, clean-up, and analytical determination by gas chromatrography (GC) as described by Bjørseth (1977).

Analysis of nicotine was performed on the original DMSO extracts, and the basic fractions dissolved in DMSO. The solutions were extracted with cyclohexane in the presence of aqueous sodium hydroxide. Nicotine was then determined in the cyclohexane phase by glass-capillary GC on SE-54 with FID, and quantification was made by treating known amounts of nicotine in DMSO in the same manner as the samples. The identity of nicotine was verified by glass-capillary GC on CPtmSil 5 with an N/P-sensitive detector.

RESULTS

Airborne Particulate Matter

Many samples collected at the two office buildings have been analyzed for mutagenicity. Results of a representative selection of samples collected simultaneously outdoors and indoors are presented in Table 1. It can be observed that outdoor air

Table 1.   Mutagenicity of Outdoor and Indoor Particulate Matter

| | | | | Revertants/m^3 Air | | | |
| | | | | TA98 | | TA100 | |
Office	Sampling Date	Sample No.	Type of Air	−S9	+S9	−S9	+S9
I	12/04/78	101	Outdoor	18	16	--	--
		102	Indoor-leaving	10	32	--	--
I	12/05/78	103	Outdoor	23	21	--	--
		104	Indoor-leaving	10	35	--	--
II	02/04/79	121	Outdoor	39	28	47	35
		122	Indoor-incoming, after filter	9	22	15	10
II	02/06/79	126	Outdoor	31	24	30	21
		125	Indoor-leaving	5	25	13	12

contains mutagens which are detectable in TA98 and TA100 in the absence of S9 and that addition of S9 either decreases or does not significantly change the mutagenic response. This result has been obtained repeatedly with ambient urban samples collected at rooftop levels (Löfroth, 1981).   Indoor air, in contrast, contains mutagens which become detectable in TA98 in the presence of S9.

Samples 103 and 104 (Table 1) were analyzed for PAH. The outdoor sample contained readily detectable concentrations: phenanthrene, 1.0; fluoranthene, 2.0; pyrene, 3.4; benz(a)anthracene, 0.7; chrysene, 1.4; benzo(e)pyrene, 0.9; benzo(a)pyrene, 0.6; indeno(1,2,3-cd)pyrene, 0.8; and benzo(ghi)perylene, 1.4 ng/m^3.   The indoor sample contained components interfering with the detectability of many PAH but it was possible to ascertain that the sample contained < 0.3, < 0.25, and < 0.06 ng/m^3, respectively, of fluoranthene, pyrene, and benzo(a)pyrene.   It was thus possible to conclude that the enhanced mutagenic effect in the presence of S9 of indoor samples was not due to a source of common PAH.

Cigarette Sidestream Smoke

The extract of cigarette sidestream particulate matter was preferentially mutagenic in TA98 in the presence of S9 (Figure 1). The observed response corresponds to about 15,000 rev/cigarette. No response has been detected in the absence of S9, i.e., less than 1,000 rev/cigarette. The response in TA100 in the presence of S9 was very weak, being less than 3,000 rev/cigarette. Several

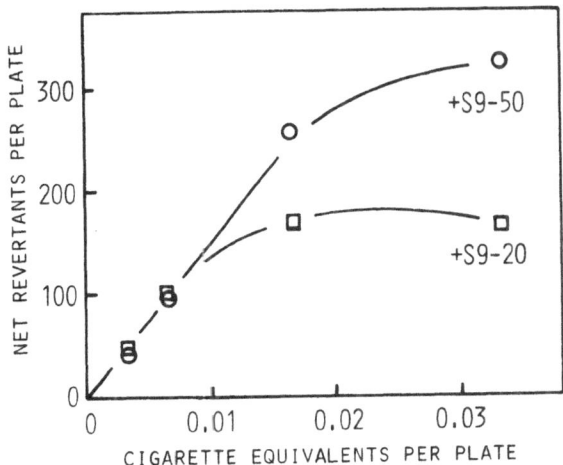

Figure 1. Mutagenicity in TA98 of an acetone extract of cigarette sidestream smoke particulate matter.

studies has been published about the mutagenicity of mainstream tobacco smoke in the **Salmonella** assay (see, e.g., Kier et al., 1974).

Fractionation

The extract of cigarette sidestream smoke particulate matter and combined extracts of outdoor and indoor samples were

fractionated with respect to polarity. Fractions as well as the
unfractionated samples were assayed for mutagenicity in TA98
(Table 2). The distribution of the mutagenic response of the
outdoor sample is very similar to that observed for other outdoor
samples (Löfroth, 1980). Among several features of outdoor
samples is a low response of the basic fraction. The indoor
sample has, in contrast, in the presence of S9 a high response
residing in the basic fraction which parallels the distribution in
sidestream smoke. The mutagenicity of the basic fraction may
partly or fully be due to a number of aza-arenes which have been
detected in both main- and sidestream tobacco smoke (Dong et al.,
1978).

Nicotine was determined in the unfractionated samples and
their basic fractions (Table 3). The differences in the
concentrations between unfractionated and basic samples may be due
to an incomplete partitioning in the fractionation or due to
differences in the extraction of nicotine by cyclohexane from the
different samples. The small amount of nicotine found in the

Table 2.   Relative Distribution of the Mutagenic Response in TA98
of Fractionated Samples

| Fraction | Outdoor 101 + 103 | | Indoor 102 + 104 | | Sidestream Smoke |
	-S9	+S9	-S9	+S9	+S9
Basic	5	5	15	65	80
Acidic	25	20	20	5	< 5
Aliphatic	< 5	< 5	< 10	< 5	< 5
Aromatic	10	35	15	10	15
Oxygenated	30	35	35	20	20
Total	70	95	85	100	115[a]
Unfractionated sample, rev/m^3 or cigarette	19	18	10	33	15,000

[a]A reconstituted sample gave a response of about
15,000 rev/cigarette.

Table 3. Nicotine Concentrations in Samples of Particulate Matter

Source	Unfractionated Sample	Basic Fraction
Outdoor: 101 + 103 ($\mu g/m^3$ air)	--	0.05
Indoor: 102 + 104 ($\mu g/m^3$ air)	1.2	0.5
Cigarette sidestream (mg/cigarette)	1.9	1.3

outdoor air sample may either be due to mixing of ventilation air
leaving the building into the rooftop air or be a general level
found in ambient air in densely populated areas with a high
proportion of smokers.

The similarity between indoor air and sidestream smoke with
respect to the distribution of the mutagenic response in different
fractions and the result of the nicotine analyses suggest that
cigarette smoke is causing the mutagenicity of the indoor air
which is detected in the presence of S9.

Nonsmoking and Smoking Conditions

The mutagenicity of outdoor and indoor particulate matter has
subsequently been measured at Office I during three consecutive
days with different work and smoking activities (Table 4). The
first day was a Sunday with no work activity. During the
following ordinary workday smoking was not allowed in the building
except for restroom areas, which were not connected to the main
ventilation system; however, a few cigarettes, about 10, were
smoked in an erroneously-designated rest area until the mistake
was discovered. During the last day, smoking was resumed. At the
end of the day, cigarettes were counted by collecting and weighing
all cigarette butts.

The weather conditions changed during the sampling period
from a rather windy situation to almost calm. This change
resulted in a variation of the outdoor air pollution levels
similar to changes observed previously (Löfroth, 1981).

For indoor samples, smoking causes a substantial increase of
the mutagenic response detected in TA98 in the presence of S9. No

Table 4. Mutagenicity of Outdoor and Indoor Particulate Matter during Different Activities

Sampling Date	Activities	Weather	Sample No.	Type of Air (Office I)	Revertants/m³ Air			
					TA98		TA100	
					-S9	+S9	-S9	+S9
02/08/81	None	+ 1.5°C 8 m/s NW	375 376 377	Outdoor Indoor-leaving Indoor-incoming, after filter	2.1 0.5 0.3	0.8 < 0.2 < 0.2	2.3 -- --	-- -- --
02/09/81	300 Persons working No smoking	- 3°C 3 m/s NW → calm	378 379 380	Outdoor Indoor-leaving Indoor-incoming, after filter	10 1.5 1.0	6 3.2 2.3	16 -- --	-- -- --
02/10/81	300 Persons working 2,000 cigarettes smoked 1,300,000 m³ fresh air exchanged	- 4°C 0.5 m/s Calm → N	381 382 383 384	Outdoor Indoor-leaving Indoor-incoming, before filter Indoor-incoming, after filter	25 7 9 4	16 24 28 11	38 18 -- --	21 10 -- --

or very little enhancement of the mutagenicity by S9 occurred when smoking was absent.

There is good quantitative agreement between the mutagenicity detected indoors and the mutagenicity expected from about 2,000 cigarettes. Assigning the indoor enhancement by S9 to about 20 rev/m^3 and multiplying this with the known air exchange of 1,300,000 m^3 (61 m^3/s x 0.25 x 24 h) results in an estimated production of mutagenicity corresponding to 26 million rev. The expected production is 2,000 cigarettes x 15,000 rev/cigarette, corresponding to 30 million rev.

The samplings were made during 24-h periods and all concentrations are consequently average values for 24 h. The mutagenic response of outdoor particulate matter is higher during daytime than during the night (Löfroth, 1981). It can be estimated that the average daytime response is about 1.35 times the average 24-h response although wide variations depending on meteorological conditions may occur. The mutagenic response from sidestream smoke is only present during office hours and the average daytime response should be about 2.6 times the average 24-h response.

Using these correction factors it can be estimated that the nonsmoking office personnel would have been exposed to about 6 rev/m^3 (4 x 1.35) during February 10 (Table 4) if everyone in the building had refrained from smoking. The presence of smokers, however, increased the exposure to about 30 rev/m^3 (11 x 2.6) for those who worked in rooms without smokers. The exposure in rooms with smokers must have been still higher.

DISCUSSION

The original purpose of the study which led to the present report was to investigate if the ventilation filter in Office I reduced the air pollution as measured by the Salmonella assay of extracts of particulate matter. It was, however, discovered at an early stage of this study that an indoor production occurred of mutagenic components which were of a type that required mammalian activation in order to elicit mutagenicity in the Salmonella assay. The origin of these components has now been firmly established to be sidestream tobacco smoke.

The original intention has been accomplished. It can be estimated that the 45-type ventilation filter removes about 50% and the 85-type 80 to 90% of the mutagenicity detected by TA98 in the absence of S9 representing the mutagenic response of ambient

outdoor particulate matter. The removal efficiency of the
mutagenicity detected in the presence of S9 representing cigarette
sidestream particulate matter is less, being 30 to 40 and 50% for
the two filter types.

The results obtained show that cigarette sidestream
particulate matter can be a substantial source of airborne
mutagenicity compared to the levels present in urban outdoor air.
It can be estimated that the smoke from one single cigarette,
corresponding to 15,000 rev, which is evenly spread in a residence
of 300 m^3 results in a mutagenic activity of 50 rev/m^3. The
annual average mutagenic activity of airborne particulate matter
collected at a rooftop site in the inner city of Stockholm has
been determined to be about 20 rev/m^3 (Löfroth, 1981).

Notwithstanding that urban air pollution and sidestream
tobacco smoke may lead to serious health problems, a comparison
with respect to general air quality indicates that sidestream
tobacco smoke may be the greater problem. Large segments of our
societies, including children of smoking parents and fellow
workers to smokers, are involuntarily exposed to sidestream
tobacco smoke with no or small possibilities of avoiding the
exposure.

ACKNOWLEDGMENT

    Studies on ambient outdoor air have been supported by grants
from the National Swedish Environment Protection Board.

REFERENCES

Alfheim, I., G. Löfroth, and M. Møller. (in press). Bioassay of
    extracts of ambient particulate matter. Environ. Health
    Perspect.

Ames, B.N., J. McCann, and E. Yamasaki. 1975. Methods for
    detecting carcinogens and mutagens with the Salmonella/
    mammalian-microsome mutagenicity test. Mutation Res.
    31:347-364.

Bjørseth, A. 1977. Analysis of polycyclic aromatic hydrocarbons
    in particulate matter by glass capillary gas chromatography.
    Anal. Chim. Acta 94:21-27.

Dong, M., I. Schmeltz, E. Jacobs, and D. Hoffman. 1978.
    Aza-arenes in tobacco smoke. J. Anal. Toxicol. 2:21-25.

Garfinkel, L. 1981. Time trends in lung cancer mortality among nonsmokers and a note on passive smoking. J. Natl. Cancer Inst. 66:1061-1066.

Hirayama, T. 1981. Non-smoking wives of heavy smokers have a higher risk of lung cancer: a study from Japan. Brit. Med. J. 282:181-185.

Kier, L.D., E. Yamasaki, and B.N. Ames. 1974. Detection of mutagenic activity in cigarette smoke condensate. Proc. Natl. Acad. Sci. USA 71:4159-4163.

Löfroth, G. 1980. Salmonella/microsome mutagenicity assays of exhaust from diesel and gasoline powered motor vehicles. In: Health Effects of Diesel Engine Emissions. W.E. Pepelko, R.M. Danner, and N.A. Clarke, eds. EPA-600/9-80-057a. U.S. Environmental Protection Agency: Cincinnati. pp. 327-342.

Löfroth, G. 1981. Comparison of the mutagenic activity in carbon particulate matter and in diesel and gasoline engine exhaust. In: Short-Term Bioassays in the Analysis of Complex Environmental Mixtures, Volume 2. M.D. Waters, S.S. Sandhu, J.L. Huisingh, L. Claxton, and S. Nesnow, eds. Plenum Press: New York. pp. 319-336.

Pitts, J.N., Jr., D. Grosjean, T.M. Mischke, V.F. Simmon, and D. Poole. 1977. Mutagenic activity of airborne particulate organic pollutants. Toxicol. Lett. 1:65-70.

Schmeltz, I., D. Hoffman, and E.L. Wynder. 1975. The influence of tobacco smoke on indoor atmospheres. Prev. Med. 4:66-82.

Talcott, R. and E. Wei. 1977. Airborne mutagens bioassayed in Salmonella typhimurium. J. Natl. Cancer Inst. 58:449-451.

Trichopoulos, D., A. Kalandidi, L. Sparros, and B. MacMahon. 1981. Lung cancer and passive smoking. Int. J. Cancer 27:1-4.

Valentin, H., H.-P. Bost, and E. Wawra. 1978. Das Passivrauchen am Arbeitsplatz - eine Gesundheitsschädigung? Zbl. Bakt. Hyg. I. Abt. Orig. B167:405-434.

ENVIRONMENTAL AROMATIC NITRO COMPOUNDS AND THEIR BACTERIAL

DETOXIFICATION

Yoshinari Ohnishi, Takemi Kinouchi, Yoshiki Manabe, and
Kazumi Wakisaka

Department of Bacteriology, School of Medicine, The
University of Tokushima, Tokushima 770, Japan

INTRODUCTION

Modern industry has produced many useful chemical agents, but
they and their by-products have also contributed to environmental
pollution. The effects of environmental chemicals on human health
are of considerable concern, since damage to DNA by environmental
mutagens, both natural and man-made, is likely to be a major cause
of cancer and other diseases (Rosenkranz, 1973; Doll, 1977; Wynder
and Gori, 1977; Ames, 1979).

Mutagenic nitroaromatic compounds have been found recently in
the urban atmosphere, automobile exhaust, and photocopies (Pitts
et al., 1978; Lofroth et al., 1980; Rosenkranz et al., 1980;
Tokiwa et al., 1981). 1-Nitropyrene (1-NP) is readily formed when
pyrene, ubiquitous in the environment, is exposed to NO_2 in the
urban atmosphere or in automobile exhaust, and is highly
mutagenic, inducing 449 his[+] rev/plate/nmol from Salmonella
typhimurium strain TA98 in the absence of S9 fraction in the
bacterial mutation test. Moreover, it is carcinogenic in rats
(Ohgaki et al., 1982).

Here, we report that wastewater from the oil-water separating
tanks of some gasoline stations in Japan contains 1-NP, and that
bacteria in the outside environment and in the human body convert

1-NP to 1-aminopyrene (1-AP) and proportionally decrease the mutagenic activity of 1-NP.

MATERIALS AND METHODS

Bacterial Strains and Cultivation

Bacterial strains used for enzyme assays are presented in Tables 1 and 2. Anaerobic bacteria were kindly supplied by Drs. T. Kawata, K. Ueno, A. Takagi, M. Ueda, M. Mutai, and M. Morotomi.

Table 1.   Production of 1-Aminopyrene from 1-Nitropyrene by Crude Extracts of Bacteria Isolated from Wastewater, and Decrease in the Mutagenic Activity

| Bacterial Strain | 1-Aminopyrene Produced | | Mutagenic Activity (%) |
	Total (nmol)	Specific Activity (nmol/h/mg protein)	
None	0	0	100[a]
Acinetobacter calcoaceticus	1.7	0.22	76.8
Aeromonas hydrophila	2.8	0.27	79.8
Enterobacter sp.	14.4	1.01	56.1
Proteus mirabilis	11.8	0.79	68.6
Proteus rettgeri	26.5	1.17	54.2
Proteus vulgaris	32.2	1.34	44.5
Pseudomonas aeruginosa	5.1	0.34	75.9
Pseudomonas alcaligenes	2.8	0.30	80.0
Pseudomonas fluorescens	4.7	0.34	79.1
Pseudomonas putida	3.3	0.34	80.2
Pseudomonas stutzeri	4.3	0.46	76.6
Bacillus sp.	13.8	0.65	57.8
Micrococcus sp.	2.2	0.19	86.2

[a]100% is 535 *his*+ rev/plate from strain TA98 in the absence of S9 mix.

Table 2.   Production of 1-Aminopyrene from 1-Nitropyrene by Crude Extracts of Bacteria, and Decrease in the Mutagenic Activity.

Bacterial Strain	1-Aminopyrene Produced		Mutagenic Activity (%)
	Total (nmol)	Specific Activity (nmol/h/mg protein)	
None	0	0	100[a]
Anaerobes			
Bacteroides fragilis	44.2	2.27	33.8
B. melaninogenicus ss. intermedius	5.6	0.27	74.1
B. melaninogenicus ss. melaninogenicus	8.5	0.63	56.2
B. oralis	6.4	0.47	73.3
B. thetaiotaomicron	69.0	2.59	16.6
B. vulgatus	54.3	2.35	28.5
Fusobactrium mortiferum	41.0	2.11	37.3
F. nucleatum	37.8	1.99	27.2
F. varium	17.5	0.93	45.5
Clostridium perfringens	41.0	1.89	33.8
C. sporogenes	35.0	1.83	31.8
Bifidobacterium adolescentis	14.0	1.68	54.6
B. bifidum	31.3	2.47	31.9
B. breve	4.6	0.42	75.1
B. infantis	< 0.4	< 0.08	93.3
Eubacterium lentum	27.6	1.59	45.9
E. limosum	19.8	1.02	46.1
Lactobacillus brevis	< 0.4	< 0.04	80.7
L. casei ss. casei	1.7	0.12	81.3
Propionibacterium acnes	< 0.4	< 0.04	80.7
Veillonella alcalescens	4.8	0.84	73.2
Peptococcus magnus	< 0.4	< 0.04	76.0
Peptostreptococcus anaerobius	36.8	2.73	34.7
Streptococcus intermedius	< 0.4	< 0.04	77.0
Aerobes			
Enterobacter aerogenes	15.7	0.84	54.6
Escherichia coli	7.2	0.38	74.6
Klebsiella pneumoniae	21.2	1.09	56.2
Proteus morganii	10.6	0.71	60.7
Salmonella typhimurium	3.0	0.19	72.0
Pseudomonas aeruginosa	3.6	0.27	75.1
Micrococcus luteus	7.3	0.44	75.3
Staphylococcus aureus	2.3	0.17	74.6
S. epidermidis	10.4	0.58	68.3
Streptococcus faecalis	5.3	0.35	74.6

[a]100% is 551 his+ rev/plate from strain TA98 in the absence of S9 mix.

For isolation of bacteria from wastewater, L broth was used.
For enzyme assays, aerobic bacteria were grown in nutrient broth
No. 2 (Oxoid) at 37°C with shaking. Anaerobic bacteria were
maintained in GAM semi-solid agar (Nissui) at room temperature and
grown in GAM medium (Nissui) at 35°C by use of the Forma
Scientific anaerobic system (model 1024) in an atmosphere of 80%
N_2, 10% H_2, and 10% CO_2.

Mutagenicity Test

Bacterial strains used for the mutation test were Salmonella
typhimurium LT-2 his strains TA98 and TA100. Mutagenesis assays
of air and automobile exhaust (Table 3) were carried out by the
procedure of Ames et al. (1975). For the mutation test on other
samples, we used the preincubation procedure (37°C, 20 min) with
50 µl of the liver S9 fraction from rats treated with
Aroclor 1254.

Table 3. Mutagenic Activity of Airborne Particles and Automobile
          Exhaust

Sample	Strain	S9	Rev/plate/m^3
Airborne particles in residential and industrial urban areas	TA98	+	1-445
Gasoline-engine exhaust from an unregulated 1972 car	TA98	+	8,500
	TA100	+	12,400
Gasoline-engine exhaust from a regulated 1975 passenger car	TA98	+	1,000
	TA100	+	4,000
Diesel-engine exhaust from a heavy-duty truck and bus	TA98	+	16,000-20,000
	TA100	+	9,000-17,000
Gasoline-engine exhaust from a 1978 passenger car	TA98	+	500
	TA100	+	800

Wastewater and Fractionation of the Extracts

One or two liters of wastewater were collected at the outlet in the oil-water separating tanks (Figure 1) of ten gasoline stations, five in the downtown area (A-E) and five in the suburbs (F-J). One liter of wastewater was mixed to 50 g ·of NaCl, 0.6 N HCl to adjust the pH to 1.0, and 200 ml of diethyl ether, and shaken at room temperature for 10 min. According to the procedure described previously (Ohnishi et al., 1980), the ether-soluble neutral, acidic, and basic fractions were separated.

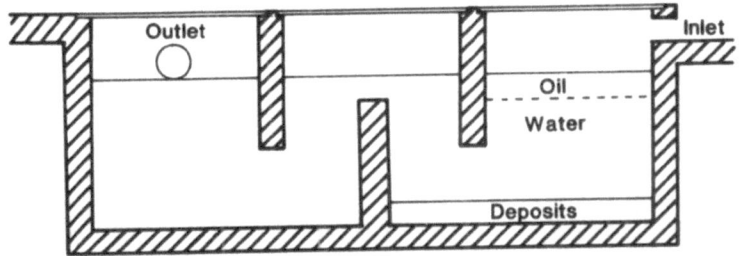

Figure 1.   Oil-water separating tank.

Chemical Analysis

To purify the mutagenic extracts and identify the mutagenic chemicals, high performance liquid chromatography (HPLC) (Shimadzu LC-3A) with a Zorbax ODS column, gas chromatography and mass spectrometry (GC-MS) (JEOL JMS-D300) were used.

Quantitative Analysis of 1-NP in Wastewater

Determination of the 1-NP content of wastewater by using HPLC was rather difficult because of contaminating materials. Therefore, we measured 1-AP converted from 1-NP by the enzyme reaction because 1-AP is fluorescent and easy to determine quantitatively. The reaction mixture (0.5 ml) contained the

ether-soluble neutral fraction or the 1-NP-corresponding fraction
eluted from the Zorbax ODS column by HPLC, nitroreductase I
partially purified from Bacteroides fragilis (Sephadex G-200
fraction), 2.5 µmol of glucose-6-phosphate (G-6-P), 2 µmol of
nicotinamide-adenine dinucleotide phosphate (HADPH), 0.5 unit of
G-6-P dehydrogenase, and 50 µmol of phosphate buffer (pH 6.8).
The mixture was incubated at 37°C for 15 h to complete the enzyme
reaction. The ethyl acetate extract of the reaction mixture was
subjected to HPLC, and 1-AP was measured quantitatively.

Preparation of Cell-Free Crude Extracts

Stationary-phase cells were harvested, washed, and suspended
in 0.1 M sodium phosphate buffer (pH 7.4). The suspension was
treated by sonic oscillation (Kubota Isonator model 200M) at
180 watts for 10 min. Cell debris was removed by centrifugation
at 9000g for 10 min. The supernatant fraction was used to
determine nitroreductase activity.

Nitroreductase Assays

The incubation mixture for 1-NP reduction contained the
following components in a total volume of 1 ml: 100 nmol of 1-NP,
4 µmol of $NADP^+$, 5 µmol of G-6-P, 33 µmol of KCl, 8 µmol of $MgCl_2$,
100 µmol of sodium phosphate buffer (pH 7.4), and enzyme fraction
(crude extract). The reaction mixture in a small tube was
incubated at 37°C for 3 h aerobically, but without shaking; 5 ml
of ethyl acetate was then added and the sample was mixed
vigorously. After low-speed centrifugation (3000g, 10 min), the
ethyl acetate phase was removed. The ethyl acetate extraction was
done twice and one portion of the ethyl acetate phase was used for
measurement of 1-AP with a Hitachi fluorescence spectrophotometer
(type MPF-2A); the other was used for the mutagenicity test.

Chemicals

$NADP^+$, NADPH, and G-6-P dehydrogenase were obtained from
Oriental Yeast Co., Ltd. G-6-P was from Sigma Chemical Co. 1-NP
and 1-AP were purchased from Aldrich Chemical Co. 1-NP was
contaminated with less than 1% 1,8- and 1,6-dinitropyrene. The
commercial 1-NP induced 1003 his+ rev/plate/250 ng from strain
TA98 in the absence of S9 mix. Pure 1-NP purified by HPLC showed
454 his+ rev/plate/250 ng. The dinitropyrene present as a
contaminant in 250 ng of the commercial 1-NP showed
453 his+ rev/plate, indicating that about 50% of the mutagenicity

of the commercial 1-NP is due to pure 1-NP and that the rest is
due to the contaminating dinitropyrene.

RESULTS

Mutagenicity of Engine Exhaust

We have determined the mutagenicity of airborne particulates,
automobile exhaust in Japan, and particulate exhaust from small
engines (Tokiwa et al., 1977, 1980; Ohnishi et al., 1980, 1982).
These data are summarized in Tables 3 and 4. The mutagenic
activity of the airborne particulates collected in residential and
industrial urban areas was 1 to 445 his[+] rev/plate/m^3 of air for
test strain TA98 incubated along with rat-liver extract in the
Ames test. Exhaust from an unregulated 1972 car provoked about
12,000 rev/plate/m^3 of exhaust gas, whereas that from a regulated
1975 passenger car showed about 4,000 rev/plate/m^3 from strain
TA100 in the presence of S9 mix. Furthermore, compared to these
gasoline engines, automobile exhaust from old heavy-duty diesel
engines showed very high mutagenic potency: about
20,000 rev/plate/m^3. The most restrictive minimal regulations of
gasoline engines of passenger cars was put into effect in Japan in
1978; exhaust from a 1978 improved passenger car equipped with a
catalyst showed about 800 rev/plate/m^3 in the Salmonella-microsome
test. On the other hand, the emissions of small engines used to
generate electric power or to run agricultural machinery are not
regulated in Japan at all. Table 4 shows that the mutagenic
activity of particulate exhaust from a small gasoline engine with
a load for electric generation of 1,000 watts was
20,500 rev/plate/m^3, and that from a small diesel engine without a
load was 14,200 rev/plate/m^3. To reduce exhaust pollutants and to
decrease fuel consumption, a newly devised combustion process
called "active thermo-atmosphere combustion" has been developed in
Japan (Onishi et al., 1979). This new lean combustion system
differs from conventional gasoline and diesel engine combustion
processes. The engine equipped with this new system is also
equipped with a catalytic muffler, because the lean combustion
permits it to have a satisfactory operating life, unlike more
conventional engines. Particulate exhaust from this engine,
called the "NiCE engine" from the initials of its company name,
Nippon Clean Engine Institute, had very little mutagenic activity:
less than 500 rev/plate/m^3.

We detected 1-NP, 3-nitrofluoranthene, and
5-nitroacenaphthene in the ether-soluble neutral fraction
extracted from diesel exhaust (Tokiwa et al., 1981).

Table 4. Mutagenic Activity of Particles in Small-Engine Exhaust

Engine	Strain	S9	Rev/plate/m^3
Gasoline engine (171 cc)	TA98	+	20,500
with 1000 watts	TA100	+	14,200
Diesel engine (269 cc)	TA98	+	14,200
without a load	TA100	+	8,480
NiCE engine (98 cc)	TA98	+	140
with 1000 watts	TA100	+	340

Mutagenicity of Wastewater from Oil-Water Separating Tanks of Gasoline Stations in Japan

In Japan, wastewater in gasoline stations, originating mainly from car-washing, must be stored in oil-water separating tanks (Figure 1). The water layer in the tank flows out to a drain and finally into the wastewater treatment plant or directly into the river. Table 5 shows the mutagenicity of extracts from wastewater of 10 tanks in 10 gasoline stations. The neutral fractions had higher mutagenicity than the other fractions. Moreover, most of the samples in neutral fractions show the highest mutagenicity for strain TA98 in the absence of S9 mix, suggesting that they contain nitroaromatic compounds. Therefore, we investigated the presence of 1-NP in the neutral fractions that induced high mutagenicity, by HPLC and GC-MS. We collected 1-NP-corresponding fractions from HPLC, subjected them to GC-MS and identified 1-NP in the neutral fractions of Samples 1 (Figure 2), 4-6, 9, and 10.

Estimation of the 1-NP Content of the Neutral Fractions Extracted from Wastewater

When we subjected the neutral fraction of sample no. 1 to HPLC, there was a peak at the retention time corresponding to 1-NP; however, the peak fraction contained the other, unknown compounds in addition to 1-NP. Therefore, we converted 1-NP in the fraction to fluorescent 1-AP with partially purified nitroreductase I, eluted 1-AP by HPLC, and measured it with the fluorescence detector. Table 6 shows the estimated 1-NP concentration of each neutral fraction of wastewater. The amount

Table 5. Mutagenic Activity of Wastewater from the Oil-Water
Separating Tanks in Gasoline Stations

| | | | \multicolumn{6}{c}{Revertants/Plate/1} | | | | | |
| | | | Neutral Fraction | | Acidic Fraction | | Basic Fraction | |
Sample Number	Gasoline Station	S9	TA98	TA100	TA98	TA100	TA98	TA100
1	A	−	13,100	843	3,450	1,120	1,530	242
		+	24,800	3,330	1,530	187	1,840	577
2	B	−	649	1,070	287	266	0	523
		+	156	0	0	225	24	288
3	C	−	567	405	409	433	320	440
		+	425	445	264	529	180	200
4	D	−	22,700	1,320	549	352	432	372
		+	13,800	3,790	0	418	0	180
5	E	−	22,700	1,500	6,940	461	240	340
		+	3,660	56	1,260	583	280	380
6	F	−	4,630	171	900	300	0	158
		+	2,510	1,490	450	1,350	113	497
7	G	−	1,530	680	2,640	1,200	56	84
		+	1,020	1,360	1,920	2,400	56	42
8	H	−	374	2,030	430	615	70	1,220
		+	107	695	123	0	0	0
9	I	−	39,800	3,610	23	552	0	476
		+	473	888	0	46	0	102
10	J	−	99,200	10,900	144	1,920	0	189
		+	3,360	679	0	96	0	0

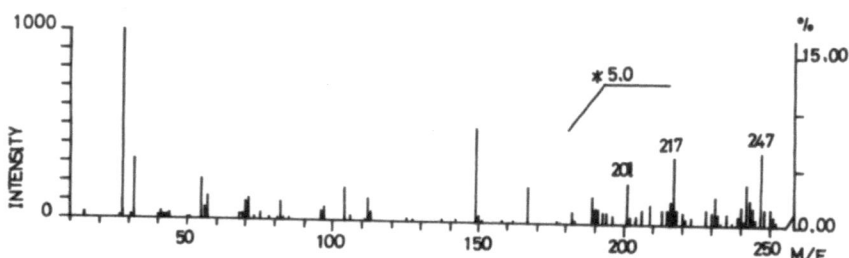

Figure 2.   Mass spectrum of 1-NP-corresponding fraction of the
            neutral fraction (wastewater no. 1) from ODS column.

Table 6.   Estimation of 1-Nitropyrene Concentration in the
           Wastewater from the Oil-Water Separating Tanks in
           Gasoline Stations

Sample Number	Gasoline Station	Quantity of 1-NP (ng/1)	Mutagenicity of Neutral Fraction TA98, -S9 (rev/plate/1)	% Mutagenicity of 1-NP in the Sample[a]
1	A	926	13,100	12.9
2	B	ND[b]	649	ND
3	C	ND	567	ND
4	D	433	22,700	3.5
5	E	101	22,700	0.8
6	F	361	4,630	14.2
7	G	ND	1,530	ND
8	H	ND	374	ND
9	I	165	39,800	0.8
10	J	323	99,200	0.6

[a]1 μg of 1-NP induces 1,816 his⁺ rev/plate from strain TA98 in the
absence of S9 mix.
[b]Not determined.

of 1-NP can explain 0.6 to 14% of the total mutagenicity of the mutagenic neutral fractions.

Bacterial Biotransformation of 1-NP to 1-AP

We now know that there is 1-NP in air and water. To determine whether 1-NP affects the human body directly or is converted to 1-AP in and outside the human body by the action of bacterial nitroreductases and whether 1-AP affects the human body, we isolated bacteria from wastewater and tested them for nitroreductase activity (Table 1). The nitroreductase activity of anaerobic bacteria in the normal flora and representative aerobic bacteria were also determined (Table 2). Bacteria in wastewater show less nitroreductase activity. Some of the normal bacterial flora of the mouth, such as Fusobacterium nucleatum, Bifidobacterium adolescentis, and Peptostreptococcus anaerobius, had very high nitroreductase activity. Crude extracts from intestinal anaerobic bacteria, including Bacteroides fragilis, B. thetaiotaomicron, B. vulgatus, Fusobacterium mortiferum, F. nucleatum, Clostridium perfringens, C. sporogenes, Bifidobacterium adolescentis, B. bifidum, Eubacterium lentum, E. limosum, and Peptostreptococcus anaerobius, all easily converted 1-NP to 1-AP and proportionally decreased the mutagenic activity of 1-NP. On the other hand, the nitroreductase activity of aerobic bacteria, except Klebsiella pneumoniae, was low.

DISCUSSION

We have detected 1-NP in engine exhaust and wastewater. 1-NP is also readily formed by exposure of pyrene, always present in city air and automobile exhaust, to NO_2. In addition, when pyrene formed in incomplete combustion processes is exposed to NO_2 in urban cooking gas, mutagenic nitro derivatives would be readily induced. We do not know how 1-NP in wastewater affects human beings and whether 1-NP in wastewater remains 1-NP in the environment or is converted to other chemicals. Biotransformation of 1-NP in bacteria in the environment or in the normal human flora may be an advantage for man because the bacteria can convert 1-NP to 1-AP and decrease the mutagenic activity. However, 1-AP is still mutagenic, though its activity is low. Therefore, it is now important to investigate the relative carcinogenicity of 1-NP and 1-AP. Since the mutagenic activity of dinitropyrenes is higher than that of mononitropyrene (Tokiwa et al., 1981), the evaluation of these dinitropyrenes and diaminopyrenes is also urgent. Carcinogenicity tests of these chemicals are under way.

ACKNOWLEDGMENT

This work was supported in part by Grant-in-Aid for
Scientific and Cancer Research from the Ministry of Education,
Science and Culture, and the Ministry of Health and Welfare of
Japan, and funds from the Nissan Science Foundation.

REFERENCES

Ames, B.N. 1979. Identifying environmental chemicals causing
    mutations and cancer. Science 204:587-593.

Ames, B.N., J. McCann, and E. Yamasaki. 1975. Methods for
    detecting carcinogens and mutagens with the Salmonella/
    mammalian-microsome mutagenicity test. Mutation Res.
    31:347-363.

Doll, R. 1977. Strategy for detection of cancer hazards to man.
    Nature 265:589-596.

Löfroth, G., E. Hefner, I. Alfheim, and M. Møller. 1980.
    Mutagenic activity in photocopies. Science 209:1037-1039.

Ohgaki, H., N. Matsukura, K. Morino, T. Kawachi, T. Sugimura,
    K. Morita, H. Tokiwa, and T. Hirota. 1982. Carcinogenicity
    in rats of the mutagenic compounds 1-nitropyrene and
    3-nitrofluoranthene. Cancer Lett. 15:1-7.

Ohnishi, Y., K. Kachi, K. Sato, I. Tahara, H. Takeyoshi, and
    H. Tokiwa. 1980. Detection of mutagenic activity in
    automobile exhaust. Mutation Res. 77:229-240.

Ohnishi, Y., H. Okazaki, K. Wakisaka, T. Kinouchi, T. Kikuchi, and
    K. Furuya. 1982. Mutagenicity of particulates in small
    engine exhaust. Mutation Res. 103:251-256.

Onishi, S., S.H. Jo, K. Shoda, P.D. Jo, and S. Kato. 1979.
    Active thermo-atmosphere combustion (ATAC)--A new combustion
    process for internal combustion engines. SAE Paper 790501.

Pitts, J.N., Jr., K.A. Van Cauwenberghe, D. Grosjean, J.P. Schmid,
    D.P. Fitz, W.L. Belser, Jr., G.B. Knudson, and P.M. Hynds.
    1978. Atmospheric reactions of polycyclic aromatic
    hydrocarbons: facil formation of mutagenic nitro
    derivatives. Science 202:515-519.

Rosenkranz, H.S. 1973. Aspects of microbiology in cancer
    research. Ann. Rev. Microbiol. 27:383-401.

Rosenkranz, H.S., E.C. McCoy, D.R. Sanders, M. Butler,
    D.K. Kiriazides, and R. Mermelstein. 1980. Nitropyrenes:
    isolation, identification, and reduction of mutagenic
    impurities in carbon black and toners. Science
    209:1039-1043.

Tokiwa, H., S. Kitamori, K. Takahashi, and Y. Ohnishi. 1980.
    Mutagenic and chemical assay of extracts of airborne
    particulates. Mutation Res. 77:99-108.

Tokiwa, H., K. Morita, H. Takeyoshi, K. Takahashi, and Y. Ohnishi.
    1977. Detection of mutagenic activity in particulate air
    pollutants. Mutation Res. 48:237-248.

Tokiwa, H., R. Nakagawa, K. Morita, and Y. Ohnishi. 1981.
    Mutagenicity of nitro derivatives induced by exposure of
    aromatic compounds to nitrogen dioxide. Mutation Res.
    85:195-205.

Tokiwa, H., R. Nakagawa, and Y. Ohnishi. 1981. Mutagenic assay
    of aromatic nitro compounds with Salmonella typhimurium.
    Mutation Res. 91:321-325.

Wynder, E.L. and G.B. Gori. 1977. Contribution of the
    environment to cancer incidence: an epidemiologic exercise.
    J. Natl. Cancer Inst. 58:825-832.

APPLICATION OF MUTAGENICITY TESTS FOR DETECTION AND SOURCE

ASSESSMENT OF GENOTOXIC AGENTS IN THE RUBBER WORK ATMOSPHERE

Agneta Hedenstedt Rannug [1] and Conny Östman[2]

[1]Department of Toxicology Genetics, Wallenberg Laboratory
University of Stockhom, S-106 91 Stockholm, Sweden
[2]Department of Analytical Chemistry, Arrhenius Laboratory
University of Stockholm, S-106 91 Stockholm, Sweden

INTRODUCTION

In 1954, the increased risks of developing urinary bladder
tumors were established among rubber workers (Case and Hosker,
1954). Later, other cancer forms were reported to appear in
excess among workers in this industry (Peters et al., 1976; Monson
and Fine, 1978; Hakama and Kilpikari, in press). The earlier use
of beta-naphthylamine as an antioxidant was probably the single
cause of the bladder tumors. Dust has been suggested as the most
probable cause of stomach cancer among workers in the weighing and
mixing departments (Maisey, 1981). Exposure to hazardous
substances in fumes generated during the vulcanization of rubber
leads to an elevated risk of developing lung disease (Peters et
al., 1976; Maisey, 1981). This exposure may also explain the
enhanced lung cancer frequencies found among workers in the curing
department (Fox et al., 1974).

The amounts and complexity of chemicals used and generated in
the rubber manufacturing processes make a risk identification very
difficult. A new approach to the problem of identifying noxious
substances in polluted working environments uses short-term
bioassays for mutagenicity and teratogenicity in combination with
human monitoring and epidemiological programs. This work,
undertaken as a Swedish-Finnish collaborative study between
Wallenberg Laboratory, the University of Stockholm and the

Institute of Occupational Health, Helsinki, can be divided into
three major areas: (1) short-term bioassays -- Ames test,
Drosophila (recessive lethals), hamster cells (HGPRT-point
mutations), micronucleus test, and chicken embryo teratogenicity;
(2) human monitoring -- urine (Ames test and tioether
concentration) and blood (SCE in lymphocytes and chromosomal
aberrations); and (3) epidemiology -- spontaneous abortions,
malformations, and cancer.

Falck et al. (1980) have observed increased mutagenic
activity in the urine of rubber workers as compared to a control
group of office clerks. A number of substances used as rubber
additives have been shown to be genotoxic in short-term bioassays
(Hedenstedt et al., 1979; Hedenstedt et al., 1981; Donner, 1981).
Mutagenic substances have been found especially among accelerators
of the dithiocarbamatic type and among antioxidants. Many of
these chemicals have melting points that admit vaporization at
mixing and curing temperatures (approximately 100 to 220°C).
Other additives, like soot and aromatic oils, may pose a risk by
emitting polynuclear aromatic hydrocarbons. Volatilized rubber
polymers seem to play an important role in the mutagenicity
detected in the fumes (Hedenstedt et al., in press).

The present investigation using the Ames test was undertaken
to determine to what extent additives and rubber materials in a
number of different polymer types contribute to the total
mutagenicity. The mutagenicity tests were conducted on condensed
curing fumes generated in a laboratory. The chemical complexity
of the condensates reported by others (Fraser and Rappaport, 1976;
Willoughby and Lawson, 1979) was confirmed by gas chromatography
(GC). Chemical separation of one chloroprene rubber condensate
was performed in order of increasing polarity, and the fractions
were tested for mutagenicity.

MATERIALS AND METHODS

Collection and Preparation of Samples

Industrial samples. Vulcanization fumes were filter
collected from chloroprene rubber (CR) cured in a continuous
process by heating to 220°C in a salt bath containing $NaNO_3$.
Fumes were collected on glass-fiber filters outside the closed
process using a high-volume sampler. The filters were
continuously extracted with acetone for 18 h. The extracts were
then concentrated by evaporation.

Laboratory-generated samples. For laboratory samples, 50 g of a rubber material was heated in a Pyrex reaction flask to temperatures corresponding to the industrial mixing and curing operations. Vapors formed by this thermal treatment were collected on a water-cooled condenser that reached down into the flask. The condenser was withdrawn every 5 min and washed in a tube containing 10 ml acetone. A total sampling period of 30 min was used for each condensate. Mutagenicity tests were usually performed within 2 h. The rubber materials were of two types: first, compounded materials with all supplements added, and second, the rubber polymers only. The first group consisted of fluor rubber (FPM), ethylene propylene rubber (EPDM), and chloroprene rubber (CR). The second group was composed of standard Malaysian rubber polymer (SMR polymer), fluor rubber polymer (FPM polymer), ethylene propylene rubber polymer (EPDM polymer), and chloroprene rubber polymer (CR polymer).

Chemical Fractionation

Materials. The column packing used for the fractionation was kieselgel 60, particle size 63 to 200 μm, from Merck, deactivated with 10% water by weight. The glass columns, with an inner diameter of 10 mm, were packed with silica to a height of 120 mm. A Teflon valve was used to regulate the solvent flow. All solvents were of analytical grade, distilled and analyzed on a capillary GC, and tested for biological activity prior to use.

Procedure. Fractionation was performed on a silica column using 300 µl of the sample. The sample consisted of the high-temperature (II) condensate from the heating of a CR sample similar in composition to that presented in Figure 4A. Four fractions were collected, in order of increasing polarity, by eluting the column with 10 ml n-hexane, 45 ml benzene, 50 ml acetone, and 50 ml ethanol. The fractions were evaporated and dissolved in suitable solvents prior to biological testing. Before fractionation, the silica column was washed with 10 ml n-hexane and this eluate was tested for biological activity.

Mutagenicity Testing

Salmonella typhimurium.TA100 and TA98 were the main strains used for the mutagenicity studies. Complementary studies were performed with TA1535 and TA1538, primarily to verify and establish the type of mutagenicity found. Only data on strains TA98 and TA100 will be presented here. The Salmonella strains were originally provided by Dr. Bruce Ames, and the mutability of

the strains and their genotypes were regularly checked by
recommended procedures (Ames et al., 1975; de Serres and Shelby,
1979). Basically, the plate-incorporation assay (Ames et al.,
1975) was followed with minor modifications. Thus, histidine
(0.1 μmol/plate) and biotin (0.1 μmol/plate) were added to the
minimal medium instead of to the soft agar. Liver preparations
(S9 fractions) from the pooled livers of five Aroclor-pretreated
male Sprague-Dawley rats were used as metabolizing systems. The
S9 mix contained 10% S9 fraction.

The concentrated filter extracts were added to the plates in
amounts corresponding to 2.13, 21.3, and 42.6 l of air.
Condensates from the laboratory curing processes were added in
amounts ranging from 0.1 μl to 100 μl. In the case of the
fractionated CR sample, 1 μl and 10 μl of the acetone-based
fractions were added.

RESULTS

Industrial Samples

The fumes generated at the curing of CR showed a
statistically significant rise in the number of mutants per plate
above the control level at all concentrations tested on TA100 in
the presence of S9 (Table 1). For TA98, a slight and
statistically nonsignificant increase was indicated. On strains
TA1535 and TA1538, an extract volume corresponding to 213 l of air
produced both direct and indirect effects that were statistically
significant (data not shown).

Laboratory-Prepared Samples

Natural rubber. Condensed gases from thermal treatments of
an SMR polymer showed a slight but not dose-dependent increase
above the controls at the 95% significance level for TA100 in the
presence of S9 at the low temperature level (I) (Figure 1). At
the high temperature level (II), this condensate produced no
increase above the controls.

Fluor rubber. Condensed gases from thermal treatments of FPM
and FPM polymers showed very similar mutagenicity patterns
(Figure 2). The condensates collected at the low temperature
level (I) had a low activity that could be due to chance
fluctuations. At the high temperature level (II), however, an
indirect effect was clearly seen on both TA98 and TA100. Parallel
experiment series (data not shown) on TA1535 and TA1538 gave a

Table 1. Mutagenicity of Filter-Collected Air Samples from
          Vulcanization of Chloroprene Rubber (CR) on
          Salmonella typhimurium TA98 and TA100 with and
          without a Metabolizing System (S9)

| | No. of Mutants/Plate ± S.E.[a] | | | |
| | TA98 | | TA100 | |
Sample Volume (1)[b]	−S9	+S9	−S9	+S9
0	31.0 ± 1.2	39.5 ± 4.3	155.3 ± 5.7	150.5 ± 6.0
2.13	30.5 ± 3.1	29.3 ± 5.0	158.8 ± 6.4	184.0 ± 7.0[*]
21.3	38.3 ± 6.6	50.8 ± 3.1	157.8 ± 2.9	199.5 ± 8.3[**]
42.6	30.3 ± 4.0	51.8 ± 3.4	165.8 ± 6.8	230.0 ± 4.1[***]

[a]Statistical analyses were performed with Student's t test.
[b]Samples were extracted with acetone.
[*]Significance level: $0.01 < p < 0.05$.
[**]Significance level: $0.001 < p < 0.01$.
[***]Significance level: $p < 0.001$.

total lack of response on strain TA1535, indicating that the
mutagenicity is due to the presence of a frameshift-inducing
substance(s).

     Ethylene propylene rubber. An EPDM polymer produced weak
effects on TA98 and TA100 in both temperature regions (Figure 3).
The results on TA1535 and TA1538 were negative (data not shown).
Sample A, a rubber material containing a curing system based on
sulfur, showed strong mutagenic effects on TA100 that were
potentiated by S9. The other sample, B, containing a
peroxide-type curing system, showed a mutagenicity pattern
reminiscent of the polymer itself.

     Chloroprene rubber. The mutagenic effects from two CR blends
(A,B) showed a striking resemblance to the mutagenicity
originating from the polymer material itself (C) (Figure 4). The
CR polymer sample produced direct-acting gaseous compounds at the
low temperature level (I) and indirect-acting compounds at the
high temperature level (II) on both TA98 and TA100. This overall
pattern could also be seen with the two compounded materials A

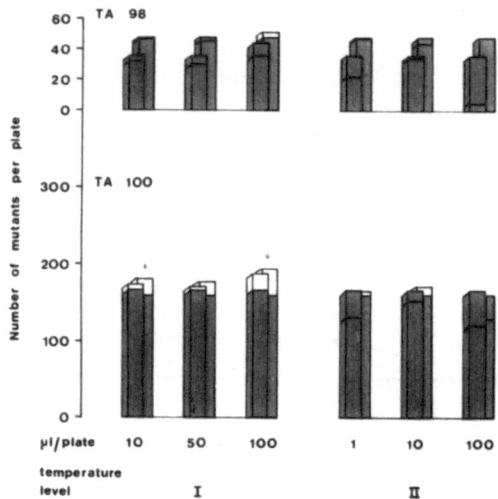

Figure 1. Mutagenicity on Salmonella typhimurium TA98 and TA100
           of condensed gases from an SMR polymer pyrolyzation at
           two temperatures corresponding to mixing (temperature
           level I) and curing (temperature level II). Filled
           bars represent control levels; bars in the front, −S9
           series; and bars in the back, +S9 series. Values in
           experiment series below control values are indicated by
           horizontal lines in the filled parts of the bars. For
           statistical analyses, see Table 1.

and B. For B, however, the presence in the condensate also of an
indirectly acting species was observed in the low temperature
region. The true base substitution character of the effects on
TA100 was established with tests on TA1535 (data not shown).

CR-Condensate Fractions

     The total condensate produced indirect effects of both base
substitution and frameshift character (Figure 5). The results on
TA98 and TA100 of the total condensate were verified on strains
TA1535 and TA1538 (data not shown). The hexane, and to a larger
extent the benzene, eluate showed these characteristic effects.
The acetone eluate was totally inactive but the most polar
solvent, ethanol, also extracted indirectly acting mutagens. A
weak, however, not dose-dependent, direct effect was also noticed
with TA100.

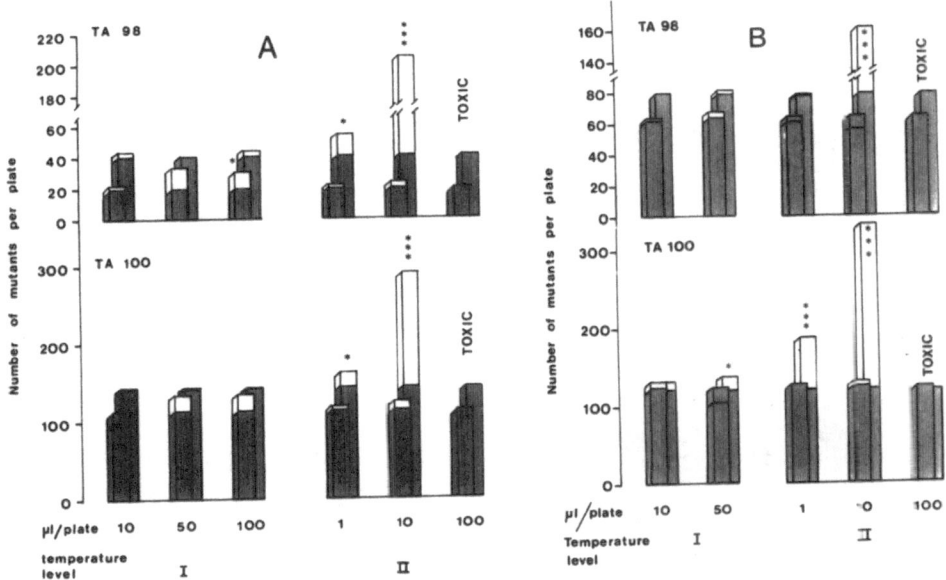

Figure 2. Mutagenicity on Salmonella typhimurium TA98 and TA100
of products from the pyrolyzation of FPM (A) and FPM
polymer (B). For further details, see Figure 1.

DISCUSSION

The data for CR and FPM pyrolyzates indicate a definite
contribution from the polymers to the mutagenicity of the
compounded materials. A natural rubber sample (SMR polymer) was
regarded as negative, and an EPDM polymer produced only weak
effects. It must be stated that the polymers were "stabilized" to
some extent with antioxidants. It is, however, exceedingly
difficult to determine the exact composition of these polymerized
materials.

The apparent difference in mutagenic activity between the two
EPDM samples (A and B in Figure 3) obviously depends on the
different curing systems used. The "low effect sample," the
peroxide-cured rubber (B), must not be considered safer based on
these data only, as the Salmonella strains used here do not
respond to mutagens of the peroxide type. Sample A consisted of a
number of sulfur-containing substances such as tetramethylthiuram
disulfide (TMTD), tellur diethyldithiocarbamate (TDEC), ethylene
thiourea (ETU), and sulfur powder. These substances or reaction

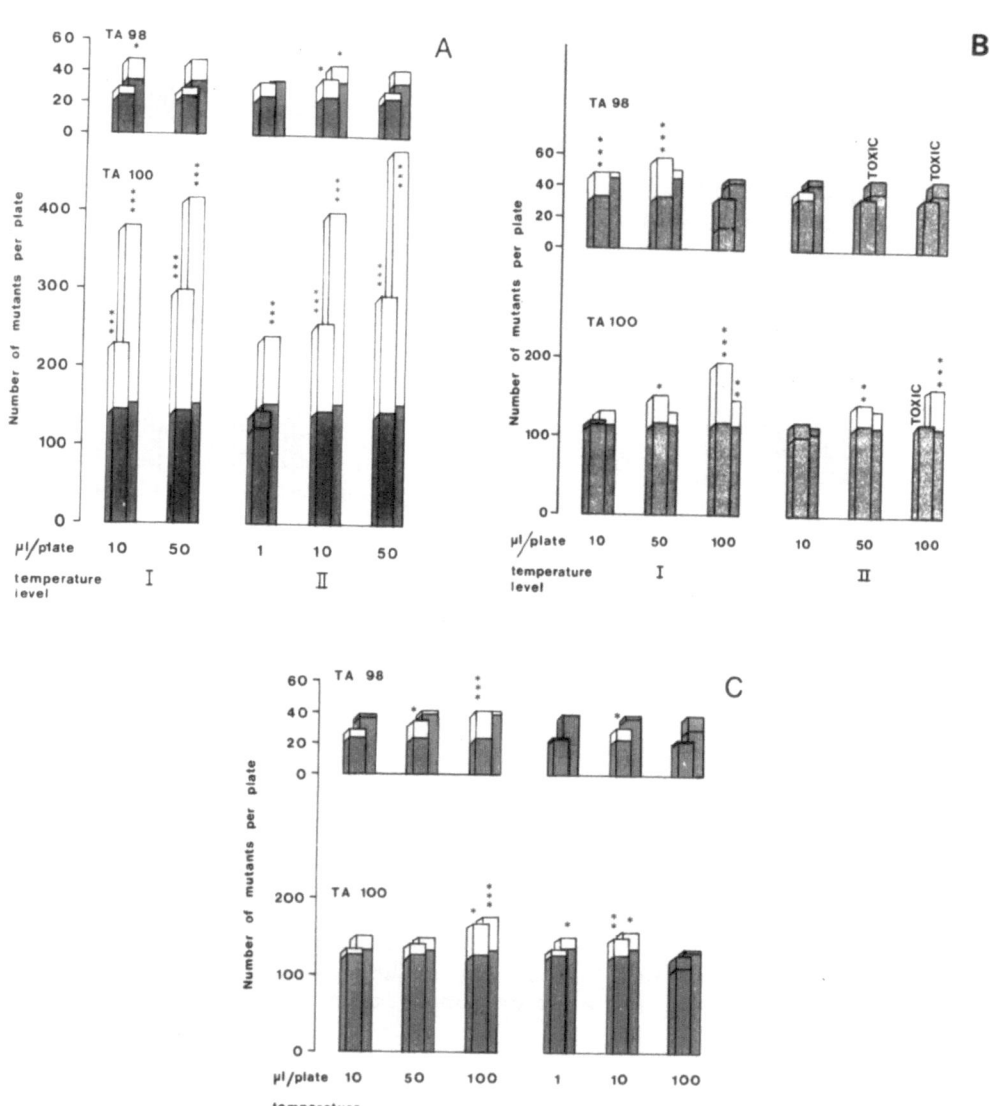

Figure 3.   Comparisons of the mutagenic effects on <u>Salmonella</u>
            <u>typhimurium</u> TA98 and TA100 of condensates collected
            from thermal treatments of two EPDM samples (A,B) and
            one EPDM polymer sample (C). For further details, see
            Figure 1.

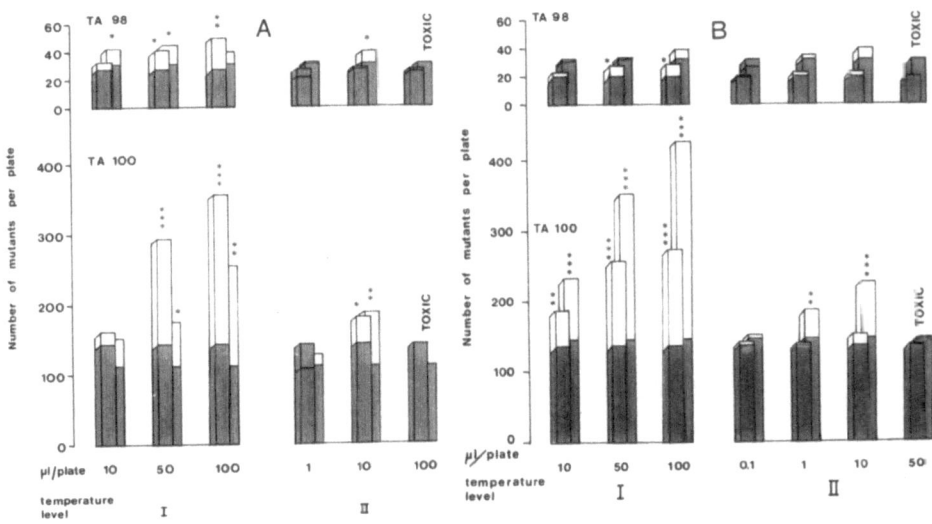

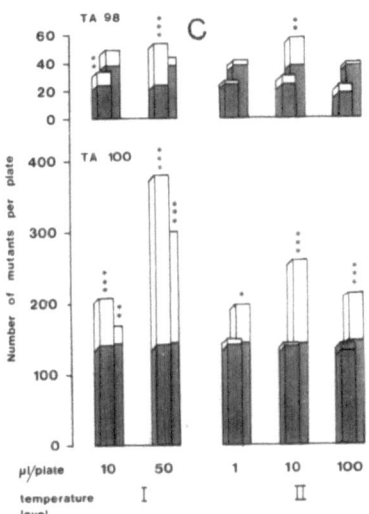

Figure 4. Mutagenicity on <u>Salmonella typhimurium</u> TA98 and TA100 of products from the pyrolyzation of two CR samples (A,B) and one CR polymer sample (C). For further details, see Figure 1.

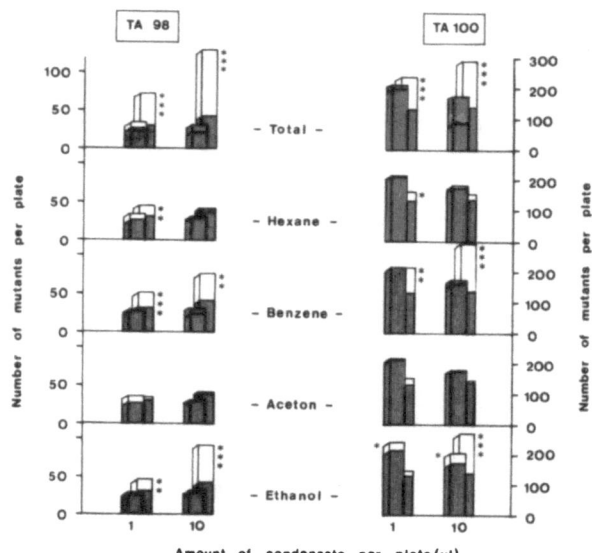

Figure 5. The total CR condensate from temperature level II and
          four fractions with increasing polarity obtained by
          separation on a silica column, tested for mutagenicity
          on Salmonella typhimurium TA98 and TA100. For further
          details, see Figure 1.

products formed during the thermal treatments probably are
responsible for the elevated mutagenicity noticed on TA100.

     One of the CR samples (Figure 4b) diverged from the
mutagenicity results for the polymer. This is likely due to the
presence of an indirectly acting base substitution-inducing
mutagen in this material, which is emitted at low temperatures.
According to the recipe, this blend contained 15 supplements
besides the CR polymer, among which soot, highly aromatic oils (HA
oils), ETU, and tetramethylthiuram monosulfide (TMTM) could
account for the additional mutagenicity. TMTM has a low melting
point (103 to 108°C) and is also an indirectly acting mutagen on
TA100 (Hedenstedt et al., 1979), which is why this compound can be
considered as the most probable candidate.

     One of the important questions with regard to the curing
fumes is whether the carbon black or HA oils contribute
significantly to the mutagenic effects observed. Among the

samples tested in this study, the FPM, EPDM (B), and CR (A, B, and fractionated) samples contained carbon black. HA oils were added only to one CR sample (B). These data give no indications of any significant contribution from these sources to the genotoxicity of the fumes.

High amounts of soot and HA oils are added to styrene butadiene rubber (SBR) to give the product (tires) the required qualities. Our data from tests on SBR pyrolyzates are difficult to interpret, and a contribution from the sources mentioned above cannot be excluded at this moment, especially at high process temperatures.

Chemical fractionation in order of polarity of one CR sample and the subsequent testing of these fractions demonstrated the complexity of the condensates. Mutagenicity was found among both nonpolar and polar substances. This sample contained small amounts of TMTM that could contribute to the mutagenicity of the chloroprene, which is known to induce reverse mutations in Salmonella typhimurium TA100 (Bartsch, 1976; IARC, 1979). Further identification of these fractions by means of capillary chromatography and GC/MS is under way. Thus, the presence of polycyclic aromatic hydrocarbons, if emitted under process conditions, will be confirmed.

REFERENCES

Ames, B.N., J. McCann, and E. Yamasaki. 1975. Methods for detecting carcinogens and mutagens with the Salmonella/ mammalian-microsome mutagenicity test. Mutation Res. 31:347-364.

Bartsch, H. 1976. Mutagenicity tests in chemical carcinogenesis. In: Environmental Pollution and Carcinogenic Risk, Volume 52. C. Rosenfeld and W. Davis, eds. INSERM Symposia Series, IARC Scientific Publications No. 13. International Agency for Research on Cancer: Lyons. pp. 229-240.

Case, R.A.M., and M.E. Hosker. 1954. Tumor of the urinary bladder as an occupational disease in the rubber industry in England and Wales. Br. J. Prev. Soc. Med. 8:39-50.

de Serres, F.J., and M.D. Shelby. 1979. Recommendation on data production and analysis using the Salmonella/microsome mutagenicity assay. Mutation Res. 64:159-165.

Donner, M., and M. Sorsa. 1981. Use of Drosophila as an indicator of risk chemicals in rubber industry. Presented at the Scandinavian Rubber Conference, Helsingfors, Finland.

Falck, K., M. Sorsa, H. Vainio, and I. Kilpikari. 1980. Mutagenicity in urine of workers in the rubber industry. Mutation Res. 79:45-52.

Fox, A.J., D.C. Lindars, and R. Owen. 1974. A survey of occupational cancer in the rubber and cable-making industries: results of five-year analysis, 1967-1971. Br. J. Ind. Med. 31:140-151.

Fraser, D.A., and S. Rappaport. 1976. Health aspects of the curing of synthetic rubbers. Environ. Health Perspect. 17:45-53.

Hakama, M., and I. Kilpikari. (in press). Cancer risk among rubber workers. J. Toxicol. Environ. Health.

Hedenstedt, A. 1981. Genetic health risks in rubber industry: Mutagenicity studies on rubber chemicals and process vapours. Presented at the Scandinavian Rubber Conference, Helsingfors, Finland.

Hedenstedt, A., C. Ramel, and C.A. Wachtmeister. (in press). Mutagenicity of rubber vulcanization gases in Salmonella typhimurium. J. Toxicol. Environ. Health.

Hedenstedt, A., U. Rannug, C. Ramel, and C.A. Wachtmeister. 1979. Mutagenicity and metabolism studies on 12 thiuram and dithiocarbamate compounds used as accelerators in the Swedish rubber industry. Mutation Res. 68:313-325.

IARC. 1979. Evaluation of the Carcinogenic Risk of Chemicals to Humans, Volume 19: some monomers, plastics and synthetic elastomers, and acrolein. International Agency for Research on Cancer: Lyons.

Maisey, L.J. 1981. Rubber vulcanization fume: its origin and control. Presented at the Scandinavian Rubber Conference, Helsingfors, Finland.

Monson, R.R., and L.J. Fine. 1978. Cancer mortality and morbidity among rubber workers. J. Natl. Cancer Inst. 61:1047-1053.

Peters, J.M., R.R. Monson, W.A. Burgess, and L.J. Fine. 1976. Occupational disease in the rubber industry. Environ. Health Perspect. 17:31-34.

Willoughby, B.G., and G. Lawson. 1979. Laboratory vulcanization as an aid to factory air analysis. Paper 28, presented at the American Chemical Society meeting in Cleveland, OH.

MUTAGENS IN AIRBORNE PARTICULATE POLLUTANTS AND NITRO DERIVATIVES

PRODUCED BY EXPOSURE OF AROMATIC COMPOUNDS TO GASEOUS POLLUTANTS

Hiroshi Tokiwa,[1] Shigeji Kitamori,[1] Reiko Nakagawa,[1] and
Yoshinari Ohnishi[2]

[1]Fukuoka Environmental Research Center, Fukuoka 818-01, and
[2]Department of Bacteriology, School of Medicine, The
University of Tokushima, Tokushima 770, Japan

INTRODUCTION

Most of the mutagenicity in airborne particulate pollutants
is found in respirable particles less than 1 μm in diameter, and
polycyclic aromatic hardrocarbons (PAHs) or their derivatives seem
to be important mutagens in these particles. In fact, the neutral
fraction extracted from airborne particulates contains direct- and
indirect-acting mutagens, and accounts for 48% of the mutagenic
activity of the whole extract (Tokiwa et al., 1980).

Many nitro-containing compounds, including nitrofurans and
nitroaromatics, are mutagenic and carcinogenic (Cohen et al.,
1973; Takemura et al., 1974; Yahagi et al., 1975). Because nitro
derivatives can be produced by the nitration of PAHs in pollutant
gases (Pitts et al., 1978), and because volatile or gaseous
compounds, including NO_2 and SO_2, are widespread air pollutants,
we wondered whether nonmutagenic chemicals might be converted to
potent mutagens when exposed to NO_2 in the environment.

We, therefore, exposed nonmutagenic aromatic or heterocyclic
compounds, which are normally detectable in the environment, to
NO_2, and determined whether potent mutagenic nitro derivatives
were produced (Tokiwa et al., 1981).

MATERIALS AND METHODS

Mutation Assay

The bacterial strains used were mutants of Salmonella typhimurium his⁻ strains TA98 and TA100, which were kindly provided by Dr. Bruce N. Ames, University of California, Berkeley, CA.

The mutation assay was performed as described by Ames et al. (1973); 0.15 ml of S9 fraction/plate was used.

Fractionation

Particulates (6.422 g) were collected from 5,646 m^3 of air at a highly polluted site in an industrial area and extracted with methanol in a Soxhlet extractor for 8 h. The methanol-extracted material (1.434 g) was then concentrated on an evaporator, dissolved in 5 ml of methanol, and separated into 10 fractions by the procedure of Swain et al. (1969).

Exposure of the Test Compounds to NO_2, SO_2, or HNO_3

Ten mg of each compound dissolved in benzene was deposited on a 23.7-cm^2 piece of filter paper (Toyo Filter Paper No. 2) and dried to remove the benzene. The specimens were then exposed in the dark to NO_2 (10 ppm) for 24 h (flow rate of 2 1/min, 30°C). When pyrene was exposed to SO_2 of HNO_3 (~ 20 ppb), it was introduced into the exposure apparatus with NO_2.

RESULTS

Indirect-Acting Mutagens in Airborne Particulate Materials

Of the 10 fractions obtained by Swain's procedure, six (BE, WAE, SAE, NME, NCH, and NNM) were found to be active (Table 1). Metabolic activity of the WAE, SAE, NME fractions, and crude extract was greater in the presence of the S9 fraction prepared from rat lung than from liver extract. The nitromethane- and cyclohexane-soluble neutral fractions were found by gas chromatography (GC) to contain 134 kinds of PAH compounds, nine of which were determined quantitatively by thin-layer chromatography and fluorometric analysis. As shown in Table 2, the amount of benzo(a)pyrene (B[a]P), chrysene, and fluoranthene corresponded to about twice that of B(a)P.

Table 1. Mutagenic Activity of Swain's Fractions of Airborne Particulate Extracts

Fraction	Quantity of Extract (mg)	Revertants/100 µg						Estimated Revertants (/m^3)	%[c]
		TA98			TA100				
		+S9[a]	+S9[b]	−S9	+S9[a]	+S9[b]	−S9		
Crude extract	1,434	175	680	720	100	102	34	444	100
Bases[d]	36.9	660	269	210	280	62	0	43.1	9.71
Bases[e]	37.9	25	4	4	60		85	1.68	0.38
Bases[f]	0.48	375		200	485	88	48	0.32	0.07
Weak acids[d]	51.1	360	1,240	330	40	220	86	32.6	7.34
Strong acids[d]	80.9	110	1,054	415	79	0	13	15.7	3.54
Weak and strong acids[f]	30.6	25		108	30	19	75	1.35	0.30
Strong acids[e]	204.2	25	180	325	5	126	133	9.04	2.04
Neutrals[g]	22.1	770	1,455	444	380	73	0	30.1	6.78
Neutrals[h]	153.1	100	150	225	52	166	154	27.1	6.10
Neutrals[i]	42.6	1,280	1,110	1,120	412	166	154	96.6	21.8

[a]Liver.
[b]Lung.
[c]The data are shown as calculated rev/m^3 of air in the presence of liver S9 fraction with strain TA98. Total recovery of mutagenicity after fractionation was 58% of that of the crude extract.
[d]Ether-soluble.
[e]Water-soluble.
[f]Insoluble.
[g]Methanol-soluble.
[h]Cyclohexane-soluble.
[i]Nitromethane-soluble.

Table 2.   Compounds Contained in the Neutral Fraction of Airborne
           Materials

Peak No. of GC	Compound	Quantity (ng/m^3)	Mutagenicity/ Carcinogenicity
3	Methylfluorene		
8	Anthracene		-/-
9	Phenanthrene		-/-
13	Methylphenanthrene		
22	Dimethylphenanthrene		
24	Fluoranthene	10	-/-
29	Pyrene	3.38	-/-
36	Methylpyrene		
	Methylfluoranthene		
39	Benzo(a)fluorene		
40	Benzo(b)fluoranthene		/+
49	Methylbenzo(a)fluorene		
55	Methylbenzo(b)fluorene		
57	Benzo(c)phenanthrene		
58	Benzo(ghi)fluoranthene		
68	Benzo(a)anthracene	5.75	+/+
	Chrysene	10.7	+/+
	Triphenylene		
81-83	Methylchrysene		
	Methylbenzo(a)anthracene		+/+
91,94,95	Benzofluoranthene		/+
101	Benzo(a)pyrene	5.33	+/+
	Benzo(e)pyrene	12.6	+/+
103	Perylene	1.34	-/
104	Dibenzophenanthrene		
	Methylbenzopyrene		
115	Anthanthrene		+/+
	Indeno(1,2,3-cd)pyrene		
118	Dibenzoanthracene		+/+
	Dibenzophenanthrene		
122	Benzo(ghi)perylene	6.43	
126	Methyldibenzoanthracene		
	Methyldibenzophenanthrene		
131-133	Dibenzopyrene		/+
134	Coronene	1.83	

Formation of Nitro Derivatives

Nitro derivatives were formed from five aromatic compounds and one heterocyclic compound, extracted with benzene, and subjected to GC analysis and mass spectrometry (MS), as described in the caption of Figure 1 and illustrated in the figure.

When pyrene was exposed to NO_2, a single nitro-substituted derivative (MW 247) was obtained. The product was determined to be 1-nitropyrene by its ultraviolet spectrum, melting point (151 to 151.5°C), and mass spectrum. The mixture of pyrene and 1-nitropyrene was mutagenic for strain TA98 in the absence of the S9 fraction (Figure 2). The 1-nitropyrene was also very mutagenic for strain TA98 (data not shown).

Fluorene contained both dihydrofluorene (MW 168) and 9-fluorenone (MW 180) in the original sample as impurities (Figure 1). Two nitro-substituted derivatives (MW 211 and 213) were produced simultaneously when fluorene was exposed to NO_2. One of them was determined to be 2-nitrofluorene (MW 211), formed from the original fluorene; the other (MW 213) was nitrodihydrofluorene, which could have been formed from a dihydrofluorene impurity. Fluorene exposed to NO_2 was also mutagenic in the absence of the rat liver S9 fraction.

Fluoranthene and carbazole are themselves nonmutagenic. Each formed two nitro-substituted derivatives (Figure 1). The two induced nitro-substituted fluoranthenes were determined to be 3- and 8-nitrofluoranthene by their mass spectra. 3-Nitrofluoranthene showed the highest mutagenic activity of the four kinds of nitro derivatives against strains TA98 and TA1538. Therefore, each nonmutagenic chemical produced some mutagenic nitro compounds when exposed to NO_2 (10 ppm for 24 h).

Chrysene, an indirect-acting mutagen, also formed a nitrochrysene derivative, determined to be 6-nitrochrysene (MW 273) by its mass spectrum, ultraviolet spectrum, and melting point (206.5 to 208°C).

Effect of SO_2 and HNO_3 on the Formation and Mutagenicity of 1-Nitropyrene

Pyrene deposited on filter paper was exposed to 7 ppm of SO_2 (flow rate, 0.5 l/min) or to a trace of HNO_3 (~ 20 ppb) in addition to 1 ppm of NO_2 (flow rate, 2.2 l/min) to determine whether the formation of 1-nitropyrene was enhanced (Table 3). After 24 h, each sample was extracted with benzene and tested for

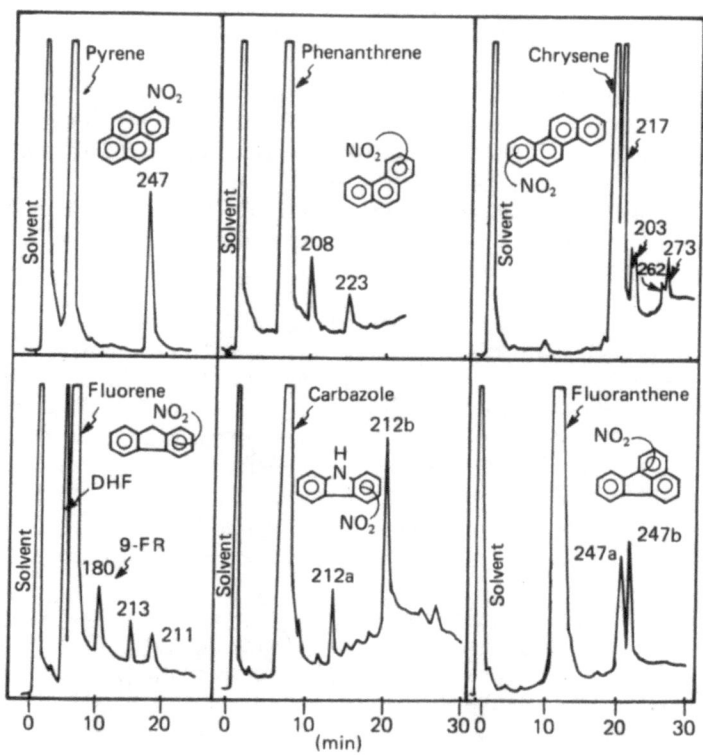

Figure 1. Gas chromatograms of nitro-substituted compounds formed
after exposure of six compounds to NO_2. Each compound
was exposed to NO_2 (10 ppm) for 24 h and extracted with
benzene. A portion of each extract was subjected to
gas chromatography and each peak induced was analyzed
by mass spectrometry. The numbers in the figure show
molecular weights determined by mass spectrometry.
Chromatographic operation conditions were as follows:
the spiral stainless column (0.3 x 300 cm) for gas
chromatography was packed with 3% dex-sil 400 GC on
60/80 mesh chromosorb W (A.W.DBCS). Carrier gas was
nitrogen, flowing at a rate of 2.0 kg/cm^2; DHF,
dihydrofluorene; 9-FR, 9-fluorenone.

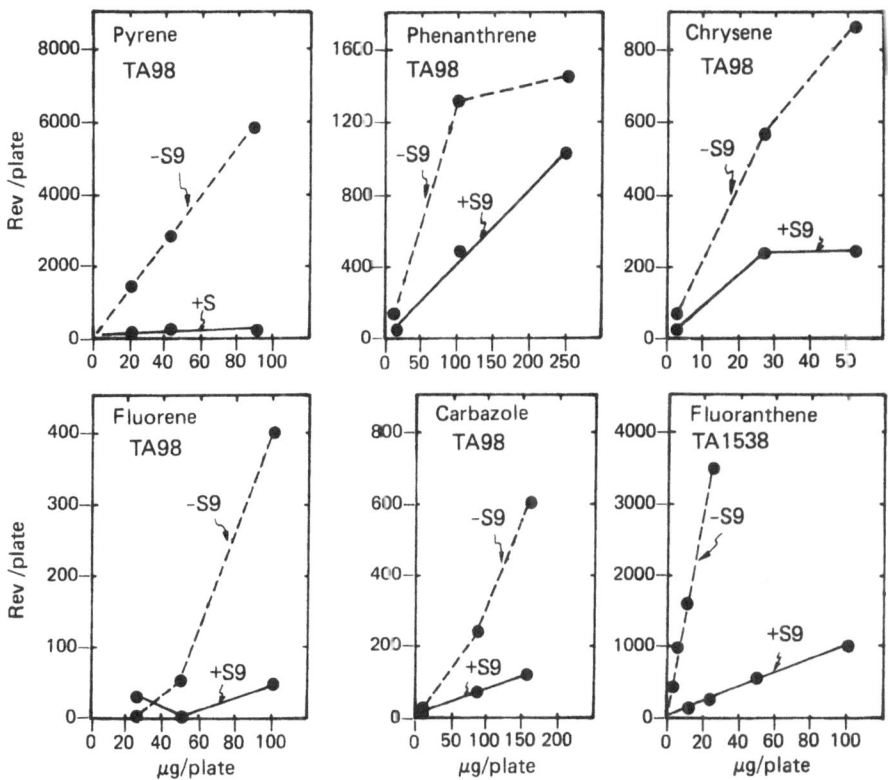

Figure 2.   Dose-response curves of the activated nitro compounds.
After exposure of 10 mg of sample to NO$_2$, a portion of
the benzene-extracted material was assayed with strains
TA98 and TA1538 with or without S9 fraction.   The
number of spontaneous revertants was subtracted as
follows: 10 (TA1538, -S9), 12 (TA1538, +S9), 39 (TA98,
-S9), 44 (TA98, +S9).   The abscissa indicates the dose
(μg/plate) of the extracted material after exposure.

mutagenicity.   The mutagenic activity induced by NO$_2$ was enhanced
by SO$_2$ or HNO$_3$.   In addition, NO$_2$ plus SO$_2$ elevated the mutagenic
activity during anaerobic incubation for 16 h early in the assay
system.   The yield of 1-nitropyrene was 0.02% with a low level of
NO$_2$ (1 ppm), 0.37% with NO$_2$ plus SO$_2$ (7 ppm), and 2.85% with NO$_2$
plus HNO$_3$ (~ 20 ppb).

Table 3. Mutagenicity and Yield of 1-Nitropyrene After Exposure of Pyrene to Various Gases[a]

	His+ Revertants/Plate After Exposure to Gases				
	NO_2 (1 ppm)	SO_2 (7 ppm)	HNO_3 (~ 20 ppb)	NO_2 + SO_2 (1 + 7 ppm)	NO_2 + HNO_3 (1 ppm + ~ 20 ppb)
Pyrene (µg/plate)					
20	70	40	9	190 (550[b])	2090
10	25	10	0	90 (260[b])	1300
5	12	2	0	40 (120[b])	900
2.5	2		0	24 (60[b])	500
Total amount of 1-nitropyrene produced (µg)	2	ND[c]	ND[c]	37	285
Yield (%)	0.02			0.37	2.85

[a]Mutagenic activity is shown as his+ revertants per plate from strain TA98 without S9 mix by the Ames test except revertants indicated by [b] in the NO_2 + SO_2 column, which show the results of anaerobiosis.
[b]These plates were incubated in anaerobic Gas Pak jars for 16 h and incubated aerobically for an additional 32 h. Spontaneous revertants were subtracted. The amount of 1-nitropyrene was determined by the peak produced in gas chromatography. The flow rate of NO_2 was 2.2 l/min, that of SO_2 was 0.5 l/min, that of HNO_3 was 0.2 l/min.
[c]Not detected.

Nitration of Pyrene in Airborne Particulate Materials

It was important to see whether nitro compounds are similarly produced when airborne particulate materials are exposed to NO_2.

Various amounts of airborne particulate material collected on the glass-fiber filter were exposed to 1 ppm of NO_2 for 24 h. As shown in Figure 3A, the mutagenic activity of the exposed sample markedly increased in the absence of S9 mix. The NO_2-exposed material was analyzed by GC, and three kinds of nitro derivatives were produced, one of which was determined to be 2-nitrofluorene by the GC-MS system (data not shown). Furthermore, when the airborne particulate material with 2 mg of pyrene added was exposed to NO_2 by the same procedure, the mutagenicity was very greatly increased (Figure 3B). The increase in mutagenicity might be due to the promotion of nitration of pyrene because this particulate sample already contained about 5.26 μg of NO_3^- and 20.3 μg of $SO_4^=/m^3$.

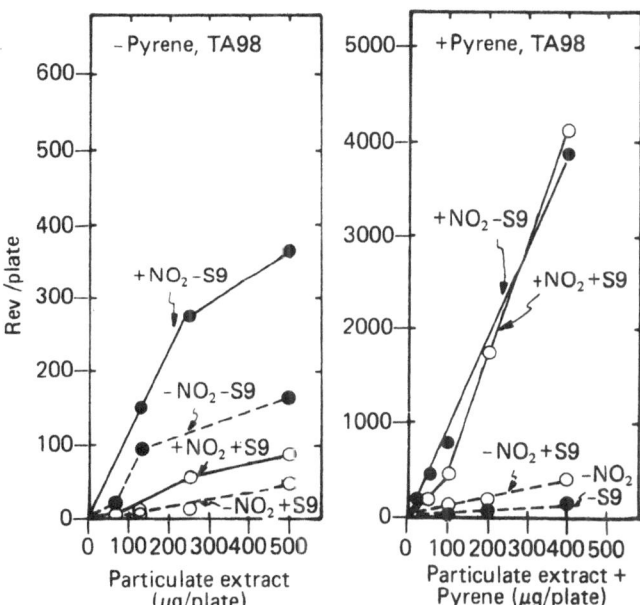

Figure 3.   Mutagenic activity of airborne particulate material exposed to NO_2 (1 ppm) with or without pyrene (2 mg).

Detection of Nitro Compounds in Diesel Exhaust Materials

It is important to know whether nitro compounds exist in environmental materials. Their detection in diesel exhaust materials was attempted by high performance liquid chromatography (HPLC) and MS.

The neutral fraction of diesel exhaust extracts was subdivided into aliphatic, polyaromatic, and oxygenated fractions by column chromatography on silica gel. As shown in Figure 4B, mutagenicity in the absence of S9 mix was observed in the oxygenated fraction. By chemical analysis of each fraction, 1-nitropyrene was detected in the polyaromatic fraction (53 $\mu g/g$ of the extracts), and 3-nitrofluoranthene and 5-nitroacenaphthene were found in the oxygenated fraction (Figure 4C-D).

DISCUSSION

To assess the contribution of PAHs (other than B[a]P) and of other chemical fractions to the mutagenicity of airborne particulates, we fractionated the crude extracts of particulate material collected in a polluted industrial site and chemically analyzed the PAHs in the neutral fraction. The concentration of B(a)P was 5.33 ng/m^3, and this is equivalent to 0.33 rev/m^3 from strain TA98 in the presence of liver S9 fraction, amounting to 2.67% of the mutagenic activity of the NCH-NNM fraction and 0.743% of the total mutagenic activity of the crude extract.

It was found that five kinds of aromatic compounds and a heterocyclic compound, most of which occur in airborne particulate matter (Daisey and Leyko, 1979; Kaden et al., 1979) and cigarette smoke condensate, readily produce nitro-substituted derivatives when exposed to 10 ppm of NO_2. All the nitro derivatives obtained were potent direct-acting mutagens.

Pitts et al. (1978) reported that B(a)P, an indirect-acting mutagen, produces both nitro derivatives and benzo(a)pyrene-quinone on exposure to NO_2. However, in our experiment, only nitro derivatives were induced from each chemical. A variety of chemical derivatives might be formed on exposure of PAH to pollutant gases under simulated atmospheric conditions (Pitts et al., 1978). On an NO_2-exposure experiment with pyrene, trace amounts of nitric acid (~ 20 ppb) or sulphur dioxide (7 ppm) were added to the exposing system in addition to NO_2 (1 ppm), and a few products in addition to nitropyrene were formed. The mutagenic activity was apparently enhanced. Furthermore, with NO_2 plus SO_2, the mutagenicity of pyrene was

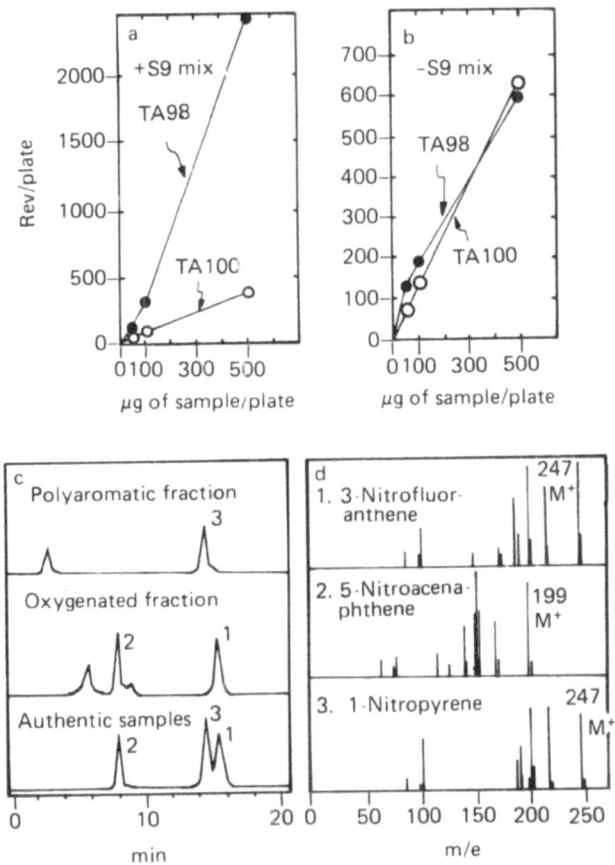

Figure 4.   Mutagenicity and nitro derivatives in diesel exhaust
             soot. Figure 4A and 4B are dose-response curves of the
             oxygenated fraction of the neutral fraction of diesel
             exhaust extract. High performance liquid
             chromatography used type LC-3A of Shimazu Co., Ltd.,
             and the data were obtained on a Zorbax ODS (35°C, 85%
             methanol, flow rate 1 ml/min, 400 nm ultraviolet
             detector). A, polyaromatic fraction; B, oxygenated
             fraction; C, high performance liquid chromatograms;
             D, mass spectra.

enhanced when anaerobic incubation (16 h) was included in the Ames test (Table 3). This promoting effect might be due to the oxygen lability of some products such as hydroxylamino intermediates (Rosenkranz and Poirier, 1979) and some nitropyrene-sulfonic acids of pyrene (Tietz and Bayer, 1939).

It is important to know whether nitro compounds exist in environmental materials. We demonstrated three kinds of nitro compounds, 1-nitropyrene, 3-nitrofluoranthene, and 5-nitroacenaphthene, by using HPLC and MS. The carcinogenicity of 5-nitroacenaphthene was shown by Takemura et al. (1974), who reported that adenocarcinoma was induced in the small intestine of rats in feeding experiments. On the other hand, Ohgaki et al. (1982) demonstrated that 1-nitropyrene and 3-nitrofluoranthene induced malignant fibrous histiocytomas at the site of injection in rats.

ACKNOWLEDGMENTS

This investigation was supported in part by a Grant-in-Aid for Scientific Research from the Ministry of Education, Science and Culture, the Ministry of Public Welfare, and the Environment Agency, Japan.

REFERENCES

Ames, B.N., W.E. Durston, E. Yamasaki, and F.D. Lee. 1973. Carcinogens are mutagens: a simple test system combining liver homogenates for activation and bacteria for detection. Proc. Natl. Acad. Sci. USA 70:2281-2285.

Cohen, S.M., E. Ertürk, A.M.V. Esch, A.J. Crovetti, and G.T. Bryan. 1973. Carcinogenicity of 5-nitrofurans, 5-nitroimidazoles, 4-nitrobenzenes and related compounds. J. Natl. Cancer Inst. 51:403-417.

Daisey, J.M., and M.A. Leyko. 1979. Thin-layer gas chromatographic method for the determination of polycyclic aromatic and aliphatic hydrocarbons in airborne particulate matter. Anal. Chem. 51:24-26.

Kaden, D.A., R.A. Hites, and W.G. Thilly. 1979. Mutagenicity of soot and associated polycyclic aromatic hydrocarbons to Salmonella typhimurium. Cancer Res. 39:4152-4159.

Ohgaki, H., N. Matsukura, K. Morino, T. Kawachi, T. Sugimura, K. Morita, H. Tokiwa, and T. Hirota. 1982. Carcinogenicity in rats of the mutagenic compounds 1-nitropyrene and 3-nitrofluoranthene. Cancer Lett. 15:1-27.

Pitts, J.N., Jr., K.A. van Cauwenberghe, D. Grosjean, J.P. Schmid, D.R. Fritz, W.L. Belser, Jr., G.B. Knudson, and P.M. Hynds. 1978. Atmospheric reaction of polycyclic aromatic hydrocarbons: facile formation of mutagenic nitro derivatives. Science 202:515-519.

Rosenkranz, H.S., and L.A. Poirier. 1979. Evaluation of the mutagenicity and DNA-modifying activity of carcinogens and noncarcinogens in microbial systems. J. Natl. Cancer Inst. 62:873-893.

Swain, A.P., J.E. Cooper, and R.L. Stedman. 1969. Large-scale fractionation of cigarette smoke condensate for chemical and biologic investigation. Cancer Res. 29:579-583.

Takemura, N., C. Hashida, and M. Terasawa. 1974. Carcinogenic action of 5-nitroacenaphthene, Br. J. Cancer 30:481-483.

Tietze, E., and O. Bayer. 1939. Die sulfosaurene des pyrens und ihre abkommlinge. Ann. Chem. 540:189-210.

Tokiwa, H., S. Kitamori, K. Takahashi, and Y. Ohnishi. 1980. Mutagenic and chemical assay of extracts of airborne particulates. Mutation Res. 77:99-108.

Tokiwa, H., R. Nakagawa, K. Morita, and Y. Ohnishi. 1981. Mutagenicity of nitro derivatives induced by exposure of aromatic compounds to nitrogen dioxide. Mutation Res. 85:195-205.

Yahagi, T., H. Shimizu, M. Nagao, N. Takemura, and T. Sugimura. 1975. Mutagenicity of 5-nitroacenaphthene in Salmonella. Gann 66:581-582.

INDEX

569